Bacteria in the Natural World

Bacteria in the Natural World

John Paul

CABI

CABI is a trading name of CAB International

CABI
Nosworthy Way
Wallingford
Oxfordshire OX10 8DE
UK

Tel: +44 (0)1491 832111
E-mail: info@cabi.org
Website: www.cabi.org

CABI
200 Portland Street
Boston
MA 02114
USA

Tel: +1 (617)682-9015
E-mail: cabi-nao@cabi.org

A catalogue record for this book is available from the British Library, London, UK.

ISBN-13: 9781836993902 (hardback)
9781836993919 (paperback)
9781836993926 (ePDF)
9781836993933 (ePub)

DOI: 10.1079/9781836993933.0000

Commissioning Editor: Ward Cooper
Editorial Assistant: Emma McCann
Production Editor: James Bishop

Typeset by Straive, Pondicherry, India
Printed and bound in the UK by CPI Group (UK) Ltd, Croydon, CR0 4YY

Contents

Preface

When I was at medical school I did my first Gram stain and saw bacteria. It was love at first sight. Bacteria fitted into a mindset that I had been cultivating for years and which had been formed as a result of my pursuits as an amateur entomologist. Just like insects there are lots of different species of bacteria, each with its own distinguishing features, growth requirements and habitat preferences and there are lots of Latin names which keep changing. Over the course of time I developed a professional interest in the bacteria that cause infection in humans but because of my strong affinity for natural history I was drawn to investigate the bacteria that live in the natural environment and to understand their place in natural history. *Bacteria in the Natural World* is the result of these efforts.

Acknowledgements

Bacteria in the Natural World would not have been written without the encouragement of my late wife, Janice Bates. When I complained to her about the paucity of authoritative but user-friendly texts about bacteria that made suitable reading for inquisitive adventurous natural historians, she encouraged me to make my own investigations and to write my own book on the subject. It happens that at around the same time I was approached by Anna Dumitriu, an artist who was interested in mapping the normal microbial flora of domestic and natural environments. Our collaborative bioart investigations continue to this day and have facilitated contact with other microbiologists from around the world. One of them, Minna Männistö, provided us with cultures of bacteria isolated from arctic soils, some of which are illustrated in this book. Many scientists, clinicians, photographers, artists, musicians, curators and naturalists helpfully responded to requests to use their figures, photographs, quotations or data or simply to answer my questions. Among them were Jasmine Bastable, David Benson, Richard Bone, Roos Brandau, Alexander Brasier, Thomas Brock, Tana Cavey, Ian Clark, Ian Clarke, Charles Cockell, Kevin Cole, Steven Edmunds, Gad Frankel, Dorian Fuller, Bill Grant, Marianne Hanzlik, Markus Heidarsson, Katie Jeffery, Barrie Johnson, Dan Joseph, Martin Kaltenpoth, Wolfgang Krumbein, Andrew Lewington, Anna Eleri Livingstone, Alex May, David Muirhead, Philippe Normand, Shannon O'Dell, D. Oesterhelt, A.S. Padmavathy, Simon Park, John Parnell, Michael Prime, James Prosser, David Pybus, Fred Rainey, Andrew Sier, David Simmons, Gordon Southam, David Streeter, David Thomas, Nigel Trewin, Cosey Fanni Tutti, Anthony Walsby, Judy Webb, Ian West, Barny Whitman, Brian Whitton and Carl Woese.

Author's biography

John Paul became acquainted with bacteria at Birmingham Medical School and later specialised in medical microbiology. After spending some time attached to a KEMRI-Wellcome Trust research programme in Nairobi he returned to the UK to work as a hospital doctor. He occupied the roles of regional microbiologist with Public Health England as well as that of honorary clinical professor at Brighton and Sussex Medical School. As a member of the Modernising Medical Microbiology research consortium he used whole genome sequencing to investigate pathogenic bacteria. He has collaborated with the artist Anna Dumitriu on a number of bioart projects over several decades. Outside of work he has pursued a lifelong interest in entomology.

1 Bacteria: their place in natural history

les plus simples que l'on puisse imaginer.
[the simplest that one could imagine.]
Louis Pasteur (1864). *Mémoire sur la fermentation acétique.*

While walking home once along a bridleway in the dead of night I encountered a little cluster of eerie cold pinpoints of light, which from a distance seemed to twinkle in the darkness in the middle of a stony path. My first impression was that the subtle glimmer had something of the look of a glow-worm about it, except that on close examination it became obvious that the glowing organisms were in fact fly larvae. Each one of their body segments was gently illuminated in the pitchy blackness of the night by a dim greenish light. The maggots were writhing about on fox faeces. It felt almost as though the little larvae were attempting to attract my attention. It was a beguiling encounter with the microbial world because a likely explanation for this apparition is that the fly larvae were infected with bioluminescent bacteria.

Bacteria are ubiquitous but seem strangely alien. Their unfamiliarity relates to the fact that we cannot get to know bacteria in the way that we know animals and plants. Animals are so like people that it is easy to understand what makes them tick. In contrast, bacteria are biologically distant and furthermore they are microscopic. Nevertheless, despite their minuteness bacteria do form an important part of our everyday experience. Bacteria are responsible for many of the natural phenomena that cause us to appreciate nature in the way that we do. For instance, the pleasurable sensibilities that are evoked by the smell of the seaside and by the smell of rain on dry soil derive from our acute sensitivity to volatile compounds made by particular kinds of bacteria. In this way the metabolic products of bacteria provide a link between modern human experience and our primal capacity to relate to the environment.

DOI: 10.1079/9781836993933.0001

Although individual bacterial cells are too small to be visible to the naked eye, bacteria en masse may decorate entire landscapes. But because such scenes seem so commonplace to us we do not generally take much notice of them. For example, when walking in the countryside it is a fairly routine experience to encounter the inky black streaks of microbial biofilms on otherwise bare rock faces. Malham Cove in Yorkshire is one of a number of well-known locations where such phenomena are writ large. The pale limestone at Malham serves to accentuate the blackish microbial streaks, which appear to run down over the surface of the rocks like dripping paint. Dancing Ledge on the Dorset coast provides another example (Figs. 1.1–1.2). There are countless other such places. The black ink streaks are actually composed of the cells of enormous numbers of microscopic cyanobacteria accompanied by other microbes. In order to survive these organisms require little more than air, sunlight and moisture. An otherwise barren rock face provides them with a suitable habitat.

Compared with most kinds of bacteria, the cells of cyanobacteria are relatively large and can be observed reasonably easily under a microscope at high magnification by examining a scraping taken from an ink-streaked cliff face (Fig. 1.3). In common with green plants, cyanobacteria are photosynthetic organisms that use sunlight as a source of energy. Cyanobacteria, however, are more tolerant than plants of extreme environmental conditions. Cyanobacteria can survive in places that are too dry, too dark, too hot or too cold for plants. As well as populating sun-baked cliff faces, cyanobacteria are able to colonise the walls of dimly lit caves, intertidal mudflats, hot springs and the pore spaces within frozen rocks. Hence, in environments where the conditions are too extreme for other life forms, cyanobacteria create natural spectacles that can be enjoyed as visible evidence of the presence of bacterial life.

The so-called iron bacteria provide another example of how bacteria can embellish a landscape and enhance our experience of nature (Fig. 1.4).

Fig. 1.1. 'Ink streaks' composed of cyanobacteria and other microbes growing on Portland stone. Dancing Ledge, Dorset, UK.

Fig. 1.2. Microbial ink streaks, detail. Dancing Ledge, Dorset, UK.

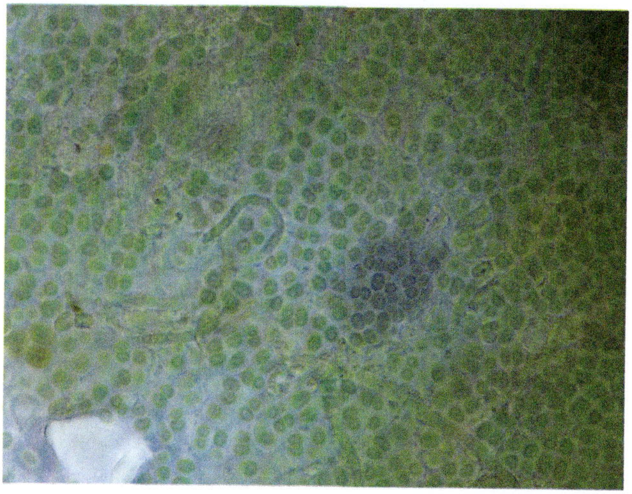

Fig. 1.3. Cyanobacteria seen on microscopy of a scraping taken from an ink-streaked rock face. Kisumu, Kenya.

Iron-oxidising bacteria are responsible for the rust-coloured slimy deposits that accumulate around iron-rich springs, bogs and seepages (Fig. 1.5). Rust-coloured iron springs stand out against the green background of nature. Their appearance seems almost alien, possibly because many of us have become accustomed to the colour of rust as being indicative of manmade iron objects, of corrosion and of industrial decay rather than as being one of the colours of nature itself. The bacteria that live in iron springs are completely natural and surprisingly beautiful. Microscopic examination of a sample of orange slime from an iron spring

Fig. 1.4. The Iron's Well, New Forest, UK, a habitat for iron-oxidising bacteria.

Fig. 1.5. Leaves coated with iron oxide and iron-oxidising bacteria. Iron's Well, New Forest, UK.

discloses the presence of the iron-oxidising bacteria that inhabit its rufescent waters (Fig. 1.6). Some iron bacteria look like bundles of straw. Others have a coiled structure and look like miniature rusty springs. Because bacteria are invisible, we may tend to overlook the telltale signs that point to their existence. However, by becoming attuned to these signals it becomes possible for any naturalist to appreciate the natural world more completely.

Individual bacterial cells are invisible to the naked eye. (Some species of bacteria are just about visible as individual cells but these are rather special exceptions.) It takes a bit of imagination to start to gain insight into what it might be like to be a bacterium and some effort to appreciate the importance of microorganisms. Bacteria do indeed continue to look microscopic (Fig. 1.7) even when viewed under a traditional light microscope. Although microscopic examination makes it possible to visualise the overall shape and arrangement of bacterial cells – some are elongated and resemble strings of sausages, while others are more globular, like bunches of grapes – the detailed morphology of individual bacteria can hardly be appreciated. Bacteria are generally so small that even at the highest magnification (×1000) they look as though they are situated far away in the distance. Looking at bacteria under a microscope is a bit like looking down at a herd of animals from a plane. The lack of discernible detail compromises our ability to develop an intuitive understanding of what these organisms really are and of what they might be doing. Imagine what it might be like for someone who had no prior knowledge and experience of other animals to be initiated into a walk in the countryside. During the course of the excursion the novice's attention might be drawn to a flock of sheep grazing on a distant hillside (Fig. 1.8). To the naive observer it would be evident that the novel organisms were roughly oval in outline and that they were mostly white in colour. It might also be possible to make out that these newly discovered entities appeared to have appendages of some sort (legs); but from a distance

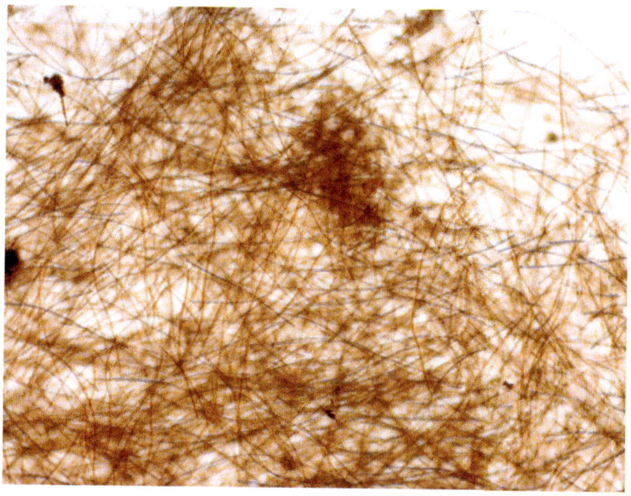

Fig. 1.6. Microscopy of microbial mat composed of iron-oxidising bacteria. Iron's Well, New Forest, UK.

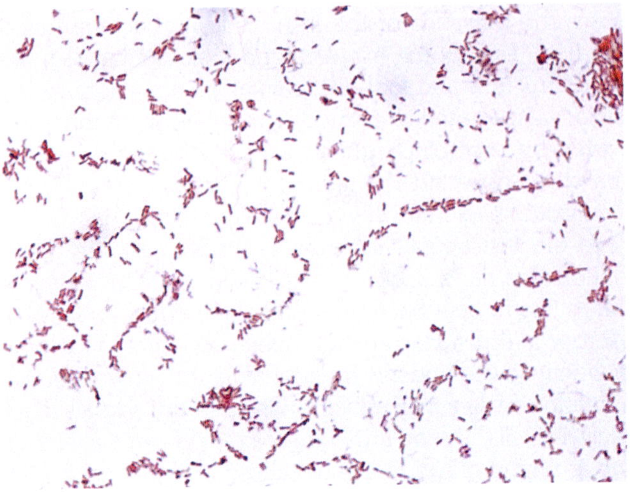

Fig. 1.7. Close up bacteria.

Fig. 1.8. Far away sheep.

that is about all that could be said about them. Microscopists face a comparable challenge when investigating the lives of bacteria.

Plant diseases caused by bacteria

One of the better known facts about bacteria is that some of them are pathogenic and have the capacity to infect plants or animals and cause disease. Because of their associations with infectious diseases, bacteria seem sinister to some minds. Bacterial infections of plants present themselves to us with characteristic physical appearances, some of which may awaken our sense of curiosity. The peculiar appearances of some plant diseases may even engender a sense of mystery. Horse chestnut trees, *Aesculus hippocastanum*, for example, are susceptible to a condition that is characterised by wine-dark exudates, which ooze down their unhealthy-looking trunks and branches (Fig. 1.9). These pigmented exudates are a sign of a disease known as bleeding canker of horse chestnut. One of the causes of bleeding canker of horse chestnut in Britain in recent years has been shown to be a bacterium called *Pseudomonas syringae* pathovar *aesculi* (Webber *et al.*, 2008). The Bleeding Canker Bacterium is just one of a large number of variants or strains of *Pseudomonas syringae*, a species of bacterium that causes a wide range of plant diseases. Each of the different pathogenic variants (pathovars) or strains of *Pseudomonas syringae*, is associated with a particular kind of plant disease. The scientific name *Pseudomonas syringae* relates to the fact that the original description of this species corresponded with a strain of the bacterium isolated from a diseased lilac, *Syringa vulgaris*, which was showing signs of lilac leaf blight. At a later date, when other variants were discovered, which caused different diseases, the strain associated with leaf blight was designated more precisely as *Pseudomonas syringae* pathovar *syringae* (the Leaf Blight Bacterium).

Another pathogenic variant of *Pseudomonas syringae* is responsible for white-colour disease of creeping thistle, *Cirsium arvense* (Fig. 1.10). Infection of thistles results in blanching of the leaves and stems, endowing them with a glossy milk-white enamelled appearance (Fig. 1.11). Canadian scientists (Zhang *et al.*, 2004) were the first to link white-colour disease of thistles to *Pseudomonas syringae* (Fig. 1.12), when they isolated a strain of the bacterium, which they designated as *Pseudomonas syringae* strain CT99B016B. The loss of green pigment has the effect of giving the plants a peculiarly non-plant like aspect, as though ghosts of their former selves. Fields haunted by

Fig. 1.9. Bleeding canker of horse chestnut. Oxford, UK.

Fig. 1.10. Creeping thistles with white thistle disease. Bramber, West Sussex, UK.

Fig. 1.11. Creeping thistle with white thistle disease. Upper Beeding, West Sussex, UK.

stands of creeping thistle affected by white-colour disease can be a noticeable feature in the countryside in some parts of Britain in late summer.

Bacteria make unnatural habitats natural

Despite the fact that much of the British landscape has been modified and managed by man for millennia, it seems natural to think of it as natural. We have got used to it the way it is and so have the plants and animals. Heathland,

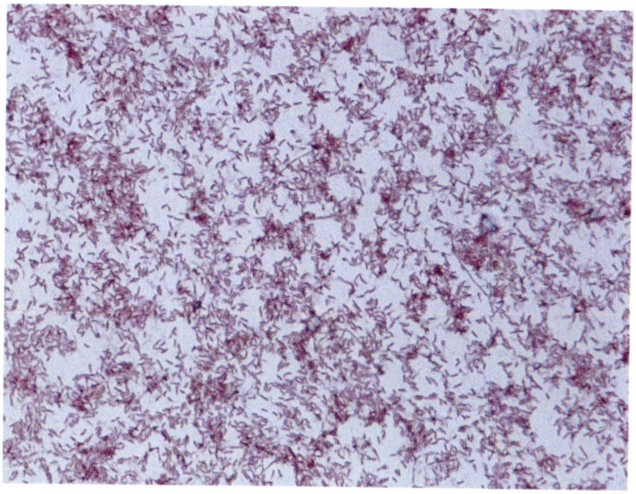

Fig. 1.12. *Pseudomonas syringae*, the agent of white thistle disease. Upper Beeding, West Sussex, UK.

for example, is venerated by naturalists not only for its beauty and timeless character but also for the rare plants, birds, reptiles and invertebrates that live on it. Heathland though is a habitat that developed on poor soils in prehistoric times after the clearance of woodland by humans. Similarly, arable fields have become valued features of our natural heritage. Although, arable landscapes are manmade and essentially unnatural, botanists treasure the specialised arable weeds that frequent them and entomologists and ornithologists prize the beetles and birds associated with them. More recently naturalists have come to appreciate the importance of so-called brownfield sites that exist in urban and former industrial areas. Derelict land soon becomes colonised by plants and animals and as a result, gradually becomes more natural. Brownfield sites are often biodiverse and may harbour species that are seldom seen elsewhere. As a consequence they are attractive to students of natural history and some sites have been recognised as being worthy of preservation.

When humans instigate an industrial process, modify an environment, build a new structure or fashion a new object, they immediately create opportunities for bacteria. Bacteria are everywhere: few places are truly sterile. It can be argued that when bacteria colonise an artefact or a manmade environment they initiate a process of making it become more natural. Such places may seem to us to be entirely artificial and may even appear to be devoid of life. Bacteria, however, live on a different scale. They see the world differently, so to speak. Because bacteria are so small, complex communities of bacteria can flourish within very small confines. A pot of yoghurt that contains living bacteria is arguably a natural habitat. A jam jar full of sludge left on a windowsill can function reasonably well as a balanced microbial ecosystem. At a microscopic scale, the outside wall of a newly constructed building will do just as well for a rock-dwelling bacterium as a natural rock face. Thermophilic bacteria that live in hot volcanic environments may feel equally at home in a smoking slag heap. At Parys Mountain (Fig. 1.13) on Anglesey, pools of acidic mining waste provide a home for bacteria (Fig. 1.14) that require conditions of high acidity. Because bacteria are everywhere, they stretch the conventional boundaries that natural historians might like to apply when they seek to define what they mean by natural history.

Bacteria are to be found in environments that humans consider to be extreme. In fact, habitats that are hostile to other forms of life often host some of the more interesting bacteria. Rubbish dumps, smoking slag heaps,

Fig. 1.13. Acidic mining waste, a home for acidophilic microbes. Parys Mountain, Anglesey, UK.

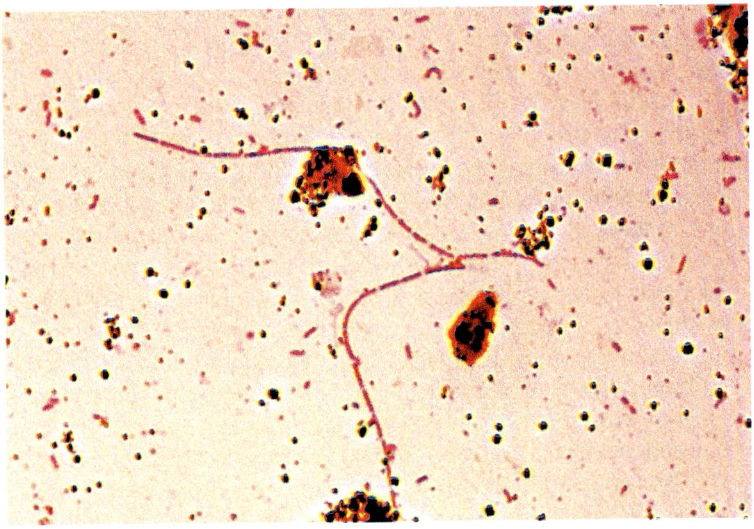

Fig. 1.14. Acidophilic bacteria seen on microscopy. Parys Mountain, Anglesey, UK.

abandoned copper mines, silage bales, anaerobic digester plants, geothermal power stations and subterranean oil reservoirs are all places that reward bacteriological study. The communities of bacteria that exist in these apparently unnatural or seemingly barren settings give them a new lease of life. Bacteria can teach us to appreciate that such places are natural habitats in their own right. Thereby we can comprehend more fully the totality of natural history. Despite an absence of sunlight, poor access to nutrients, high pressures and near-boiling temperatures, a petroleum reservoir situated deep beneath the floor of the North Sea may harbour a wonderful variety of specialised bacteria. A pristine example of such an environment is more natural than places that are actively managed by humans, such as the New Forest or the Norfolk Broads.

Bacteria are everywhere

Bacteria are everywhere and many species of bacteria are very widely distributed. Every habitat is associated with its own distinctive microbial flora. Each bacterium has its own environmental requirements. For growth, some bacteria require high temperatures, some require high concentrations of salt and some favour acidic environments. The nutritional needs of different species of bacteria vary from those that require a rich mixture of organic compounds to those which are able to subsist on a diet of as little as hydrogen and carbon dioxide. Many bacteria show resilience to adverse conditions and are able to colonise the empty spaces left vacant by other organisms. Because some bacteria like to consume the same foodstuffs as are favoured by humans and because some bacteria act as agents of wound infection, we have made efforts to create bacteria-free zones by cooking and canning our food supplies and by sterilising surgical equipment. Few places though are naturally sterile.

Similar kinds of bacteria can often be found, even in widely separated locations, in places where the environmental conditions suit their particular requirements. The microbial floras of salt lakes, for example, are remarkably similar, even on different continents. For this reason it makes more sense to consider bacteria as globally distributed organisms rather than to study them on a regional or national basis. In order to understand the bacteria of the British Isles it is necessary to perform an investigation of the bacteria of the world. This is not only because many kinds of bacteria show cosmopolitan patterns of distribution but also because studies conducted in other parts of the world help to fill in the gaps in our knowledge concerning our own microbial flora.

The geographical distribution patterns of bacteria were studied by the Dutch microbiologists Martinus Beijerinck (1851–1931) and Lourens Baas Becking (1895–1963). In response to the discontinuous distribution patterns shown by certain species of bacteria, Lourens Baas Becking formulated the expression: 'everything is everywhere, but the environment selects' (Baas Becking, 1934). For example, thermophilic bacteria that flourish only at very high temperatures can be found at widely separated locations wherever suitable habitat exists. Bacteria found in hot springs in Iceland are similar to bacteria found in hot springs in Yellowstone National Park. This suggests that thermophilic bacteria may be capable of dispersal over long distances. Further evidence for global dispersal is provided by the recovery of thermophiles from manmade habitats.

One thermophile, *Thermus scotoductus* (the Hot Water Pipe Bacterium), was described as new to science in 1994, based on a strain cultured from a hot water pipeline at a geothermal facility in Iceland (Fig. 1.15). The optimal temperature for growth of the Hot Water Pipe Bacterium is 70°C. Strains of the Hot Water Pipe Bacterium have also been found in hot springs in New Mexico and in Portugal (Tenreiro *et al.*, 1995). It seems likely that geothermal features serve as the natural habitat for this thermophilic bacterium. Bacteria identified as *Thermus scotoductus* have been cultured from fissure water in a South African gold mine at a depth of two miles (3.2 km) below the surface. The Hot Water Pipe Bacterium has also been recorded in Britain. Bacteria cultured from a hot water tap at the London Hospital (now the Royal London Hospital) in Whitechapel (Fig. 1.16) in the 1970s were subsequently identified as *Thermus scotoductus* (Pask-Hughes and Williams, 1975; Pask-Hughes and Shaw, 1982).

The enchanting fragrance of bacteria

People are attracted to the smell of the seaside and to the scent produced by rain on dry soil. The beguiling smell of the seaside is the odour of dimethylsulfide (C_2H_6S). Dimethylsulfide yields so seductive an aroma that it is sometimes used as an ingredient in perfumes. Dimethylsulfide is only agreeable, however, when present in extremely low concentrations. At higher concentrations its odour

Fig. 1.15. Geothermal hot water pipes provide a habitat for the thermophilic bacterium, *Thermus scotoductus*. Iceland.

Fig. 1.16. The Royal London Hospital, UK, where the thermophilic bacterium, *Thermus scotoductus* was isolated from hot tap water.

become unpleasant. In fact, dimethylsulfide is toxic to humans. Dimethylsulfide (C_2H_6S) is a breakdown product of dimethylsulfoniopropionate ($C_5H_{10}O_2S$). Dimethylsulfoniopropionate is produced by particular kinds of bacteria, algae and plants and from this molecule certain kinds of marine bacteria produce odorous dimethylsulfide. The major producers of dimethylsulfoniopropionate in salt marshes, estuaries and in deep oceanic sediments are bacteria.

When you pay a visit to a coastal marsh you enter a zone of intense microbial productivity. It has been estimated that a gram of sediment from a salt marsh on the Norfolk coast (Fig. 1.17) can be expected to contain roughly 100 million dimethylsulfoniopropionate-producing bacteria. The compound is synthesised by many species of salt marsh bacteria, including members of the genera *Oceanicola* and *Marinobacter* (Williams *et al.*, 2019). Concentrations of the molecule in coastal sediments may greatly exceed those found in open water. The degradation of dimethylsulfoniopropionate by bacteria yields dimethylsulfide (Kiene *et al.*, 2000). Numerous species of marine bacteria are known to be involved in the demethylation of dimethylsulfoniopropionate and the release of dimethylsulfide (Liu *et al.*, 2018). Some marine bacteria use dimethylsulfoniopropionate as a source of carbon and sulphur and some species incorporate it into their cells to protect them against osmotic stress in conditions of variable salinity. Planktonic algae as well as bacteria store dimethylsulfoniopropionate in their cells to protect them against osmotic stress. Each year phytoplankton synthesise about eight billion tonnes of dimethylsulfoniopropionate (Yoch, 2002). Salt marsh plants such as Common Cordgrass, *Spartina anglica* also synthesise and store dimethylsulfoniopropionate. The scent of aerosolised dimethylsulfide tells us that we are by the sea. It is also a climatically active molecule that promotes cloud formation in the atmosphere.

Fig. 1.17. Dimethylsulfide produced by bacteria in coastal sediments largely accounts for the distinctive smell of sea air. Stiffkey Marshes, Norfolk, UK.

Humans are extremely sensitive to geosmin ($C_{12}H_{22}O$). Geosmin is the scent of freshly turned soil and of parched ground after a shower. Geosmin has an attractive odour and highly diluted geosmin is used as an ingredient in perfumes. Geosmin is produced by filamentous soil bacteria that belong to the genus *Streptomyces* (Polak and Provasi, 1992). Geosmin is also produced by some species of fungi and by the beetroot, *Beta vulgaris*. The beetroot, known for its robust earthy taste, is an unusual vegetable in that its flavour is derived from geosmin. Geosmin is present also in water, where it is produced by aquatic cyanobacteria (Lu *et al.*, 2003; Suurnäkki *et al.*, 2015). Human responses to geosmin vary according to the context. Whereas the scent of geosmin rising from damp earth after a drought is delightful, the taste of geosmin in corked wine is repulsive. Similarly we find geosmin-tainted fish to be unpalatable and we avoid drinking water that tastes of geosmin (Darriet *et al.*, 2000; Jüttner and Watson, 2007; Houle *et al.*, 2011; Lukassen *et al.*, 2019). Although we find the smell of freshly turned soil to be pleasant, we find the smell of damp buildings to be musty and uninviting. When we enter a damp building, we interpret the musty odour as a sign of decay. *Streptomyces albidoflavus*, a geosmin-producing bacterium that gives damp buildings their musty odour is a member of the genus of bacteria that produce geosmin in soil (Claeson and Sunesson, 2005). When we enter a musty building we know that something is wrong. We know somehow that a bacterium has found its way indoors and made itself at home in the wrong place.

There appears to be something primordial and atavistic about our ability to detect and respond to geosmin. When we venture out between the showers on a summer's day and pick up the scent of geosmin on the gentle breeze, it makes us feel as though we are heading in the right direction, whereas food tainted with geosmin's earthy tang tends to make us think twice about ingesting it; somehow we seem to know instinctively that what we are consuming may do us harm. Other animals as well as humans react to geosmin. Geosmin attracts mosquitoes to egg-laying sites but it repels fruit flies (Melo *et al.*, 2020). Springtails (*Collembola*) are attracted to geosmin. Springtails are minute highly mobile soil arthropods. *Streptomyces* bacteria are dispersed by spores. When they sporulate, *Streptomyces* bacteria release geosmin, which attracts springtails. Springtails not only feed on the streptomycetes but also aid in the dispersal of their spores (Becher *et al.*, 2020).

Luminous bacteria

If you venture out into the field after dark you may witness beautiful and mysterious nocturnal phenomena. In suitable locations where light pollution is not excessive it is not unusual to encounter the occasional glow-worm, *Lampyris noctiluca*; and with luck perhaps other kinds of natural luminescence such as luminous bacteria and fungi In coastal areas you may witness the bioluminescent marine dinoflagellate, *Noctiluca scintillans*. There are also luminous marine bacteria such as the bacteria that belong to the genus *Photobacterium* (Seafood Glow Bacteria). Among other things, these luminous marine bacteria are associated with the decomposition of seafood. A neglected fridge with a broken light that contains the leftovers from a *plateau de fruits de mer* would not be an unreasonable place to start looking for them. Consumer complaints received by the US Food and Drug Administration have included mention of red snapper fillets that emitted green iridescent light, shrimps that glowed in the dark and luminous imitation crab meat. When investigated, all of these cases were found to be culture-positive for a glow-in-the-dark marine bacterium, *Photobacterium phosphoreum* (Sado, 1998). Bioluminescent bacteria occur widely in nature, although they generally go unnoticed unless seen in otherwise very dark conditions (Figs. 1.18–1.19). Luminous bacteria must have been more widely experienced in past ages before electric lighting blinded us to the subtlety of microbial luminescence. In *A Christmas Carol* (1843), Charles Dickens described Jacob Marley's ghost as having '*a dismal light about it, like a bad lobster in a dark cellar*'.

In an address to the South London Entomological and Natural History Society, the entomologist Kenneth G. Blair suggested that some of the observations recorded as examples of will o' the wisp might be accounted for by the presence of bioluminescent bacteria on the bodies of insects (Blair, 1921–1922). Kirby and Spence in their *Introduction to Entomology* noted that in 1780 at Ickleton, Cambridgeshire a farmer had brought to the Reverend Doctor Sutton of Norwich a mole cricket 'and told him that one of his people, seeing a *Jack-o'lantern*, pursued it and knocked it down, when it proved to be this insect, and the identical specimen shown to him'. W.J. Lucas, an authority on crickets and other Orthoptera, speculated that the Reverend Doctor Sutton's mole cricket might have been attacked by a bioluminescent fungus (Kirby and

Fig. 1.18. Cultures of luminous bacteria (dark): “Bioluminescent Bacteria Light Lab” by Dr Simon Park and Anna Dumitriu. Part of “Bacteria Light Lab” at The Wellcome Collection, London, UK (Alex May).

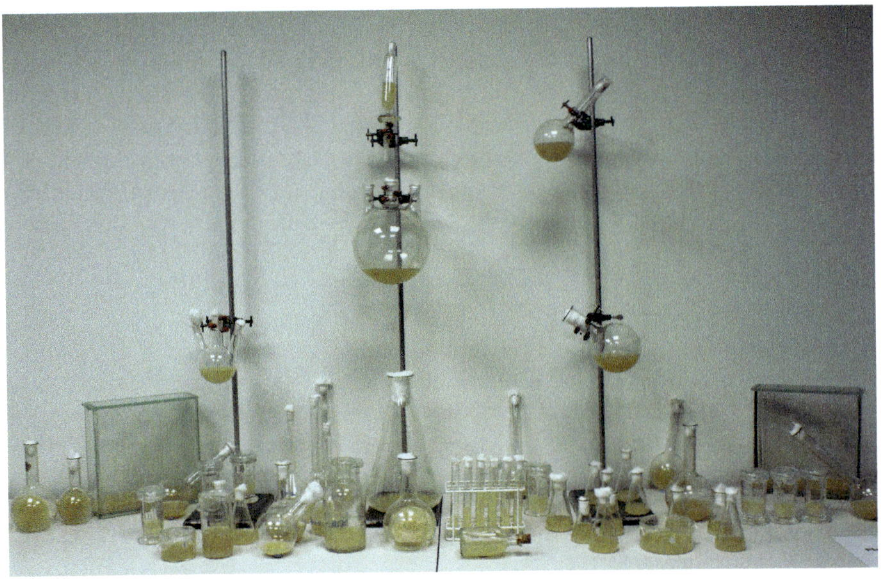

Fig. 1.19. Cultures of luminous bacteria (light): “Bioluminescent Bacteria Light Lab” by Dr Simon Park and Anna Dumitriu. Part of “Bacteria Light Lab” at The Wellcome Collection, London, UK (Alex May).

Spence, 1816–1826; Lucas, 1920). It seems more likely in fact that the mole cricket had been made phosphorescent by the presence of glow-in-the-dark bacteria.

Bacteria as insect pathogens

The mole cricket, *Gryllotalpa gryllotalpa*, is now vanishingly rare in Britain. Populations of mole crickets still occur on Guernsey and in the New Forest and occasionally mole crickets turn up in association with plants imported from southern Europe (Fig. 1.20). In contrast to the current situation in the British Isles, in some parts of the world mole crickets are so common that they are regarded as plant pests. In the USA, *Steinernema scapterisci*, a species of nematode worm, is marketed as a biological control agent of mole crickets. The worms act as vectors for a pathogenic bacterium *Xenorhabdus innexi* (the Cricket Slayer). The genus *Xenorhabdus* includes dozens of species, all of which are deadly insect pathogens. Numerous species of insect are susceptible to infection. *Xenorhabdus* bacteria live as symbionts of *Steinernema* worms. The worms transport the bacteria to their insect victims in a special pouch. Later on, the nematodes reproduce inside the deceased insect.

Xenorhabdus has a sister genus, *Photorhabdus*. The bacteria that belong to the genus *Photorhabdus* are bioluminescent. Their insect victims glow in the dark. *Photorhabdus* bacteria live as symbionts in the intestines of nematode worms belonging to the genus *Heterorhabditis*. The worms enter their prey through natural openings or by incising the cuticle. Once inside, the worms release their cargo of bacteria into the body cavity of the insect. The bacteria secrete a lethal toxin. The nematodes feast on the culture of bacteria inside the dead insect and reproduce before marching off in search of another victim. Numerous insects are known to be susceptible to infection by *Photorhabdus*,

Fig. 1.20. Mole cricket. Big Plant Nursery, Sussex, UK.

including at least one species of cricket, the house cricket, *Acheta domesticus*. One species, *Photorhabdus asymbiotica* (the Angels' Glow Bacterium) is able to grow at human body temperature and is known to cause luminous wound infections in humans (Forst *et al*., 1997; Ciche and Ensign, 2003; Waterfield *et al*., 2009; Mulley *et al*., 2015; Sajnaga and Kazimierczak, 2020).

Bacteria in folklore: falling stars and will o' the wisp

Although bacteria generally go unnoticed, some species form distinctive, readily visible colonies. After a spell of heavy rain, conspicuous olive-green, jelly-like blobs called nostoc may suddenly appear, as though from out of the blue. Field naturalists may not bother to take much notice of them but these blobs are of interest to a microbiologist because they are actually colonies of *Nostoc commune*, which is a kind of cyanobacterium (Fig. 1.21). When the weather is dry, nostoc colonies exist inconspicuously in the form of desiccated crusts (Fig. 1.22). When exposed to moisture, nostoc crusts absorb water and swell into vibrant blobs of jelly. Nostoc blooms are a feature of various kinds of habitat, including chalk grassland and limestone pavement. Nostoc adds an element of drama to calcareous landscapes (Fig. 1.23). Nostoc blobs may appear also on artificial structures made of stone or concrete, such as walls, flat roofs and driveways (Fig. 1.24).

Nostoc provides one explanation for the substance (or substances) known colloquially as star jelly. Star jelly is a term that has been used to envelop a whole constellation of mysterious blobs and slimes which occur in the natural world. Star jelly is a perplexing phenomenon. It has become a part of folklore and it has attracted serious scientific enquiry. Nostoc does not hold a monopoly on star jelly: it can be noted that slime moulds, for example, provide one of a number of other plausible explanations for sightings of jelly-like

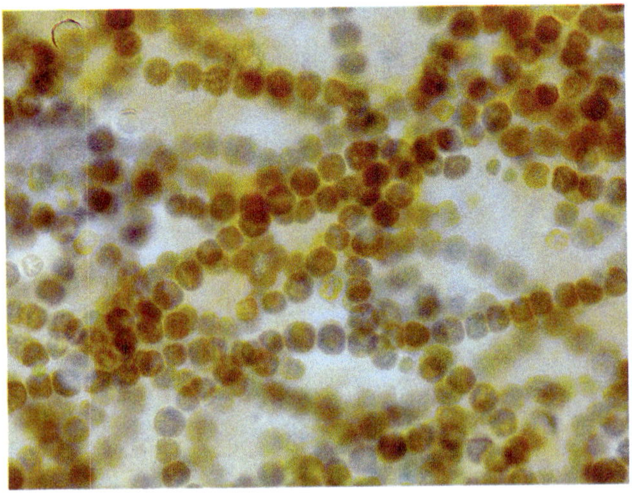

Fig. 1.21. Cells of *Nostoc commune* seen on microscopy. Anchor Bottom, Sussex, UK.

Fig. 1.22. Partly desiccated *Nostoc commune* on chalk downland at Anchor Bottom, Sussex, UK.

Fig. 1.23. Downland habitat for *Nostoc commune* at Anchor Bottom, Sussex, UK.

apparitions in nature. Curiously, the abrupt appearance of star jelly has been linked in folklore to meteor showers. It was supposed that the jelly-like masses represented the remnants of fallen stars. However, showers of common or garden rain will suffice to make nostoc visible. Originally spelt *Nostoch*, the name was coined by the Swiss physician, botanist and astrologer Paracelsus

Fig. 1.24. Nostoc on damp concrete after heavy rain at Prora, Germany.

(1493–1591). Paracelsus does not seem to have left a record of the word's derivation. Malcolm Potts, an authority on the cyanobacteria, proposed that *Nostoch* may represent a combination of the English and German words for nostril in reference to nostoc's green mucilaginous appearance (Potts, 1997). Malcolm Potts collected a long list of popular terms for nostoc; as well as star jelly, these include star shot, star slime and fairy butter in English, *pwdr sêr* (star rot) in Welsh, *beurre magique*, *fleur du soleil* and *merde du coucou* in French and *Himmelsblume* and *Schleimling* in German.

Will o' the wisp, or *ignis fatuus* (foolish fire), is another folklore oddity of interest to bacteriologists. Typical descriptions relate to a form of luminescence, often in boggy terrain, that would appear to move from one spot to another (Fig. 1.25). Although will o' the wisp seems to be out of fashion these days, sightings do not seem to have been particularly rare events in the past. Will o' the wisp has been a source of inspiration for poets and scientists. Numerous literary references to it exist. Goethe's fairytale *The green snake and the beautiful lily* features two will o' the wisps (Goethe, 1795), whilst one of the characters in D.H. Lawrence's novel, *Women in love* (Lawrence, 1920) has qualms about '*....lilies and snakes, and the ignis fatuus....*' (Lawrence, 1920). Will o' the wisp is known as *ellyldân* (elf fire) in Welsh, *tannos* (night fire) in Cornish, *teine sìth* (fairy light) in Gaelic, *feu follet* (goblin fire) in French and *Irrlicht* (wandering light) in German. There are many alternative terms for will o' the wisp in English, including jack o' lantern, friar's lantern and hobby lantern. Such a multiplicity of popular names creates an impression that an encounter with will o' the wisp was once a not uncommon, if albeit a rather weird experience for significant numbers of ordinary people. Star jelly and will o' the wisp both occupy common ground in the human psyche. The body of folklore that encompasses star jelly and will o' the wisp provides insights

Fig. 1.25. *The ignis-fatuus, or will'o the wisp: a mirage in a marsh.* Coloured wood engraving by C. Whymper (Wellcome Collection). Public Domain Mark.

into ways of thinking about the natural world that pre-date the Enlightenment, the invention of the microscope, the discovery of bacteria and the systematic application of the scientific method as a mode of inquiry.

Louis Blesson, a nineteenth century investigator of mysterious natural events described observations that he had made of star jelly as well as will o' the wisp when he contributed a paper to *The Edinburgh New Philosophical Journal* (1832), entitled: *Observations on the Ignis Fatuus, or Will-with-the Wisp, Falling Stars, and Thunderstorms*. With regard to star jelly, Blesson noted that he had: '*frequently observed on meadows and fields that slimy, leek-green matter, which is commonly taken for the product of falling-stars, fire-balls, andc*'. Speculating on a terrestrial explanation he noted that: '*plants, particularly the cryptogamic ones in the vicinity, ought to be examined. I may add that I observed this jelly, in a forest under a fir tree, where there was no possibility of its having fallen from the sky*'. Louis Blesson also had first-hand experience of will o' the wisp. He recorded his observations of the *ignis fatuus* in a marshy valley in Germany where: '*During the day bubbles of air were seen rising from it, and in the night blue flames were observed shooting from and playing over its surface*'. Blesson performed an experiment to test a conjecture that bubbles seen rising during the day might be connected to the mysterious lights seen at night: '*I first extinguished the flame, and then hastened with a torch to the spot from whence the gas bubbled up, when instantaneously a kind of explosion was heard, and a red light was seen over eight or nine square feet of the surface of the marsh, which diminished to a small blue flame, from two and a half to three feet in height, that continued to burn with an unsteady motion. It was therefore no longer doubtful that this ignis fatuus was caused by the evolution of inflammable gas from the marsh.'*

Louis Blesson's nineteenth century experiments support a widely held view that at least some, if not the majority of the observed lighting effects that have been described under the will o' the wisp umbrella relate to the spontaneous combustion of marsh gas or methane, a flammable gas released as a waste product by methane-producing microbes. Nevertheless, spontaneously combusting marsh gas is not the only natural cause of nocturnal luminescence: there are bioluminescent insects, bacteria and fungi. They too might account for some of the mysterious incidents that have been labelled as will o' the wisp, jack o' lantern or the *ignis fatuus*. A number of factors might have led to will o' the wisp's decline in modern times, including light pollution, the draining of bogs and fens and a tendency for people to travel by car rather than to take a short cut home at night through swampy terrain.

Methane-producing microbes are just as much at home in the airless depths of a landfill site full of rotting rubbish as they are in a stagnant bog of decaying vegetation. The same species of microbes that occur in bogs can also be found in rubbish dumps. Nowadays a magnificent modern-day alternative to the *ignis fatuus* can be witnessed in the form of the heat haze that rises above an actively managed landfill site where methane released by decaying rubbish is being burnt in a controlled manner to power a methane-burning generating station (Figs. 1.26–1.27). When standing at the boundary fence of a rubbish dump and contemplating the scene one has to be impressed by the thought of

Fig. 1.26. Methane-burning generating plant. Small Dole, Sussex, UK.

the enormous numbers of methane-producing microorganisms that are living underground and out of sight of the observer in their peculiar manmade habitat amongst the tonnes of discarded organic matter. Although *ignis fatuus* (foolish fire) may be a thing of the past, this does not stop us in this modern age from enjoying will o' the wisp's contemporary counterpart, *ignis quisquiliarum* (trash torchlight) (Fig. 1.28). Bacteria have important roles in decomposition and more generally in soils and sediments in the recycling of carbon, nitrogen

Fig. 1.27. The heat haze above a methane-burning generating plant. Small Dole, Sussex, UK.

Fig. 1.28. Methane flare warning sign at a landfill site. Port Solent, Hampshire, UK.

and other elements. Whole landscapes can be shaped and coloured by bacteria, especially in environments that are too hot, too cold, too dry, too salty or too acidic for other forms of life.

Bacteria, *Archaea* and the eukaryotes: three domains of life

Methane-producing microbes (methanogens) are not 'true bacteria'. They belong to a sister group of microbes known as the archaea. At one time, the archaea, or archaebacteria as they were once known, were considered simply to constitute a group of specialised bacteria. Many of the archaea (singular, archaeon) are found in extreme environments such as salt lakes and hot springs. To our eyes, archaea and true bacteria look much the same: all of them are microscopic and many of them are similar in size and shape. Although when seen under the microscope, archaea and bacteria may look superficially alike, significant differences exist between them in their genetic makeup and biochemistry. Archaeal ribosomes, for example, the cellular organelles involved in protein synthesis, are structurally different to those of bacteria. Awareness of the differences between bacteria and archaea led to the classification of the archaea as a separate group. Initially, they were classified as a distinctive group of bacteria and were called archaebacteria. The name archaebacteria is derived from Greek and means ancient (ἀρχαῖος, archaíos) bacteria as it was supposed that archaea such as the methanogens might be representative of the kind of organisms that existed during the early stages of life on earth when the planet's atmosphere lacked significant amounts of oxygen. Later, a scheme of classification was proposed that recognised the fundamental distinctiveness of the archaea as a group in its own right by giving the archaea the status of a 'domain' that was of equal rank to the bacteria. At a molecular level, the archaea are more closely related in some respects to eukaryotes than they are to the true bacteria. This book is about two major groups or domains of microorganisms: the Domain *Bacteria*, which accommodates the bacteria and the Domain *Archaea*, which accommodates the archaea.

According to the three-domain system of classification, which was formally proposed by the American biologist Carl Woese (1928–2012) and colleagues in 1990, all cellular organisms belong to one of three major groups or domains: *Bacteria*, *Archaea* and the eukaryotes (Woese *et al.*, 1990). The eukaryotes (Eukarya or Eukaryota) are organisms whose cells have a nucleus, a special organelle that contains the genetic blueprint of the organism encoded in strands of deoxyribonucleic acid (DNA). The term eukaryote is derived from the Greek word for a nut or kernel, κάρυον (karyon). Karyon is an alternative but less commonly used term for nucleus. The eukaryotic organisms include the animals, plants, fungi and also many single-celled organisms such as amoebae and green algae. The cells of bacteria and archaea are generally much smaller than eukaryotic cells (Figs. 1.29–1.30). There is much less DNA in bacterial and archaeal cells than there is in a eukaryotic cell and bacterial and archaeal DNA is not enclosed in a nucleus or comparable special structure. Because archaea and bacteria lack a nucleus they can be referred to collectively as prokaryotic organisms, in contrast to the eukaryotic organisms that do have one.

Fig. 1.29. Bacteria. A watercolour by Anna Dumitriu.

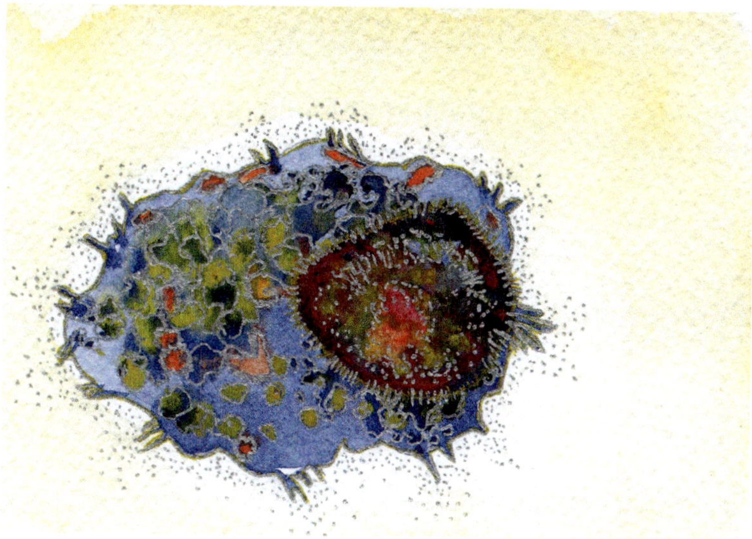

Fig. 1.30. Eukaryotic cell. A watercolour by Anna Dumitriu.

The term prokaryote is a useful umbrella term when discussing collections of microbes that might consist of archaea as well as bacteria. Some biologists like to use the terms eubacteria, or true bacteria, to emphasise that they really mean bacteria as opposed to archaea. Meanwhile other biologists seem content to continue to apply the term bacteria informally to any prokaryotic organism. Much of the subject matter of this volume concerns members of the Domain *Bacteria*, simply because of the fact that on a global basis bacteria account for significantly more individual cells, significantly more biomass and more described species than the archaea.

Replication and sporulation

Bacteria are able to tolerate a wider range of environmental conditions than plants and animals. For some of them, the optimal temperature for growth is above 80°C. No plant or animal can tolerate such temperatures. Bacteria are highly adaptable. One of the reasons for their success relates to the ease with which they are able to exchange chunks of their genetic material, even between distantly related species. Another reason for their success relates to their high replication rates. Given a supply of nutrients and suitable conditions, a population of a species of bacterium such as the well-known *Escherichia coli* (Escherich's or the Common Gut-bug) can double at roughly hourly intervals. Conversely, in more hostile conditions, such as in nutrient-depleted deep oceanic sediment, the bacteria entombed there survive by ticking over at a very slow rate and may replicate as infrequently as once every few thousand years (Jørgensen and Marshall, 2016; Starnawski *et al.*, 2017; Bradley *et al.*, 2018). When conditions become unfavourable for them, some species of bacteria ensure their survival by producing resilient spores. The spores of a strain of the soil bacterium *Bacillus amyloliquefaciens* recovered from a desert region in Morocco can withstand exposure to a temperature of 420°C (Beladjal *et al.*, 2018).

Relationships with plants and animals: symbionts, commensals and pathogens

Bacteria have a long history. Putative microfossils found in rock samples from northern Quebec suggest that bacteria may have existed as early as 4 billion years ago (Dodd *et al.*, 2017). Physical evidence of the former presence of ancient bacteria exists in the form of microfossils and also in the form of mineralised laminated microbial mats known as stromatolites (Fig. 1.31). Stromatolites occur widely in ancient rock formations. An ancient stromatolite reef found at Pilbara in Western Australia has been dated as being almost 3.5 billion years old. Stromatolites are still being formed today by mats of bacteria in a few select sites around world.

Multicellular animals do not start to appear in large numbers in the fossil record until about 530 million years ago. The first land plants began to appear at about the same time. The bodies of plants and animals provide habitats for bacteria. Bacteria that live in association with other organisms are known as symbionts and the term symbiosis is used as a general term to describe situations in which different kinds of organism live together in close proximity. The term symbiosis encompasses a wide range of different kinds of relationship. At one extreme is mutualism, an intimate kind of symbiosis in which all partners derive direct benefit from the relationship. A well-known example of mutualism is the symbiosis formed between plants of the pea family and nitrogen-fixing bacteria in which the bacteria supply the plant with nitrogenous chemicals in exchange for a place to live and access to nutrients (Fig. 1.32). Nitrogen-fixing bacteria live inside plant cells situated within specialised root nodules.

Fig. 1.31. Devonian (c. 390 million year old) stromatolite. Stromness, Orkney, UK.

Fig. 1.32. Sea pea, *Lathyrus japonicus*, a leguminous plant whose root nodules accommodate nitrogen-fixing bacteria. Rye Harbour, Sussex, UK.

The term commensalism describes a less intimate kind of relationship. Commensal bacteria may be thought of as messmates, eating at the same table as their host. Commensal bacteria derive benefit by forming relationships with other organisms. Commensal bacteria are generally neither harmful nor particularly beneficial to their host. A typical flowering plant can accommodate thousands of species of bacteria on the surfaces of its leaves, shoots, flowers and seeds, as well as inside root tissues and within the internal spaces of stems and leaves. Enormous numbers of commensal bacteria occur also on the external surfaces and within the body cavities of animals. Compartments of the mammalian body, including the oral cavity, large intestine and upper respiratory tract each host a distinctive commensal microbial flora. Different animal species are associated with their own particular species of commensal bacteria. Some species of bacteria, however, can be found living as commensals of many different kinds of animals. *Escherichia coli* is a characteristic and well-known member of the gut flora of man and other animals. Most strains of *E. coli* exist as harmless members of the intestinal flora but there are also pathogenic strains of *E. coli* that cause enteric disease. Pathogens cause harm by invading and destroying tissues or by releasing toxins. Pathogens can be defined as organisms that cause disease.

Symbiotic bacteria involved in long-term relationships with other organisms but whose actions are detrimental to the host are generally referred to as parasitic bacteria. An example of parasitic symbionts are the male-killing bacteria that induce the death of male ladybirds during their development. The bacteria are passed onto the next generation via the eggs of infected female ladybirds. By inducing the death of male insects, the bacteria increase the proportion of infected ladybirds (Majerus, 1994).

The mammalian gut provides accommodation for a complex normal flora of microbes. Hence, the gut is an example of an internal body compartment that can be described as being naturally non-sterile. A disadvantage of such an arrangement is that it renders the host susceptible to the ingress of potential pathogens. These may include species of bacteria associated with food poisoning, for example. It has long been customary to refer to the bacteria that normally reside in the human gut simply as the normal intestinal flora. There is, however, a growing appreciation that gut bacteria are more than just accidental passengers. We have coevolved with our gut flora and intestinal microbes are actually of importance to our wellbeing. There is a theory that a major driving force behind the evolution of the highly developed mammalian immune system is the selective advantage proffered by the ability to accommodate a beneficial gut flora. Our immune system not only provides a defence against invasion of the body by disease-causing bacteria but also allows us to manage and tolerate a community of gut bacteria which facilitates our access to nutrients extracted from ingested food.

The best known microbial resident in the mammalian gut flora is probably *Escherichia coli* (Escherich's or the Common Gut-bug). *E. coli* is a member of the *Enterobacteriaceae*, a family of bacteria particularly associated with the gut flora of animals. Microbiologists often use the informal term 'coliform' to refer to *E. coli* and the related species in the family *Enterobacteriaceae* that resemble it (Fig. 1.33). Beside the coliform group, many other major phylogenetic groups of bacteria contribute to the communities of bacteria that occur in the guts of animals. Many of these specialised gut bacteria are anaerobes which are intolerant of oxygen. One of the reasons that *E. coli* is so well known is that it is tolerant of a wide range of growth conditions. Hence, this organism

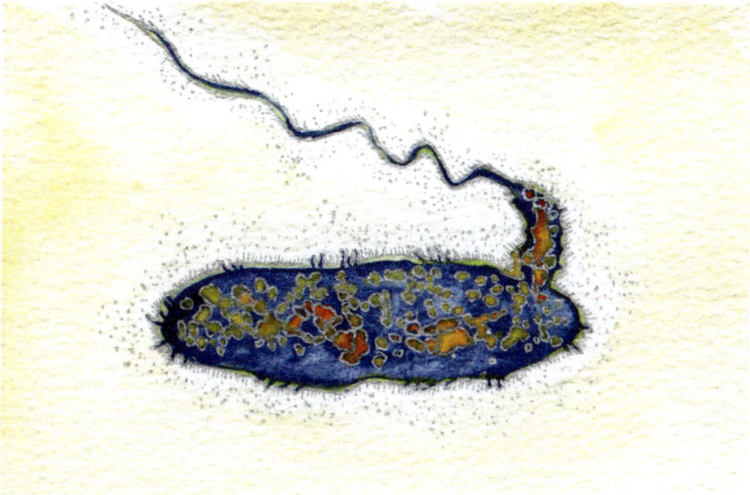

Fig. 1.33. *Escherichia coli.* A watercolour by Anna Dumitriu.

lends itself to study in the laboratory. In contrast, the strictly anaerobic bacteria, which make up a large proportion of the gut flora must be cultured in anoxic conditions in specialised microbiology laboratories.

Human interest in bacteria

Bacteria play key roles in soil ecology in decomposition and the recycling of minerals and nutrients. They occur throughout land, freshwater and saltwater environments and move through the air attached to particles of dust, in droplets or as spores. Bacteria live where plants and animals cannot: they have been recovered from deep below the earth's surface, from the stratosphere high above the earth and from extremely hot, acidic or saline environments. They can survive and replicate, albeit slowly, at very low temperatures. Bacteria thrive even in the Antarctic Dry Valleys where desert dryness and freezing temperatures combine to form one of the most hostile environments on the earth's surface. Some bacteria survive high doses of ionising radiation. Nevertheless, there is little popular interest in these organisms. Bacteria most often make us sit up and take notice when they cause disease, even though only a tiny percentage of the many known species of bacteria are capable of infecting humans. Bacteria were present billions of years before animals and plants appear in the fossil record. Much of the history of life on earth concerns bacteria but much of the (fairly short) history of bacteriological study relates to the investigation of infectious diseases and the association of particular diseases with their causative bacteria.

The Dutch microscopist, Antonie van Leeuwenhoek (1632–1723) is credited as being the first person to visualise individual bacterial cells. However, microbiology did not really take off as a scientific discipline until the latter half of the nineteenth century. Researchers such as Robert Koch (1843–1910) and his colleagues in Germany developed culture techniques that included the use of dishes of nutrient agar, making it possible to isolate different species of bacteria in pure culture. Many studies conducted by the pioneers of bacteriology concentrated on the very select group of species that cause diseases of people, other animals and plants. This sort of thing gives bacteria a bad name and encourages a popular view that all bacteria are pathogens that are trying to kill us. In more recent decades the importance of bacteria in global ecology, in the history of life on earth and in geological history has been increasingly appreciated. Also, it may be noted that bacteria have their uses. For example, enzymes that withstand high temperatures, derived from thermophilic bacteria, are included in biological washing powders and are used in starch processing. Heat-stable enzymes obtained from thermophilic bacteria also form the basis of the diagnostic PCR (polymerase chain reaction) tests that are performed in medical laboratories. Bacteria have applications in the recovery of metals from ore deposits and in the bioremediation of polluted sites. Others are used in the manufacture of foods and drinks. The distinctive flavours of some foods including particular kinds of cheese, beer, salt and perhaps even strawberries are due to the presence of substances made by bacteria. Bacteria had something like a 2 billion year head start on the more complex forms of life. Bacteria tell you just about everything you need to know about life on earth and maybe about carbon-based life anywhere. Bacteria make humans look small.

A brief description

little sticks, were flying about …

Joseph Conrad (1899). *Heart of Darkness.*

Bacterium, Latinised form of the Greek, βακτεριον, *bakterion*, diminutive of βακτρον, *baktron*, a stick.

Bacteria are huge: they occur just about anywhere on the planet where life can exist, including land, sea, air and subterranean rocks. As a group they constitute a large proportion of the earth's biomass; but as individuals they are microscopic. A much-studied bacterium such as *Escherichia coli*, which forms part of the intestinal flora, measures about 5 μm (micrometres) or five thousandths of a millimetre in length. To give some idea of scale, this is roughly a thousandth the diameter of a small printed full stop. Even when viewed at the highest magnification (×1000) under a light microscope, *E.coli* still looks small (a bit like looking at a tiny printed hyphen). Many of the well-known bacteria are similar in size to *E. coli*. Cells of '*Nanoarchaeum equitans*', (the Pygmy Archaeon-rider) a thermophilic archaeon found in a hydrothermal vent near Iceland are a mere 400 nm (nanometre) or 400 millionths of a millimetre in diameter and less than a tenth the size of *E. coli* (Huber *et al.*, 2002). '*Nanoarchaeum equitans*' lives attached to a larger species of archaeon, possibly as a mutualistic symbiont or possibly as a parasite (Fig. 1.34).

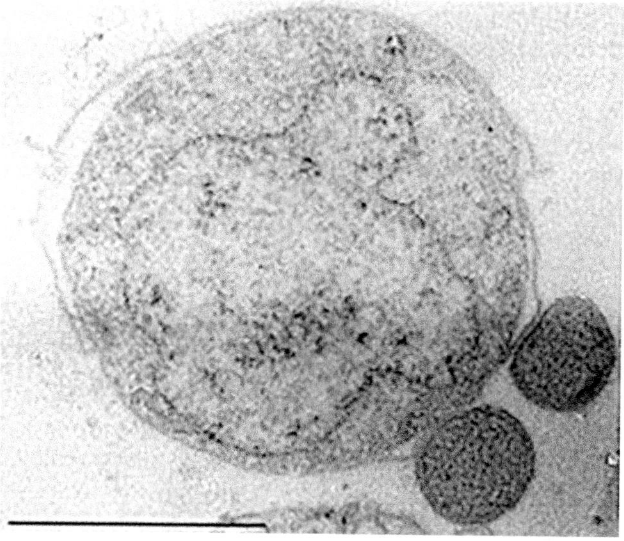

Fig. 1.34. Cells of '*Nanoarchaeum equitans*' dwarfed by its host archaeon, *Ignicoccus hospitalis* (Karl Stetter). Public Domain.

The largest known bacterium, '*Thiomargarita magnifica*', (the Big Bacterium) was discovered in a mangrove swamp on the island of Guadeloupe in the French West Indies as recently as 2009, although the discovery was not published until 2022 (Volland *et al.*, 2022). The filamentous cells of the Big Bacterium grow to a length of up to a centimetre and are visible to the naked eye. The previous record holder for the world's largest bacterium, *Thiomargarita namibiensis* (the Pearl of Namibia) was discovered living off the coast of Namibia in 1997. The cells of *Thiomargarita namibiensis* are roughly spherical in shape and may reach a diameter of as much as 750 μm (about the size of a small printed full stop), making them just about visible to the naked eye. The Pearl of Namibia and its relatives are known as giant sulphur bacteria because they feed on hydrogen sulphide. Giant sulphur bacteria use nitrate to convert hydrogen sulphide to sulphur. Nitrate is a scarce commodity in marine sediments and the enormous size of these bacteria is related to a huge vacuole which is used to store nitrate. Giant sulphur bacteria also have smaller storage vacuoles which contain phosphate. A bacterium related to *Thiomargarita namibiensis* but with smaller cells (diameter up to 375 μm) inhabits marine sediments in the Gulf of Mexico. Other kinds of giant sulphur bacteria occur widely in marine environments (Salman *et al.*, 2013).

Another (just about) visible bacterium is "*Epulopiscium fishelsoni*" (Fishelson's Fish-feaster) a species that occurs as part of the gut flora of the brown surgeonfish, *Acanthurus nigrofuscus*, in the Red Sea. Surgeonfish in various parts of the Pacific and off South Africa also harbour related bacteria. "*E. fishelsoni*" is known to grow up to 600 μm in length. Giants such as the Pearl of Namibia and Fishelson's Fish-feaster though are exceptional. *Thiomargarita namibiensis* and "*Epulopiscium fishelsoni*" are like Jupiter and Saturn among the planets of the bacterial system.

The cyanobacteria are unusual among the more widely distributed groups of bacteria in being represented by many species that have large cells by bacterial standards. The relatively large size of cyanobacterial cells, their green colour due to the presence of chlorophyll and the tendency of many species to grow as visible colonies as blobs or as long filaments or streamers caused them for many years to be regarded by many scientists as primitive plants or mini-algae rather than bacteria. Confusingly, cyanobacteria are sometimes called blue-green algae.

Gram-positive and Gram-negative bacteria

Bacteria occur in a variety of shapes. Many are approximately spherical and called cocci (from the Greek, κόκκος, kokkos, berry) or rod-shaped and called bacilli (from the Latin, *bacillus*, little rod) or simply rods. Other shapes include curved rods and spirals. As well as describing bacteria according to their shape, bacteriologists often like to categorise bacteria as being Gram-positive or Gram-negative. Gram's stain is a traditional laboratory tool, used for differentiating bacteria into two major groups according to the composition and structure of their cell walls (Fig. 1.35). Ones that stain dark blue with Gram's stain are said to be Gram-positive and ones that stain pink are described as Gram-negative (Figs. 1.36–1.37). During the staining process, a microscope slide coated with bacteria is exposed to a dark blue stain (crystal violet), then washed with acetone and then dipped in a pink counterstain (such as safranin or neutral red). Gram-staining characteristics of bacteria relate to cell wall structure. Gram-positive cell walls retain the dark blue crystal violet stain, even

Fig. 1.35. Gram-staining equipment.

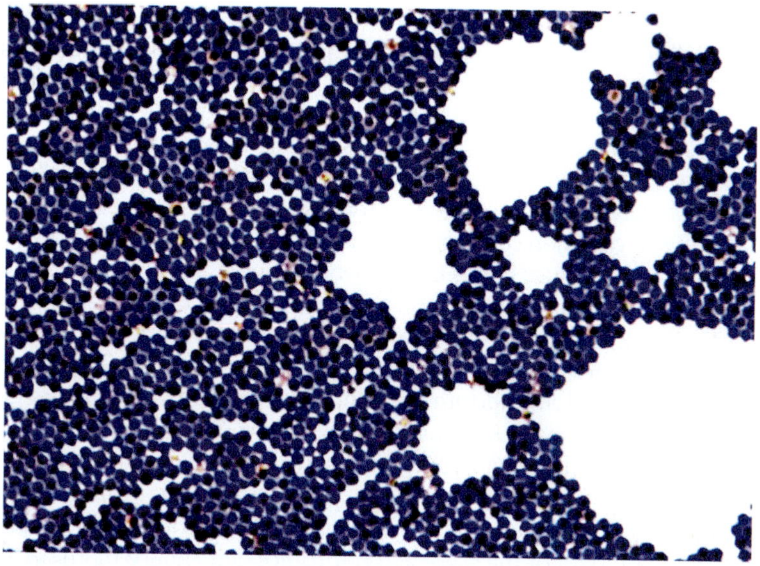

Fig. 1.36. Bacteria with 'Gram-positive' cell walls stain violet with Gram's stain.

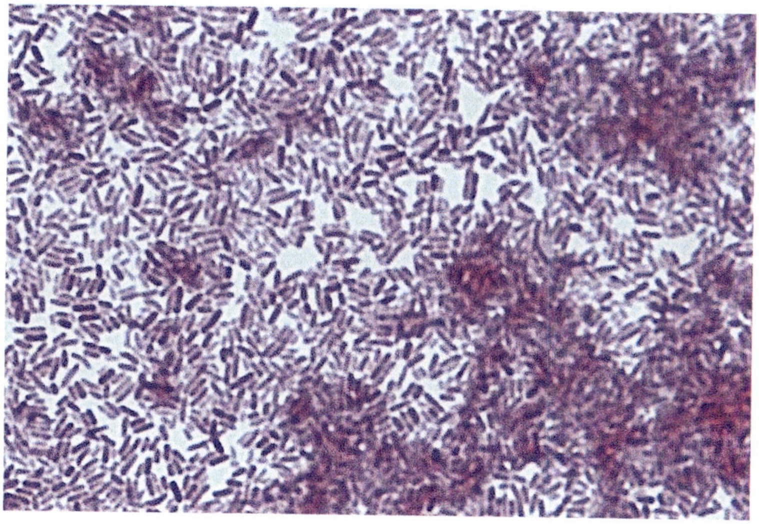

Fig. 1.37. Bacteria with 'Gram-negative' cell walls stain pink with Gram's stain.

when washed in acetone. Acetone has the effect of washing the crystal violet stain out of Gram-negative cells.

When looked at under a microscope, Gram-negative bacteria are made visible by the more delicate colour of the pink counterstain and appear pink rather than dark blue. Gram-positive cell walls are composed of a thick layer of peptidoglycan whereas Gram-negative walls have a thin peptidoglycan layer and also an outer membrane. The staining technique is named after Hans Christian Joachim Gram (1850–1938), a Danish physician and bacteriologist.

While working in Berlin, he developed a staining technique that was published in 1884 and has become known as the Gram stain or Gram's stain.

There are two major groups of Gram-positive bacteria, the Phylum *Bacillota* (previously *Firmicutes*) and the Phylum *Actinomycetota* (previously *Actinobacteria*). Numerous species of harmless environmental bacteria belong to the Phylum *Bacillota*. In addition, it happens that the *Bacillota* include a number of species of medical and veterinary importance, including *Staphylococcus aureus*, a common cause of skin and wound infections, *Streptococcus pyogenes*, which among other things is the agent of scarlet fever, and *Streptococcus pneumoniae*, an agent of pneumonia. The *Actinomycetota* are common inhabitants of soils and include bacteria that produce mould-like colonies that resemble fungi.

There are numerous taxonomic groups of Gram-negative bacteria. The Phylum *Pseudomonadota* accounts for many of the Gram-negative bacteria that can easily be cultured in the laboratory. The *Pseudomonadota* (formerly known as the Phylum *Proteobacteria*) are partitioned into classes (the names of which end in *-proteobacteria*). The Class *Alphaproteobacteria* includes a number of bacteria that form symbiotic relationships with eukaryotic cells and some members of the Class *Alphaproteobacteria* are mammalian pathogens. The Class *Betaproteobacteria* includes diverse environmental bacteria, as well as bacteria associated with animals. The Class *Gammaproteobacteria* includes enteric bacteria, such as *Escherichia coli* as well as numerous infectious agents of animals and plants.

Nomenclature

Scientific names for bacteria superficially resemble the scientific names of plants and animals. For example, *Escherichia coli*, a name which consists of a generic name (starting with an upper case letter) and a specific name (written entirely in lower case). As is the convention for plants and animals it is customary to print the scientific names of bacteria in italics. Many naturalists will not feel much need to go any deeper into bacterial nomenclature, although one practical point of difference from botanical and zoological nomenclature that should be noted is that with bacterial nomenclature it is customary to print the names of higher ranking taxa in italics. Hence: Domain: *Bacteria*, Domain: *Archaea*, Phylum: *Cyanobacteriota*, Family: *Corynebacteriaceae* and so on. Zoologists and botanists would usually write such higher ranking names in plain text. A helpful consequence of this convention is that an author can choose to distinguish between whether they mean to use the name of a higher rank precisely in a formal sense (by printing it in italics, starting with a capital letter), or imprecisely in an informal sense (by printing it entirely in lower case in plain text). For example, when we see the term *Bacteria* printed thus, we know that the author is writing unambiguously about members of the Domain *Bacteria*.

A number of requirements must be met for a prokaryotic name to be considered as being validly published. As some of these requirements are fairly exacting, the bacteriological literature is littered with names which have not been validly published. Some of these names are quite widely used. It is conventional to write names of organisms that have not been validly published in inverted commas. "*Epulopiscium fishelsoni*" is an example of a widely used name that has not been validly published. One of the requirements for valid publication of a name is the submission of a type culture of the organism to a reference collection. For organisms that are recalcitrant to isolation in culture but which have been well-characterised by molecular methods, such as genetic sequencing, authors can opt to propose a new name as a candidate taxon under the designation of *Candidatus* (abbreviated as *Ca.*). An example is "*Candidatus* Nanohalarchaeum antarcticus" (the Antarctic Salt-bug). The names of *Candidatus* taxa are written in inverted commas and the generic and specific names are printed in plain text. For additional information about nomenclature including a note about the names of phyla, see Chapter 10.

In addition to scientific names, microbiologists often find it convenient to use informal laboratory names that refer to the results of commonly used laboratory tests, such as Group A streptococcus, for *Streptococcus pyogenes* or Group B streptococcus, for *Streptococcus agalactiae*, in reference to Rebecca Lancefield's (1895–1981) serological typing method for streptococci.

Abundance and biomass

Although bacteria are microscopic they do occur in vast numbers and bacteria en masse may be visible and may even be conspicuous in some habitats as colourful mats or as glutinous strings that environmental microbiologists call bacterial streamers. There may be a million bacteria in a millilitre of pond water and many millions in a gram of soil but researchers were sceptical that a credible estimate for the total number of bacteria and archaea in the world could be calculated until William Whitman, a microbiologist at the University of Georgia realised that a plausible estimate could be reached by sampling a range of representative environments, counting the bacterial cells in each sample, multiplying each count to calculate the number of cells in the world volume of each environment and adding the totals. This approach relies on the assumption that worldwide average densities of organisms can be determined for each habitat with reasonable accuracy. The seas, fresh water and soil contain enormous numbers of bacteria but according to Whitman's findings, the most important environments appeared to be subsurface sediments (Whitman *et al.*, 1998). We are accustomed to thinking of the biosphere as the thin layer of the earth's surface that is inhabited by plants and large animals. However, bacteria have been recovered from subsurface water from boreholes and mines down to depths of several kilometres below the earth's surface. Hence, it can be supposed that bacteria may occupy a huge volume of the subsurface. Whitman's research provided a foundation, which other scientists have subsequently built upon, by incorporating results obtained from more recent surveys of microbial life in deep oceanic sediments (Kallmeyer *et al.*, 2012). According to a recent estimate, the global population of bacteria has been estimated to be somewhere in the order of one nonillion (equivalent to one followed by 30 zeros or 10^{30}) cells, most of which dwell in the subsurface (Bar-On *et al.*, 2018).

This equates to approximately 70 gigatonnes (equivalent to 70 trillion metric tonnes) of carbon. Archaea are less abundant that bacteria. It has been estimated that the global population of archaea amounts to approximately 100 octillion (equivalent to one followed by 29 zeros or 10^{29}) cells, which equates to 7 gigatonnes (equivalent to 7 trillion metric tonnes) of carbon. As with bacteria, most archaea are thought to live in the subsurface.

Geographical distribution and dispersal

A curious thing about bacteria and other microorganisms as compared with animals and plants is that their patterns of geographical distribution do not appear to correspond well with the biogeographical zones favoured by zoologists and botanists. The distribution patterns of many animals can be mapped to a set of zoogeographical zones (Afrotropical, Australasian etc.), each with its own distinctive fauna, although it is true that some animals have cosmopolitan distributions and it is not surprising that synanthropic species will tend to show up wherever there are people. However, in the case of microorganisms even non-synanthropic members of the same species, or at least closely related species seem to turn up anywhere in the world where there is suitable habitat to support their needs. Discontinuous global distribution patterns have been demonstrated for bacteria and archaea as well as for algae, rotifers and other microorganisms. Even highly specialised bacteria that thrive only in very localised geographically restricted habitats can be found in similar places at widely separated sites. An example of a distinctive species with a markedly discontinuous geographical distribution is Walsby's Square Archaeon, *Haloquadratum walsbyi*, which thrives only in nearly saturated salt solution in conditions where few other organisms exist (Fig. 1.38). It has been found in brine pools as far

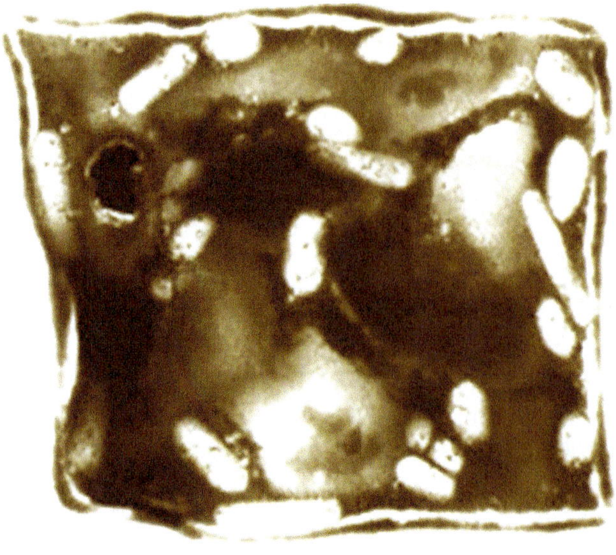

Fig. 1.38. Walsby's Square Archaeon. (Image kindly provided by Anthony Walsby.)

apart as Sinai, Spain and Australia. Walsby's Square Archaeon has even been detected in a sample of commercially available table salt sourced from a solar saltern on the island of Ibiza (Satari *et al.*, 2021). Strains from different continents have been reported to show only minor genetic differences (Dyall-Smith *et al.*, 2011). Hence, Walsby's Square Archaeon is a good example of an organism that supports the Baas Becking hypothesis: that everything is everywhere but the environment selects.

The geographical distribution of Walsby's Square Archaeon suggests that it may have the capacity to disperse in a non-vegetative state, even over great distances. Viable bacteria of many kinds can be detected in air samples and it is apparent that microorganisms attached to particles of dust and in droplets may be transported far and wide. Some bacteria are known to be resistant to desiccation and ultraviolet radiation, whilst others produce resistant spores that are resistant to desiccation and very high temperatures. The survival of even a small number of cells or spores will allow colonisation of new sites. Bacteria that find favourable conditions for growth are able to replicate and may soon give rise to enormous numbers of cells. Of course, many bacterial cells dispersed by the wind may fail to reach a place in which the environmental conditions meet their requirements. When airstreams from Africa sweep Saharan dust and bacteria northwards from the Sahara Desert to Europe one wonders how many of these desert microbes will find conditions in which they can flourish. A research group based in Grenoble, France reported the detection of molecular sequences indicative of bacteria originating from the Sahara Desert from dust samples collected from the snows of Mont Blanc (Chuvochina *et al.*, 2011). Naturalists interested in migrating birds and insects like to keep an eye on the long-range weather forecast and in particular on the wind direction. On occasions when Saharan dust clouds reach the UK (Figs. 1.39–1.40) there is the chance that the dust will be accompanied by unusual migrating insects such

Fig. 1.39. Saharan dust on a parked car. Upper Beeding, Sussex, UK.

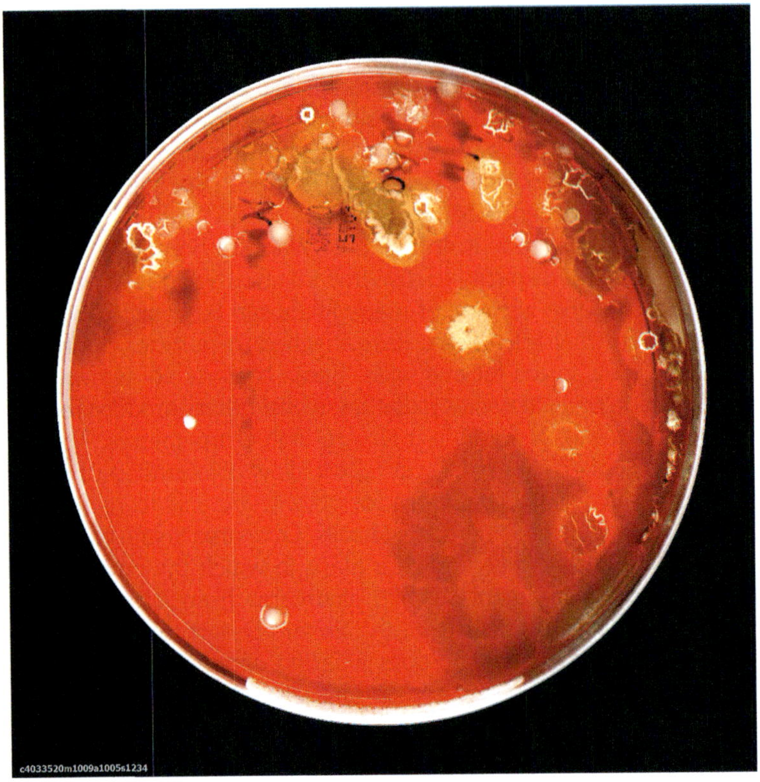

Fig. 1.40. Bacterial colonies cultured from Saharan dust, Oxford, UK (Tana Cavey).

as a Vagrant Emperor Dragonfly, *Hemianax ephippiger*, or perhaps the odd Desert Locust, *Schistocerca gregaria.* But with the Saharan dust also come the Saharan bacteria.

Microbes can disperse by water as well as by air. Advances in DNA sequencing technology have made it possible to compare the genetic relatedness of whole communities of bacteria and archaea from different locations without the need to culture individual strains of bacteria. Such an approach was used to investigate the microbial communities associated with methane seeps situated on the seafloor at widely separated locations in the world's oceans (Ruff *et al*., 2015). Methane seeps, also known as cold seeps, are spots on the seafloor where methane and other hydrocarbons emerge from the subsurface into the ocean's waters. Methane seeps offer nutrient-rich islands of opportunity to methane-feeding microbes on an otherwise relatively barren seafloor. An archipelago of methane seeps extends around the globe. Methane seeps harbour a characteristic microflora, consisting of particular kinds of methanotrophic archaea and sulphate-reducing bacteria, which utilise methane as a food source. Samples taken from twenty-three methane seeps from locations as widely separated as the North Sea, the South Atlantic and the Pacific Ocean show that methane seeps share a characteristic microflora. Many of the methanotrophic archaea and sulphate-reducing bacteria associated with methane seeps are globally distributed and have been detected at multiple isolated seep sites. Methane seep microbes are distinctive. They differ from the microbes found elsewhere in the world's oceans. The geographical distribution of methane seep microbes supports the Baas Becking hypothesis, that everything is everywhere but that the environment selects. Methane seeps are anoxic environments. The methanotrophic archaea that dwell in methane seeps are anaerobic organisms that do not thrive in the presence of oxygen. Nevertheless, it would appear that despite their intolerance of oxygen, methanotrophic archaea possess the capacity to traverse great distances through the world's oceans from one seep to another.

Something to consider when investigating the geographical distribution of bacteria relates to the level of resolution that can be applied when comparing the genetic relatedness of bacteria found at different locations. Careful comparison may reveal distinctive differences between superficially similar bacteria found at different locations. Such is the case with bacteria that belong to the thermophilic genus *Sulfurihydrogenibium* (sulphur gobblers). Sulphur gobblers can be found in hot springs situated as far apart as Iceland, the Azores and Yellowstone (Figs. 1.41–1.43). The scientific name *Sulfurihydrogenibium* refers to the capacity of these organisms to oxidise elemental sulphur using a process that yields water as a waste product. Sulphur gobblers thrive only in hot (65–70°C) sulphurous waters. Accordingly their geographical distribution is discontinuous and highly localised. The bacteria found in Iceland, the Azores and Yellowstone are closely related organisms that share a common ancestor. However, there are differences and the bacteria found in Iceland, the Azores and Yellowstone have been described as different species: *Sulfurihydrogenibium kristjanssonii*, *Sulfurihydrogenibium azorense* and *Sulfurihydrogenibium yellowstonense* (Nakagawa *et al.*, 2005; Flores *et al.*, 2008). At Yellowstone distinctive strains

Fig. 1.41. Hot spring habitat of *Sulfurihydrogenibium kristjanssonii*. Hveragerði, Iceland.

Fig. 1.42. Geothermal site for *Sulfurihydrogenibium azorense*. Furnas, Azores (Kevin Cole).

Fig. 1.43. Hot spring habitat of *Sulfurihydrogenibium yellowstonense*. Mammoth, Yellowstone, USA.

of *Sulfurihydrogenibium* have evolved in different hot springs. The distribution of the different strains of *Sulfurihydrogenibium* correlates with the geological record of previous volcanic eruptions at Yellowstone. Each new eruption creates novel geothermal sites which are seeded with *Sulfurihydrogenibium*. From these pioneering populations of bacteria new strains subsequently evolve (Takacs-Vesbach, 2008).

Whilst large parts of the globe are bacteriologically connected, some regions are less well connected than others. The polar regions are a special case in that the Arctic and Antarctic regions seem able to maintain their own distinctive communities of bacteria. Many of the bacteria found in association with polar ice are known to be specialised psychrophiles which are adapted to life at low temperatures. Although Arctic sea ice and Antarctic sea ice would appear to provide essentially the same type of habitat a review of data on polar bacteria has shown that the species mix varies from pole to pole (Staley and Gosink, 1999). It may be the case that the bacterial communities at the two poles have evolved in relative isolation and that the Baas Becking hypothesis does not apply at very high latitudes.

Diversity

Bacteria and archaea constitute a significant proportion of the world's biomass (Whitman *et al.*, 1998; Bar-On *et al.*, 2018). Prokaryotes are involved in the recycling of elements and organic molecules and influence the composition of the atmosphere and global climate. An understanding of the ecology of an ecosystem or habitat requires knowledge of its microbial diversity. It is more than a trivial question to ask how many species of bacteria there are: species counts are measures of prokaryotic diversity that reflect the structure of ecosystems and allow comparisons to be made between different locations and different kinds of habitat. Species catalogues provide information that help us to understand what bacteria and archaea are doing, how they interact, how they affect the regulation of nutrient cycles, how they underpin the health of ecosystems and how they respond to environmental disturbance.

In a review of global species diversity published as recently as 2011, bacteria were shown to account for as little as 0.5% of described living organisms. Bacteria were represented by only about 10,000 species, compared with roughly a million species of arthropods, 100,000 species of fungi and 200,000 species of flowing plants (Mora *et al.*, 2011). However, these figures do not accurately reflect the actual total numbers of living species for any of these groups. This statement is especially true for bacteria because the review included only those species which have been formally described in the scientific literature. The formal description of new species of bacteria is a highly specialised business, which in some ways is even more demanding than the process involved in describing a new species of bird or insect. As a consequence, formally described species of bacteria are somewhat underrepresented when compared with other groups of organisms.

New species of bacteria are conventionally described on the basis of cultured isolates and comparison with the type cultures of known species. However, most of the bacteria and archaea that live as members of complex microbial communities, such as exist in soils or marine sediments cannot be formally described because they have not been cultured. Most of the bacteria that can be seen on microscopy in samples obtained from such environments cannot be grown easily, or grown at all using standard culture techniques.

When environmental samples are stained with acridine orange, a fluorescent dye, they can be seen to contain many more microbial cells than the number of colonies that can be cultured from them using conventional culture media. Many bacteria have exacting conditions for growth that make them recalcitrant to culture in the laboratory. Nevertheless, it is possible to obtain information about microbial diversity by analysing DNA that has been extracted from environmental samples.

Because most species of bacteria remain undescribed and have not been given scientific names, published investigations of microbial diversity that are based on molecular methods often substitute the term 'species' with the term 'OTU' (Operational Taxonomic Unit). OTUs are based on the DNA sequence relatedness of 16s rRNA genes. For practical purposes OTUs are usually defined as groups of organisms whose DNA sequences share greater than 97% identity. Although this method probably leads to underestimates of the numbers of actual species, an advantage of using this approach to study microbial diversity is that it can be applied to large global databases of archived DNA sequence data which have been obtained from a wide range of environmental samples. These databases can be interrogated to investigate levels of microbial diversity locally and worldwide (Yarza *et al.*, 2014).

A team of scientists based at the University of Oregon and University of British Columbia examined 1.7 billion 16s rRNA sequences, derived from 492 studies from around the world (Louca *et al.*, 2019). Sampled habitats included soils, permafrost, salt marshes, seawater, deep sediments, lakes, rivers, hydrothermal vents, methane seeps, hot springs, animals and plants. The diversity represented by the entire collection amounted to a total of 739,880 separate OTUs, or 'species' in a broad sense. Most OTUs were represented by at least three sample sites and most of them were globally distributed. In order to reach an estimate of the actual world total of OTUs, the researchers examined the shape of accumulation curves that were obtained by adding increasing numbers of samples to the study set. An accumulation curve, or collector's curve, plots the cumulative number of species detected against the cumulative number of sampled individuals. It is a method that was developed to study the diversity of Lepidoptera in Malaya (Fisher *et al.*, 1943). Analysis of the shape of the curve can be used to predict the actual total number of species. In this way it was estimated that the global total number of 'species' of bacteria may lie somewhere within a range of 800,000 to 1.6 million OTUs.

The following chapters explore various aspects of bacteria in the context of the natural history of the British Isles and elsewhere.

References

Baas-Becking, L.G.M. (1934) Geobiologie of inleiding tot de milieukunde, W.P. Van Stockum and Zoon, The Hague, Netherlands.

Bar-On, Y. M., Phillips, R. and Milo, R. (2018) The biomass distribution on Earth. *Proceedings of the National Academy of Sciences of the United States of America*, 115, 6506–6511.

Becher, P.G., Verschut, V., Bibb, M.J., Bush, M.J., Molnár, B.P. *et al.* (2020) Developmentally regulated volatiles geosmin and 2-methylisoborneol attract a soil arthropod to *Streptomyces* bacteria promoting spore dispersal. *Nature Microbiology*, 5, 821–829.

Beladjal, L., Gheysens, T., Clegg, J.S., Amar., M. and Mertens, J. (2018) Life from the ashes: survival of dry bacterial spores after very high temperature exposure. *Extremophiles*, 22, 751–759.

Blair, K.G. (1921–1922) Annual Address to the Members. *Proceedings of the South London Entomological and Natural History Society*, 9–29.

Blesson, L. (1832) Observations on the ignis fatuus, or will-with-the-wisp, falling stars, and thunder storms. *The Edinburgh New Philosophical Journal*, 14, 90–94.

Bradley, J.A., Amend., J.P. and LaRowe, DE. (2019) Survival of the fewest: microbial dormancy and maintenance in marine sediments through deep time. *Geobiology*, 17, 43–59.

Chuvochina, M.S., Marie D., Chevaillier, S., Petit, J-R., Normand, P. *et al.* (2011) Community variability of bacteria in alpine snow (Mont Blanc) containing Saharan dust deposition and their snow colonisation potential. *Microbes and Environments*, 26, 237–247.

Ciche, T.A. and Ensign, J.C. (2003) For the insect pathogen *Photorhabdus luminescens*, which end of a nematode is out? *Applied and Environmental Microbiology*, 69, 1890–1897.

Claeson, A.S. and Sunesson, A.L. (2005) Identification using versatile sampling and analytical methods of volatile compounds from *Streptomyces albidoflavus* grown on four humid building materials and one synthetic medium. *Indoor Air*, 15, 41–47.

Darriet, P., Pons, M., Lamy, S. and Dubourdieu, D. (2000) Identification and quantification of geosmin, an earthy odorant contaminating wines. *Journal of Agricultural and Food Chemistry*, 48, 4835-4838.

Dodd, MS, Papineau, D, Grenne, T., Slack, J.F., Rittner, M, Pirajno, F., O'Neil, J. and Little, C.T.S. (2017) Evidence for early life in Earth's oldest hydrothermal vent precipitates. *Nature*, 543, 60–64.

Dyall-Smith, M.L., Pfeiffer, F., Klee, K., Palm, P., Gross, K. *et al.* (2011) *Haloquadratum walsbyi*: limited diversity in a global pond. *PLoS ONE*, 6, e20968.

Fisher, R.A., Corbet, A.S., and Williams, C. B. (1943) The relation between the number of species and the number of individuals in a random sample of an animal population. *Journal of Animal Ecology*, 12, 42–58.

Flores, G.E., Liu, Y., Ferrera, I., Beveridge, T.J. and Reysenbach, A.L. (2008) *Sulfurihydrogenibium kristjanssonii* sp. nov., a hydrogen- and sulfur-oxidizing thermophile isolated from a terrestrial Icelandic hot spring. *International Journal of Systematic and Evolutionary Microbiology*, 58, 1153–1158.

Forst, S., Dowds, B., Boemare N. and Stackebrandt, E. (1997) *Xenorhabdus* and *Photorhabdus* spp. bugs that kill bugs. *Annual Revues in Microbiology*, 51, 47–72.

Goethe, J.W.v. (1795) *Märchen* [Tale]. In Schiller, F. (ed.) *Die Horen* [Horae]. Cottaische Buchhandlung, Tübingen, Germany.

Houle, S., Schrader, K.K., Le François, N.R., Comeau, Y., Kharoune, M. *et al.* (2011) Geosmin causes off-flavour in arctic charr in recirculating aquaculture systems. *Aquaculture Research*, 42, 360–365.

Huber, H., Hohn, M.J., Rachel, R., Fuchs, T., Wimmer, V.C. *et al.* (2002) A new phylum of *Archaea* represented by a nanosized hyperthermophilic symbiont. *Nature*, 417, 63–67.

Jørgensen, B.B. and Marshall, I.P.G. (2016) Slow microbial life in the seabed. *Annual Review of Marine Science*, 8, 311–332.

Jüttner, F. and Watson, S.B. (2007) Biochemical and ecological control of geosmin and 2-methylisoborneol in source waters. *Applied Environmental Microbiology*, 73, 4395–4406.

Kallmeyer, J., Pockalny, R., Adhikari, R.R., Smith, D.C. and D'Hondt, S. (2012) Revised estimate of subseafloor microbial life. *Proceedings of the National Academy of Sciences of the United States of America*, 109, 16213–16216.

Kiene, R.P., Linn, L.J. and Bruton, J.A. (2000) New and important roles for DMSP in marine microbial communities. *Journal of Sea Research*, 43, 209–224.
Kirby, W. and Spence, W. (1816–1826) An introduction to entomology, or, elements of the natural history of insects: with plates. Longman, Hurst, Rees, Orme, Brown and Green, London, UK.
Lawrence, D.H. (1920). The rainbow. Methuen, London, UK.
Liu, J., Liu, J., Zhang, S. H., Liang, J., Lin, H. *et al.* (2018) Novel insights into bacterial dimethylsulfoniopropionate catabolism in the East China Sea. *Frontiers in Microbiology*, 9, 3206.
Louca, S., Mazel, F., Doebeli, M. and Parfrey, L.W. (2019) A census-based estimate of Earth's bacterial and archaeal diversity. *PLOS Biology*, 17, e3000106.
Lu, G., Edwards, C.G., Fellman, J.K., Mattinson, D.S. and Navazio, J. (2003) Biosynthetic origin of geosmin in red beets (*Beta vulgaris* L.). *Journal of Agricultural and Food Chemistry*, 51, 1026–1029.
Lucas, W.J. (1920) A monograph of the British Orthoptera. Ray Society, London, UK.
Lukassen, M. B., de Jonge, N., Bjerregaard, S. M., Podduturi, R., Jørgensen, N. *et al.* (2019) Microbial production of the off-flavor geosmin in tilapia production in Brazilian water reservoirs: importance of bacteria in the intestine and other fish-associated environments. *Frontiers in Microbiology*, 10, 2447.
Majerus, M. (1994) Ladybirds. New Naturalist. Collins, London, UK.
Melo, N., Wolff, G.H., Costa-da-Silva, A.L. and Arribas, R. (2020) Geosmin attracts *Aedes aegypti* mosquitoes to oviposition sites. *Current Biology*, 30, 127–134.e5.
Mulley, G., Beeton, M. L., Wilkinson, P., Vlisidou, I., Ockendon-Powell, N. *et al.* (2015) From insect to man: *Photorhabdus* sheds light on the emergence of human pathogenicity. *PLoS ONE*, 10 (12), e0144937.
Nakagawa, S., Shtaih, Z., Banta, A., Beveridge, T.J., Sako, Y. *et al.* (2005) *Sulfurihydrogenibium yellowstonense* sp. nov., an extremely thermophilic, facultatively heterotrophic, sulfur-oxidizing bacterium from Yellowstone National Park, and emended descriptions of the genus *Sulfurihydrogenibium*, *Sulfurihydrogenibium subterraneum* and *Sulfurihydrogenibium azorense*. *International Journal of Systematic and Evolutionary Bacteriology*, 55, 2263–2268.
Pask-Hughes, R. and Williams, R.A.D. (1975) Extremely thermophilic gram-negative bacteria from hot tap water. *Journal of General Microbiology*, 88, 321–328.
Pask-Hughes, R.A. and Shaw, N. (1982) Glycolipids from some extreme thermophilic bacteria belonging to the genus *Thermus*. *Journal of Bacteriology*, 149, 54–58.
Polak, E.H. and Provasi, J. (1992) Odor sensitivity to geosmin enantiomers. *Chemical Senses*, 17, 23–26.
Potts, M. (1997) Etymology of the genus name *Nostoc* (Cyanobacteria). *International Journal of Systematic Bacteriology*, 47 (2), 584.
Ruff, S.E., Biddle, J.F., Teske, A.P., Knittel, K., Boetius, A. *et al.* (2015) Global dispersion and local diversification of the methane seep microbiome. *Proceedings of the National. Academy of Sciences of the United States of America*, 112, 4015–4020.
Sado, P.N. (1998) "Glowing" seafood. *California Journal of Environmental Health*, 21, 15–18.
Sajnaga, E. and Kazimierczak, W. (2020) Evolution and taxonomy of nematode-associated entomopathogenic bacteria of the genera *Xenorhabdus* and *Photorhabdus*: an overview. *Symbiosis*, 80, 1–13.
Salman V., Bailey J.V. and Teske A. (2013) Phylogenetic and morphologic complexity of giant sulphur bacteria. *Antonie Van Leeuwenhoek*, 104, 169–186.
Satari, L., Guillén, A., Latorre-Peréz, A. and Porcar, M. (2021) Beyond archaea: the table salt bacteriome. *Frontiers in Microbiology*, 12, 714110.

Starnawski, P., Bataillon, T., Ettema, T.J.G., Jochum, L.M., Schreiber, L. *et al.* (2017) Microbial community assembly and evolution in subseafloor sediment. *Proceedings of the National Academy of Sciences of the United States of America*, 114, 2940–2945.

Suurnäkki, S., Gomez-Saez, G.V., Rantala-Ylinen, A., Jokela, J., Fewer, D.P. *et al.* (2015) Identification of geosmin and 2-methylisoborneol in cyanobacteria and molecular detection methods for the producers of these compounds. *Water Research*, 68, 56–66.

Takacs-Vesbach, C., Mitchell, K., Jackson-Weaver, O. and Reysenbach, A.-L. (2008) Volcanic calderas delineate biogeographic provinces among Yellowstone thermophiles. *Environmental Microbiology*, 10, 1681–1689.

Tenreiro, S., Nobre, M.F., Hoste, B., Gillis, M., Kristjansson, J.K. and da Costa, M.S. (1995). DNA:DNA hybridization and chemotaxonomic studies of *Thermus scotoductus*. *Research in Microbiology*, 146, 315–324.

Volland, J.M., Gonzalez-Rizzo, S., Gros, O., Tyml T, Ivanova, N. *et al.* (2022) A centimeter-long bacterium with DNA contained in metabolically active, membrane-bound organelles. *Science*, 376, 1453–1458.

Waterfield, N.R., Ciche, T. and Clarke, D. (2009) *Photorhabdus* and a host of hosts. *Annual Review of Microbiology*, 63, 557–574.

Webber, J.F., Parkinson, N.M., Rose, J., Stanford, H., Cook, R.T.A. *et al.* (2008) Isolation and identification of *Pseudomonas syringae* pv. *aesculi* causing bleeding canker of horse chestnut in the UK. *Plant Pathology*, 57, 368–368.

Whitman, W.B., Coleman, D.C. and Wiebe, W.J. (1998) Prokaryotes: the unseen majority. *Proceedings of the National Academy of Sciences of the United States of America*, 95, 6578–6583.

Williams, B.T., Cowles, K., Bermejo Martínez, A., Curson, A.R.J., Zheng, Y. *et al.* (2019) Bacteria are important dimethylsulfoniopropionate producers in coastal sediments. *Nature Microbiology*, 4, 1815–1825.

Woese, C.R., Kandler, O. and Wheelis, M.L. (1990) Towards a natural system of organisms: proposal for the domains Archaea, Bacteria, and Eucarya. *Proceedings of the National Academy of Science of the United States of America*, 87, 4576–4579.

Yarza, P., Yilmaz, P., Pruesse, E., Glöckner, F., Ludwig, W. *et al.* (2014) Uniting the classification of cultured and uncultured Bacteria and Archaea using 16S rRNA gene sequences. *Nature Reviews. Microbiology*, 12, 635–645.

Yoch, D.C. (2002) Dimethylsulfoniopropionate: its sources, role in the marine food web, and biological degradation to dimethylsulfide. *Applied and Environmental Microbiology*, 68, 5804–5815.

Zhang, W., Sulz, M., Mykitiek, T., Li, X., Yanke, L. *et al.* (2004) A Canadian strain of *Pseudomonas syringae* causes white-colour disease of *Cirsium arvense* (Canada thistle). In: Cullen, J.M., Briese, D.T., Kriticos, D.J., Lonsdale, W.M., Morin, L. *et al.* (eds.). Proceedings of the XI International Symposium on Biological Control of Weeds, April 27 – May, Canberra, Australia. 215–220.

2 Bacteria in earth history

First forms minute, unseen by spheric glass,
Move on the mud, or pierce the watery mass;
Erasmus Darwin (1803). *The Temple of Nature.*

Bacteria build, erode and transform. Some bacteria promote the precipitation of minerals and the accretion of sediments, while the presence of other kinds of bacteria leads to weathering and erosion. In so doing bacteria remould the world. Bacteria influence the places in which they live in many ways. The ongoing transformative effects of bacteria on their surrounding environments can be seen today. Furthermore, evidence of their past involvement in modifying the world can be detected in the geological record. The term geomicrobiology is used to encompass the study of the geochemical and geological phenomena that are mediated by bacteria and other microbes. Observations made of ongoing geomicrobiological processes can be used to assess the influence of past generations of microbes on the shaping of land forms. Britain's geology preserves a varied account of ancient bacteria and of the roles that they played in earth history. During the course of a tour of the British Isles the field naturalist can explore locations where the rocks harbour visible and tangible bacterially induced structures and deposits and visit sites where the geological record is known to include the actual remnants of bacteria that have been preserved as microfossils. By exploring different geological horizons it is possible to appreciate the roles of bacteria in different kinds of environments at different times in earth history. The geological record preserves evidence of environments that no longer occur in the British Isles, such as deserts, salt lakes, hot springs and tropical seas. These lost habitats once supported thriving populations of specialised bacteria. Geological sites also allow the field naturalist to investigate environments of bacteriological importance that would otherwise be practically inaccessible. Examples of these exotic places include ocean floor methane

DOI: 10.1079/9781836993933.0002

seeps and hydrothermal vents. When exploring the geological vestiges of ancient environments it is possible to imagine the communities of bacteria that lived in them and sometimes to witness signs of their former influence.

Palpable evidence for the existence of ancient bacteria is provided by trace fossils known as stromatolites. Stromatolites have a characteristic laminated structure consisting of the mineralised layers of former microbial mats made by successive generations of bacteria. Fossilised stromatolites can be seen in the field at a number of locations in the British Isles. It is also possible to find more subtle signs of ancient bacteria in the form of bedding planes in rock formations that display the mineralised surfaces of sediments that were once coated and stabilised by bacterial mats. Such signs are known as 'microbially induced sedimentary structures'. Bacteria themselves are sometimes preserved as microfossils and sometimes even identifiable components of bacterial cells, in the form of the tiny magnetic organelles formed inside the cells of magnetic bacteria, can be found by microscopic examination of sedimentary rocks.

Fossils and fossilisation

The fossilisation of animals and plants is mediated by the action of bacteria (Carpenter, 2001). As soon as an organism dies its resident bacterial flora can take advantage of the available nutrients and a process of decomposition begins. Remains of organisms that become buried in sediments become further colonised by bacteria of environmental origin. Furthermore, the remains of the dead organism become exposed to minerals supplied in groundwater. Often only the hard parts, such as shells and bones are preserved. Other structures may decompose completely and leave only an impression on the surface of a sediment. Alternatively, the body of an animal may be replaced with minerals to leave a cast. Bacteria induce the precipitation of calcium carbonate and create conditions that favour the precipitation of silica. Permineralisation refers to the infiltration of minerals into the spaces of an object such as the minute cavities within bones. Bacteria induce the precipitation of calcium carbonate and other minerals and transform bones into rocks. Fossils become heavy as a result of mineralisation and may be become darker in tone as a result of the precipitation of particles of black iron sulphide. During the fossilisation process, sulphate-reducing bacteria release sulphide which combines with iron to form iron pyrite (FeS_2), an iron sulphide compound. Often the pyrite is finely dispersed through the fossil matrix (Carpenter, 2001). This gives many fossils their characteristically blackened appearance which makes them stand out from the surrounding rocks. However, fossils that are composed of crystalline iron pyrite, or fool's gold, glitter with a brilliant metallic lustre (Fig. 2.1). Bacterial biofilms that colonise the surfaces of dead organisms can become mineralised. Soft features such as the outlines of muscle fibres may be preserved in exceptional detail. Even the outlines of mineralised bacteria may be visible on the fossil's surface (Briggs *et al*., 2005). Plant fossils become preserved according to the same principles as apply to animals (Dunn *et al*., 1997). A feature of plant fossils and plant semifossils relates to the fact that in the absence of oxygen, lignin, a constituent of woody plant tissue, is resistant to microbial decomposition. In anoxic waterlogged conditions, trunks and stems build up as layers of peat. Eventually, after a period of sustained compression beneath layers of sediment, all that remains is coal, which may preserve the outlines of former plants (Fig. 2.2).

Fig 2.1. Fossilised ammonite composed of pyrite. Hock Cliff, Gloucestershire, UK.

Fig 2.2. Fossilised fern from the Somerset coal measures at Radstock, UK. Coal is a negative marker of bacterial decomposition.

Earth processes and bacteria

During the course of a journey around Britain the peripatetic naturalist can witness present-day signs of bacterially induced mineral deposition and erosion. Geologists argue that much of the geological record can be explained in terms of processes that can be witnessed in action today and that the earth is subject to continuous subtle modifying processes that bring about gradual structural changes over the course of time. In 1788 the Royal Society of Edinburgh published James Hutton's *Theory of the earth; or an investigation of the laws observable in the composition, dissolution, and restoration of land upon the globe*. James Hutton (1726–1797) studied the surface of the earth in the vicinity of his home in the Scottish Borders. He observed the ongoing processes of weathering, erosion, sedimentation and accretion and reasoned that these processes could explain the origin and composition of rocks. Hutton examined the juxtaposition of different rock formations and concluded that the origin of land forms could be traced back to the deposition of sediments on the seafloor and their subsequent upward displacement. His idea of a continually evolving earth which was subjected to continuous cycles of consolidation and destruction became a principle of geological science known as uniformitarianism. A full understanding of this geological principle is incomplete unless the effects of microbial processes on the natural environment are considered. In contrast to the principle of uniformitarianism, catastrophism emphasises the role of abrupt catastrophic events to explain the origin of land forms. Contemporary geological ideas combine uniformitarianism with catastrophism in portraying earth history as a story of gradual change interrupted by episodes of disruption.

Geological evidence suggests that bacteria have been directly involved in contributing to some of the upheavals in earth history and also indicates that bacteria have flourished in the aftermath of events that have been catastrophic for other organisms. Bacteria can cause environmental change. Such changes can in turn affect populations of other kinds of bacteria. By releasing oxygen as a waste product of photosynthesis, bacteria brought about the gradual oxygenation of the earth. The geological record shows that this process began at least 2.4 billion years ago. Pyrite (iron sulphide) undergoes spontaneous oxidation in the presence of oxygen. Rocks formed from sediments deposited in shallow water from later than about 2.4 billion years ago lack grains of pyrite. From about 2.2 billion years ago, sandstone deposits begin to appear in which individual grains of sand are coated with a red layer of iron oxide, showing that oxygen was by then present in the atmosphere (Schopf, 2011). The oxygenation of the planet would have been deleterious to strictly anaerobic bacteria which were intolerant of oxygen. Nowadays, strictly anaerobic bacteria are confined to anoxic environments such as stagnant sediments and the guts of animals.

The oxygenation of the earth can be linked to a period of global glaciation, sometimes referred to as 'snowball earth', which lasted until about 2.1 billion years ago, when icy conditions reached as far south as the tropics. A possible explanation for the drop in global temperatures relates to the instability of methane in the presence of oxygen. When the earth was populated by anaerobic microbes it seems reasonable to suppose that the atmosphere would have been rich in methane which is one of the principal waste products of such organisms. Methane is a greenhouse gas that would have promoted high temperatures globally. Oxygen reacts with methane to yield carbon dioxide, a less potent greenhouse gas, resulting in global cooling. A later period of global glaciation also occurred more recently, approximately 720–635 million years ago and evidence for this event is preserved in the geological record on the Isle of Islay and at a few other sites in Scotland and Ireland. The visitor to Islay has the opportunity to examine glacial sediments as well as beds of fossilised stromatolites which occur in association with layers of dolomitic limestone known as cap carbonates (Fig. 2.3). Cap carbonates are so called because they lie on top of glacial sediments (Harland, 2007). Vast accumulations of cap carbonates formed in the fallout of the eventual meltdown when temperatures increased globally. Cap carbonate deposits rich in stromatolites are known from a number of locations around the world (Hoffman *et al.*, 2017).

Constructive bacteria

Bacteria help to stabilise substrates and form soils which encourage the growth of plants and promote further stabilisation. Bacteria and other microbes form dune crusts in coastal areas and in deserts. Nitrogen-fixing bacteria enrich impoverished sandy soils with nitrogen. In China, cultures of nitrogen-fixing cyanobacteria have been applied to sandy soils in an effort to impede desertification

Fig. 2.3. An outcrop of dolomitic rocks (known as 'cap carbonates'). Islay, UK.

(Park *et al.*, 2017). Sediments are colonised by enormous numbers of micro-organisms. Bacteria and other microbes secrete slimes, more correctly known as extracellular polymeric substances, which trap and bind particles of sediment together. The physical presence of bacteria together with their secreted extra-cellular polymeric substances make sediments sticky and promote the stabilisation of sedimentary surface features such as ripple marks and slow down the migration of sediments (Lichtman *et al.*, 2018). By stabilising sediments, bacteria help to impede the erosion of coastal land forms. Bacteria can induce the precipitation of minerals, including carbonates in the form of calcite, dolomite, tufa (Fig. 2.4) and travertine, as well as phosphates, sulphates, iron oxide and other metallic compounds.

Microbially induced sedimentary structures

Among the more subtle signs of the presence of bacteria are so-called 'microbially induced sedimentary structures'. Bacteria grow across surfaces as layers of cells known as biofilms. Bacteria secrete sticky extracellular polymeric substances which promote adherence. On sediment surfaces, bacterial biofilms develop into resilient microbial mats that consist of networks of microbial filaments supplemented by secreted slimy extracellular polymeric substances which trap particles of sediment. Microbial mats provide sediments with a protective elasticated coating which impedes the movement and modelling of the sediments by physical forces. The interplay between the physical forces that act on sedimentary surfaces and the resistance to those forces by microbial mats

Fig. 2.4. Aquatic *Nostoc verrucosum* colonies, other cyanobacteria and eukaryotic algae growing on a tufa cascade. Ventnor, Isle of Wight, UK.

can result in a variety of morphologies that are collectively described as microbially induced sedimentary structures (Noffke *et al.*, 2001). Microbial mats are widespread and easily visible in coastal environments and around estuaries, on mudflats, shorelines and inland in shallow pools and beside rivers. Microbial mats are composed of microbial communities of bacteria and single-celled eukaryotes, such as algae and diatoms. Beneath a microbial mat, communities of bacteria in the underlying sediment may be stratified at different depths below the surface according to the environmental conditions. Mats are made more noticeable when fragmented by the shearing forces of wind and water or when they have been disturbed by animals. When multicellular organisms became abundant during the Cambrian Period from about 540 million years ago, microbial mats became increasingly subjected to bioturbation or disturbance by grazing and burrowing animals. There is a school of thought that Precambrian times might have been something of a golden age for microbial communities. The term 'microbially induced sedimentary structures' can be applied to trace fossils found in rock formations that represent the mineralised remains of ancient mats as well as to present-day sedimentary structures made by living bacterial mats. When fossilised microbial mats are found in the geological record they provide clues to geologists about ancient environments and their microbial communities.

Microbially induced sedimentary structures can be observed on intertidal flats and in similar environments. The ecology and morphology of microbial mats have been systematically investigated in diverse environments in several places around the world, including Canada, the USA, Tunisia, Egypt, Saudi Arabia and Germany. Microbially induced sedimentary structures have been studied in detail at Mellum Island on the North Sea coast of Germany (Gerdes and Klenke, 2007). Environmental conditions on the mudflats at Mellum Island are not too much unlike those of some intertidal sites around the British Isles. At Mellum Island microbial mats are dominated by two species of cyanobacterium, *Oscillatoria limosa* (the Slimy Swinger) and *Coleofasciculus chthonoplastes* (the Plastic Carrier Bug). At Mellum Island, sediment surfaces that are coated with microbial mats display a variety of structural features, which include multidirectional ripple marks, mat chips, erosional remnants and erosional pockets, gas domes, collapsed gas domes, tufted mats and polygonal cracks (Figs. 2.5–2.9). These features will be so familiar to anyone who spends time walking along coastal flats as to be almost unnoticeable. Remarkably though, the mineralised traces of similar structures can be seen in ancient geological deposits, suggesting that microbial mats have a heritage which dates back in earth history for more than 3 billion years.

When used by geologists as tools to make inferences about past environmental conditions and ancient life, microbially induced sedimentary structures are of most value when they can be distinguished from structures formed by non-biological processes. The differences between sedimentary structures formed as a result of wholly physical processes and structures formed under the influence of microbes and their products can be subtle and relative. At least 17 different kinds of microbially induced sedimentary structure have been identified (Noffke, 2010). Among the more easily recognisable microbially induced

Fig. 2.5. Bioturbated cyanobacterial mat with animal prints. Norfolk, UK.

Fig. 2.6. Desiccated, torn cyanobacterial mat. Parys Mountain, Anglesey, UK.

sedimentary structures are multidirectional ripple marks. Observations of modern mud flats show that during the course of a season, ripple marks become stabilised by the growth and adherence of microbes in the upper layer of sediment. As tides and currents shift, new ripple patterns are formed in the vicinity. They too become stabilised such that a multidirectional array of ripple marks

Fig. 2.7. Desiccated, fragmented microbial mat. Rye Harbour, Sussex, UK.

Fig. 2.8. Desiccated, rehydrated, displaced cyanobacterial mat with oxygen bubbles. Salthouse, Norfolk, UK.

is formed (Fig. 2.10). The significance of multidirectional ripple marks is that they suggest that microbial stabilisation of sediment has taken place. It can be argued that without such stabilisation the ripple pattern would quickly become refashioned when the flow changes direction whereas when the substrate is

Fig. 2.9. Bioturbated cyanobacterial mat with ruptured gas domes. Widewater, Sussex, UK.

Fig. 2.10. Multidirectional ripple marks, a feature of sediments stabilised by extracellular polymeric substances produced by bacteria. Holme, Norfolk, UK.

stabilised by microbial action a secondary set of ripple marks can develop that is superimposed on the previous pattern. Although multidirectional ripple marks can be seen to develop in the absence of visible mats, this does not mean to say that microbes have not contributed to the stability of the sediments. Seawater contains extracellular polymeric substance, a microbial slime produced by microbes. It has been shown that even low concentrations of extracellular polymeric substance will stabilise ripple marks on sediment surfaces (Malarkey *et al.*, 2015; Davies *et al.*, 2016).

Britain's oldest microbially induced sedimentary structures

Microbially induced sedimentary structures can be seen among exposures of Britain's oldest sedimentary rocks in northwest Scotland where they are accompanied by microfossils of an extinct cyanobacterium, which has been given the scientific name *Eohalothece lacustrina* (Strother and Wellman, 2016). *Eohalothece lacustrina* (the Old Lake Bug), was a planktonic cyanobacterium, colonies of which probably floated about in the water column. It resembles modern planktonic cyanobacteria. Microfossils of *Eohalothece lacustrina* (Fig. 2.11) have been found in rocks of the Nonesuch Formation in Michigan State and at various sites in Scotland, including Ob Gauscavaig on the Isle of Skye and Cailleach Head. At one locality, Lower Diabaig, a coastal village situated at the end of a single track road not far from Torridon, there are slanting beds of grey shale on the shoreline that show an abundance of ripple marks and desiccation cracks (Figs. 2.12–2.14). The rocks at Diabaig were formed from sediments which accumulated approximately 1 billion years ago in a shallow lake. Microbially induced sedimentary structures, including wrinkle marks show that microbial mats formed around the lake margins in spots that were prone to desiccation. Tony Prave, a geologist based at the University of St Andrews described these microbially induced sedimentary structures at Lower Diabaig as early evidence of the emergence of early life onto land (Prave, 2002). The Torridonian rocks of Scotland pre-date the emergence of plants and multicellular organisms. One billion years ago, Lower Diabaig was a microbial

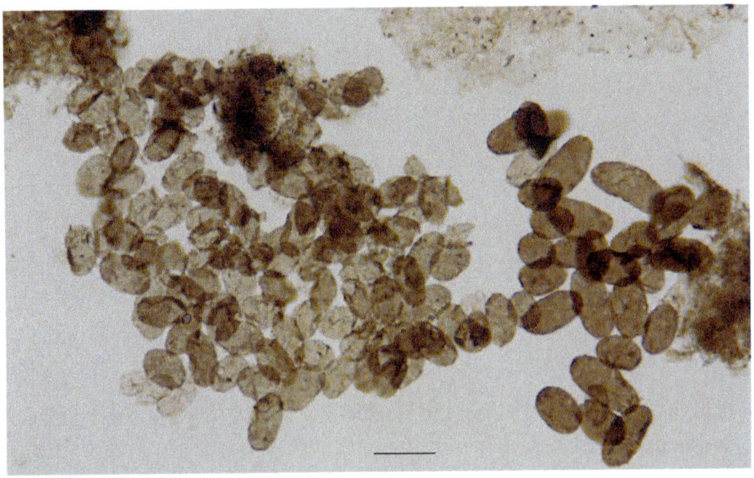

Fig. 2.11. Ovoid cells of *Eohalothece lacustrina* from the Precambrian Nonesuch Formation. Scale bar represents 10 μm (Image cropped from figure 3 from Strother and Wellman, 2016). Creative Commons Attribution Licence.

Fig. 2.12. Precambrian (1 billion year old) ripple marks. Lower Diabaig, Northwest Highlands, UK.

Fig. 2.13. Traces of microbial mat surfaces on Precambrian shale. Lower Diabaig, Northwest Highlands, UK.

Fig. 2.14. Beds of shale containing microbially induced sedimentary structures. Lower Diabaig, Northwest Highlands, UK.

world of bacteria and unicellular eukaryotes. To the north of Diabaig at Stoer Bay and in neighbouring areas, many rock surfaces are adorned with ancient, ripple marks indicating the former existence of a shallow water environment (Fig. 2.15). Microfossils as well as mineralised layers suggestive of microbial mats have been found in rocks of the Stoer Group. The rocks at Stoer are richly coloured red sandstones, which have been dated as being almost 1.2 billion years old.

In some parts of the world microbially induced sedimentary structures have been discovered which are amazingly ancient. The geological record demonstrates that microbes have been in existence for at least 3.5 billion years. Stable parts of the earth's crust are known as cratons. Really ancient unmodified cratons complete with geological evidence of their ancient environments occur in Western Australia as the Pilbara Craton and in South Africa as the Kaapvaal Craton. In South Africa, the Nhlazatse Section of the 2.9 billion year old Pongola Supergroup includes a large assemblage of unusually well preserved microbially mediated features. Despite the antiquity of the rocks their preserved sediment surfaces show ripple marks and cracks just like those seen on modern sandy coasts (Noffke *et al.*, 2008). Also in South Africa, an even older 3.4 billion year old microbial mat was detected in the fine-grained Josefsdal Chert deposits (Westall *et al.*, 2006). The Pilbara region of Western Australia is renowned among geomicrobiologists for its ancient microbially induced sedimentary structures, its stromatolites and its ancient microfossils. The 3.48 billion year old Dresser Formation includes polygonal desiccation cracks and evidence of chipped microbial mats (Noffke *et al.*, 2013). Microscopic examination of fine-grained chert samples from the Dresser Formation has revealed microfossils

Fig. 2.15. 1.2 billion year old multidirectional ripple marks. Stoer, Northwest Highlands, UK.

that resemble the modern microorganism *Methanocaldococcus jannaschii* (Jannasch's Gas-ball) a hyperthermophilic methanogen and member of the *Archaea* (Glikson *et al.*, 2008). Jannasch's Gas-ball grows at high temperatures on a diet of hydrogen and carbon dioxide and produces methane as a waste product. Microscopy of rock samples from the Dresser Formation also shows structures which have been interpreted as the remains of gas bubbles trapped in extracellular polymeric substance or microbial slime. Dome shaped and conical stromatolites occur in the same deposits. The Dresser Formation includes deposits of geyserite, a mineral composed of silica, formed in the vicinity of geysers and hot springs (Djokic *et al.*, 2017). Overall, the evidence points to the former existence of a 3.48 billion year old geothermal environment with hot springs populated by communities of thermophilic microbes, which in some respects are not unlike those that can be seen today in locations such as Yellowstone National Park (Fig. 2.16) and in Iceland.

Biomineralisation

The term biomineralisation is used to describe the production or induction by living organisms of structures or deposits that are composed of minerals. The mere presence of bacteria can promote biomineralisation. Because bacteria are very small their surface area to volume ratios are very high. Large numbers of bacterial cells can make a significant impact on an environment. Bacterial cells are surrounded by cell walls and by a thin proteinaceous coat (called the S-layer). The cells of many kinds of bacteria are additionally surrounded by a capsule of slime. Attached to the outside of bacterial cells are various

Fig. 2.16. Geyserite deposits that include the remains of silicified microbial mats. Excelsior Geyser, Yellowstone, USA.

molecular structures, such as amino acids, which give the outer surfaces of bacteria an overall negative charge. The negative charge attracts metal ions that in turn bind other substances such as carbonate or silicates. Bacteria promote the precipitation of many other minerals (Douglas and Beveridge, 2006). Biomineralisation can be classified either as being controlled biomineralisation or induced biomineralisation (Lowenstam, 1981). Controlled biomineralisation relates to situations where an organism plays an active role in mineral deposition. In contrast, induced biomineralisation relates to situations where mineral deposition occurs passively, for example in response to the release of waste products that influence the chemistry of the environment in which an organism lives. Biologically controlled mineralisation generates organised crystalline mineral deposits, such as the shells of marine organisms, whereas induced mineralisation results in deposits whose crystalline structure is disorganised. Examination of the crystalline structure of calcium carbonate deposits provides clues about the kinds of organisms involved in their deposition (Päßler *et al.*, 2018). Cyanobacteria for example induce the precipitation of disorganised deposits of calcium carbonate (Figs. 2.17–2.19). Fixation of carbon dioxide during photosynthesis increases acidity and promotes mineral precipitation. Also, bacterial cells act as sites for precipitation and may become encrusted with carbonate minerals. In addition to the precipitation of calcium carbonate ($CaCO_3$) as calcite or aragonite, cyanobacteria can also induce the precipitation of magnesium carbonate as magnesite ($MgCO_3$) and calcium magnesium carbonate ($CaMg(CO_3)_2$) as dolomite. The chemical composition of the medium in which the bacteria live influences the composition of the precipitate. Around springs and waterfalls where the water is saturated with dissolved minerals,

Fig. 2.17. Carbonate deposits on granite with cyanobacteria. St Michael's Mount, Cornwall, UK.

Fig. 2.18. Mineralised cyanobacteria (*Rivularia*) colonies. Whale Chine, Isle of Wight, UK.

bacteria and their secreted extracellular polymeric substance enhance carbonate deposition, leading to the formation of tufa deposits. At carbonate-rich hot springs, bacteria mediate the precipitation of calcium carbonate as travertine. Travertine deposits may include impressions of colonies of bacteria that once lived in them (Fig. 2.20). These are discussed as specific examples of microbial habitat in Chapter 3, *Meet the bacteria*. Mineralised structures formed by microbes are called microbialites. Some authors use the term in a narrow sense for mineralised structures that develop in water bodies. Others use a broader

Fig. 2.19. Cross section of partially mineralised *Rivularia* colony. Whale Chine, Isle of Wight, UK.

Fig. 2.20. Impressions made by bacterial streamers in travertine. Mammoth Hot Springs, Yellowstone, USA.

definition and refer to carbonate precipitates in caves and other kinds of microbially induced structures as examples of microbialites. Although microbialites occur widely in nature their presence often goes unnoticed. Laminated microbialites formed by microbial mats are called stromatolites, whereas those whose microstructure is disorganised in appearance are known as thrombolites.

Whiting events, ooids and oolite

Blooms of cyanobacteria as well as blooms of planktonic eukaryotes sometimes trigger spectacular large-scale 'whiting events'. Whiting events happen when calcium carbonate precipitation is induced by microbes en masse. Whiting events occur not only in the seas but also sometimes in freshwater lakes. The precipitated calcium carbonate sinks to the bottom to form a layer of lime sediment. Lime sediments eventually consolidate to form limestone. Limestone deposition occurs especially in warm shallow seas such as occur today in the Persian Gulf, the Gulf of Mexico and around the Bahamas. In experimental conditions using bacteria isolated from Bahaman seawater, several species of bacteria were shown to promote calcium carbonate precipitation (McCallum and Guhathakurta, 1970). Not all limestones are formed as a result of the precipitation of calcium carbonate from solution: some consist largely of shells or the skeletal remains of organisms. Corallian limestone for example contains the remains of corals. Oolitic limestone is a type of limestone that is composed of ooids (Fig. 2.21). Ooids are ovoid. Ooids consist of concentric layers of precipitated minerals that form around a nucleus, such as a shell fragment. Oolitic limestones are well-represented in the British geological record. Oolite deposition today occurs in the seas around the Bahamas, where banks of oolite sand accumulate from the vast numbers of developing ooids. The ooids are polished to a smooth finish by shifting currents. Oolite sand from the Bahamas is commercially available, complete with bacteria, to provide healthy conditions for tropical fish in marine aquaria. Diverse kinds of bacteria, which include species engaged in various kinds of metabolic activity, including photosynthesis, sulphate reduction, sulphate oxidation and denitrification grow as biofilms over the surfaces of developing ooids. All of them promote calcium carbonate

Fig. 2.21. Microscopic view of oolitic limestone composed of tiny ooids. Chedworth, Gloucestershire, UK.

precipitation (Diaz *et al.*, 2017; Batchelor *et al.*, 2018). The Bahamas are of geomicrobiological interest not only for oolite deposition but also for stromatolites. 'Living' stromatolites are known from the shorelines of several islands that border Exuma Sound. An association between oolite and stromatolites was first noted by the German geologist Ernst Kalkowsky (1851–1938). He studied sediments from the Triassic Period in northern Germany and published a scientific paper entitled: '*Oolith und Stromatolith im norddeutschen Buntsandstein*' (Kalkowsky, 1908). Kalkowsky coined the term '*Stromatolith*' (stromatolite). Kalkowsky reasoned that both ooids and stromatolites might be microbially mediated structures. He speculated that ooids might be generated by single colonies of organisms and that stromatolites might be built by layers of organisms.

Many localities of special interest to naturalists are associated with oolitic limestone, including important calcareous grassland sites in the Cotswolds, Somerset, Dorset and elsewhere (Fig. 2.22). Oolite is quarried for use as a building stone. Portland stone, for example, is a kind of oolite. Disused oolite quarries become valuable sites for wildlife. Oolitic limestones in Britain include Hunts Bay oolite in South Wales, Burrington oolite in the Mendip Hills and Goblin Coombe oolite in Somerset, which date from the Carboniferous Period and Osmington oolite, inferior oolite and great oolite which all date from the Jurassic Period. Bath stone is an example of the Great oolite which is quarried near Bath.

Methane seep carbonate mounds

Unless trapped by impervious rocks, hydrocarbons such as methane and oil will naturally find their way to the surface. Methane seeps are a feature of the seafloor at continental shelf margins where large volumes of methane gas

Fig. 2.22. Oolitic limestone scenery. Selsley Common, Gloucestershire, UK.

escape into the atmosphere. The escape of natural gas from methane seeps situated on the UK Continental Shelf has been estimated to account for a significant proportion of Britain's methane emissions (Judd *et al.*, 1997). However, much of the methane that seeps onto the seafloor from methane seeps is utilised as a food source by methanotrophic archaea and by sulphate-reducing bacteria. Sulphate-reducing bacteria use sulphate to oxidise the methane that emerges from methane seeps and produce hydrogen sulphide as a waste product (Wegener *et al.*, 2015). The processes associated with methane oxidation promote the precipitation of carbonates, leading to the formation of carbonate mounds. The carbonate mounds and the microbial mats associated with them provide a habitat for other microbes, coelenterates, molluscs and fish.

The fossilised remains of methane seep carbonate mounds that date back to the Jurassic Period can be seen at a number of sites in the British Isles (Allison *et al.*, 2008). On the shores of the Bristol Channel near the village of Kilve in Somerset, coastal erosion has exposed a nest of carbonate mounds which are visible at low tide (Fig. 2.23). They can be seen from the cliff top coast path, or can be experienced more directly by picking one's way for some distance along the foreshore though a jumble of coastal rocks. The conical mounds measure several metres in diameter and contain the fossilised remains of foraminifera, bivalves and ammonites. Three hundred miles away from Kilve, carbonate mounds can also be seen at Ravenscar on the Yorkshire coast. The fossilised carbonate mounds at Kilve and Ravenscar overlie thick deposits of black shale (Fig. 2.24). The shale beds represent former marine sediments, rich in organic matter. Methane, released as a waste product by methanogenic archaea in those ancient sediments, would have risen through fault lines to emerge on the Jurassic seafloor at seep sites.

Fig. 2.23. Methane seep carbonate mounds. Kilve, Somerset, UK.

Silica mineralisation

Bacteria are indirectly involved in the genesis of flint nodules (Fig. 2.25). Flint is a form of chert, which is a general term for rocks that consist of fine-grained silica, or silicon dioxide (SiO_2). Flint commonly occurs in association with chalk,

Fig. 2.24. Beds of limestone and black shale. Kilve, Somerset, UK.

Fig. 2.25. Flint wall. Upper Beeding, West Sussex, UK.

a variety of limestone composed of calcium carbonate derived from unicellular planktonic eukaryotes called coccolithophores. Conspicuous beds of nodular flint are a common feature of chalk deposits. Flints add visual interest to chalk cliffs. Pebble beaches composed of flints that have been eroded from the chalk and fashioned by wave action form the basis of important littoral habitats for specialised plants and animals. During the Cretaceous Period when coccolithophores abounded in warm seas, marine sediments consisted of a kind of ooze composed of the calcium carbonate plates shed by coccolithophores as well as the decayed remains of organisms such as sponges, which acted as sources of silica. Sulphate-reducing bacteria in the deeper anoxic sediments produced hydrogen sulphide which migrated upwards through the sediment and reacted with oxygen to release hydrogen ions which dissolved the calcium carbonate and also promoted silica precipitation in any cavities in the upper sediment layers (Clayton, 1986). Some flints are interestingly shaped and are considered to represent the casts of the cavities made by burrowing organisms which might have acted as preferential sites for silica precipitation. Flint nodules may contain fossilised sponges or other organisms (Fig. 2.26). These fossils often contain pyrite (FeS_2) deposits, brought about by the interaction of iron ions with hydrogen sulphide released by sulphate-reducing bacteria in the ancient sediments.

Rhynie, Aberdeenshire: bacteria preserved in silica from a hot spring

Very few sites for geothermal springs have been detected in the geological record worldwide. Only one such site is known in Britain (Trewin, personal

Fig. 2.26. Sponge fossil containing pyrite in flint nodule. Seaford Head, Sussex, UK.

communication, 2009). It is close to the village of Rhynie in Aberdeenshire (Fig. 2.27). The Rhynie chert, a finely grained silica deposit, formed around hot springs during the Devonian Period 407 million years ago. The chert preserves the remains of plants, animals and even individual microbial cells in fine detail. The Rhynie chert documents an ecosystem from the Devonian period (Edwards *et al.*, 2017). Geothermal systems channel boiling water from deep in the earth to geysers and hot springs at the surface. As the water cools it becomes supersaturated with dissolved silica, which precipitates out as sinter or geyserite. The finely grained silica preserves the detailed structure of trapped organisms. Thermophilic bacteria and archaea grow abundantly in hot springs. Although microbial cell walls and slime may act as templates for the precipitation of sinter, its precipitation appears to be largely spontaneous (Yee *et al.*, 2003; Lalonde *et al.*, 2005). The Rhynie chert was discovered by William Mackie (1856–1932) a doctor and amateur geologist. Mackie found chunks of Rhynie chert scattered on the surface in fields near Rhynie in 1912. The chert deposits have since been investigated by digging trenches to excavate samples and by painstakingly cutting sections from them for microscopic examination (Figs. 2.28–2.29) (Edwards *et al.*, 2017). A new species of cyanobacterium that formed microbial mats and grew as tuft-shaped colonies was described in 2007 as *Croftalania venusta* (Krings *et al.*, 2007). The name *Croftalania* is a combination of the surname and middle name of William Noble Croft and Eric Alan George, who were the authors of a monograph on the cyanobacteria of the Rhynie chert (Croft and George, 1959). Examination of sectioned *Croftalania* mats from Rhynie has even yielded a 407 million year old testate

Fig. 2.27. Rhynie, Aberdeenshire, UK, close to the site of a fossilised geothermal spring dating from the Devonian Period.

Fig. 2.28. Rhynie chert sample (Nigel Trewin, School of Geosciences, Aberdeen University, UK, with permission).

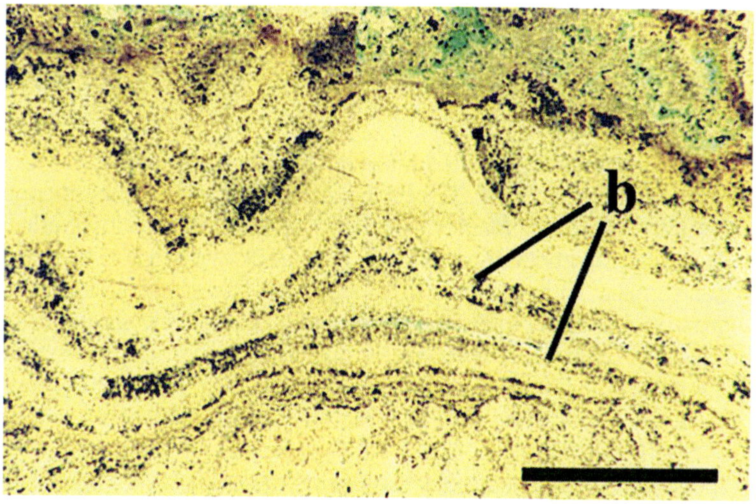

Fig. 2.29. Cyanobacteria microfossils in a sample of Rhynie chert (Nigel Trewin, School of Geosciences, Aberdeen University, UK, with permission).

amoeba, a unicellular shelled eukaryote (Strullu-Derrien, 2019). The scene at Rhynie 407 million years ago was that of a floodplain with rivers and lakes, situated between mountains (Wellman, 2019). There was volcanic activity and there were hot springs similar to those which occur today at Yellowstone and elsewhere (Fig. 2.30). Today there is nothing to see at surface level except for an unremarkable grassy field situated close to the outskirts of the village.

Fig. 2.30. A geothermal spring in Yellowstone National Park, USA.

Living stromatolites

Stromatolites are laminated mineralised structures formed by bacteria or other microbes (Figs. 2.31–2.32). Living stromatolites were first recognised as 'algal mounds' by Philip Playford in 1954 at Hamelin Pool, Shark Bay, Western Australia (Cockbain *et al.*, 2017). As a geologist Phillip Playford (1931–2017) was familiar with fossilised stromatolites. The stromatolites at Hamelin Pool grow in a shallow inlet where the water is relatively stagnant and more salty than the open sea. They are impressively bulky structures that grow up to 1.5 m (almost 5 feet) in height. The age of living stromatolites can be determined by using radiocarbon dating. This has shown that some living stromatolites at Shark Bay may be almost 1000 years old (McNamara, 1992). Very large specimens may be even older. By using radiocarbon dating and other methods their growth rate has been estimated to vary between approximately 0.1 and 0.5 mm per year (Jahrnert and Collins, 2012). Living stromatolites in Western Australia are composed largely of calcium carbonate ($CaCO_3$) and are themselves a form of limestone. Many fossilised stromatolites also consist largely of limestone. In some fossilised stromatolite formations the original calcium carbonate has been replaced by other minerals such as siderite (iron carbonate) or silica (Petrash *et al.*, 2016). In his description of fossilised stromatolites in 1908 Ernst Kalkowsky noted an association between stromatolites and oolite in Triassic rocks in Germany. The same relationship is seen at Shark Bay and in the Bahamas where living stromatolites occur today in association with

Fig. 2.31. Stromatolites. Shark Bay, Western Australia.

Fig. 2.32. Stromatolitic structures growing in a hot spring. Yellowstone National Park, USA.

oolite sand. Common factors that promote oolite formation and stromatolites are waters rich in calcium carbonate and environmental conditions favourable to the bacteria that induce the precipitation of calcium carbonate. At Shark Bay, stromatolites occur at different depths and exhibit different morphologies. Shapes include domes, columns, clubs and cones. Other microbial deposits occur with the stromatolites including tabular mats, pustular mats and tufted mats. As the internal microstructure of some of the microbialites at Shark Bay is clotted as opposed to laminated they can more correctly be referred to as thrombolites rather than stromatolites. Smooth mats of filamentous cyanobacteria are associated with finely laminated stromatolites in the subtidal zone. Coccoid bacteria and diatoms are associated with coarsely laminated stromatolites and with thrombolites in deeper water. Thrombolites show signs of disruption by shelled molluscs (Jahrnert and Collins, 2012). The Shark Bay specimens have come to be treated as standard examples of living stromatolites. Since the discovery of the Shark Bay stromatolites, an increasing number of varieties of mineralised microbially mediated structures have been referred to as stromatolites. In addition to Shark Bay and the Bahamas, locations for stromatolites include Lake Thetis in Western Australia, Giblin River in Tasmania, Lake Socompa and Laguna Negra in Argentina and Pavilion Lake, British Columbia, Canada (Farías *et al.*, 2013; Proemse *et al.*, 2017; Mlewski *et al.*, 2018). Some of these structures are quite small: the Tasmanian examples measure only about 1 mm in thickness and 10 cm in diameter. Some stromatolites grow in saline conditions and others in fresh water. Some of them grow in hot springs.

Microstromatolites and shore platform stromatolites in the British Isles

Subtle signs of microbially mediated carbonate precipitation can often be seen in places where the local conditions encourage microbes to flourish in a carbonate-rich environment. The surfaces of stones in coastal rock pools can become laminated with layers of calcium carbonate (Figs. 2.33). Some species of cyanobacteria in the genus *Rivularia* grow in the form of hemispherical colonies which may reach a diameter of several centimetres in diameter (Fig. 2.34). *Rivularia* colonies become calcified and laminated with successive seasons of growth. These little calcified pimples are sometimes referred to as microstromatolites. Microstromatolites formed by colonies of *Rivularia biasolettiana* (Biasoletto's Brook Blister) and *R. haematites* (the Bloodstone Brook Blister) are a feature of carbonate-rich streams in the British Isles. At localities such as Gordale Beck and Malham Tarn in the Craven District of Yorkshire, submerged rocks and pebbles may become conspicuously encrusted with *Rivularia* colonies. *Rivularia* colonies can withstand periods of desiccation (Pentecost and Franke, 2010).

Microbialites found at a number of locations on the South African coast and in Northern Ireland and the Isle of Harris in Scotland have been referred to as 'shore platform stromatolites'. In 2011, shore platform stromatolites were discovered close to the Giant's Causeway in pools towards the landward margin of a basalt rock platform (Cooper *et al.*, 2013). Rock surfaces in the pools provide a suitable habitat for a cyanobacterium which has been identified as belonging to the genus *Rivularia*. The cyanobacterial colonies are encrusted with a thin layer of carbonate, 2–4 mm in thickness (Fig. 2.35). Microscopic

Fig. 2.33. Cyanobacteria and green algae with calcium carbonate precipitates in a coastal rock pool on dolomitic limestone. Islay, UK.

Fig. 2.34. Pebbles encrusted with mineralised colonies of cyanobacteria (*Rivularia*). Malham Tarn, Yorkshire, UK.

examination of encrusted pebbles shows layers of carbonate as well as trapped particles and diatoms. The pools are situated in a zone that appears to be unusually favourable for the growth of cyanobacteria and for mineralisation. The local geology provides a source of calcium carbonate and the variable salinity of

Fig. 2.35. Microstromatolites formed by cyanobacteria in a coastal rock pool near the Giant's Causeway, County Antrim, Northern Ireland.

the pools renders them unfavourable for marine and freshwater organisms that might compete with or disrupt the growth of cyanobacteria. In South Africa at some locations shore platform stromatolites may flourish as florid encrustations which reach a thickness of several centimetres. Shore platform stromatolites have been observed to thrive in proximity to sites where active tufa formation is taking place. Where carbonate-rich water flows at a gradient, such as down a cliff face, tufa deposits accumulate, whereas in adjacent rock pools the same source of carbonate creates conditions that allow shore platform stromatolites to develop (Smith *et al.*, 2018).

Ancient stromatolites

Ancient fossilised 3.3 billion year old stromatolites are known from South Africa. The oldest known undisputed fossilised stromatolites currently known are those of the Pilbara Craton in Western Australia. Some of them have been dated as being almost 3.5 billion years old. It seems remarkable that fossils of such antiquity can still be detected on the earth's surface. Over the course of earth history land masses have formed, migrated around the globe, whilst rocks have been eroded, buried, heated, metamorphosed and recycled. Western Australia and South Africa include some unusually old and unusually stable masses of rock. Even older 3.7 billion year old sedimentary rocks occur in Isua supracrustal belt in Greenland. Wavy structures seen in the Isua rocks have been reported as putative stromatolites but because the Isua rocks have been deformed this interpretation has not been universally accepted (Nutman *et al.*, 2016; Allwood *et al.*, 2018). The earth is about 4.5 billion years old.

The oldest rocks in the British Isles are the 3 billion year old Lewisian gneiss deposits of northwest Scotland but these are rocks that have spent much of their time situated miles underground where they have been cooked, folded and recrystallised. The Pilbara stromatolites of Australia, as well as putative microfossils found in the same deposits, are regarded by many geologists to be the world's oldest known fossils. A detailed survey of the Pilbara deposits has allowed reconstruction of ancient environmental conditions. Different kinds of stromatolite exist in the Pilbara region. There are dome-shaped laminates that encrust the underlying rock, wavy laminates, iron-rich laminates, conical laminates and egg-carton shaped structures. Some stromatolitic forms lie sandwiched between layers of volcanic rocks (Allwood *et al.*, 2006). Stromatolites became most abundant and diverse during the Proterozoic aeon (2.5 billion to 540 million years ago). Evidence from the morphology, mineralogy and biochemistry of stromatolites from this time suggests that complex communities of bacteria and archaea lived together, resulting in the formation of stromatolitic reefs. Impressive examples of stromatolite formations are known from Australia, Canada and elsewhere. The British geological record includes stromatolite fossils from the late Precambrian as well as from the Cambrian, Devonian and later periods. Impressive living stromatolites can be seen in a few select sites around the world, including Lake Thetis and Shark Bay in Western Australia. Details on where to see examples are given in Chapter 9, *Bacteria tourism*.

By the beginning of the Cambrian Period, stromatolites were in decline, possibly as a result of disturbance or competition caused by emerging populations of multicellular or unicellular eukaryotes. Since then they have had their moments of resurgence, which have been linked to mass extinction events that caused the depletion of populations of eukaryotes. In particular, the emergence of the foraminifera has been proposed as a possible reason for the decline in abundance of stromatolites. Foraminifera are unicellular protists which became abundant in the fossil record during the Cambrian Period. Some foraminifera have calcite shells in which they farm green algae. Others creep around feeding on bacteria and diatoms using an appendage called a reticulopod which disturbs sediment. Observations made at Highborne Cay in the Bahamas where both thrombolites and stromatolites occur, suggest that foraminifera are disruptive to the lamination of stromatolites (Bernhard *et al.*, 2013). There are numerous different kinds of stromatolite-building bacteria. They simply need the right conditions in order to build stromatolites. Where a suitable environment exists, stromatolites can be expected to appear. Apart from lack of competition and disturbance by other organisms another prerequisite for the development of stromatolites is an adequate supply of carbonate. The abundance of stromatolites in the geological record correlates with the preponderance of dolomitic sediment deposition (Peters *et al.*, 2017). Microbes promote the precipitation of dolomite and dolomite provides material for the growth of stromatolites.

Fossil stromatolites in the British Isles

The search for stromatolites is rather like hunting for rare plants. They blend into their surroundings. Stromatolites are rocks that tend to resemble other rocks present in the same location. Unlike birds and insects stromatolites do not betray their presence by movements or by making any noise. Even when armed with a map and location details, after arriving at the right spot it may take some moments to get one's eye in. Then suddenly, amongst the surrounding rocks, a stromatolite will reveal itself through its distinctive laminated appearance. These layers record the cyclical growth pattern of past generations of microbes. Because the narrow bands of mineralisation tend to be more resistant to erosion than the adjacent layers they may project slightly from the surface of the rock and feel sharp to the touch. Stromatolites also stand out visually from the adjacent rocks on account of their wavy patterns and curved outlines which signify organic growth. Many stromatolite localities in the British Isles are beautiful and remote places that are rich in other aspects of natural history (Fig. 2.36). Stromatolites occur among the Precambrian rocks of Anglesey

Fig. 2.36. Map showing sites for fossil stromatolites (red stars) and microfossils (blue stars) in the British Isles (adapted from Hellerick, Blank topographic map of the British Isles). Creative Commons Attribution-Share Alike 3.0 Unported licence.

and Scotland. More recent examples include those from the Ordovician in Sutherland, the Devonian in Orkney and the Permian in Yorkshire and County Durham. In each situation the geological evidence tells a story of opportune environmental conditions for microbial life and of circumstances which were less favourable for more complex organisms. Quarries and eroding shorelines are good places to see exposed, recognisable stromatolites. Stromatolites had their heyday during Precambrian times. To see really extensive fossilised formations of Precambrian stromatolites *in situ* one would need to travel to those parts of the world where there are large chunks of ancient exposures of sedimentary Precambrian rocks, as occur in Canada, Australia or South Africa. Specimens from some of these places may be seen in museums in the UK.

Precambrian stromatolites on the Isle of Islay

Precambrian stromatolites are to be seen on the east coast of the Isle of Islay on a stretch of the shoreline not far from the Bunnahabhain distillery (Maltman *et al.*, 2000). To the south of the distillery, a coastal path leads for a short distance towards a headland, Rhubha a' Mhill. This is a pleasant location (Fig. 2.37). As well as the interest provided by the distillery, there are good views across the Sound of Islay towards the glistening quartzite hills of Jura. The sun was out when I went in search of the stromatolites at Bunnahabhain and Dark-green fritillaries and Ringlet butterflies were flying about on a sheltered grassy slope above the beach. Outcrops of greyish and brownish dolomitic

Fig. 2.37. Bunnahabhain, Islay, UK, a locality for Precambrian stromatolites.

limestone decorate the hillside. Near the base of the slope, not far from the path I encountered the eroded face of a dome-shaped stromatolite. Partly covered with moss, its significance could easily have been overlooked during the course of a more casual visit. A black patina revealed the laminated structure of the stromatolite to good effect (Fig. 2.38). In the intertidal zone, closer to the headland, the shattered dome of another stromatolite became apparent. On first glance the nature of this structure was not immediately obvious. From a distance it looked like any other coastal rock, sculpted by wave action and exposure to the elements (Fig. 2.39). Clumps of thrift had gained a root hold on its fissured and fractured surface. However, on close examination, fine curved laminations characteristic of organic growth were clearly visible. The stromatolites at Bunnahabhain occur as part of a layer of dolomitic rocks that are known as cap carbonates because they cap or cover a deeper layer of glacial rocks. Dolomite deposition is an indirect marker of microbial activity. Dolomite is the name of a mineral and is also the name of a type of rock that consists largely of dolomite (calcium magnesium carbonate). Dolomitic rocks usually consist of a mixture of dolomite and calcium carbonate and are also known as dolomitic limestone or dolostone. Cap carbonates can be seen in various places around the world, in similar geological circumstances to the thick layer of rocks that form the hillside above the beach at Bunnahabhain.

Precambrian glacial deposits and snowball earth

Islay is of special interest to people who study the history of life on earth. Early in the history of the study of geology, the island yielded clues about a

Fig. 2.38. Precambrian stromatolite. Bunnahabhain, Islay, UK.

Fig. 2.39. Eroded stromatolite. Bunnahabhain, Islay, UK.

period in Precambrian times when it is suspected that a worldwide glaciation event occurred. This time is known as the Cryogenian Period (720–635 million years ago). The Cryogenian was characterised by multiple cycles of freezing which alternated with warmer interglacial periods. Despite what might seem to have been trying environmental conditions, microbial life thrived during the interglacial periods as evidenced by the discovery of stromatolites from different geological horizons in the Cryogenian rocks of Islay as well as in the Garvellachs, a group of islands situated to the north of the Isle of Jura. The end of the Cryogenian was marked by a time when stromatolites bloomed. In addition to the ones at Bunnahabhain, stromatolites from the Cryogenian period have been found at multiple sites from around the world.

In 1870 and 1871, James Thomson presented papers to the British Association for the Advancement of Science entitled: *On the occurrence of pebbles and boulders of granite in schistose rocks of Islay, Scotland* and *On the stratified rocks of Islay* (Thomson, 1871a; Thomson, 1871b). Thomson noted the presence of granite boulders embedded in Precambrian sediments and suggested that glaciers had been responsible for transporting the boulders to what is now the island of Islay. James Thomson (1823–1900) was a self-educated geologist. He travelled extensively. He published an account of the geology of Islay and he became an authority on fossilised corals. Following Thomson's observations of Precambrian glacial deposits on Islay, similar rock formations containing drop stones were discovered in Africa, Asia, Australia and the Americas. Later, by studying the magnetic signatures recorded by the rocks when they were formed, it became possible to show that glacial sediments were deposited during the Cryogenian Period at all latitudes, including the tropics.

More than a billion years before the Cryogenian Period, an earlier phase of global cooling occurred known as the Huronian glaciation, which has been attributed to the oxygenation of the planet by photosynthetic cyanobacteria. In contrast, the reasons for global cooling during the Cryogenian Period are less clear. The Huronian glaciation (2.4–2.1 billion years ago) and the later Cryogenian Period are sometimes referred to as 'snowball I' and 'snowball II'.

The Precambrian glacial deposits on Islay are known as the Port Askaig tillite. Tillite is a term used to describe rock formed from glacial till, which is the debris deposited by a glacier. Till consists of an unsorted mixture of clay, sand and rock fragments. At Port Askaig on the east coast of Islay, quarried rock faces composed of tillite form the walls of a geological amphitheatre that backs onto the car park situated behind the ferry terminal. This makes it possible to study Cryogenian rocks within minutes of disembarking from a mainland ferry. Suspended in the grey matrix of the tillite lie rounded clasts of pink granite and other rocks, some the size of small pebbles, others as big as boulders (Fig. 2.40). They are known as drop stones because they were dropped from floating ice into the water column and sank into the underlying sediment. Cryogenian boulder deposits which have been correlated with the Port Askaig tillite occur at a few other places in Scotland and also in Ireland in County Donegal and at Doogort Bay on Achill Island in County Mayo. At Doogort Bay an exposure of tillite known as the Doogort Boulder Bed exists in close proximity to a bed of metamorphosed limestone. Although much altered from its original state and frequently crossed by veins of calcite, the presence of alternating paler and darker bands suggests a stromatolitic origin for sections of the Doogort Limestone (Muir, 1978). At Doogort Bay at low tide it is possible to pick one's way rather precariously along the boulder-strewn foreshore in

Fig. 2.40. Granitic drop stone in the Loch Askaig tillite, Islay, UK.

search of traces of Precambrian microbial life. Here and there one encounters a greyish outcrop of putative stromatolitic limestone (Fig. 2.41) whilst from the boulders on the foreshore protrude the differentially eroded clasts of granitic drop stones (Fig. 2.42).

It has been reasoned that the Cryogenian period came to an end as a result of increasing levels of atmospheric carbon dioxide released from volcanoes, according to which hypothesis the resulting greenhouse conditions would

Fig. 2.41. Metamorphosed stromatolitic limestone, Doogort Bay, Achill Island, County Mayo, Ireland.

Fig. 2.42. Granitic drop stone protruding from eroded tillite boulder, Doogort Bay, Achill Island, County Mayo, Ireland.

eventually have led to a rapid meltdown of ice, accompanied by erosion of glacially shattered rocks by carbonic acid, saturation of the oceans with calcium carbonate and widespread deposition of calcium carbonate on the ocean floor. The Port Askaig tillite is capped by dolomitic limestone. Dolomite precipitation is known to be promoted by several different groups of bacteria and archaea, including photosynthetic cyanobacteria, sulphide-oxidising bacteria, sulphate-reducing bacteria and methanogenic archaea (Moore *et al.*, 2004; Kenward *et al.*, 2009; Xu *et al.*, 2019). In the case of post-Cryogenian dolomite it has been suggested that methanogenic archaea might have mediated the conversion of calcium carbonate to dolomite. Although unsuitable for many other forms of life, the extreme conditions would have been ideal for the growth of microbial mats and for the development of stromatolites over large areas of the globe (Hoffman *et al.*, 2017). The cap carbonates mark the beginning of the Ediacaran Period (635–542 million years ago), named after an enigmatic assemblage of fossils from the Ediacara Hills in Australia. The Ediacaran Period was followed by the Cambrian Period (541–485 million years ago), which is when complex multicellular animals really began to blossom and diversify. Stromatolites gradually declined in abundance over the course of the Ediacaran and Cambrian Periods (Riding, 2006).

Stromatolites from the early Ordovician at Durness, Sutherland

Durness has been described as the most north-westerly village in mainland Britain. To the west of the village, the coastline at Balnakeil Bay between Eilean Dubh and Balnakeil preserves a geological record of the conditions that occurred from the late Cambrian Period into the early Ordovician Period. This time of transition has been identified as a time of resurgence for stromatolites (Riding, 2006). Numerous traces of microbial life have been described from Durness, including microbial mats, columnar and dome-shaped stromatolites and thrombolites. Unlike stromatolites, which have a laminated morphology, thrombolites are less organised in appearance. Living thrombolites can still be seen today at a few select locations around the world, such as Lake Clifton in Western Australia. The late Cambrian was a time of change that saw the extinction and decline of some groups of animals. The succession of rocks that is now exposed along the present-day coastline at Durness was deposited between about 510 and 470 million years ago. During those times, the setting was that of a tropical continental margin, situated about 20° south of the equator (Cocks and Torsvik, 2006). The Durness rocks consist mostly of limestone and dolomite. The presence of oolite shows that some deposits were formed in shallow seas similar to those of the present-day Bahamas. In such conditions bacteria induce the precipitation of carbonate as ooids. Stabilised ooid deposits consolidate to form oolitic limestone. The conditions also favoured the growth of stromatolites. It is likely that they flourished in hypersaline waters as occur today at Hamelin Pool in Western Australia. Above the tidal zone extensive mudflats became colonised by microbial mats. Desiccation cracks are evident on preserved sediment surfaces. This was an arid environment, similar to coastal plains known as sabkhas that occur today around the shores of the Persian Gulf.

Five hundred million years later, different conditions prevailed when I visited Balnakeil Bay. It was a cool windy afternoon. Along the shoreline an expanse of massive dolomitic rocks separated the sandy beach from the green of the golf course. While scrambling over the rocks I suddenly realised that one particular dome-shaped rock was in fact a weather-beaten stromatolite. From a distance the dome resembled the surrounding grey dolomitic rocks but when seen up close, the surface showed the characteristic mineralised layers of stromatolitic growth (Fig. 2.43). The rocks at Balnakeil Bay have been sculpted by the elements. Weathering causes the mineralised laminations of the stromatolites to stand proud from the rock surface as minute grey lines of points and ridges that feel spiky to the touch like the teeth of a comb (Fig. 2.44). The jagged pattern formed by the laminations resembles Peter Saville's art work for the record sleeve of *Unknown Pleasures*, an album by Joy Division. Even dolomite itself is of microbiological significance. Dolomite precipitation is promoted by microbial activity. Furthermore dolomite precipitation does not readily happen spontaneously. Sulphate-reducing bacteria are suspected to have been the inducers of dolomite precipitation at Durness (Wright, 1997; Wright, 1999). An expanse of mudflats with stagnant sediments and decaying mats of cyanobacteria would have allowed large populations of sulphate-reducing bacteria to flourish. By reducing magnesium sulphate, sulphate-reducing bacteria make magnesium ions available for dolomite (calcium magnesium carbonate) precipitation. Additionally, sulphate reduction increases alkalinity and this also induces dolomite precipitation.

Fig. 2.43. Stromatolite dome. Durness, Sutherland, UK.

Fig. 2.44. Stromatolite laminations. Durness, Sutherland, UK.

Devonian stromatolites in Orkney

Much of Orkney dates from the Devonian period when the ancient land masses of Laurentia ('North America') and Baltica ('northern Europe') were joined together and situated near the equator. About 390 million years ago, a shallow lake (Lake Orcadie) situated near the middle of this tropical land mass was subjected to periods of inundation, drying and refilling, resulting in the characteristic red and orange sedimentary rocks that grace Orkney's landscape. The Stromness flagstone beds contain the fossilised remains of Devonian fish. During periods of drought and high salinity, stromatolites flourished along the edge of Lake Orcadie. Examples of them can be seen on the coast near Stromness and also on the cliff top at Yesnaby. The Stromness stromatolites can be reached fairly easily by walking westward along the coast from the campsite at Ness, past the gun emplacement and onto the shoreline. The scene here is an ancient coastline that happens once again to be exposed at sea level. The rocks show ripple marks and cracks, indicating past environmental conditions of shallow water and mudflats which were prone to desiccation. Nestled among these features there are some beautiful rust-coloured, domed stromatolites. Chert nodules associated with the stromatolite mounds have been interpreted as Magadi-type cherts (Parnell, 1987). Magadi-type chert is a form of silica that forms in soda lakes as typified by Lake Magadi, situated in the Rift Valley in Kenya (Fig. 2.45). Lake Magadi is highly alkaline but also microbially highly productive. It is the haunt of photosynthetic bacteria and archaea that can tolerate extremely alkaline conditions. The presence of Magadi-type chert nodules on Orkney shows that the ancient Lake Orcadie was subjected to periods of extreme alkalinity, aridity and desiccation. A visit to the shoreline near Stromness allows the naturalist to witness traces left by an ancient soda lake and visit one of Britain's lost habitats. Ten miles north of Stromness there are spectacular west-facing cliffs at Yesnaby and an abundance of stromatolites. In order to scrutinise the rock exposures and search for stromatolites there is no need to venture too close to the cliff edge because the cliff top at Yesnaby has been battered and scoured bare by stormy weather for a considerable distance inland (Fig. 2.46). Situated between the parking area and the

Fig. 2.45. Modern chert deposits in Lake Magadi, an alkaline lake in the Rift Valley, Kenya, resemble those seen among the Devonian rocks of Orkney, UK.

Fig. 2.46. Yesnerby, Orkney, UK, a locality for Devonian stromatolites.

cliff top an interpretative board illustrates Lake Orcadie accompanied by depictions of Orcadian stromatolites and their modern counterparts from Shark Bay in Australia. The stromatolites at Yesnaby have a curious columnar shape. Locally they are known as horse tooth rock because when examined in cross section they can be seen to consist of concentric laminations (Fig. 2.47). Examples of them can be seen in Stromness Museum. Ripple-marked stones at Yesnaby show that the sediments were deposited in shallow water.

Carboniferous stromatolites on the Firth of Forth

In the Carboniferous Period (359–299 million years ago) the British Isles were situated close to the equator. During the early Carboniferous, limestone deposition occurred in marine basins. Many of Britain's limestone formations date from the early Carboniferous, including those in South Wales, the Mendips, the Avon Gorge, the Great Orme, Morecambe Bay, the Pennines, the Midland Valley in Scotland and also the Burren in Ireland. Carboniferous stromatolites occur in association with the Randerston limestones in Fife and the Clifton Down limestone in South West England (Fig. 2.48). Near Queensferry on the Firth of Forth some stromatolites can be seen that appear to have developed in the shallows of a tropical saline lake, named Lake Cadell after the Scottish geologist Henry Moubray Cadell (1860–1934). The stromatolites are situated on a sloping shelf of stone among a jumble of slippery rocks near the end of a little spur of land that juts out from the coast (Fig. 2.49). There are ripple-marked slabs of sandstone in the vicinity, suggesting that the stromatolites grew in conditions similar to those that occur today in Lake Thetis, a saline lake with living stromatolites in Western Australia (Fig. 2.50). Microfossils which have been interpreted as possible cyanobacteria have been found in rocks adjacent

Fig. 2.47. Devonian 'horse tooth' stromatolites. Yesnerby, Orkney, UK.

Fig. 2.48. Carboniferous stromatolites in the Clifton Down limestone. Barnhill Quarry, Gloucestershire, UK.

Fig. 2.49. Carboniferous stromatolites at Queensferry on the Firth of Forth, UK.

to stromatolites at Queensferry (Follows and Tyson, 1998). Next to the stromatolitic rocks there is a sliver of dolomitic rock, known as Long Rib, close to which there are beds of oil shale. The oil in these beds is a hydrocarbon residue that resulted from the accumulation, decay and transformation of the organic matter of photosynthetic microbes that populated Lake Cadell.

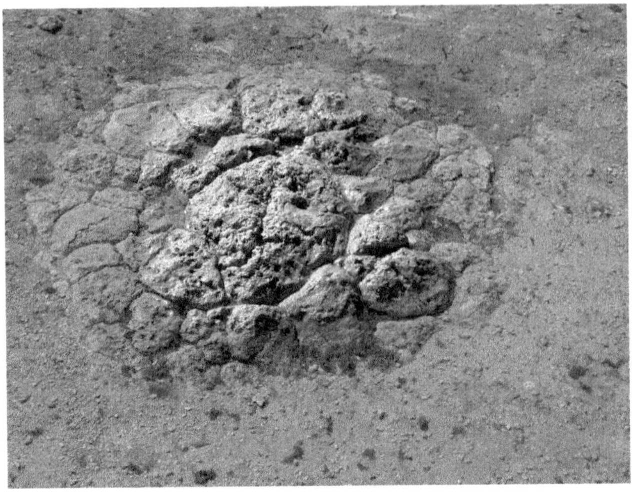

Fig. 2.50. A living stromatolite at Lake Thetis, Western Australia.

Permian stromatolites at Blackhall Rocks and South Elmsall Quarry

During the late Permian Period Britain was situated in the tropics to the north of the equator. About 255 million years ago County Durham was on the edge of a salty inland water body, called the Zechstein Sea, which extended across the North Sea into northern Europe. An arid climate favoured the deposition of evaporites, including rock salt and potash. These are now mined at Boulby in North Yorkshire. Coastal reefs formed by the growth of stromatolites and bryozoans developed in the shallows of the Zechstein Sea. Bryozoans are small, colonial marine animals, also known as moss animals. Bryozoans are most abundant in the tropics. Today the stromatolite reefs are incorporated among masses of dolomitic limestone, informally known as the Permian magnesium limestone. Exposures of the magnesian limestone can be seen at a number of sites in Yorkshire and County Durham. The magnesium limestone is of special interest to botanists and entomologists. Coastal outcrops of magnesian limestone and old quarries are often rich in natural history. In County Durham, a subspecies of the Northern brown argus butterfly, *Arica artaxerxes* ssp. *salmacis*, also known as the Durham argus or Castle Eden argus is closely associated with herb-rich magnesian limestone grassland. Not far south of Castle Eden Dene, some magnificent domed stromatolites can seen on the coast at Blackhall Rocks. After an interval of 255 million years it happens that these structures have come to the surface once again on an east-facing shoreline. In Permian times conditions were probably not too much unlike those that prevail at Shark Bay today. On a sunny day with the right perspective it is just about possible to imagine what it was like when the cyanobacteria that built the domes were still alive (Fig. 2.51). The domes are visible at low tide and are partially obscured by the growth of seaweed (Fig. 2.52). For decades

Fig. 2.51. Permian stromatolite on the Durham coast, at Blackhall rocks, UK.

Fig. 2.52. Stromatolite dome at Blackhall rocks, Durham, UK.

the locality received spoil from an adjacent coal mine. An aerial flight used to convey mining spoil towards the beach where it was dumped. The cliff top and the now dismantled aerial flight feature in the closing scene of the film *Get Carter* (1971), where Carter, a character played by the actor Michael Caine is taken out by a hit man. More recently the Durham coast has been cleaned up and promoted for recreation. There is now a Blackhall Rocks local nature reserve. I had visited this part of the Durham coast a number of times to look for plants and insects before I learnt about the fossil stromatolites. Viewed from the cliff top the stromatolites are not at all conspicuous. They blend in with the surrounding outcrops of limestone. On close inspection, however, their wavy, laminated organic structure makes them stand out from the other rocks. There is a large, partly eroded dome at the upper end of the shore just south of Gin Cave, whilst another outcrop is visible among the jumble of rocks closer to low water mark. Grazing limpets make indentations in the surfaces of the domes. Even as fossils, stromatolites are not safe from the predations of browsing eukaryotes.

Permian stromatolites can be seen in cross section at South Elmsall Quarry in Yorkshire (Fig. 2.53). South Elmsall Quarry is promoted as part of the UKRIGS (UK regionally important geological and geomorphological sites) initiative. The quarry is a short walk from an adjacent minor road and is only a few miles from the A1. A stromatolite reef is clearly visible as a set of wavy laminated structures stretched out along the upper face of the northeast wall of the quarry (Fig. 2.54). The stromatolite reef is situated on top of a bryozoan reef. This suggests that conditions which initially promoted the growth of eukaryotic bryozoans deteriorated in favour of the growth of bacteria. It seems likely that this former shallow margin of the Zechstein Sea became progressively saltier at the expense of the bryozoans which died off. The reef they built provided a platform suitable for the growth of a magnificent assemblage of stromatolites.

Fig. 2.53. Permian stromatolite seen in cross section. Elmsall Quarry, Yorkshire, UK.

Fig. 2.54. Permian stromatolites at Elmsall Quarry, Yorkshire, UK.

Cotham marble

Rather like Dr Who, stromatolites turn up during troubled times for life on earth. Cotham marble is actually a kind of stromatolitic limestone, a thin layer of which can be seen at exposures here and there in southwestern England and south Wales. Cotham marble deposits correlate with an extinction event which occurred at the end of the Triassic Period about 201 million years ago. At this time Britain was situated about 35° north of the equator. The Cotham marble was deposited in carbonate-rich water in a shallow subtropical channel. The fine-grained deposits contain filamentous microfossils and spherical microfossils which have been identified as a green alga. The marble rests on a bed rich in pyrite (iron sulphide) suggesting that development of the stromatolitic layer was mediated by a microbial mat populated by eukaryotic green algae and bacteria associated with a deeper anoxic layer of sulphate-reducing bacteria. The Triassic–Jurassic extinction might have been caused by climate change related to volcanic activity. Meteor impact has also been proposed as a possible cause, although a plausible impact site has not been identified (Simms, 2007; Ibarra *et al.*, 2016). Whatever the reasons, the conditions provided a brief window of opportunity for stromatolite-building microbes. Cotham marble was once quarried at Cotham, a suburb of Bristol. Natural exposures of the stromatolitic limestone are scattered over a wide area of South West England. Localities include Culverhole Point on the Devon coast, Stowey Quarry in Somerset, Manor Farm near Aust Cliff and the banks of the Little Avon River at Lower Woods Nature Reserve in Gloucestershire. Before the woods became a nature reserve I would cycle to the Lower Woods to collect insects. Decades later, after

I had developed an interest in stromatolites I returned once again to the Lower Woods to investigate Cotham marble *in situ*. When seen in the wild, Cotham marble is a pale-coloured knobbly material, which is sometimes described as vermiculated limestone (Figs. 2.55–2.56). Cotham marble does not really

Fig. 2.55. A loose broken chunk of Cotham marble found in the Lower Woods, Gloucestershire, UK.

Fig. 2.56. A moss-covered lump of Cotham marble in the Little Avon River, Gloucestershire, UK.

come to life until cut and polished. In cross section, the contrasting dark and pale layers and dendritic shapes suggest a rural landscape of fields and trees. It is known also as landscape marble and is appreciated for its decorative appeal. Strips of unfinished vermiculated limestone grace the walls of Cotham House, Cotham Hill, Bristol (Fig. 2.57). Cotham House currently accommodates the Department of Philosophy of the University of Bristol. Cotham House is marked on Rocque's 1742 *A Plan of the City of Bristol*. A former resident, Sir George White (1854–1916), founder of the British and Colonial Aeroplane Company refurbished Cotham House at great expense. Appreciation of this elegant and historic building is enhanced by the knowledge that its ornamental features were created by communities of microbes.

Portland stone, stromatolites and thrombolites

When I first visited the Isle of Portland, my quarry was the scaly cricket, *Pseudomogoplistes vicentae*, an elusive insect that lives amongst the shingle of Chesil Beach, between the island and the mainland. During later visits to Portland I turned my attention to the limestone quarries which host a distinctive population of the Silver-studded blue butterfly, *Plebejus argus*, as well as many other interesting insects. Eventually my interests matured to those of the bacteria tourist and I revisited the Portland quarries to see Jurassic stromatolites and thrombolites. During the late Jurassic Period, about 145 million years ago, a warm shallow carbonate-rich sea covered the area that later became the Dorset coast. Microorganisms mediated the creation of large banks of oolite sand in conditions that probably resemble those of the Bahamas today. The banks

Fig. 2.57. Cotham marble frieze, Cotham House, Bristol, UK.

became stabilised and consolidated to form thick beds of Portland stone. Lumps of Portland stone can still be admired in the disused quarries that provided the stones used to construct St Paul's Cathedral, the British Museum and other historic buildings. An advantage of Portland stone and of other oolitic limestones is that they can be cut and fashioned in any direction. Portland stone weathers well and has a pleasing bright appearance. The versatility of Portland stone can be appreciated at Tout Quarry Sculpture Park and Nature Reserve. This former quarry contains dozens of sculptures that have been carved in blocks of stone that remain on the site. The collection includes one of Sir Antony Gormley's earlier works, *Still falling*, which depicts the outline of a diving figure, carved into a vertical quarried face of Portland stone.

The Purbeck beds, which overlie the Portland stone, record a history of changing environmental conditions. Three cap deposits, the Skull Cap, Hard Cap and Soft Cap provide a record of brackish to hypersaline lake environments (Gallois *et al.*, 2018). All three layers contain stromatolites and thrombolites. The cap deposits are interrupted by the Lower Dirt Bed and Great Dirt Bed, which are the lithified remains of ancient soils. Forests of trees grew in the dirt beds, most of which have been identified as a species of conifer, *Protocupressinoxylon purbeckensis* (Purbeck Proto-cypress). *Protocupressinoxylon purbeckensis* was described as a new species on the basis of fossils found in Dorset. Specimens of Purbeck Proto-cypress have also been identified in late Jurassic deposits in Spain and Switzerland. The dominance and wide distribution of a single species of tree in the late Jurassic indicates a lack of botanical diversity, possibly related to stressful environmental conditions (Philippe *et al.*, 2010). A petrified tree trunk, retrieved from a Portland quarry, can be seen in the grounds of the Heights Hotel (Fig. 2.58). During episodes when the water

Fig. 2.58. Petrified tree. Portland, Dorset, UK.

levels rose and coalesced into saline lakes the waterlogged trees died and stromatolites began to grow on the lake bed. The dead trunks and branches of dead trees were colonised by communities of microbes which mediated the growth of thrombolites. Unlike stromatolites, thrombolites have a porous, disorganised microstructure. The thrombolites seen in Portland deposits have in fact been likened to modern deposits of tufa (Perry, 1994). Often the remains of dead trees decayed completely to leave rot holes (Fig. 2.59). In a disused quarry at Freshwater Bay on the east coast of Portland numerous rot holes can be seen in Hard Cap deposits, that mark spots where the trunks and branches of submerged dead trees acted as foci for the growth of cyanobacteria and the formation of thrombolites.

A spectacular collection of microbial mounds and the silicified remnants of dead trees known as the Fossil Forest can be seen near Lulworth Cove. Lulworth Cove is well-known to naturalists as being the place where in 1832, the entomologist J.C. Dale (1791–1872) discovered the Lulworth skipper butterfly, *Thymelicus acteon*, as a species new to Britain. The Fossil Forest is situated inside the army ranges in a dramatic position, perched on a ledge above the sea (Fig. 2.60). An interpretative board provides information about the local geology and displays some nice pictures of dinosaurs that used to roam the area. A row of thrombolite-encrusted trunks marks the seaward edge of the ledge (Fig. 2.61). The trees once grew in Jurassic soil that is now represented by the Great Dirt Bed. The thrombolites constitute part of the Soft Cap which was deposited after the trees became submerged. An additional stratum of microbial importance, the Perryfield stromatolite bed, can be seen above the cap deposits in quarries on the Isle of Portland. The Perryfield stromatolite bed (Fig. 2.62) includes numbers of individual stromatolites that resemble the living stromatolites at Lake Thetis in Western Australia. The proximity of evaporite deposits suggests an arid hypersaline environment (West, 1975). Although

Fig. 2.59. A rot hole marking the site of a former tree trunk, surrounded by thrombolite. Portland, Dorset, UK.

Fig. 2.60. The Fossil Forest at Lulworth, Dorset, UK.

Fig. 2.61. Thrombolite-encrusted former trunk. Lulworth, Dorset, UK.

Perryfield Quarry is private property, a public footpath provides views of the exposed face of Perryfield North Quarry, also known as Windmill Quarry after a disused windmill that overlooks the quarry. The stromatolite bed is close to the top (Fig. 2.63). Portland stone continues to be excavated from Perryfield Quarry. An adjacent disused quarry is now a butterfly reserve.

Fig. 2.62. Stromatolite, from Portland, Dorset, UK.

Fig. 2.63. Perryfield Quarry, a stromatolite site. Portland, Dorset, UK.

Microstromatolites of the Hunstanton red chalk

To anyone accustomed to the idea that chalk is white, the cliffs at Hunstanton are a peculiar sight. At Hunstanton on the Norfolk coast stands a line of west-facing cliffs composed of sandstone and of red and white chalk. Three contrastingly coloured layers of rock stretch from one end of the cliffs to the other (Fig. 2.64). Resembling as it does an elaborate piece of confectionery, from a distance the scene almost looks edible. At the bottom lies the Carstone Formation, a layer of orange-brown sandstone, which contains numerous rounded pebbles, indicating that this layer of rock was deposited in shallow water. The top layer is made up of beds of white chalk. In between these two layers is the Hunstanton Formation, formerly known as the Red Chalk. The term red chalk is still used informally when referring to the Hunstanton Formation. The red and white chalk layers are deep marine deposits that date from the Cretaceous Period. Red chalk can be seen at other sites on the east coast and also in Germany. Hunstanton provides what is possibly the best known example of red chalk in Britain. The redness of the red layer is caused by the presence of finely dispersed particles of iron oxide. In addition to its distinctive red colouration the red chalk at Hunstanton is known for crust-like deposits situated in the boundary zone between the red chalk and the white chalk (Fig. 2.65). Microscopic stromatolitic structures seen in the red chalk crust have been interpreted as being of possible microbial origin (Andrews *et al.*, 2015). The putative microstromatolites at Hunstanton have been tentatively identified as *Frutexites*. *Frutexites* is a generic name for stromatolitic microfossils that show tree-like growth patterns and which are rich in iron and manganese

Fig. 2.64. The Hunstanton red chalk. Hunstanton, Norfolk, UK.

oxides. *Frutexites* microstromatolites have been recorded from various geological settings, including limestones, fracture surfaces and from fossilised seafloor surfaces, known by geologists as carbonate hardgrounds. *Frutexites* has been portrayed as something of an enigma (Jeans *et al.*, 2016). Efforts to find evidence of a link between *Frutexites* and any particular kind of microbe remained elusive until 2017 when researchers in Sweden discovered living *Frutexites*-like biofilms growing on the walls of a tunnel in the Äspö Hard Rock Laboratory situated at a depth of 160 m beneath an island in the Baltic Sea (Heim *et al.*, 2017). *Frutexites*-like structures were found in association with iron-rich seepages of brackish water (Fig. 2.66). The biofilms were found

Fig. 2.65. Red crust-like deposits coating detached blocks of chalk below the cliffs at Hunstanton, Norfolk, UK.

Fig. 2.66. Iron-oxidising biofilms in the Äspö Hard Rock Laboratory, Sweden. (Image cropped from figure 2 from Heim *et al.*, 2017). Creative Commons Attribution Licence.

to contain iron-oxidising bacteria as well as species of bacteria that oxidise ammonia and nitrite. It has been suggested that oxidation of ferric iron by bacteria may account for the red colour of the Hunstanton red chalk (Andrews *et al.*, 2015).

Fossil bacteria

Multiple strands of evidence in the geological record demonstrate that microbial life has existed for more than 3 billion years. In addition to stromatolites and microbially induced sedimentary structures made by microbial mats, these various strands of evidence include the fossilised remains of individual bacterial cells and magnetite and greigite particles derived from the intracellular organelles found in magnetic bacteria. Further supportive evidence is provided by carbon isotope profiles that are known to be associated with particular groups of microbes and other biomarkers indicative of an organic origin for geological deposits (Boschker and Middelburg, 2002). Microfossils are the compressed or mineralised remains of individual microbes. Confusingly, however, inorganic chemical processes can generate filamentous and tubular crystalline microstructures that can resemble microorganisms (Brasier *et al.*, 2015; McMahon, 2019). This makes it important to develop criteria by which actual microfossils can be distinguished from artefacts. The detection of biomarkers and spatial association with stromatolites and microbial mat structures as well as general considerations regarding the geological context help to establish the credibility of putative fossil microbes. Because bacteria pre-date the emergence of multicellular plants and animals, microfossil bacteria are of particular interest to those who study the early history of life on earth. One of the problems that Darwin identified in the *Origin of Species* was the lack of a fossil record from the Precambrian, despite a wealth of fossil material dating from the Cambrian onwards. Nevertheless, Darwin was aware of the discovery of trace fossils in Precambrian rocks at Long Mynd in Shropshire (Salter, 1856; Salter, 1857; Darwin, 1859). The Long Mynd deposits include wrinkle marks, indicative of microbial mats and filamentous microfossils which resemble cyanobacteria (Callow and Brasier, 2009). The Long Mynd rocks, which are approximately 560 million years old belong to the Ediacaran Period of the late Precambrian. The sandstones and mudstones of the Long Mynd have been interpreted as delta and river sediments. Curious ring-shaped, dome-shaped or parallel ridge-shaped structures exist in some of the bedding planes. They measure several millimetres in diameter and are laminated in cross section. They have been interpreted as torus-shaped microstromatolites created by mats of cyanobacteria (Menon *et al.*, 2017). In the British Isles, Precambrian microfossil cyanobacteria are known also from 1 billion year old Torridonian deposits from Diabaig and other localities in Scotland (Strother and Wellman, 2016). Some groups of bacteria, such as cyanobacteria and sulphide-oxidising bacteria have reasonably distinctive morphological features. The interpretation of microfossils as cyanobacteria and sulphide-oxidising bacteria provides information about ancient ecosystems. Even remotely related species of bacteria, however, often look much the same. This means that microfossil bacteria are not especially useful for the study of bacterial evolution or for dating geological horizons. In contrast, fossilised plants and animals are morphologically distinctive, often to the point of being recognisable as individual species.

The first assemblage of Precambrian microfossils to gain wide acceptance as such by the scientific community was detected in the Gunflint chert deposits of Ontario, Canada in 1953 by Stanley A. Tyler (1906–1963), a geology professor at the University of Wisconsin (Tyler and Barghoorn, 1954; Barghoorn and Tyler, 1965). The Gunflint Chert Formation is roughly 2 billion years old. It contains stromatolites which are associated with well-preserved filamentous and spheroid microfossils. The deposits even contain identifiable biomarkers that can be associated with those of living bacteria. Some protein component profiles from Gunflint chert microfossils resemble those of modern cyanobacteria (Alleon *et al.*, 2016). Following the discovery of microfossils in the Gunflint chert, comparably well-preserved microfossils were detected in the 850 million year old Bitter Springs chert in the Northern Territory of Australia (Barghoorn and Schopf, 1965). Microfossils have since been discovered in numerous Precambrian deposits from Australia, North America, Africa and India. Many of them were found in association with stromatolites and some were supported by the detection of biomarkers (Schopf, 2006). The photosynthetic pigment okenane is a biomarker of photosynthetic purple sulphur bacteria. Purple sulphur bacteria require anoxic, sulphide-rich environments. Okenane has been detected in the 1640 million year old Barney Creek Formation in Northern Australia (Brocks and Schaeffer, 2008). The 3465 million year old Apex chert deposits in the Pilbara region of Western Australia preserve the remains of an ancient hydrothermal vent ecosystem (Schopf *et al.*, 2018). The chert contains a variety of differently shaped microstructures. These structures have been variously interpreted as representing the remains of different species of microbes or as chemical artefacts (Brasier *et al.*, 2015). The ratio of naturally occurring carbon isotopes in living organisms varies depending on their metabolism and food sources. Measurements of carbon isotopes associated with five morphologically distinguishable Apex chert microfossils suggest that one of the microbial species was a member of the methanogenic archaea, two were methanotrophic bacteria and two others were photosynthetic bacteria (Schopf *et al.*, 2018).

Tubular and filamentous microstructures seen in hydrothermal vent precipitates from the Nuvvuagittuq Greenstone Belt in Quebec, Canada currently hold the record for the earth's oldest putative microfossils (Dodd *et al.*, 2017). Uranium-lead radiometric dating of zircons found in association with the precipitates gives a minimum age of 3770 million years old. Samarium–neodymium radiometric dating of neighbouring rocks gives a maximum possible age of 4280 million years old. The Nuvvuagittuq samples come from a volcanic ocean floor environment with pillow lava and carbonate-rich iron formations. Similar iron-rich environments exist near hydrothermal vents on the seafloor today. They provide a habitat for iron-oxidising bacteria, such as *Mariprofundus ferrooxydans* (the Deep Sea Iron-bug) and ocean floor bacteria related to *Gallionella ferruginea* (the Rusty Iron-bug), which mediate the precipitation of iron oxide deposits (Li *et al.*, 2012). The geological record preserves evidence of such environments throughout earth history. Putative microfossils which represent the remains of presumptive iron-oxidising bacteria have been found at multiple sites worldwide (Little *et al.*, 2004). Such a site exists in Wales at Llanddwyn Island. From the carpark at Newborough Warren a walk along the beach leads to a sand bar that connects Anglesey to Llanddwyn Island. Even before reaching the island strikingly pigmented mounds of pillow lava rise out of the sand. The greenish lava heaps are colonised by lichens and incorporate lumps of red jasper (Fig. 2.67). Jasper is a fine-grained chert, strained red by haematite, a form of iron oxide. Microscopic thread-like structures can be seen in thin sections of Llanddwyn jasper. It has been claimed that they might represent the fossilised remains of iron-encrusted filamentous microorganisms (Peat, 1984; Peat, 1988). Alternatively, they could be chemical artefacts. The Llanddwyn rocks are roughly 550 million years old. It is uncertain if they date from the late Precambrian or early Cambrian. Sections of more recent jasper deposits from a 352 million year old seafloor hydrothermal vent deposit at the Ballynoe Mine, near Silvermines in County Tipperary contain twisted ribbon-like microfossils that convincingly resemble the iron-oxidising bacterium *Gallionella ferruginea* (the Rusty Iron-bug). The Ballynoe vent deposits date from the early Carboniferous Period. Also in Ireland, another early Carboniferous hydrothermal vent, which has yielded worm fossils, has been discovered at Tynagh in County Galway (Banks, 1985; Boyce *et al.*, 2003).

Gypsum

Gypsum ($CaSO_4{\cdot}2H_2O$) is a form of calcium sulphate that incorporates water molecules. Gypsum occurs as fine-grained deposits and also in a variety of crystalline forms, including selenite, satin spar and desert rose. Fine-grained

Fig. 2.67. Pillow lava with red jasper. Newborough Warren, Anglesey, UK.

gypsum is known as alabaster. Alabaster mined historically at Chellaston near Derby was once used to supply an international trade in mediaeval alabaster carvings. Gypsum mining continues today at a number of locations in England. Many gypsum deposits are evaporites. Evaporites are accumulations of minerals derived from water bodies which have evaporated in arid conditions. Evaporite pans provide a distinctive habitat associated with specialised bacteria. Such a place is the Chott el Djerid, a huge evaporite pan on the edge of the Sahara Desert in southern Tunisia (Drake *et al.*, 2004). A causeway runs across this remarkably experiential landscape (Fig. 2.68). Evaporite pans are rich microbial worlds but they are hostile to most other forms of life.

Geological gypsum deposits document the former existence of lost habitats and have yielded microfossils that provide information about ancient ecosystems. Gypsum deposits are well-represented in Britain where they are associated especially with the Triassic Mercia mudstone (Fig. 2.69). These sediments date from a time when hot arid conditions prevailed in Britain. Natural exposures of bands of satin spar and nodules of alabaster can be seen at Aust Cliff beside the River Severn. At Aust and other localities near Bristol, gypsum occurs in association with strontium sulphate ($SrSO_4$) as the rare mineral celestine, or celestite (Nickless *et al.*, 1976). Several kinds of bacteria can induce celestine precipitation, including cyanobacteria and purple sulphur bacteria (Sanz-Montero *et al.*, 2009). In Ireland in the counties of Cavan, Fermanagh and Leitrim, evaporite beds containing nodules of gypsum and celestine occur in association with deposits of dolomite and stromatolitic laminated limestones. The laminations record the growth of microbial mats which once flourished on tidal flats in a desert climate during the Carboniferous Period (West *et al.*, 1968).

Fig. 2.68. Active gypsum deposition at Chott el Djerid, Tunisia.

Fig. 2.69. Aust Cliff showing Triassic gypsum deposits. Gloucestershire, UK.

Fossil microbes have been detected in gypsum deposits in Italy, Spain, Australia, the USA, Mexico and Peru. Microbes seen in these deposits have been interpreted as the mineralised remains of sulphur-oxidising bacteria (Schopf *et al.*, 2012; Andreetto *et al.*, 2019). Gypsum deposits from Spain and Italy are associated with the desiccation of the Mediterranean Sea during the Miocene epoch approximately 5.5 million years ago, an episode referred to as the Messinian Salinity Crisis or the Messinian Event. Microfossils found in association with these evaporite deposits have been interpreted as representing the remains of the filamentous sulphide-oxidising bacteria *Thioploca*, *Beggiatoa* and the giant globular bacterium *Thiomargarita*. At Fayetteville Green Lake, a freshwater lake in New York State, the cells of cyanobacteria have been shown to act as templates for gypsum precipitation (Konhauser, 2007).

Phosphate

In their spare time, connoisseurs of the London Clay Formation gravitate to the fossil-rich northern shoreline of the Isle of Sheppey in Kent (Fig. 2.70). The sediments were deposited in the shallows of a subtropical ocean 52–51 million years ago during the early Eocene (Balson, 1990). Scattered along the beach between Minster and Warden Point one finds the fossilised remains of sea creatures and the seeds of plants that grew in the neighbouring mangrove swamps and forests (Collinson, 1984; Clouter, 2000). The scene was probably not much unlike some coastal areas of Africa and Asia today (Fig. 2.71). Fossils at Sheppey include the remains of crabs, sharks and the seeds of magnolia and mangrove trees. Many of the fossils have been converted

Fig. 2.70. Warden Point, Sheppey, UK.

Fig. 2.71. Mangrove swamp. Manda Island, Kenya.

to pyrite, while others are entombed in cream-coloured phosphate nodules (Figs. 2.72–2.73). The conversion of organic matter to pyrite (iron sulphide) is mediated by sulphate-reducing bacteria. Sulphate-reducing bacteria oxidise the remains of dead organisms and release hydrogen sulphide as a waste product. In iron-rich sediments this stimulates the formation of iron sulphide which can replace carbonate compounds within organic matter and result in fossils that are composed of pyrite (Thiel *et al.*, 2019). Sulphate-reducing bacteria have

Fig. 2.72. Fossilised plant stems composed of pyrite. Sheppey, UK.

Fig. 2.73. Fossiliferous phosphate nodules. Sheppey, UK.

also been implicated in the genesis of phosphate deposits. Phosphorus is an essential nutrient for living organisms. Among other things phosphorus is a component of deoxyribonucleic acid (DNA). In the modern world phosphate deposition is associated with coastal waters where nutrient-rich upwelling currents promote the growth of plankton. The remains of these organisms rain down to the seafloor where they accumulate in sediments and decompose. Sulphate-reducing bacteria liberate hydrogen sulphide, which encourages the growth of thriving populations of sulphur-oxidising bacteria (Crosby and Bailey, 2012). Some sulphur-oxidising bacteria, such as *Thiomargarita namibiensis* (the Pearl of Namibia), accumulate phosphate which is stored as an energy reserve in the form of polyphosphate. The subsequent breakdown of polyphosphate saturates the immediate environment with phosphate, which mineralises as apatite ($Ca_5(PO_4)_3(F,Cl,OH)$), leading to the generation of phosphate deposits. It is possible that something like this might have been happening in the seas around southeastern England in the early Eocene. The Pearl of Namibia is an unusually large species of bacterium. Some cells attain a diameter of 750 μm. In China, the phosphatic rocks of the 600 million year old Doushantuo Formation provide the basis for a commercial mining operation. The fine granular structure of the Doushantuo deposits makes them suitable for the detection of microfossils. Enigmatic globular microfossils present in the deposits have been interpreted as representing the microscopic eggs of some unknown animal. But as the remains of no suitable candidate animal species have yet been detected in these Precambrian rocks, a simpler interpretation might be that these microfossils represent the fossilised remains of giant sulphur-oxidising bacteria which were actively engaged in the accumulation and storage of phosphate (Bailey *et al.*, 2007).

Serpentine

The mineral serpentine ($(Mg,Fe)_3Si_2O_5(OH)_4$) is a product of the hydration of olivine ($(Mg,Fe_2)SiO_4$), a green mineral that occurs abundantly in the earth's upper mantle but much less frequently on the earth's surface. Serpentine is created when water comes into contact with the rocks of the earth's mantle, as may occur at tectonic plate boundaries when a portion of the mantle is uplifted or when a descending slab of crust carries water and sediment down with it into the mantle. The chemical reactions involved in the formation of serpentine generate a certain amount of heat and more importantly result in the release of hydrogen and methane molecules as byproducts. By acting as a source of hydrogen and hydrocarbons, serpentinisation zones situated below the earth's surface create environments that have the potential to sustain microbial life (Brazelton *et al.*, 2012).

Numerous zones of active serpentinisation exist in the world today in the vicinity of tectonic plate boundaries. Regions where serpentinisation is currently taking place include the Mid-Atlantic Ridge and the western margin of the Pacific Plate. The Mid-Atlantic Ridge marks the boundary between the North American and Eurasian plates. The two plates are slowly drifting apart. During the process, rocks of the earth's mantle come into contact with seawater, resulting in the formation of serpentine. At the Lost City hydrothermal field situated near the Mid-Atlantic Ridges samples taken from hydrothermal chimneys demonstrate the presence of diverse subsurface microbes. Among them are methanogenic archaea, including *Methanocaldococcus jannaschii* (Jannasch's Gas-ball), as well as the hydrogen-oxidising bacteria, *Hydrogenophaga atypica* (the Unusual Hydrogen Eater), *Hydrogenophaga palleroni* (Palleroni's Hydrogen Eater), *Hydrogenophaga taeniospiralis* (the Squiggly Hydrogen Eater) and the hydrogen-producing bacteria, *Heliobacillus mobilis* (the Deep-sea Sun Bug) and *Elusimicrobium minutum* (the Elusive Microbe) (Brazelton *et al.*, 2012). Also discovered amongst the Lost City DNA sequence data were sequences that matched those of "*Ca.* Desulforudis audax-viator" (the Bold Traveller), a globally distributed subsurface bacterium which has been retrieved from numerous deep locations, including a South African gold mine, a borehole in Finland and a nuclear test site in Nevada.

Stretched out along the western margin of the Pacific Plate between Japan and the island of Guam lie a series of massive submarine serpentine mud volcanoes which convey methane, mud, fluids and chunks of serpentine upwards towards the seafloor. The presence of organic matter in lumps of serpentine sampled from the South Chamorro mud volcano points to the likely existence of a deep microbial ecosystem. The summit of the South Chamorro mud volcano is itself situated at an ocean depth of 3 km. Deep below it exists a zone in which serpentinisation of the Earth's mantle is taking place. Serpentinisation zones associated with descending plate margins may in fact be the deepest places on earth that can sustain life. As they descend, plate margins remain relatively cool in comparison with the surrounding rocks. It has been estimated that descending plate margin temperatures may remain within the tolerances of hyperthermophilic microorganisms, even at depths of 8–15 km (Plümper *et al.*, 2017). At the time of writing, the maximum reported temperature at which microbial cells can actively grow is 122°C (Takai *et al.*, 2008).

In the British Isles outcrops of serpentine are few and far between. They represent ancient elevated fragments of the earth's mantle which underwent serpentinisation long ago as a consequence of contact between the hot mantle and seawater. Terrestrial serpentine outcrops are the last vestiges of former deep subsurface microbial habitats. Bacteria tourists who visit exposures of serpentine rocks in Britain today can use their imagination to picture these lost suboceanic microbial worlds. Serpentine itself is an intrinsically beautiful material (Fig. 2.74). A serpentine outcrop near Holyhead was once worked as the source of decorative Mona marble, whilst Cornish serpentine continues to serve in the production of ornaments and jewellery. Serpentine rocks contribute to the singular appeal of the Lizard peninsula. At Kynance Cove the rocks are red and green and sculpted by the elements to a reptilian finish (Fig. 2.75). Serpentine soils are generally alkaline and are associated with special communities of plants. The serpentine soils of the Lizard peninsula provide a home for a number of local species, including Cornish heath, *Erica vagans*, a plant which is locally common on the Lizard but not found elsewhere in Britain. Serpentine scenery can also be experienced at the Keen of Hamar on Unst, Shetland (Fig. 2.76). One of the flowering plants at the Keen of Hamer, Edmondston's chickweed, *Cerastium nigrescens* is considered to be endemic to Unst. While wandering among the shattered stones at the Keen of Hamar in search of rare flowers, the visiting naturalist has the opportunity to contemplate an earlier episode from earth history when the serpentine rocks were

Fig. 2.74. Serpentine. Keen of Hamar, Unst, UK.

Fig. 2.75. Serpentine outcrop at Kynance Cove, Cornwall, UK.

Fig. 2.76. Serpentine flora. Keen of Hamar, Unst, UK.

coming into being below an ancient seafloor and when chemical byproducts of serpentinisation sustained hyperthermophilic microbes in the deep biosphere.

Metals

Metals, including iron, manganese, copper and zinc are not only essential for living organisms but they are also potentially toxic to them. To perform oxygenic photosynthesis cyanobacteria require magnesium, iron, copper and manganese. Metals act as a cofactor in almost half of known enzymes. A nickel-dependent enzyme allows the thermophilic bacterium *Carboxydothermus hydrogenoformans* (the Hot Swamp Bug) to grow on a diet of carbon monoxide and water. The Hot Swamp Bug releases hydrogen as a waste product. There are also nickel-dependent microbes that oxidise hydrogen and release carbon dioxide and water as waste products. Bacteria have uptake mechanisms that enable them to obtain essential metals from the environment. Bacteria also have mechanisms that allow them to resist metal toxicity. In the presence of oxygen, metal-oxidising bacteria mediate the precipitation of iron, manganese and other metal oxides. In iron-rich spring water bacteria such as *Leptothrix ochracea* (the Ochre Iron-bug) passively induce the precipitation of ferric iron. Meanwhile, other iron bacteria, such as *Gallionella ferruginea* (the Rusty Iron-bug) actively exploit the chemical disequilibrium in waters rich in ferrous iron to obtain energy for growth by catalysing the conversion of ferrous to ferric iron. While oxygenated conditions favour bacteria that utilise oxidation reactions, anoxic environments are populated by microbes that make use of reduction reactions. In anoxic environments, metal-reducing bacteria obtain chemical energy for growth by reducing metal oxides. Scores of species of bacteria can catalyse the reduction of insoluble ferric to soluble ferrous iron. A byproduct of the dissolution of iron oxides is phosphorus, an essential nutrient, which tends to bind to iron oxides. Whereas reduced iron and manganese are soluble, the opposite is true of uranium and vanadium. Oxidised uranium (VI) precipitates out of solution when reduced to uranium (IV) by uranium-reducing bacteria. Uranium-reducing bacteria remove soluble uranium from water sources and are important sources of uranium ores. Similarly, sulphate-reducing bacteria play an important role in reducing soluble gold (III) to metallic gold and account for some gold deposits, such as those associated with the Easdale Subgroup black shales in Ireland and Scotland (Parnell *et al.*, 2017). Bacteria are of value in reducing metal toxicity at contaminated sites and in the leaching of metal compounds from ores during mining operations.

Banded iron formations

Banded iron formations consist of layers of reddish-brown ironstone which alternate with paler bands of chert. The alternating bands represent intermittent periods of iron oxide precipitation and its subsequent deposition as dark layers onto paler silica-rich sediments. One theory is that banded iron formations are a consequence of the oxygenation of the planet by photosynthetic cyanobacteria which released oxygen as a waste product. When in contact with oxygen, dissolved ferrous iron in the oceans would have oxidised to ferric oxide, precipitated out and sunk to the bottom. The iron-rich bands may represent cycles of photosynthetic activity (Klein, 2005). The earliest date for oxygenic photosynthesis is not known but significant changes in earth geochemistry that can be attributed to the oxygenation of the seas and at a later date to the atmosphere are not clearly evident until about 2.4–2.2 billion years ago (Schopf, 2011). Most banded iron formations, however, date to a much earlier time in earth history, from between about 3.5–2 billion years ago.

Another theory is that photosynthetic iron-oxidising bacteria might have been responsible for creating the ancient banded iron formations. Unlike cyanobacteria, iron-oxidising bacteria do not produce oxygen as a waste product. Microfossils seen in banded iron chert deposits at Cape Vani on the Greek island of Milos resemble a species of living photosynthetic iron bacterium, *Rhodomicrobium vannielii* (van Niel's Iron-bug). In the presence of light, van Niel's Iron-bug oxidises ferrous iron to ferric iron (Fru *et al.*, 2013). Cultures of *Rhodomicrobium vannielii* are pink in colour and require light, carbon dioxide and anoxic conditions for growth. Some cells become entombed within encrustations of precipitated iron oxide. The Cape Vani chert was formed relatively recently, about 2 million years ago in shallow water in association with a hydrothermal vent. Precambrian Banded iron formations differ from the Cape Vani chert in that they contain very little organic matter and appear to lack microfossils. This suggests that their iron-oxide content was precipitated remotely as a result of the activities of microorganisms living much higher up in the water column. Such an organism is *Chlorobium phaeoferrooxidans* (the Grey Iron-bug), a photosynthetic non-oxygenic iron-oxidising bacterium which inhabits the waters of an iron-rich Rift Valley lake in Africa. The Grey Iron-bug floats in the water column. It does not become encrusted with precipitated minerals. The Grey Iron-bug provides a plausible model for microbially mediated banded iron deposition by pelagic bacteria in the Precambrian oceans (Thompson *et al.*, 2019).

The largest banded iron deposits are associated with areas of ancient unaltered Precambrian crust, such as occur in Western Australia, South Africa and North America. Britain's oldest rocks, the Lewisian gneiss of north Scotland are about 3 billion years old but these rocks are highly distorted metamorphosed versions of their former selves. However, in Wester Ross the 2 billion year old Loch Maree group of meta-sedimentary rocks includes some modified but recognisable banded iron formations (Park, 2002). These rocks are only partially metamorphosed. They have been cooked but are not too badly burnt. Chert has been converted to quartzite and iron oxide to magnetite. An accessible banded iron outcrop exists at Flowerdale on the Gairloch Estate. Seen from a distance the outcrop does not look out of the ordinary (Fig. 2.77). The banded iron deposits form a low grassy ridge marking a field boundary beside the Cherry Hill Walk, a way-marked path. On close inspection the rocks can be seen to show alternating layers of white quartzite and dark brown magnetite (Fig. 2.78). Loose fragments have a heavy metallic feel about them. It is a remarkable experience to hold one of those fragments and speculate that it might have been a bacterium, an ancient antecedent of the Grey Iron-bug perhaps, that caused that rock to be the way it is today.

Fig. 2.77. Site of banded iron outcrop at Flowerdale, Gairloch, UK.

Fig. 2.78. Banded iron. Flowerdale, Gairloch, UK.

Desert varnish

Desert varnish is a thin blackish patina of manganese and iron oxides and clay particles that builds up on rock surfaces in arid climates. Desert varnish is microbiologically interesting because it constitutes a special kind of habitat that is associated with a distinctive desert varnish microbial flora consisting of communities of bacteria and fungi. Although some of the microbes associated with desert varnish are known to be able to oxidise manganese and iron, the precise roles that bacteria may play in the creation of desert varnish are uncertain (Dorn and Oberlander, 1981; Lang-Yona *et al.*, 2018). Desert varnish develops slowly, not only on rocks but also on other objects such as abandoned World War II fuel cans in the Sahara Desert (Eastes *et al.*, 1988). This indicates that airborne dust deposits probably provide the source materials for desert varnish rather than the rocks themselves. Desert varnish forms underneath stones as well as at more exposed sites. Desert dwelling peoples have used desert varnish as a medium in which to carve petroglyphs by chipping at the black layer to reveal the underlying rock. Images incised in desert varnish may still look fresh after hundreds of years of exposure to the elements, suggesting that layers of desert varnish may take comparably long periods of time to develop. The petroglyphs depicted in Figs. 2.79–2.80 were seen during the course of a visit to Cub Creek, Dinosaur National Monument in Utah. They have been attributed to people belonging to the Fremont culture who lived at Cub Creek more than 1000 years ago (Cole, 1995).

The British geological record includes some exposures of rocks that are coated with ancient desert varnish. Desert varnish-coated rocks present among

Fig. 2.79. Petroglyphs in desert varnish at Cub Creek, Dinosaur National Monument, Utah, USA. Fremont culture, c. 650 AD or older.

Fig. 2.80. Petroglyph in desert varnish at Cub Creek, Dinosaur National Monument, Utah, USA. Fremont culture, c. 650 AD or older.

the Brockram deposits in Cumbria record the existence of a stony desert that existed during the Permian Period. A line of desert varnish that runs along the cliff face at Budleigh Salterton in Devon marks the surface of a stony desert from the Triassic Period. The layer of desert varnish can easily be seen running between the lumpy Budleigh Salterton pebble bed and the overlying Otter sandstone (Figs. 2.81–2.83).

Fig. 2.81. Thin layer of pebbles with dark coating of desert varnish dating from the Triassic Period. Budleigh Salterton, Devon, UK.

Fig. 2.82. The desert varnish layer is situated below a band of yellow sandy deposits. Budleigh Salterton, Devon, UK.

Fig. 2.83. Triassic cliffs at Budleigh Salterton, Devon, UK.

Uranium nodules

Situated at the foot of slumping cliffs, Littleham Cove can be reached by taking a long stroll along the beach to the west of Budleigh Salterton. The geology at Littleham Cove offers a glimpse into the deep biosphere. The Littleham mudstone has been dated to the late Permian Period, close to the Permian–Triassic boundary about 250 million years ago. In common with many other Permian deposits from East Devon, much of the Littleham mudstone is coloured red by oxidised iron. The mudstone sediments are thought to have been deposited in ephemeral shallow lakes in an arid environment that was subjected to dust storms and flash floods (Gallois, 2019). Green patches show where ferric iron has been reduced to ferrous iron. Here and there, one encounters curious spheroids of grey matter, encircled by reduction zones of green mudstone (Fig. 2.84). These grey nodules are composed of uranium, vanadium and organic matter. Known as reduction spheroids they came into being more than a kilometre below the earth's surface and more than 10 million years later than the time of deposition of the mudstone sediments. It is likely that colonies of sulphate-reducing or iron-reducing bacteria mediated the creation of the nodules deep in the earth. Iron-reducing bacteria are also capable of reducing soluble uranium (VI) to insoluble non-crystalline uranium (IV). Concentrations of uranium in the reduction spheroids are much higher than in the surrounding sediments. A similar process can be observed today in a subterranean sandstone deposit in Wyoming where microbially mediated non-crystalline uranium (IV) is being precipitated from soluble uranium (VI) supplied from groundwater (Bhattacharyya *et al.*, 2017; McMahon *et al.*, 2018). In addition to Littleham

Fig. 2.84. Uranium nodules, Littleham Cove, Devon, UK.

Cove, uranium-rich reduction spheroids can also be seen in Devonian mudstones at Dingwall in Scotland and in other red beds in the British Isles.

Weathering

Weathering is a microbial phenomenon. Exposed rocks become colonised by bacteria, algae, fungi and lichens, which degrade rock surfaces to obtain nutrients. Bacteria colonise rock surfaces as biofilms and clothe themselves with extracellular polymeric substances that help to retain moisture and provide suitable habitat for additional species of bacteria. Bacteria burrow into and dissolve rocks. Porous and cracked rocks provide homes for bacteria. Bacteria synthesise organic acids as metabolic byproducts that have the potential to dissolve carbonates, potassium and silica and cause stones to disintegrate. Liberated mineral grains and fragments of organic matter become soil. Even hard-wearing igneous and metamorphic rocks such as granite and gneiss are susceptible to leaching of minerals and gradual degradation by bacteria (Styriaková *et al.*, 2012). In nutrient-poor soils, rocks provide a source of essential elements. For example, on the moors of southwestern England bacteria that live in the vicinity of rock outcrops can thrive on a simple diet of glucose, ammonium chloride and granite (Fig. 2.85). Oxalic acid produced by the bacteria mediates the release of essential minerals and causes weathering (Olsson-Francis *et al.*, 2016).

Being composed largely of acid-soluble calcium carbonate, limestones are especially susceptible to erosion by attack from carbonic acid generated from carbon dioxide dissolved in rain water and also by organic acids released as metabolic byproducts by colonising bacteria. Whilst some bacteria mediate the dissolution of carbonate rocks other bacteria induce the precipitation of carbonates. The continuous erosion of limestone and precipitation of carbonates account for the dynamism of limestone landscapes, much of which can be traced to processes that involve bacteria. In the Craven District of Yorkshire examples of weathered limestone, limestone pavement with clints and grykes, ink streaks on cliff faces, microbially mediated tufa formation and calcified microstromatolites can be seen during the course of a single walk (Fig. 2.86).

Fig. 2.85. Bacteria contribute to the weathering of granite. Eroded granite outcrop, Cornwall, UK.

Fig. 2.86. Bacteria contribute to the weathering of limestone. Limestone pavement, Malham, Yorkshire, UK.

Cyanobacteria are pioneering organisms. They colonise newly exposed limestone surfaces both above and below the water surface. Some bacteria colonise the minute spaces within the rock while others actively bore into the surface to create living space. Filamentous cyanobacteria of the genus *Schizothrix* colonise submerged limestone surfaces and progressively penetrate the rock by exploiting weaknesses between calcite crystals, leading eventually to fragmentation (Pentecost and Whitton, 2000). The spectacular bacterial ink streaks which decorate Malham Cove in Yorkshire contribute to its gradual erosion (Fig. 2.87). Cyanobacteria have been calculated to erode the cliff face at Malham at a rate of 3 mm per 100 years (Pentecost and Whitton, 2000). Surface-dwelling microbes form biofilms, secrete extracellular polymeric substances and adopt a sedentary existence. Motility is a more valuable trait for microbes that colonise the nutrient-poor lacunae that exist deeper inside rocks. *Variovorax paradoxus* (the Awkward Omnivore) can utilise a wide range of nutrients and occurs in a variety of habitats including the spaces inside rocks. The Awkward Omnivore is a motile bacterium that becomes less motile when supplied with a source of nutrients (Samuels *et al.*, 2020).

Shales are flaky, fine-grained sedimentary rocks that fracture along bedding planes. Black shales are blackened by variable amounts of carbon and by iron pyrite derived from the decomposition of organic matter. Iron-oxidising

Fig. 2.87. Cyanobacterial ink streaks at Malham Cove, Yorkshire, UK.

bacteria influence the weathering of pyrite-bearing rocks. Pyrite oxidation generates oxyhydride deposits that track along the seams between rock planes in cliff faces. Fallen rock fragments below receding shale cliffs on the North Yorkshire coast near Staithes are coated with rust-coloured oxides (Cockell *et al.*, 2011). The shale cliffs near Staithes are populated by the iron-oxidising bacterium *Acidiferrobacter thiooxydans* (the Kentucky Acid-bug), an acidophile that was originally described from coal spoil in Kentucky. *Acidiferrobacter thiooxydans* oxidises sulphur as well as ferrous iron. Weathered pyrite-bearing black shales are acidic environments. Other shale-dwelling microbes on the Yorkshire coast that share the acidic habitat of *Acidiferrobacter thiooxydans* include acid-tolerant *Methylocella* species. *Methylocella* species can subsist on a variety of simple carbon compounds, including methane. Shales with low pyrite content are associated with neutral rather than acid environments. The low iron content is reflected by the microbial flora. Iron-poor environments are populated by bacteria such as members of the genus *Pseudomonas* that secrete iron-chelating molecules called siderophores. Siderophore-secreting bacteria also contribute to the weathering of shales and of other rocks.

Bacteria cause weathering of statues and manmade structures. Stone walls become colonised by cyanobacteria and green algae (Fig. 2.88). Photosynthetic microbes fix carbon dioxide, which leads to the accumulation of organic matter, increased water retention and the creation of opportunities for heterotrophic bacteria and plants. These processes lead to deterioration of stonework. Nitrifying bacteria also become established in building stones. Nitrifying bacteria

Fig. 2.88. Lund Cathedral, Sweden, whose walls are colonised by the cyanobacterium *Microcoleus vaginatus*.

oxidise atmospheric ammonia and produce nitrous or nitric acid, which contributes to the weathering of limestone. In built up areas, industrial pollution exposes stonework to sulphur dioxide and nitrogen dioxide. Colonising bacteria can oxidise these compounds to produce sulphuric and nitric acid as waste products which exacerbate the erosion of buildings and monuments (Graue *et al.*, 2013).

Diverse taxonomic groups of bacteria are associated with the decay of stone buildings (Dakal and Cameotra, 2012). Samples of eroded sandstone taken from Portchester Castle in Hampshire were shown to be colonised by at least seven different bacterial phyla (Fig. 2.89). Archaea belonging to the genus *Halococcus* were also present, showing that conditions of high salinity exist within the stonework that provide suitable habitat for halophilic archaea (Zanardini *et al.*, 2016). Portchester Castle faces the sea and is exposed to the prevailing wind. Archaea have also been implicated in the deterioration of ancient wall paintings. Microbially degraded fourteenth century frescoes in the Catherine Chapel at Castle Herberstein in Austria and the painted walls of a Roman tomb at Carmona in Spain were found to be colonised with extremophile archaea that could take advantage of the conditions of high salinity and alkalinity (Piñar *et al.*, 2001). The weathered walls of Lincoln Cathedral are colonised by carbonate-dissolving bacteria, including *Bacillus sporothermodurans* (the Heat Resistor) and *Arthrobacter agilis* (the Agile Arthropod). Lincoln Cathedral was built using locally sourced oolitic limestone (Skipper *et al.*, 2016). Microbial action can lead to a build-up of patina and change in colour of the rock surface as well as to overall deterioration (Fig. 2.90). In Scotland, pigmented biofilm samples taken from the sandstone walls of historic buildings were shown to be harbouring pigment-producing bacteria, including *Streptomyces microflavus* (the Yellow Chain-bug), *Streptomyces candidus* (the White Chain-bug) and *Arthrobacter roseus* (the Pink Arthropod) (Suihko

Fig. 2.89. Exfoliated stonework, eroded by bacteria at Richard II's Palace, Portchester Castle, Hampshire, UK.

Fig. 2.90. Microbially pigmented, encrusted sandstone at Richard II's Palace, Portchester Castle, Hampshire, UK.

et al., 2007). In Austria and Germany reddish biofilms on historic buildings were found to be associated with bacteria belonging to the genus *Rubrobacter*. *Rubrobacter bracarensis* (the Braga Rubble-bug) was described as a new species that was isolated from a pigmented biofilm in a Portuguese church.

In *Pleasure of Ruins*, the writer Rose Macaulay (1881–1958) comprehensively surveyed mankind's enduring love of decayed architecture (Macaulay, 1953). Humans seem to have taken a lugubrious delight in buildings that are damaged beyond repair for as long as anyone can remember. The imperceptibly slow rate of deterioration enhances our reverence for an imperfectly known past. But the bacteriocentric observer is more fortunate than the romantic poets and artists. In the countryside and in the city bacteria tourists can savour an even more sublime vision. We can enjoy weathered rocks and the damaged stonework of old buildings in the knowledge that what we are witnessing is a work in progress of transformation performed by bacteria and archaea. It is happening so slowly that we cannot see it. Nevertheless, we can still take our time to appreciate the scene and to show respect for the flourishing communities of specialised microbes that have made a home for themselves within the rocks.

References

Alleon, J., Bernard, S., Le Guillou, C., Marin-Carbonne, J., Pont, S. *et al.* (2016) Molecular preservation of 1.88 Ga Gunflint organic microfossils as a function of temperature and mineralogy. *Nature Communications*, 7, 11977.

Allison, P., Hesselbo, S. and Carlton, B. (2008) Methane seeps on an Early Jurassic dysoxic seafloor. *Palaeogeography, Palaeoclimatology, Palaeoecology*, 270, 230–238.

Allwood, A.C., Walter, M.R., Kamber, B.S. and Burch, I.W. (2006) Stromatolite reef from the Early Archaean era of Australia. *Nature*, 414, 714–718.

Allwood, A.C., Rosing, M.T., Flannery, D.T., Hurowitz, J.A. and Heirwegh, C.M. (2018) Reassessing evidence of life in 3,700-million-year-old rocks of Greenland. *Nature*, 563, 241–244.

Andreetto, F., Dela Pierre, F., Gibert, L.,Natalicchio, M. and Ferrando, S. (2019) Potential fossilized sulfide-oxidizing bacteria in the upper miocene sulfur-bearing limestones from the Lorca Basin (SE Spain): paleoenvironmental implications. *Frontiers in Microbiology*, 10, Article number 1031, 1–15.

Andrews, J., Kendall, A. and Hall, A. (2015) Microbial crust with Frutexites(?) and iron staining in chalks: Albian–Cenomanian boundary, Hunstanton, UK. *Geological Magazine*, 152, 1–11.

Bailey, J.V., Joye, S.B., Kalanetra, K.M., Flood, B.E. and Corsetti, F.A. (2007) Evidence of giant sulphur bacteria in neoproterozoic phosphorites. *Nature*, 445, 198–200.

Balson, P.S. (1990) Episodes of phosphogenesis and phosphorite concretion formation in the North SeaTertiary. In Notholt, J.G. and Jarvis, I. (eds.), Phosphorite Research and Development. *Geological Society Special Publication*, 52, 125–137.

Banks, D.A. (1985) A fossil hydrothermal worm assemblage from the Tynagh lead-zinc deposit in Ireland. *Nature*, 313, 128–131.

Barghoorn, E.S. and Schopf, J.W. (1965) Microorganisms from the Late Precambrian of Central Australia. *Science*, 150, 337–339.

Barghoorn, E.S. and Tyler, S.A. (1965) Microorganisms from the Gunflint Chert. *Science*, 147, 563–577.

Batchelor, M.T., Burne, R.V., Henry, B.I., Li, F. and Paul J. (2018) A biofilm and organomineralisation model for the growth and limiting size of ooids. *Scientific Reports*, 8 (1), 559.

Bernhard, J.M., Edgcomb, V.P., Visscher, P.T., McIntyre-Wressnig, A., Summons, R. E. *et al.* (2013) Insights into foraminiferal influences on microfabrics of microbialites at Highborne Cay, Bahamas. *Proceedings of the National Academy of Sciences of the United States of America*, 110, 9830–9834.

Bhattacharyya, A., Campbell, K. M., Kelly, S. D., Roebbert, Y., Weyer, S. *et al.* (2017) Biogenic non-crystalline U (IV) revealed as major component in uranium ore deposits. *Nature Communications*, 8, 15538.

Boschker, H.T.S. and Middelburg J.J. (2002) Stable isotopes and biomarkers in microbial ecology. *FEMS Microbiology Ecology*, 40, 85–95.

Boyce, A.J., Little, C.T.S. and Russell, M.J. (2003) A new fossil vent biota in the Ballynoe barite deposit, Silvermines, Ireland: Evidence for intracratonic sea-floor hydrothermal activity about 352 Ma. *Economic Geology*, 98, 649–656.

Brasier, M., Antcliffe, J., Saunders, M. and Wacey, D. (2015) Changing the picture of Earth's earliest fossils (3.5–1.9 Ga) with new approaches and new discoveries. *Proceedings of the National Academy of Sciences of the United States of America*, 112, 4859–4864.

Brazelton, W.J., Nelson, B. and Schrenk, M.O. (2012) Metagenomic evidence for H2 oxidation and H2 production by serpentinite-hosted subsurface microbial communities. *Frontiers in Microbiology*, 2, 268.

Briggs, D.E.G., Moore, R.A., Shultz, J.W. and Schweigert, G. (2005) Mineralization of soft-part anatomy and invading microbes in the horseshoe crab *Mesolimulus* from the Upper Jurassic Lagerstätte of Nusplingen, Germany. *Proceedings of the Royal Society B*, 272, 627–632.

Brocks, J. J. and Schaeffer, P. (2008) Okenane, a biomarker for purple sulfur bacteria (Chromatiaceae), and other new carotenoid derivatives from the 1640 Ma Barney Creek Formation. *Geochimica et Cosmochimica Acta*, 72, 1396–1414.

Callow, R.H.T. and Brasier, M.D. (2009) A solution to Darwin's dilemma of 1859: exceptional preservation in Salter's material from the late Ediacaran Longmyndian Supergroup, England. *Journal of the Geological Society*, 166, 1–4.

Carpenter, J. (2001) How to Make a Fossil: Part 1 – Fossilizing Bone. *Journal of Paleontological Sciences*, JPS.C.07.0001, 1-10.

Clayton, C.J. (1986) The chemical environment of flint formation in Upper Cretaceous Chalks. *In* Sieveking, G. de G. and Hart, M.B. (eds). The scientific study of flint and chert: Proceedings of the Fourth International Flint Symposium held at Brighton Polytechnic, 10–15 April 1983. Cambridge University Press, Cambridge, UK.

Clouter, F. (2000) London Clay Fossils from the Isle of Sheppey: A collector's guide to the fossil animals from the London Clay between Minster and Warden Point, Sheppey. Medway Lapidary and Mineral Society, UK.

Cockbain, A.E., Hocking, R. and Reid, P. (2017) Vale Phillip Elliot Playford: a life of geology, exploration and history, 27 December 1931 – 12 July 2017. *Journal of the Royal Society of Western Australia*, 100, 46–48.

Cockell, C.S., Pybus, D., Olsson-Francis, K., Kelly, L., Petley, D.N. *et al.* (2011) Molecular characterization and geological microenvironment of a microbial community inhabiting weathered receding shale cliffs. *Microbial Ecology*, 61, 166–181.

Cocks, L.R.M. and Torsvik, T.H. (2006) European geography in a global context from the Vendian to the end of the Palaeozoic. *Geological Society, London, Memoirs*, 32, 83–95.

Cole, S.J. (1995) Legacy on stone: rock art of the Colorado Plateau. Four Corners Books, Boulder, Colorado.

Collinson, M.E. (1984) Fossil plants of the London Clay. Field Guides to Fossils Series No. 1. The Palaeontological Association, London, UK.

Cooper, J.A.G., Smith, A.M. and Arnscheidt, J. (2013) Contemporary stromatolite formation in high intertidal rock pools, Giant's Causeway, Northern Ireland: preliminary observations. *Journal of Coastal Research,* Special Issue No. 65, 1675–1680.

Corstjens, P.L., de Vrind, J.P., Westbroek, P. and de Vrind-de Jong, E.W. (1992) Enzymatic iron oxidation by *Leptothrix discophora*: identification of an iron-oxidizing protein. *Applied and Environmental Microbiology*, 58, 450–454.

Croft, W.N. and George, E.A. (1959). Blue-green algae from the Middle Devonian of Rhynie, Aberdeenshire. *Bulletin of the British Museum (Natural History) Geology*, 3, 341–353.

Cronin, A.J. (1937) The Citadel. Gollancz, London, UK.

Crosby, C.H. and Bailey, J.V. (2012) The role of microbes in the formation of modern and ancient phosphatic mineral deposits. *Frontiers in Microbiology*, 3, 241.

Dakal, T.C. and Cameotra, S.S. (2012) Microbially induced deterioration of architectural heritages: routes and mechanisms involved. *Environmental Sciences Europe*, 24, (36), 1–13.

Darwin, C.R. (1859) On the origin of species by means of natural selection, or the preservation of favoured races in the struggle for life. John Murray, London, UK.

Davies, N.S., Liu, A.G., Gibling, M.R. and Miller, R.F. (2016) Resolving MISS conceptions and misconceptions: A geological approach to sedimentary surface textures generated by microbial and abiotic processes. *Earth-Science Reviews*, 154, 210–246.

Diaz, M., Eberli, G. P., Blackwelder, P., Phillips, B. and Swart, P. K. (2017) Microbially mediated organomineralization in the formation of ooids. *Geology*, 45 (9), 771–774.

Djokic, T., Van Kranendonk, M., Campbell, K., Campbell, K.A.,Walter, R.R. *et al.* (2017) Earliest signs of life on land preserved in ca. 3.5 Ga hot spring deposits. *Nature Communications*, 8, 15263.

Dodd, MS, Papineau, D, Grenne, T., Slack, J.F., Rittner, M. *et al.* (2017) Evidence for early life in Earth's oldest hydrothermal vent precipitates. *Nature*, 543, 60–64.

Dorn, R.I. and Oberlander, T.M. (1981) Microbial origin of desert varnish. *Science*, 213, 1245–1247.

Douglas, S. and Beveridge, T.J. (2006) Mineral formation by bacteria in natural microbial communities. *FEMS Microbiology Ecology*, 26, 79–88.

Drake, N.A., Eckardt, F.D. and White, K.H. (2004) Sources of sulphur in gypsiferous sediments and crusts and pathways of gypsum redistribution in southern Tunisia. *Earth Surface Processes and Landforms*, 29, 1459–1471.

Dunn, K.A., McLean, R.J.C., Upchurch, G.R., Jr. and Folk, R.L. (1997) Enhancement of leaf fossilization potential by bacterial biofilms. *Geology*, 25, 1119–1122.

Eastes, J.W., Hearn, P.P., Jr., Breed, C.S. and McCauley, J.F. (1988) Weathering of a metal artifact in a Saharan environment: evidence for a novel form of desert varnish. *Applied Spectroscopy*, 42, 827–831.

Edwards, D., Kenrick, P. and Dolan, L. (2017) History and contemporary significance of the Rhynie cherts – our earliest preserved terrestrial ecosystem. *Philosophical Transactions of the Royal Society B*, 373, 20160489, 1–7.

Farías, M.E., Rascovan, N., Toneatti, D.M., Albarracín, V.H., Flores, M.R. *et al.* (2013) The discovery of stromatolites developing at 3570 m above sea level in a high-altitude volcanic Lake Socompa, Argentinean Andes. *PLoS ONE*, 8 (1), e53497.

Follows, B. and Tyson, R.V. (1998) Organic facies of the Asbian (early Carboniferous) Queensferry Beds, Lower Oil Shale Group, South Queensferry, Scotland, and a brief comparison with other Carboniferous North Atlantic oil shale deposits. *Organic Geochemistry*, 29, 821–844.

Fru, C.E., Ivarsson, M., Kilias, S.P., Bengtson, S., Belivanova, V. *et al.* (2013) Fossilized iron bacteria reveal a pathway to the biological origin of banded iron formation. *Nature Communications*, 4, 2050.

Gallois, A., Bosence, D. and Burgess, P.M. (2018) Brackish to hypersaline facies in lacustrine carbonates: Purbeck Limestone Group, Upper Jurassic–Lower Cretaceous, Wessex Basin, Dorset, UK. *Facies*, 64 (12), 1–39.

Gallois, R.W. (2019) The stratigraphy of the Permo-Triassic rocks of the Dorset and East Devon Coast World Heritage Site, U.K. *Proceedings of the Geologists' Association*, 130, 274–293.

Gerdes, G. and Klenke, T. (2007) States of biogenic bedding as records of the interplay of ecologic time and environment (a case study of modern siliciclastic sediments, Mellum Island, southern North Sea). Senckenbergiana Maritima, 37, 129–144.

Glikson, M., Duck, L.J., Golding, S.D., Hofmann, A., Bolhar, R. *et al.* (2008) Microbial remains in some earliest Earth rocks: Comparison with a potential modern analogue, *Precambrian Research*, 164, 187–200.

Graue, B., Siegesmund, S., Oyhantcabal, P., Naumann, R., Licha, T. *et al.* (2013) The effect of air pollution on stone decay: the decay of the Drachenfels trachyte in industrial, urban, and rural environments – a case study of the Cologne, Altenberg and Xanten cathedrals. *Environmental Earth Sciences*, 69, 1095–1124.

Harland, W.B. (2007) Origins and assessment of snowball Earth hypotheses. *Geological Magazine*, 144, 633–642.

Heim, C., Quéric, N.V., Ionescu, D., Schäfer, N. and Reitner, J. (2017) *Frutexites*-like structures formed by iron oxidizing biofilms in the continental subsurface (Äspö Hard Rock Laboratory, Sweden). *PLoS ONE*, 12, e0177542.

Hoffman, P.F., Abbot, D.S., Ashkenazy, Y., Benn, D. I., Brocks, J. J. *et al.* (2017) Snowball Earth climate dynamics and Cryogenian geology-geobiology. *Science Advances*, 3, e1600983.

Hutton, J. (1788) Theory of the Earth; or an investigation of the laws observable in the composition, dissolution, and restoration of land upon the Globe. *Transactions of the Royal Society of Edinburgh*, 1, 209–304.

Ibarra, Y., Corsetti, F.A., Greene, S.E. and Bottjer, D.A. (2016) A microbial carbonate response in synchrony with the end-Triassic mass extinction across the SW UK. *Scientific Reports*, 6, 19808.

Jahrnert, R.J. and Collins, L.B. (2012) Characteristics, distribution and morphogenesis of subtidal microbial systems in Shark Bay, Australia. *Marine Geology*, 303–306, 115–136.
Jeans, C.V., Turchyn, A.V. and Hu, X-F. (2016) Sulfur isotope patterns of iron sulfide and barite nodules in the Upper Cretaceous Chalk of England and their regional significance in the origin of coloured chalks. *Acta Geologica Polonica*, 66, 227–256.
Judd, A., Davies, G., Wilson, J., Holmes, R., Baron, G. *et al.* (1997). Contributions to atmospheric methane by natural seepages on the UK continental shelf. *Marine Geology*, 137, 165–189.
Kalkowsky, E. (1908) Oolith und Stromatolith im norddeutschen Bundsandstein. *Zeitschrift der Deutschen Geologischen Gesellschaft*, 60, 68–125.
Kenward, P.A., Goldstein, R.H., González, L.A. and Roberts, J.A. (2009) Precipitation of low-temperature dolomite from an anaerobic microbial consortium: the role of methanogenic *Archaea*. *Geobiology*, 7, 556–565.
Klein, C. (2005) Some Precambrian banded iron formations (BIFs) from around the world: their age, geologic setting, mineralogy, metamorphism, geochemistry, and origin. *The American Mineralogist*, 90, 1473–1499.
Konhauser, K.O. (2007) Introduction to Geomicrobiology. Blackwell, Oxford, UK.
Krings, M., Kerp, H., Hass, H., Taylor, T.N. and Dotzler, N. (2007) A filamentous cyanobacterium showing structured colonial growth from the Early Devonian Rhynie chert. *Review of Palaeobotany and Palynology*, 146, 265–276.
Lalonde, S.V., Konhauser, K.O., Reysenbach, A.-L. and Ferris, F.G. (2005) Thermophilic silicification: the role of *Aquificales* in hot spring sinter formation. *Geomicrobiology*, 3, 41–52.
Lang-Yona, N., Maier, S., Macholdt, D.S., Müller-Germann, I., Yordanova, P. *et al.* (2018) Insights into microbial involvement in desert varnish formation retrieved from metagenomic analysis. *Environmental Microbiology Reports*, 10, 264–271.
Li, J., Zhou, H., Peng, X., Wu, Z., Chen, S. *et al.* (2012) Microbial diversity and biomineralization in low-temperature hydrothermal iron–silica-rich precipitates of the Lau Basin hydrothermal field. *FEMS Microbiology Ecology*, 81, 205–216.
Lichtman, I.D., Baas, J.H., Amoudry, L.O., Thorne, P.D., Malarkey, J. *et al.* (2018) Bedform migration in a mixed sand–mud intertidal environment and implications for bed material transport predictions. *Geomorphology*, 315, 17–32.
Little, C.T.S., Glynn, S.E.J. and Mills, R.A. (2004) Four-hundred-and-ninety-million-year record of bacteriogenic iron oxide precipitation at sea-floor hydrothermal vents, *Geomicrobiology Journal*, 21, 415–429.
Lowe, J. (1937) Rat leprosy. A critical review of the literature. *International Journal of Leprosy*, 5, 311–328; 463–481.
Lowenstam, H.A. (1962) Magnetite in denticle capping in recent chitons (Polyplacophera). *Geological Society of America Bulletin*, 73, 435.
Lowenstam, H.A. (1981) Minerals formed by organisms. *Science*, 211, 1126–1131.
Macaulay, R. (1953) Pleasure of Ruins. Weidenfeld and Nicolson, London, UK.
Malarkey, J., Baas, J.H., Hope, J.A., Aspden, R.J., Parsons, D.R. *et al.* (2015) The pervasive role of biological cohesion in bedform development. *Nature Communications*, 6, 6257.
Maltman, A., Elliot, R., Muir, R. and Fitches, W. (2000) (2nd ed.) Guide to the Geology of Islay. Institute of Geography and Earth Sciences, University of Wales, Aberystwyth, UK.
McCallum, M.F. and Guhathakurta, K. (1970) The precipitation of calcium carbonate from seawater by bacteria isolated from Bahama bank sediments. *Journal of Applied Bacteriology*, 33, 649–655.
McMahon, S., Hood, A.V.S., Parnell, J. and Bowden, S. (2018) Reduction spheroids preserve a uranium isotope record of the ancient deep continental biosphere. *Nature Communications*, 9, 4505.
McMahon, S. (2019) Earth's earliest and deepest purported fossils may be iron-mineralized chemical gardens. *Proceedings of the Royal Society B*, 286, 20192410.

McNamara, K. (1992) Stromatolites. Western Australian Museum, Perth, Australia.

Menon, L.R., McIlroy, D. and Brasier, M.D. (2017) 'Intrites' from the Ediacaran Longmyndian Supergroup, UK: A new form of microbially-induced sedimentary structure (MISS). *Geological Society Special Publication*, 448, 271–283.

Mlewski, E.C., Pisapia, C., Gomez, F., Lecourt, L., Soto Rueda, E. *et al.*(2018) Characterization of pustular mats and related rivularia-rich laminations in oncoids from the Laguna Negra lake (Argentina). *Frontiers in Microbiology*, 9, 996.

Moore, T.S., Murray, R.W., Kurtz, A.C. and Schrag, D.P. (2004) Anaerobic methane oxidation and the formation of dolomite. *Earth and Planetary Science Letters*, 229, 141–154.

Muir, M.D. (1978) Possible stromatolitic origin of the Dalradian Doogort Limestone, Achill Island, County Mayo. *Journal of Earth Sciences*, 1, 49–52.

Nickless, E.F.P., Booth, S.J. and Mosley, P.N. (1976) The celestite resources of the area north-east of Bristol. Mineral Assessment Report 25. Institute of Geological Sciences. HMSO, London, UK.

Noffke, N., Gerdes, G., Klenke, Th. and Krumbein, W.E. (2001) Microbially induced sedimentary structure: a new category within the classification of primary sedimentary structures. *Journal of Sedimentary Research*, 71, 649–656.

Noffke, N., Beukes, N., Hazen, R. and Swift, D. (2008) Exceptionally preserved microbial mats of Meso-Archean age: the Sinqueni Formation, Pongola Supergroup, South Africa. *Geobiology*, 6, 5–20.

Noffke, N. (2010) Geobiology: microbial mats in sandy deposits from the Archean Era to today. Springer Verlag, Heidelberg, Germany.

Noffke, N., Christian, D., Wacey, D. and Hazen, R.M. (2013) Microbially induced sedimentary structures recording an ancient ecosystem in the ca. 3.48 billion-year-old Dresser Formation, Pilbara, Western Australia. *Astrobiology*, 13, 1103–1124.

Nutman, A.P., Bennett, V.C., Friend, C.R.L., Van Kranendonk, M.J. and Chivas, A.R. (2016) Rapid emergence of life shown by discovery of 3,700-million-year-old microbial structures. *Nature*, 537, 535–538.

Olsson-Francis, K., Pearson, V., Schofield, P., Oliver, A. and Summers, S. (2016) A study of the microbial community at the interface between granite bedrock and soil using a culture-independent and culture-dependent approach. *Advances in Microbiology*, 6, 233–245.

Park, R.G. (2002) The Lewisian Geology of Gairloch, NW Scotland. Geology Society, London, *Memoirs*, 26.

Park, C.-H., Li, X.R., Zhao, Y., Jia, R.L. and Hur, J.-S. (2017) Rapid development of cyanobacterial crust in the field for combating desertification. *PLoS ONE*, 12 (6), e0179903.

Parnell, J. (1987) Mineralized Magadi-type cherts in the Devonian of Scotland; support for a model of sulfide deposition in low-temperature alkaline conditions. *Economic Geology*, 82, 1053–1056.

Parnell, J., Perez, M., Armstrong, J., Bullock, L., Feldmann, J. *et al.* (2017) A black shale protolith for gold-tellurium mineralisation in the Dalradian Supergroup (Neoproterozoic) of Britain and Ireland. *Applied Earth Science*, 126, 161–175.

Päßler, J.-F., Jarochowska, E., Bestmann, M. and Munnecke, A. (2018) Distinguishing biologically controlled calcareous biomineralization in fossil organisms using electron backscatter diffraction (EBSD). *Frontiers in Earth Science*, 6. Article number 16.

Peat, C. (1984) Comments on some of Britain's oldest microfossils. *Journal of Palaeontology*, 3, 65–71.

Peat, C. (1988) Precambrian microfossils of the British Isles. *Proceedings of the Shropshire Geological Society*, 7, 2–3.

Pentecost, A. and Franke, U. (2010) Photosynthesis and calcification of the stromatolitic freshwater cyanobacterium *Rivularia. European Journal of Phycology*, 45, 345–353,

Pentecost, A. and Whitton, B.A. (2000) Limestone. In: *The ecology of cyanobacteria*. (eds.) Whitton, B.A. and Potts, M. Kluwer, Dordrecht, Netherlands.

Perry, C.T. (1994) Fresh-water tufa stromatolites in the basal Purbeck Group (Upper Jurassic), Isle-of-Portland, Dorset. *Geological Journal*, 29, 119–135.
Peters, S.E., Husson, J.M. and Wilcots, J. (2017) The rise and fall of stromatolites in shallow marine environments. *Geology*, 45, 487–490.
Petrash, D.A., Robbins, L.J., Shapiro, R.S., Mojzsis, S.J. and Konhauser, K.O. (2016) Chemical and textural overprinting of ancient stromatolites: timing, processes, and implications for their use as paleoenvironmental proxies. *Precambrian Research*, 278, 145–160.
Philippe, M., Billon-Bruyat, J-P., Garcia-Ramos, J.C., Bocat, L., Gomez, B. *et al.* (2010) New occurrences of the wood *Protocupressinoxylon purbeckensis* Francis: implications for terrestrial biomes in southwestern Europe at the Jurassic/Cretaceous boundary. *Palaeontology*, 53, 201–214.
Piñar, G., Saiz-Jimenez, C., Schabereiter-Gurtner, C., Blanco-Valera, M.T., Lubitz, W. *et al.* (2001) Archaeal communities in two disparate deteriorated ancient wall paintings: detection, identification and temporal monitoring by denaturing gradient gel electrophoresis. *FEMS Microbiology Ecology*, 37, 45–54.
Plümper, O., King, H.E., Geisler, T., Liu, Y., Pabst, S. *et al.* (2017) Forearc serpentinites as incubators for life. *Proceedings of the National Academy of Sciences of the United States of America*, 114, 4324–4329.
Prave, A.R. (2002) Life on land in the Proterozoic: Evidence from the Torridonian rocks of northwest Scotland. *Geology*, 30, 811–814.
Proemse, B.C., Eberhard, R.S., Sharples, C., Bowman, J.P., Richards, K. *et al.* (2017) Stromatolites on the rise in peat-bound karstic wetlands. *Scientific Reports*, 7, 15384.
Riding, R. (2006) Microbial carbonate abundance compared with fluctuations in metazoan diversity over geological time. *Sedimentary Geology*, 185, 229–238.
Salter, J.W. (1856) On fossil remains in the Cambrian rocks of the Longmynd and North Wales. *Quarterly Journal of the Geological Society of London*, 12, 246–251.
Salter, J.W. (1857) On annelid burrows and surface markings from the Cambrian rocks of the Longmynd. *Quarterly Journal of the Geological Society of London*, 13, 199–207.
Samuels, T., Pybus, D., and Cockell, C. S. (2020) Casamino acids slow motility and stimulate surface growth in an extreme oligotroph. *Environmental Microbiology Reports*, 12, 63–69.
Sanz-Montero, M.E., Rodríguez-Aranda, J.P. and García del Cura, M.A. (2009) Bioinduced precipitation of barite and celestite in dolomite microbialites Examples from Miocene lacustrine sequences in the Madrid and Duero Basins, Spain. *Sedimentary Geology*, 222, 138–148.
Schopf, J.W. (2006) Fossil evidence of Archaean life. *Philosophical Transactions of the Royal Society B*, 361, 869–875.
Schopf, J.W. (2011) The paleobiological record of photosynthesis. *Photosynthesis Research*, 107, 87–101.
Schopf, J.W. and Kudryavtsev, A.B. (2012) Biogenicity of Earth's earliest fossils: a resolution of the controversy. *Gondwana Research*, 22, 761–771.
Schopf, J.W., Kitajima, K., Spicuzza, M.J., Kudryavtsev, A.B. and Valley, J.W. (2018) SIMS analyses of the oldest known assemblage of microfossils document their taxon-correlated carbon isotope compositions. *Proceedings of the National Academy of Sciences of the United States of America*, 115, 53–58.
Simms, M.J. (2007) Uniquely extensive soft-sediment deformation in the Rhaetian of the UK: Evidence for earthquake or impact? *Palaeogeography, Palaeoclimatology, Palaeoecology*, 244, 407–423.
Smith, A., Cooper, A., Misra, S., Guastella, L., Bharuth, V. *et al.* (2018) The extant shore platform stromatolite (SPS) facies association: A glimpse into the Archean? *Biogeosciences*, 15, 2189–2203.
Strother, P.K. and Wellman, C.H. (2016) Palaeoecology of a billion-year-old non-marine cyanobacterium from the Torridon Group and Nonesuch Formation. *Palaeontology*, 59 (1), 89–108.
Strullu-Derrien, C.L., Kenrick, P., Goral, T. and Knoll, A.H. (2019) Testate amoebae in the 407-million-year-old Rhynie chert. *Current Biology*, 29, 461–467.

Styriaková, I., Styriak, I. and Oberhänsli, H. (2012) Rock weathering by indigenous heterotrophic bacteria of *Bacillus* spp. at different temperature: A laboratory experiment. *Mineralogy and Petrology*, 105, 3–4.

Suihko, M-L., Alakomi, H-L., Gorbushina, A., Fortune, I., Marquardt, J. *et al.* (2007) Characterization of aerobic bacterial and fungal microbiota on surfaces of historic Scottish monuments. *Systematic and Applied Microbiology*, 30, 494–508.

Takai, K., Nakamura, K., Toki, T., Tsunogai, U., Miyazaki, M. *et al.* (2008) Cell proliferation at 122°C and isotopically heavy CH4 production by a hyperthermophilic methanogen under high-pressure cultivation. *Proceedings of the National Academy of Sciences of the United States of America*, 105, 10949–10954.

Thiel, J., Byrne, J. M., Kappler, A., Schink, B. and Pester, M. (2019) Pyrite formation from FeS and H2S is mediated through microbial redox activity. *Proceedings of the National Academy of Sciences of the United States of America*, 116, 6897–6902.

Thomson, J. (1871a) On the occurrence of pebbles and boulders of granite in schistose rocks of Islay, Scotland. Report of the British Association for the Advancement of Science, 40th Meeting, Liverpool, UK.

Thomson, J. (1871b) On the stratified rocks of Islay. Report of the 41st Meeting of the British Association for the Advancement of Science, Edinburgh. John Murray, London, UK.

Thompson, K.J., Kenward, P.A., Bauer, K.W., Warchola, T., Gauger, T. *et al.* (2019) Photoferrotrophy, deposition of banded iron formations, and methane production in Archean oceans. *Science Advances*, 5, eaav2869.

Tyler, S.A. and Barghoorn, E.S. (1954) Occurrence of structurally preserved plants in Precambrian rocks of the Canadian Shield. *Science*, 119, 606–608.

Wegener, G., Krukenberg, V., Riedel, D., Tegetmeyer, H.E. and Boetius, A. (2015) Intercellular wiring enables electron transfer between methanotrophic archaea and bacteria. *Nature*, 526, 587–590.

West, I.M., Brandon, A. and Smith, M. (1968) A tidal flat evaporitic facies in the Visean of Ireland. *Journal of Sedimentary Petrology*, 8 (4), 1079–1093.

West, I.M. (1975) Evaporites and associated sediments of the basal Purbeck Group (Upper Jurassic) of Dorset. *Proceedings of the Geologists' Association, London*, 86, 205–225.

Westall, F., de Ronde, C.E., Southam, G., Grassineau, N., Colas, M. *et al.* (2006) Implications of a 3.472–3.333 Gyr-old subaerial microbial mat from the Barberton greenstone belt, South Africa for the UV environmental conditions on the early Earth. *Philosophical Transactions of the Royal Society B: Biological Sciences*, 361, 857–1876.

Wright, D.T. (1997) An organogenic origin for widespread dolomite in the Cambrian Eilean Dubh Formation, north western Scotland. *Journal of Sedimentary Research*, 67, 54–64.

Wright, D.T. (1999) The role of sulphate-reducing bacteria and cyanobacteria in dolomite formation in distal ephemeral lakes of the Coorong region, South Australia. *Sedimentary Geology*, 126, 147–157.

Xu, F., You, X., Li, Q. and Liu, Y. (2019) Can primary ferroan dolomite and ankerite be precipitated? Its implications for formation of submarine methane-derived authigenic carbonate (MDAC) chimney. *Minerals*, 9, 413.

Yee, N., Phoenix, V.R., Konhauser, K.O., Benning, L.G. and Ferris, F.G. (2003) The effect of cyanobacteria on silica precipitation at neutral pH: Implications for bacterial silicification in geothermal hot springs. *Chemical Geology*, 199, 83–90.

Zanardini, E., May, E., Inkpen, R., Cappitelli, F., Murrell, J.C. *et al.* (2016) Diversity of archaeal and bacterial communities on exfoliated sandstone from Portchester Castle (UK). *International Biodeterioration and Biodegradation*, 109, 78–87.

Zanardini, E., May, E., Purdy, K.J. and Murrell, J.C. (2019) Nutrient cycling potential within microbial communities on culturally important stoneworks. *Environmental Microbiology Reports*, 11, 147–154.

3 Meet the bacteria

> *Comme les moindres corps sont pour ainsi dire des petits mondes pleins de creatures merveilleuses il ne faut pas s'imaginer, qu'il y ait des deserts absolument parlant, quoyqu'ils soyent deserts pour nous.*
> [Just as the smallest bodies are in fact little worlds full of wonderful creatures, we should not think of deserts in an absolute sense, however deserted they may seem to us.]
>
> Gottfried Wilhelm Leibniz (1698). Letter to André Morell.

Looking at bacteria is rather like looking at modern art: when somebody explains what you are looking at, bacteria begin to make a lot more sense. Nevertheless, it remains hard to get to grips with bacteria, partly because they are invisible to the naked eye and partly because the lives of bacteria are so different from the lives of humans. Although bacteria are very small, it is easy to become familiar with the physical appearance of bacterial cells simply by looking at images of them, while access to a suitable microscope, ideally one with an oil immersion lens that allows magnification of ×1000, makes it possible to visualise actual bacteria.

Among the features that distinguish bacteria from animals and plants is the fact that bacteria as a group employ multiple nutritional strategies. In contrast, all animals rely essentially on only one nutritional strategy. Animals use organic matter obtained from the tissues of other organisms as a source of food and animals extract energy from food by using oxygen from the atmosphere to oxidise it. In fact there are many kinds of bacteria that do the same. Unlike animals, plants use sunlight as an energy source coupled with carbon dioxide from the atmosphere as a source of carbon to manufacture organic carbon compounds. There are also photosynthetic bacteria that do what plants do. However, there are many other kinds of bacteria that utilise numerous

DOI: 10.1079/9781836993933.0003

alternative nutritional strategies that are shared neither with plants nor with animals. Instead of using oxygen as an oxidising agent to liberate energy from organic matter, some bacteria use other substances as oxidising agents, including sulphur, sulphate (SO_4^{2-}), arsenic, selenium, ferric iron (Fe III, or Fe^{3+}) and nitrate (NO_3^-). Whereas plants (and photosynthetic bacteria) rely on sunlight as a source of energy to fix carbon dioxide, some bacteria (such as the iron-oxidising bacteria) are able to harness the energy released from chemical reactions (e.g. the oxidation of iron) and use it to synthesise organic compounds. Overall, this means that in any given habitat a multitude of different kinds of bacteria will be present and all of them will be busy shuffling chemical compounds from one state to another. Compared with the food webs which can be mapped out for plants and animals, the food webs for communities of environmental bacteria are complex.

Each type of habitat is associated with its own characteristic communities of bacteria. The composition of the communities of bacteria that exist in any particular place is determined among other things by the supply of nutrients, the other organisms that are present, the physical conditions and the local chemistry. Places that provide access to calcium carbonate and sunlight, for example, such as cliff faces and water bodies in limestone country are habitats where one might expect to encounter an abundance of photosynthetic cyanobacteria. Coastal marshes rich in organic matter and sulphur compounds are characterised by communities of bacteria involved in the sulphur cycle. Stagnant bogs and swamps provide homes for the archaea and bacteria that are involved in the generation of methane and its degradation. Awareness of the bacteria associated with different kinds of habitat makes it possible to view these environments through new eyes. In order to understand what is going on in these environments at a microscopic level it helps if you try to focus on the things that matter to the bacteria. The American microbiologist Betsey Dexter Dyer uses the term bacteriocentric (Dyer, 2003). The world starts to look quite different when you imagine what it might be like to be a bacterium.

Iron-oxidising bacteria

A wetland where the runnels and seepages run red with iron oxide deposits is as good a place as any to experience bacteria in the wild. Masses of actual bacteria are often visible and it is not particularly difficult to develop an understanding of what they are doing. The elements iron and oxygen account for approximately 35% and 28%, respectively of the mass of the earth. Chemical reactions involving these two elements figure prominently in environmental bacteriology. Rust-coloured mats, slimes, filaments and streamers signify the presence of iron-oxidising bacteria. Minor examples of such signs are commonplace and occur widely in wetlands. Particularly florid examples of iron deposition may occur in bogs, on coastal landslips, in old quarries, around disused mine workings and in other localities where as a result of the underlying geology and local conditions large volumes of iron-rich waters flow to the surface. Iron bacteria lend themselves to microscopic examination. Deposits of iron oxide naturally impart iron bacteria with a delightful rust colour and there is no need to use special stains in order to see them.

In common with most metallic elements, iron rarely occurs in nature as a metal. In the presence of oxygen and water, metallic iron spontaneously turns to rust. The name rust can be used to refer to a variety of oxidised states of iron that incorporate water molecules. The technical term for rust is hydrated iron oxide ($Fe_2O_3{\cdot}H_2O$). In nature, iron commonly occurs in one of two oxidised states. In airless conditions such as in the depths beneath a subterranean spring, iron usually occurs in a water-soluble state known as ferrous iron (Fe II, or Fe2+). In the presence of oxygen, ferrous iron (Fe2+) will naturally react with oxygen to become insoluble ferric iron (Fe III, or Fe3+). Specialised iron-oxidising bacteria increase the rate of this oxidation reaction from ferrous to ferric iron and as a result, rust-coloured deposits composed of bacterial cells and iron oxide begin to accumulate.

At sites where groundwater rich in ferrous iron wells up to the surface, iron-oxidising bacteria use enzymes to catalyse the oxidation of ferrous iron and convert it to ferric iron. Iron-oxidising bacteria put the energy released from the oxidation reaction to their own use. This is one of many examples that illustrate how bacteria are able to extract an advantage from a situation of chemical disequilibrium. Several species of so-called iron bacteria are able to use the energy liberated by the oxidation of ferrous iron to fix carbon dioxide to build solid organic carbon compounds. In so doing they act as primary producers and at the same time they also enhance the process of iron deposition. This is an example of carbon dioxide fixation or primary production based on the energy of a chemical reaction. This may seem quite strange when we are so much more familiar with green plants that use sunlight as a source of energy for carbon fixation. The significance of bacteria in iron precipitation has been shown experimentally by comparing rates of deposition in unfiltered or filtered (bacteria-free) water or by comparing deposition rates in systems before and after the addition of antibiotics to remove the effects of bacteria.

In some habitats, iron-oxidising bacteria are so numerous that they form visible aggregations at sites of iron precipitation (Figs. 3.1–3.3). Iron oxide deposits and mats of iron-oxidising bacteria may clog water channels and modify the local hydrology. Over the course of time, the growth of microbial mats or slimes combined with iron precipitation and the collection of debris and sediment may lead to significant deposits of bog iron or iron pan (consisting of the minerals limonite or goethite) or iron-rich clay (ochre). By impeding drainage, accumulations of iron pan encourage the development of bogs and wetlands. The accumulation of iron pan is an ongoing process as iron leached from the nearby hills continues to reach the bog by percolating through subterranean channels. Bright ferruginous seepages, accumulations of rust-coloured slime and iridescent surface films render wild boggy landscapes even more beautiful. A typical example of such a place is the Teal Lough Special Area of Conservation (H7388), a large area of undisturbed blanket bog in Northern Ireland (Fig. 3.4). The boggy habitat supports specialised insects that typically occur in upland mires, such as the Large Heath butterfly, *Coenonympha tullia* and moorland beetles, such as the ground beetle *Carabus glabratus*. Microscopic

Fig. 3.1. Iron-oxidising bacteria. Coastal seepage at Clashnessie, Sutherland, UK.

Fig. 3.2. Iron-oxidising bacteria at Whale Chine, Isle of Wight, UK.

examination of a sample of orange sludge taken from the bog reveals the presence of iron-oxidising bacteria. Iron-oxidising bacteria actively promote iron oxidation. Bog mosses may also contribute to the formation of iron pan by acting as passive substrates for iron precipitation. At Dersingham Bog in Norfolk (Fig. 3.5), deposits of bog iron have been shown to occur in association with the bog moss, *Cladopodiella fluitans* (Stevenson and Masson, 2016).

Fig. 3.3. Iron-oxidising bacteria at Whale Chine, Isle of Wight, UK.

Deposits of bog iron have been exploited by humans for millennia as easily accessible sources of raw material for the smelting of iron and the construction of iron artefacts. Deposits of bog iron are generally covered by layers of peat and by bog vegetation and are not normally visible at the surface of a bog. However, at Lowland Point in Cornwall, where coastal erosion has cut a low sea cliff into a marshy coastal plain, the anatomy of a bog iron deposit can be scrutinised in detail and within the context of the local geology (Scott *et al.*, 2011). Rock-hard lumps of bog iron are clearly visible beneath layers of rock and soil and detached chunks of bog iron may be collected from the top of the beach (Fig. 3.6). The source of the iron is a massive outcrop of iron-rich gabbro which constitutes the bedrock of this particular corner of the Lizard Peninsula. A short distance away from Lowland Point, the coast path passes an old gabbro quarry at Dean Point. Ochreous seepages glissade down the quarry's wall to the quarry floor where they discharge into slimy stagnant buff-coloured pools, strewn with gigantic boulders of the local gabbro (Fig. 3.7). It is a most pleasing spectacle.

Bacteria associated with iron oxidation include *Leptothrix ochracea* (the Ochre Iron-bug) and *Gallionella ferruginea* (the Rusty Iron-bug). *Leptothrix* grows as a chain of cells surrounded by a sheath (Fig. 3.8). *Gallionella* grows

Fig. 3.4. Iron-oxidising bacteria. Seepage in peat bog. Mill Lough, Northern Ireland.

Fig. 3.5. Passive iron precipitation on bog moss. Dersingham Bog, Norfolk, UK.

Fig. 3.6. Bog iron deposits. Lowland Point, Cornwall, UK.

Fig. 3.7. Iron-rich seepages and swamp. Dean Point, Cornwall, UK.

Fig. 3.8. Iron-oxidising bacteria (*Leptothrix*) seen on microscopy. Clashnessie, Sutherland, UK.

as a twisted corkscrew structure (Fig. 3.9). Sheaths of *Leptothrix ochracea* can easily be observed on microscopy and this organism was described, as long ago as 1797, originally as *Conferva ochracea* (Roth, 1797) (Fig. 3.10). The Ochre Iron-bug uses iron oxidation as a means of obtaining energy and also for fixing carbon dioxide. In addition, the Ochre Iron-bug supplements its diet with organic carbon compounds discarded by other organisms (Fleming *et al.*, 2018). Whereas *Leptothrix ochracea* generates a straw-like sheath of iron oxide as it grows, *Gallionella ferruginea* produces a twisted stick-like structure which becomes encrusted with precipitates of iron oxide. The Rusty Iron-bug is widespread in freshwater and marine environments. Iron-rich geothermal springs are associated with specialised thermophilic iron-oxidising bacteria and archaea (Fig. 3.11). One of them is *Metallosphaera yellowstonensis* (the Yellowstone Iron-bug), a member of the archaea whose optimum temperature for growth is 65°C.

When investigating boggy terrain it is not unusual to encounter puddles and seepages that are coated with iridescent films that superficially resemble oil spills, except that whereas oil films are viscous and liquid, these films are frangible and when gently tapped with a stick the film will crack apart like a caramelised crème brûlée. *Leptothrix discophora* (the Rainbow Iron-bug) is a relative of the Rusty Iron-bug. The Rainbow Iron-bug attaches itself to the water surface with a holdfast mechanism. The Rainbow Iron-bug secretes a matrix of extracellular polysaccharides and proteins that coat the interface between water and air with a brittle light-refracting layer (Fig. 3.12). In turbulent well-oxygenated streams the Rainbow Iron-bug can also be found on the surface of pebbles, which become coated with dark encrustations of manganese oxide.

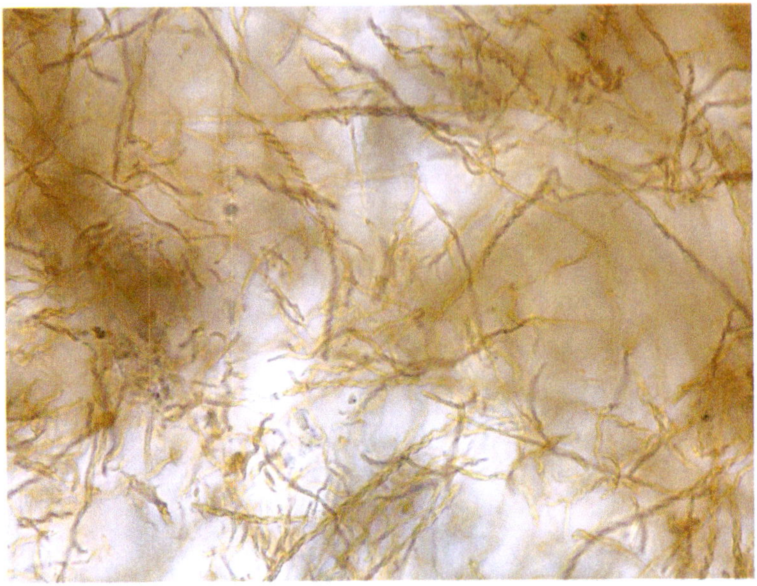

Fig. 3.9. Iron-oxidising bacteria (*Gallionella*) seen on microscopy. Whale Chine, Isle of Wight, UK.

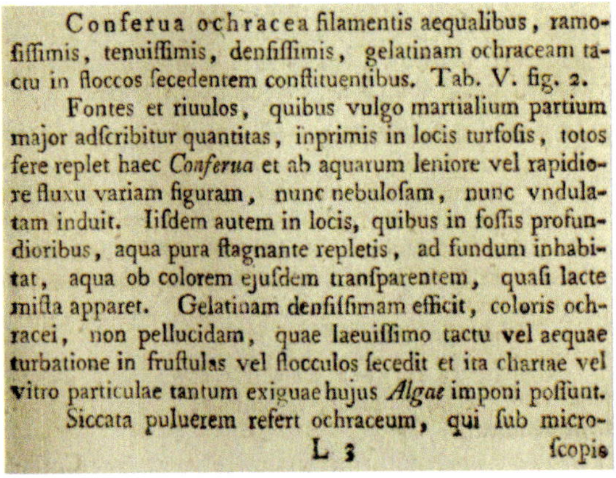

Conferua ochracea filamentis aequalibus, ramoſiſſimis, tenuiſſimis, denſiſſimis, gelatinam ochraceam tactu in floccos ſecedentem conſtituentibus. Tab. V. fig. 2.

Fontes et riuulos, quibus vulgo martialium partium major adſcribitur quantitas, inprimis in locis turfoſis, totos fere replet haec *Conferua* et ab aquarum leniore vel rapidiore fluxu variam figuram, nunc nebuloſam, nunc vndulatam induit. Iiſdem autem in locis, quibus in foſſis profundioribus, aqua pura ſtagnante repletis, ad fundum inhabitat, aqua ob colorem ejuſdem tranſparentem, quaſi lacte miſta apparet. Gelatinam denſiſſimam efficit, coloris ochracei, non pellucidam, quae laeuiſſimo tactu vel aequae turbatione in fruſtulas vel flocculos ſecedit et ita chartae vel vitro particulae tantum exiguae hujus *Algae* imponi poſſunt. Siccata puluerem refert ochraceum, qui ſub micro-

L 3 ſcopio

Fig. 3.10. Albrecht Roth's (1797) original description of the iron bacterium *Leptothrix ochracea* (as *Conferva ochracea*).

Leptothrix discophora is a manganese-oxidising bacterium as well as an iron oxidiser (Corstjens *et al.*, 1992).

Iron oxides may be deleterious to aquatic organisms. High iron concentrations have been shown to be toxic to aquatic invertebrates such as water fleas, *Daphnia magna*, and water lice, *Asellus aquaticus*. Ochre deposits are harmful to fish. Particles of ochre can clog the pores of fish eggs and accumulate in the gills of fish (Larsen and Olsen, 1950). The microbial mats generated by iron bacteria often smother stream beds and displace algae and vascular plants.

Fig. 3.11. Thermophilic iron bacteria and archaea growing at 60°C. Whirligig Geyser, Yellowstone National Park, USA.

Fig. 3.12. Refractive film made by the manganese-oxidising bacterium *Leptothrix discophora*. Cothill National Nature Reserve, UK.

Hence, compared with other aquatic habitats, iron springs often appear to have an impoverished flora and fauna. Therefore, iron springs are ideal places to see bacteria. They are places where colonies of bacteria are readily visible to the naked eye and where bacterial colonies are attractively highlighted by bright orange deposits of ferric iron. Furthermore, iron springs are places where bacteria are not severely obscured by the presence of dense populations of animals and plants. For an account of some of Britain's iron springs and their cultural associations, see Chapter 9, *Bacteria tourism*.

Magnetic bacteria

Once during the course of a visit to the Isle of Skye I encountered a solitary geologist who was hammering against an exposure of magnetite. Magnetite is a magnetic mineral composed of iron oxide. There is such an abundance of magnetic rocks in some parts of Skye as to render compass readings unreliable. The rocks that compose the Black Cuillin and other parts of the island are igneous in origin. As they cooled they became magnetised (Whetton and Myers, 1951). Until the 1960s it was assumed that all natural sources of magnetite in nature were derived from igneous or metamorphic rocks. Then in 1962 Heinz Lowenstam (1912–1993), a scientist based at the California Institute of Technology showed that the radula teeth of chitons contain magnetite of biological origin (Lowenstam, 1962). Chitons are limpet-like marine molluscs. The occurrence of biological magnetite was later found to be widespread in living organisms. Birds, bats and turtles use magnetite to orient themselves by sensing the earth's magnetic field. Although magnetite crystals occur in the human brain it is uncertain if humans can sense magnetic fields. There is, however, little doubt that some bacteria can sense and respond to magnetic fields. Bacteria that can orient themselves according to magnetic fields are called magnetotactic bacteria. Magnetotactic bacteria were reported in 1975 by Richard Blakemore, a scientist working at the Woods Hole Oceanographic Institution in Massachusetts, USA, although it subsequently emerged that as early as the 1960s, an Italian researcher, Salvatore Bellini had been distributing manuscripts to colleagues that documented his observations of what he described as magnetosensitive bacteria (Blakemore, 1975; Bellini, 2009; Frankel, 2009). It has since been shown that the cells of bacteria incorporate specialised organelles called magnetosomes that contain crystals of magnetite (Fe_3O_4) or greigite (Fe_3S_4). Bacteria actively control magnetite precipitation inside their cells. This is unusual in that most cases of bacterially mediated mineral deposition are passive processes that are influenced by the release of metabolic waste products or the innate properties of bacterial cells. Magnetosomes consist of chains of magnetic crystals. Bacteria use these miniature magnets to navigate against magnetic fields.

Magnetotactic bacteria are widely distributed and occur both in fresh water and salt water. They belong to several of the major phylogenetic groups, including the Phylum *Pseudomonadota*, the Phylum *Nitrospirota* and the Phylum *Planctomycetota*, as well as two recently proposed candidate phyla, "*Candidatus* Omnitrophota" and the "*Candidatus* Latescibacterota". The genes encoding the synthesis of magnetosomes that are found in all of these diverse magnetotactic bacteria exhibit common features and appear to have been derived from a common ancestor. Magnetotaxis appears to be an ancient trait that can be traced back to a time that pre-dates the diversification of some of the major lineages of bacteria (Lefèvre and Bazylinski, 2013; Lin *et al*., 2018).

Most magnetic bacteria are microaerophiles that thrive in low concentrations of oxygen in the transition zone between anoxic and oxygenated layers of sediment. Others are aerobes that reside in the higher oxygenated zone or anaerobes that live in deeper anoxic sediments. Magnetic bacteria are able to navigate along geomagnetic field lines to maintain a position in the water column or in sediments where the oxygen concentration is suitable for them (Hanzlik *et al.*, 1996). It is possible to recover magnetic bacteria from a sediment sample kept in a container simply by attaching the south pole of a magnet to the outside of the container. The container should be kept in the dark for at least several hours after which a sample may be withdrawn by pipette from the area adjacent to the magnet for study. Magnetotactic bacteria are abundant in some habitats. In lake sediments in Germany, a bacterium provisionally named "*Candidatus* Magnetobacterium bavaricum" (the Bavarian Magnet-bug) has been shown to be the dominant species of bacterium (Fig. 3.13). Magnetotactic bacteria that live in sulphide-rich anoxic sediments synthesise magnetosomes that are composed of greigite (iron sulphide) instead of magnetite (iron oxide). An example is the sulphate-reducing bacterium, *Solidesulfovibrio magneticus* (the Magnetic Sulphur).

Magnetosomes may become preserved as microfossils in sediments and sedimentary rocks. Conditions that favour the preservation of bacterial magnetite occur during periods of rapid sedimentation. During periods of slow sedimentation, magnetite is more likely to dissolve. Hence, the ancient sediments of deep water bodies preserve a record of varying climatic and sedimentary conditions. Climate researchers have used bacterial magnetite from the sediments of Lake Baikal in Siberia (Fig. 3.14) to investigate the lake's depositional conditions and the climatic history of the surrounding country. Lake Baikal is ancient and its sediments are several kilometres deep. It the deepest and most voluminous freshwater lake in the world (Peck and King, 1996; Demory *et al.*, 2005).

Fossilised bacterial magnetosomes impart sedimentary rocks, such as chalk and other varieties of limestone, with a palaeomagnetic record. Identifiable

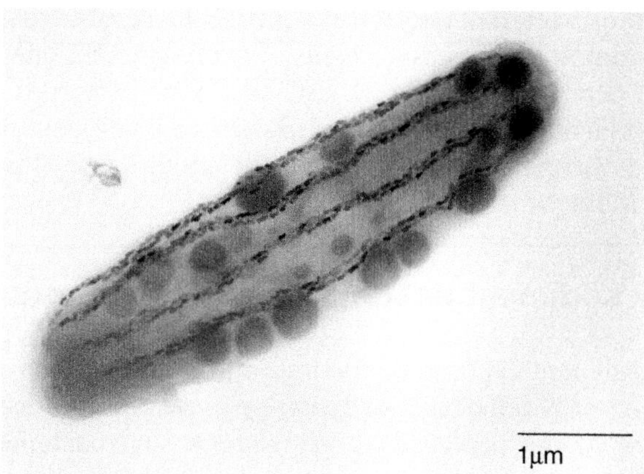

Fig. 3.13. Electron micrograph showing magnetosomes of "*Candidatus* Magnetobacterium bavaricum". (Image kindly provided by Marianne Hanzlik.)

Fig. 3.14. Lake Baikal, Siberia, where magnetic microfossils in deep sediments preserve a magnetostratigraphic record of varying climatic conditions.

magnetosomes can be seen on microscopic examination of prepared samples of chalk from the cliffs of southern England. At intervals that vary between tens of thousands to tens of millions of years the earth's magnetic field undergoes reversal and the magnetic poles exchange polarity. Magnetic minerals, such as bacterial magnetite, preserve a record of the direction of the earth's magnetic field through geological time. The measurement of magnetic signals preserved in the geological record forms the basis of a method known as magnetostratigraphy. The magnetostratigraphic record can be used to align rock sequences from different locations. Magnetostratigraphy complements biostratigraphy, which is based on the identification of assemblages of fossils that characterise different geological horizons. Geologists based at the University of Southampton used measurements of geomagnetic signals preserved in the sea cliffs at Scathell's Bay (Fig. 3.15) and Culver Cliff on the Isle of Wight and at Seaford Head (Fig. 3.16) in East Sussex to investigate the alignments of geological horizons in the chalk along the south coast of England (Montgomery, 1994; Montgomery *et al.* 1998).

Black sediment bacteria: the sulphate-reducing bacteria

Black sediment exposed by the imprint of your boot when you walk along the shoreline of a pond, a river bank, or a coastal mud flat, is generally a good indicator of the presence of sulphate-reducing bacteria. Depending on local conditions, the black sediment layer may be situated at some depth or may lie just beneath the surface (Figs. 3.17–3.18). The bacteria themselves are not actually black in colour but there is such a close association between black sediments and sulphate-reducing bacteria that I sometimes like to think of them

Fig. 3.15. Scatchell's Bay, Isle of Wight, UK, where magnetic microfossils in chalk preserve a record of the Earth's magnetic field.

Fig. 3.16. Seaford Head, Sussex, UK, where magnetic microfossils were used to investigate geological alignments.

as 'black sediment bacteria', or 'black sulphur bacteria'. Pond margins and salt marshes are productive habitats that contain copious amounts of organic matter. Stagnant sediments, however, are anoxic environments and in the absence of oxygen, bacteria must use other substances to oxidise organic compounds. Instead of using oxygen, sulphate-reducing bacteria use sulphate as an

Fig. 3.17. Black iron sulphide sediment in boot print. River Blyth, Suffolk, UK.

Fig. 3.18. Black iron sulphide sediment exposed by paw prints. East Head, Sussex, UK.

oxidising agent. Sulphate-reducing bacteria release hydrogen sulphide as a waste product which reacts with iron ions to form iron sulphide. Iron sulphide is black in colour. It is iron sulphide that accounts for the inky blackness of anoxic sediments, whilst any excess hydrogen sulphide that escapes into the atmosphere is responsible for the unpleasant sulphurous odour of stagnant wetlands.

Diverse sulphate-reducing bacteria

There are numerous different species of sulphate-reducing bacteria. The generic names of many of the sulphate reducers indicate that this is what they do. Examples include: *Desulfobacter*, *Desulfobacterium*, *Desulfobacula*, *Desulfobulbus*, *Desulfococcus*, *Desulfoluna*, *Desulfonema*, *Desulfosarcina*, *Desulfosporo sinus*, *Desulforhabdus*, *Desulfurosoma*, *Desulfurivibrio*, *Desulfovibrio* and *Pseudodesulfovibrio* (Fig. 3.19). The sulphate-reducing bacteria do not constitute a single phylogenetic group. Sulphate reduction is practiced by members of distant branches of the bacterial family tree. Many of the sulphate-reducers are Gram-negative bacteria that belong to the Phylum *Pseudomonadota*. Others are Gram-positive bacteria that belong to the Phylum *Bacillota* (previously *Firmicutes*). Members of the genus *Thermodesulfovibrio* belong to a third division of bacteria, the Phylum *Nitrospirota*. *Thermodesulfovibrio* species have been found in wastewater treatment plants and in thermal springs. As indicated by the generic name *Thermodesulfovibrio*, these sulphate-reducers are thermophiles. They grow optimally at temperatures of 55–60°C.

Widewater lagoon and its sulphate-reducing bacteria

Widewater Lagoon is a body of brackish water that stretches along the coast for about half a mile or so behind the seafront at Lancing in West Sussex (Fig. 3.20). On the landward side, the lagoon is flanked by a row of houses and the A259. On the seaward side, a shingle bank separates the lagoon

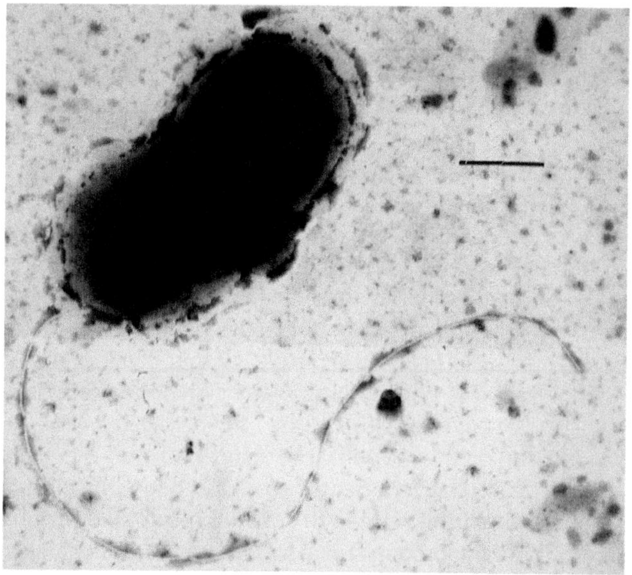

Fig. 3.19. Transmission electron micrograph of the bacterium *Desulfovibrio vulgaris* (bar equals 0.5 microns). (Graham Bradley). Public Domain.

Fig. 3.20. Saltmarsh habitat on mudflats at Widewater, West Sussex, UK

from the English Channel. Widewater provides a home for Lagoon Cockles, *Cerastoderma glaucum*, a rare hydrozoan, *Thieliana navis* and a rare sea anemone, *Edwardsia ivelli*. The surrounding mud flats and shingle are colonised by salt-tolerant plants, including Purple Glasswort, *Salicornia ramosissima* and Sea Heath, *Frankenia laevis*. The lagoon sediments and the adjacent mudflats provide suitable habitat for sulphate-reducing bacteria. It is possible to enjoy Widewater's microbes simply by standing on the shoreline and gazing at the sediments at the bottom of the lagoon. Not far below the water's mirror-like surface and beyond the reflected sky, a complex multicoloured microbial mat extends across the surface of the sediments like an enormous felt blanket. Different shades of green, grey, brown and red merge into one another. Dingy green signifies the presence of cyanobacteria and eukaryotic algae. Minute bubbles seen clinging to the surface of the mat contain oxygen released as a waste product by these photosynthetic organisms. Silvery grey tints hint at the presence of so-called white sulphur bacteria, whereas vaguely purple-coloured patches of sediment point to the existence of purple sulphur bacteria. The white sulphur bacteria and the purple sulphur bacteria are all feeding on hydrogen sulphide that is percolating up through the microbial mat from an anoxic zone of black sediment situated not far below the surface. The anoxic zone is the domain of the sulphate-reducing bacteria. They extract energy from organic matter in the sediments by oxidising it with sulphate, a chemical reaction that results in the release of hydrogen sulphide. Iron sulphide, formed by the interaction of hydrogen sulphide with iron, accumulates in the sediments and paints them black. One day I visited Widewater and encountered a mysterious black square, cut into the surface of the sediment (Fig. 3.21). The image reminds me of Black Square (1913), a painting by the artist Kazimir Malevich (1879–1935), which depicts a black square.

Fig. 3.21. Black square of sediment, rich in iron sulphide. Widewater, Sussex, UK.

Whilst some of the hydrogen sulphide that reaches the surface of the sediment is utilised by white sulphur bacteria and purple sulphur bacteria as a food source, the rest of it diffuses into the atmosphere. The smell of hydrogen sulphide is somewhat repulsive, although at a breezy location such as Widewater its odour is not usually noticeable. The depth and salinity of the water at Widewater vary in response to variations in rainfall and rates of evaporation. During the long hot summer of 1995 the lagoon shrank in size. In the settled anticyclonic weather, Widewater and its exposed mudflats became unusually sulphurous. In 2013 a pipe was installed which connected the lagoon to the open sea. Widewater's water level is now less erratic. Widewater is just one example of a place where a visiting naturalist can experience sulphur bacteria. Widewater exhibits features that are typical of many coastal sites around the British Isles. The thing that makes Widewater special is Ivell's sea anemone, *Edwardsia ivelli*, which is known only from Widewater.

Hydrogen sulphide production by sulphate-reducing bacteria

The human nose is very sensitive to hydrogen sulphide. Despite its unpleasant odour, naturalists who enjoy exploring stagnant coastal lagoons and salt marshes learn to react positively to the scent of hydrogen sulphide because it is a sign that one has arrived in a biologically interesting place. Salt marshes provide a combination of conditions that allow sulphate-reducing bacteria to proliferate. Mats of cyanobacteria and green algae as well as salt marsh plants are able to take advantage of unobstructed access to sunlight. As primary producers, photosynthetic organisms enrich the underlying sediments with organic carbon compounds. Furthermore, these locally grown provisions are supplemented by waves of organic matter which are regularly washed in by the tide. The sediments themselves are stagnant and waterlogged and relatively anoxic at even a short depth below the surface. In the airless shallow sediments, numerous fermentative bacteria feast on the fragmented remains of plants, algae and other organisms. The discarded metabolic byproducts of fermentative bacteria act as a source of low-molecular-weight hydrocarbons which can sustain an abundance of sulphate-reducing bacteria and also methanogenic archaea. Coastal marshes are often rich in sulphate, which the sulphate-reducers use as an oxidising agent to extract energy from the dribs and drabs of organic detritus that permeate the sediments.

Hydrogen sulphide, released as a waste product diffuses towards the surface, to serve as a stimulus to other communities of sulphur bacteria. Hydrogen sulphide is denser than air. The invisible haze of hydrogen sulphide that lingers over the surface of a salt marsh in high summer is an integral part of the coastal aesthetic and a major contributor to the singular appeal of coastal environments. Nevertheless, you can have too much of a bad thing. Hydrogen sulphide is toxic. Exposure to relatively small concentrations of hydrogen sulphide can be lethal. Although hydrogen sulphide seems unlikely to pose much of a hazard to naturalists who visit the more blustery coasts of the British Isles, dangerous concentrations of the gas have been recorded at certain spots on the French coast. In 2009 a horse was reported to have died after being poisoned by exposure to hydrogen sulphide on a beach in Brittany (Le Nen, 2009). Brittany's hydrogen sulphide trouble spots are places that are subject to unusually exuberant growths of sea lettuce, *Ulva lactuca*. These extravagant blooms of seaweed have been linked to excessive volumes of polluting agricultural runoff. During the summer months enormous heaps of rotting seaweed accumulate on the shoreline. Summer is the season of maximum productivity for sulphate-reducing bacteria in coastal environments and large volumes of hydrogen sulphide are released into the summer air, leading to warning signs and beach closures.

A sedimentary journey

The Severn Estuary is a murky and turbulent place (Fig. 3.22). This can easily be witnessed by looking down into its spiralling turbid currents from the footpath which stretches along the length of the original Severn Bridge. Impedance of light penetration by suspended particulate matter limits the growth of phytoplankton and restricts productivity. Nevertheless, the Severn Estuary receives an abundance of nutrients that are washed in on the tide from the Bristol Channel and from multiple rivers, including the Severn, Wye, Avon and Usk, which together drain a huge area of England and Wales. The estuarine sediments teem with bacteria. Samples obtained at Woodhill Bay (Fig. 3.23) near Portishead in Somerset have been reported to contain as many as 12–15 billion bacteria per millilitre of sediment (Webster *et al.*, 2010). The banks of the Severn Estuary and the Bristol Channel into which the estuary merges are bounded by expansive tidal flats which are replete with nutrients. Copious amounts of

Fig. 3.22. The River Severn, Aust Warth, Gloucestershire, UK.

Fig. 3.23. Estuarine sediment at Woodhill Bay, Somerset, UK.

organic matter are continuously and systematically degraded by microbes. The two main groups of microbes that complete its dissolution are the sulphate-reducing bacteria and the methanogenic archaea. Sulphate-reducing bacteria require a source of sulphate and abound in sulphate-rich environments such as estuarine sediments (Fig. 3.24). Beneath the Severn Bridge at Aust Warth, in Gloucestershire, sulphate-reducing bacteria have been shown to be responsible for the breakdown of almost two-thirds of the simple organic compounds that occur in estuarine sediment (Wellsbury *et al.*, 1996).

One of the pleasures of exploring Britain's coastline is to enjoy the contrasting scenery at different locations. But it is not just the scenery that varies from place to place: there are differences in the flora and fauna and these differences extend as far as the microbial realm. In coastal sediments, sulphate-reducing bacteria are largely responsible for completing the last steps in the degradation of organic matter. Other kinds of bacteria break down the larger molecules. Sulphate-reducing bacteria feed on the leftovers. Sulphate reducers subsist on simple organic compounds, such as acetate, lactate, propionate, butyrate and amino acids. Sulphate reducers can also utilise hydrogen as a food source. Different species of sulphate-reducing bacteria feed on different substrates and the species of sulphate-reducing bacteria and the substrates used by them vary from one locality to another. A study of sulphate-reducing bacteria at coastal locations in Scotland showed that a wide range of molecular substrates was used by sulphate reducers at Kingoodie Bay (Fig. 3.25) near Dundee on the

Fig. 3.24. Estuarine sediment at Aust Warth, Gloucestershire, UK.

Fig. 3.25. Estuarine habitat for sulphate-reducing bacteria at Kingoodie Bay, UK.

east coast of Scotland and also at Loch Eil, an arm of Loch Linnhe (Fig. 3.26) near Fort William on the west coast. At Kingoodie Bay and Loch Eil the water column is well connected to the open sea, sulphate and organic substrates are available in abundance and high rates of sulphate reduction occur. Loch Etive, in contrast, a sea loch on the the west coast, is relatively isolated from the open sea. At Loch Etive, low rates of sulphate reduction occur and a single compound, acetate, was shown to account for 100% of the organic matter utilised by the loch's resident sulphate-reducing bacteria (Parkes *et al.*, 1989).

The composition of the sulphate-reducing flora varies according to such factors as salinity, the availability of organic substrates and the availability of the sulphate ions used by the bacteria to oxidise these substrates. The River Colne in Essex is a tidal river that extends inland from Colne Point towards Colchester and beyond. At Colne Point a narrow spit of shingle meets a line of sand dunes, tidal channels and a coastal salt marsh (Fig. 3.27). The diverse habitats make Colne Point attractive to birds, insects, birdwatchers and entomologists and also to bacteria tourists. In the high-salinity environment of Colne Point, populations of sulphate-reducing bacteria are sustained by ample supplies of sulphate and organic molecules. The sulphate-reducing flora is dominated by bacteria related to members of the genus *Desulfovibrio*. Further upstream, where it becomes less salty and where supplies of sulphate diminish, bacteria related to *Desulfobacterium* predominate. This is possibly because as well as using sulphate as an oxidising agent, these more versatile bacteria are able to make do with nitrate as an alternative (Nedwell *et al.*, 2004; Kondo *et al.*, 2007).

Fig. 3.26. Sea loch and salt marsh. Habitats for sulphate-reducing bacteria. Loch Linnhe, UK.

Fig. 3.27. Coastal salt marsh at Colne Point, Essex, UK.

In addition to salinity and the availability of organic substrate and sulphate, sulphate-reducing bacteria may be influenced also by temperature. Not far from the town of King's Lynn in Norfolk, where the Great Ouse reaches the sea, sulphate-reducing bacteria occur in abundance. They are sustained by ample supplies of organic molecules and sulphate ions. The productivity of these bacteria has been shown to correlate with temperature. Rates of sulphate reduction peak during the warm summer months. Further upstream at Denver Sluice (Fig. 3.28) where the river reaches its tidal limit 15 miles from the sea, for reasons that are unclear this seasonal pattern disappears (Trimmer *et al.*, 1997).

Purple sulphur bacteria

A bloom of purple sulphur bacteria, seen on a sultry day in high summer is a natural spectacle of lavish beauty. Massive blooms of purple sulphurs, however, are seldom seen in most parts of the British Isles. Purple sulphur bacteria are usually fairly unobtrusive, although one might notice the odd patch of purple-coloured sediment during the course of a coastal walk (Fig. 3.29). Some localities are more predisposed to blooms of purple sulphur bacteria than others. Orkney is one such place, whilst in mainland Europe reported hotspots for purple sulphurs include Brittany and the Baltic (Tank *et al.*, 2011; Hubas *et al.*, 2017). In Orkney, blooms of purple sulphurs are a feature of sheltered beaches around Scapa Flow (Herbert, 1985; van Gemerden *et al.*, 1989). Although I have

Fig. 3.28. Upper tidal limit of the River Ouse at Denver Sluice, Norfolk, UK.

Fig. 3.29. Purple sulphur bacteria. Killard National Nature Reserve, County Down, Northern Ireland.

not been fortunate to witness blooms of purple sulphur bacteria during the course of my sojourns in the Orkney Islands, I have had better luck on the Baltic coast (Figs. 3.30–3.31). Once, while on holiday on the island of Öland I was delighted to discover that the beach adjacent to the campsite where we were staying was ablaze with purple sulphurs and on another occasion I encountered a vibrant bloom of purple sulphurs in a sheltered bay on the island of Gotland. To eyes accustomed to the conventional greens and blues of nature, the shocking pink of a bloom of purple sulphurs is an arresting visual experience.

Photosynthetic organisms, including plants and purple sulphur bacteria use the captured energy of sunlight to fix carbon dioxide and synthesise organic carbon compounds. But whereas plants use carbon dioxide (CO_2) and water (H_2O) as raw materials from which to build more complex organic hydrocarbons, purple sulphur bacteria use hydrogen sulphide (H_2S) instead of water. Whereas plants release oxygen as a waste product, purple sulphur bacteria produce elemental sulphur. Plants, green algae and cyanobacteria use varieties of the green pigment chlorophyll to capture light energy but purple sulphur bacteria use purple carotenoid pigments. The microbial mats that grow on the surface of coastal mudflats and intertidal beaches are made up of layers of different kinds of bacteria. The uppermost tier usually consists of a layer of cyanobacteria and green algae. These green-pigmented microbes often sit on top of a layer of purple sulphur bacteria. Green microbes do not absorb green light: it passes straight through them to reach the purple sulphur bacteria that live below them. Whereas chlorophyll absorbs the blue and red wavelengths of light, the purple carotenoid pigments of purple sulphur bacteria capture green light. Purple sulphur bacteria help to stabilise the sandy coastal sediments on

Fig. 3.30. Bloom of purple sulphur bacteria. Öland, Sweden.

Fig. 3.31. Bloom of purple sulphur bacteria, detail. Öland, Sweden.

which they live by producing copious quantities of particulate-binding adhesive slimes, known more correctly as extracellular polymeric substances (Hubas *et al.*, 2013; Hubas *et al.*, 2017). Purple sulphur bacteria are photosynthetic organisms but they can also live as scavengers by feeding directly on decomposing organic matter. During the summer months purple sulphurs sometimes venture up onto the surface to take advantage of deposits of dead seaweed. A glut of decomposing seaweed can trigger a massive bloom of purple sulphurs.

Historic interest in purple sulphur bacteria

When microbiology as a science was in its infancy and bacteria were still thought of as being primitive plants, one of the pioneers of sulphur bacteria research was the English surgeon and naturalist Edwin Lankester (1814–1874). Lankester took an interest in sulphur springs, their chemical properties and the plants and animals associated with them. He visited sulphur springs in England and Scotland and wrote a book about the sulphur springs at Askern in Yorkshire (Lankester, 1842). In all of the springs that he visited, Lankester observed a mysterious flocculent pigmented substance, which he referred to as *glairine*, a term used by continental authors for deposits found in mineral springs. *Glaire* means phlegm, or egg white in French. In the sulphur wells at Harrogate, Lankester noted the existence of '*red animalculae*'. Lankester '*also found a rose-coloured glairine at Moffat, depending on the existence of infusory animalcules; but these were of a much smaller size, and evidently of a different family from those found by me in the Askern and Harrogate waters*'. Clearly, these animalcules must have been purple sulphur bacteria. Lankester commented that: '*The existence of these animalcules, under the influence of a gas destructive to most other forms of life, is a striking instance of the wonderful adaption of organic life to the varied conditions in which organic matter is disposed.*' In his description of Askern and its sulphur springs Edwin Lankester shows an appreciation for all aspects of natural history. He describes the local geology. He gives an account of ongoing geomicrobiological processes, including the formation of tufa and other carbonate deposits in the Askern district. He provides lists of the infusoria (microbes), plants and animals. He notes a close association between '*pink deposits*', *very small animalcules*' and '*sulphuretted hydrogen*' (hydrogen sulphide). He also notes that '*the Mole-Cricket's song is often heard on summer evenings*'.

Edwin Lankester passed on the baton to his son, the zoologist Sir Edwin Ray Lankester (1847–1929). Lankester junior led an interesting life. He became, among other things, an authority on invertebrate animals, a professor of zoology at University College London, a director of the London Natural History Museum and a friend of the German philosopher and political theorist Karl Marx (1818–1883). While working in Oxford, Lankester's attention was drawn to some neglected jars of river water that had been left standing on a window sill. The jars contained the decomposing bodies of caddis fly larvae. He noted a '*purple-red growth … which spread itself over the decaying matters at the bottom of the water, and of patches forming an irregular coating on the glass of that side which was most exposed to the light*'. Furthermore, Lankester noted that the '*coloured products characterised by definite absorption-bands, by the agency of Bacteria, offers new data for speculation as to the chemical antecedents of such products, and the significance of colour in organisms generally*' (Lankester, 1873). Although he observed a multiplicity of growth habits and morphologies, which he described at length, Lankester came to the conclusion that a single species of '*peach-coloured*' bacterium was involved, which he named "*Bacterium rubescens*". There are in fact numerous different kinds of purple sulphur bacteria.

Whilst many species of bacteria are almost colourless, there is no need to use special stains in order to visualise purple sulphur bacteria under the microscope. On microscopic examination unstained cells of these bacteria are usually purple. They can be seen to incorporate conspicuous vacuoles that contain elemental sulphur, a waste product of photosynthesis. Purple sulphur bacteria are Gram-negative bacteria that belong to the Phylum *Pseudomonadota*. There are many species of purple sulphur bacteria which come in a variety of shapes and sizes (Figs. 3.32–3.33). A study of purple sulphurs from a coastal lagoon at Kiel Fjord on the Baltic coast detected dozens of species of purple sulphur bacteria (Tank *et al.*, 2011). *Thiocapsa roseopersicina* (the Peach-coloured Purple Sulphur) was reported to be the dominant species of purple sulphur bacterium on beaches in Orkney where it is accompanied by *Chromatium vinosum* (the Wine-red Purple Sulphur) (van Gemerden *et al.*, 1989).

Green sulphur bacteria and green iron bacteria

Green sulphur bacteria are photosynthetic anaerobes that manage to eke out a living by subsisting at very low light intensities in anoxic environments. If you examine a microbial mat in cross section you may be lucky and spot a thin green layer of green sulphur bacteria situated at the base of the mat. Green sulphurs live in places that are dark but not completely dark. In coastal

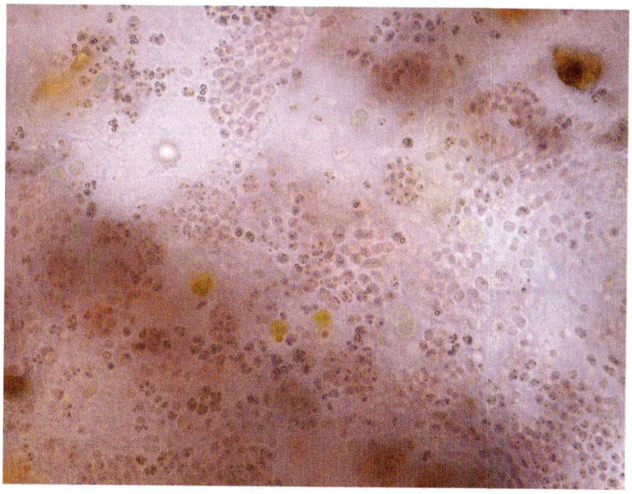

Fig. 3.32. Purple sulphur bacteria seen on microscopy. Possibly *Thiocapsa.* Killard National Nature Reserve, County Down, Northern Ireland.

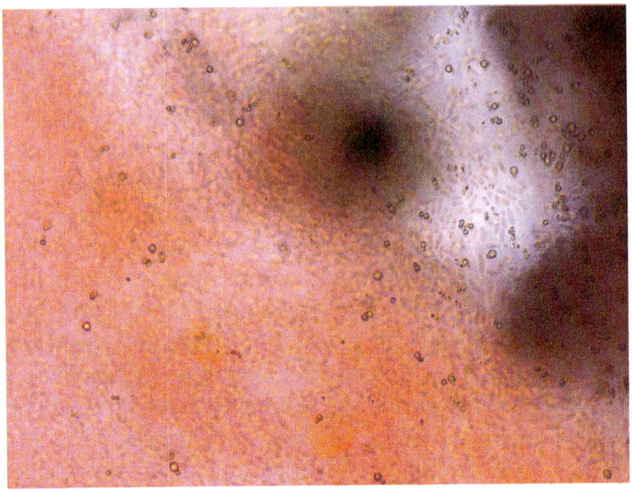

Fig. 3.33. Purple sulphur bacteria seen on microscopy. Possibly *Chromatium*. Widewater, Sussex, UK.

sediments green sulphur bacteria inhabit a twilight zone, underneath the more conspicuous layers of purple sulphur bacteria, cyanobacteria and green algae. Green sulphur bacteria occur in stagnant lakes, in sulphurous hot springs, in dimly lit caves and inside mine entrances, as well as in coastal sediments. Green sulphurs contain a variety of the green pigment chlorophyll with which they capture the energy of sunlight to perform photosynthesis. They use hydrogen sulphide to fix carbon dioxide in order to synthesise complex hydrocarbons, as do purple sulphur bacteria; and just like the purple sulphurs, green sulphurs produce elemental sulphur as a waste product. The green sulphur bacteria belong to a distinct phylum of bacteria, the Phylum *Chlorobiota* (previously known at the *Chlorobi*).

As well as being able to fix carbon dioxide and synthesise organic carbon compounds, green sulphur bacteria can also fix nitrogen. Green sulphur bacteria are important primary producers in the waters of murky deep stagnant lakes. Green sulphur bacteria occupy the deeper anoxic layers of the water column. Priest Pot, a small lake in the Lake District, contains high concentrations of phosphate and other nutrients and is known to support a diverse community of microbes, including green sulphur bacteria and purple sulphur bacteria (Goddard *et al*., 2005). One of the green sulphurs found in Priest Pot was tentatively identified as "*Clathrochloris hypolimnica*" (the Bottom-feeding Green Sulphur). Another species showed close DNA sequence relatedness to "*Chlorobium ferrooxidans*" (the Green Iron-bug). The Green Iron-bug differs from the typical green sulphur bacteria: it engages in a version of photosynthesis known as photoferrotrophy, which involves the oxidation of ferric iron (Fe II).

Green sulphur bacteria were found to represent a large proportion of the bacteria detected in sediment samples obtained from inside the entrance of a flooded coal mine at Ynysarwed, near Neath in South Wales (Blackwell *et al*.,

2019). The South Wales coalfield is perforated by old tunnels and disused mine shafts. Following the collapse of Britain's coal mining industry in the relatively recent past, abandoned mines now provide accommodation for enterprising bacteria. The sediments at Ynysarwed are rich in ferrous iron. It seems likely that the mine's resident *Chlorobiota* are green iron bacteria.

White sulphur bacteria

Heaps of rotting seaweed are a common sight on the coast in summer. As well as providing a habitat for beach-dwelling beetles and kelp flies, piles of decomposing seaweed sustain communities of purple sulphur bacteria and white sulphur bacteria. Unlike the photosynthetic (or phototrophic) purple and green sulphur bacteria, which use sunlight as an energy source, the so-called white sulphur bacteria use energy released from chemical reactions to fix carbon dioxide. White sulphur bacteria are known as chemoautotrophic organisms. They capture the chemical energy that is released when hydrogen sulphide reacts with oxygen. Hence, white sulphurs can be found at sites where hydrogen sulphide comes into contact with oxygen, such as in sulphur springs and at the surface of sediments that are rich in organic matter (Figs. 3.34–3.36). Large mats of white sulphur bacteria can also occur in well-oxygenated water on the seafloor. It is not unusual to encounter heaps of decomposed seaweed that are coated with white sulphur bacteria. When hydrogen sulphide reacts with oxygen, elemental sulphur is generated as a waste product. Elemental sulphur's milky appearance gives white sulphur bacteria their popular name. As the bacteria themselves are not actually pigmented, some microbiologists

Fig. 3.34. White sulphur bacteria on seaweed. Lower Largo, Fife, UK.

Fig. 3.35. White sulphur bacteria. Widewater, Sussex, UK.

Fig. 3.36. White sulphur bacteria and purple sulphur bacteria. Killard National Nature Reserve, County Down, Northern Ireland.

prefer to call them colourless sulphur bacteria. Examples of white sulphur bacteria include the genera *Beggiatoa* and *Thioploca* which have relatively large filamentous cells. White sulphurs are motile and will adjust their position in a column of water or sediment depending on the availability of oxygen and hydrogen sulphide. Success in finding white sulphur bacteria in labile intertidal habitats depends on the conditions being just right for them to be visible at the surface. Occasionally, at sites that are well supplied with hydrogen sulphide one encounters eye-catching mixtures of purple and white sulphur bacteria. These bacteria can create extravagant abstract compositions in purple and white.

Sulphur bacteria at home

A Winogradsky column, makes it possible to enjoy sulphur bacteria in the comfort of your own home (Fig. 3.37). In a paper published in 1887, Sergei Winogradsky described how to make what became known as a Winogradsky jar, or Winogradsky column. Winogradsky (1856–1953) was born in Kiev, then part of the Russian Empire. Following an education in Saint Petersburg he worked in Switzerland and at the Pasteur Institute in France where he conducted studies in soil microbiology. He investigated the involvement of bacteria in the cycling of sulphur and nitrogen in soils and sediments. He pioneered a culture technique, which makes it possible to study communities of interdependent environmental bacteria *in vitro*. The microbes in a Winogradsky column segregate into zones according to different physicochemical conditions. Bacteria that thrive in the absence of oxygen proliferate in the anoxic depths of the column. Their waste products stimulate the growth of other species higher up in the column. Production of hydrogen sulphide by sulphate-reducing bacteria at the base of the column will trigger a spectacular bloom of purple sulphur

Fig. 3.37. A Winogradsky jar, containing sulphur bacteria, cyanobacteria and eukaryotic algae.

bacteria that provides a beautiful colour contrast to blue-green cyanobacteria and eukaryotic green algae in the upper reaches of the column.

A Winogradsky column consists of a tall glass container to which are added a sample of soil, or sediment, water and nutrients. Winogradsky added rhizomes of Flowering Rush, *Butomus umbellatus* (Fig. 3.38), which he chopped up and threw into the column. He tried using other unspecified varieties of rhizome but they gave less satisfactory results. The rhizomes of *Butomus* decayed very slowly and acted as a viable nutrient source for many months. As an alternative he tried using boiled hay which caused a putrid fermentation and an annoying proliferation of 'banal bacteria'. Winogradsky added 1–2 grams of calcium sulphate to his column. Within 5–7 days, the system began to generate hydrogen sulphide, the base of the column blackened and the water became cloudy. The first bacteria to gain Winogradsky's attention were filamentous white sulphur bacteria, followed by the emergence of purple sulphurs. Eventually the walls of the column become coated by a purple lawn of bacterial growth which harboured all of the different kinds of sulphur bacteria that Winogradsky described in his paper. An edition of Winogradsky's collected works was published in 1949. It includes Winogradsky's recipe for what become known as a Winogradsky column (Winogradsky, 1949).

By adding different ingredients, Winogradsky columns can be modified to create conditions that favour different bacteria (Grau *et al.*, 1991; Dyer, 2003). Despite Winogradsky's appetite for Flowering Rush, chopped *Butomus* rhizomes are not essential ingredients. Screwed up paper can be substituted for *Butomus* as a source of carbohydrates. A pipette can be used to take samples

Fig. 3.38. Flowering rush, *Butomus umbellatus*. Chopped *Butomus* rhizomes served as an ingredient in Winogradsky's original recipe.

from different depths in the column for microscopic examination. Fermentative and sulphate-reducing bacteria can be observed in a sample of black sludge taken from the bottom of the column. The middle of the column provides a home for purple, green and white sulphur bacteria, while mats of cyanobacteria and green algae develop at the top of the column. Winogradsky columns do not necessarily read the textbooks: some of them misbehave, with the result that chaotic mosaics, consisting of patches of purple sulphur bacteria, green sulphur bacteria, cyanobacteria and eukaryotic algae develop, in a haphazard fashion, on the walls of the jar (Fig. 3.39). A Winogradsky column requires little more in the way of maintenance than a need to replace the water lost through evaporation. A well-prepared Winogradsky column is a colourful self-contained ecosystem that will give its custodian years of pleasure.

To observe the zones of differently coloured bacteria in a Winogradsky column is a bit like looking down at the ground from a plane at high altitude. Variations in the colour and texture of the landscape below allow the observer to infer the existence of different kinds of vegetation and land use. However, from such a great distance actual plants and animals remain invisible, just as the bacteria in a Winogradsky jar remain invisible to the naked eye. At best our appreciation of the ways of bacteria is detached. Much of what is known about bacteria comes from laboratory-based studies that employ the methods

Fig. 3.39. Detail from a 'disorganised' Winogradsky column.

of biochemistry and molecular biology. The results of such studies show that many aspects of the lives of bacteria are rather strange when compared with those of the larger organisms that we know from routine experience.

The degradation of nitrogenous waste by bacteria

A home aquarium is a bit like a Winogradsky column, except that a Winogradsky column is more self-contained. Whereas, a Winogradsky column is sustained by no more than sunlight, exposure to the atmosphere and the occasional top up with tap water, a home aquarium receives daily supplements of organic matter (fish food). Commercially available fish foods for use in home aquaria are proteinaceous concoctions that contain large amounts of organic nitrogen. Listed ingredients include processed fish meal, poultry, wheat, soy and spirulina. (Spirulina is a food product made from farmed cyanobacteria.) Fish excrete nitrogenous waste as ammonia (NH_3), a substance that is toxic to fish. The health of an aquarium depends on the conversion of ammonia to compounds which are less toxic; to nitrite (NO_2^-) and then to nitrate (NO_3^-), by nitrifying bacteria, a process known as nitrification. Aquaria are microcosms that can be used as models to study the nitrification processes that occur in the wider world.

In a series of papers published in 1890 and 1891, entitled *Recherches sur les organisms de la nitrification*, Winogradsky described the two steps that constitute the classic nitrification process (Winogradsky, 1949). The first step involves the oxidation of ammonia by ammonia-oxidising bacteria (or archaea). The second step involves the oxidation of nitrite by nitrite-oxidising bacteria. In 1892 Winogradsky isolated and described an ammonia-oxidising bacterium: *Nitrosomonas europaea* (The European Ammonia-bug). At a later date the first nitrite-oxidising bacterium was isolated and described in honour of Winogradsky as *Nitrobacter winogradskyi* (Winogradsky's Nitrite-bug). Both steps of the nitrification process result in the release of small amounts of chemical energy that are captured by nitrifying bacteria and used by them to fix carbon dioxide in order to build organic carbon compounds.

The bacteria that dispose of nitrogenous waste by converting ammonia to nitrate occur widely in soils, sediments, seawater and fresh water. They are inconspicuous organisms. Whereas the denizens of a Winogradsky column delight the observer with a vibrant multicoloured spectacle, nitrifying bacteria are invisible. By comparison with the world of art, if the sulphur bacteria in a Winogradsky column are the abstract expressionists of the microbial world, then nitrifying bacteria are the purveyors of cool minimalism. You can gaze through the glass of a fish tank and see the fish but you cannot see the nitrifying bacteria.

There are many species of ammonia-oxidising bacteria. The European Ammonia-bug is just one of a number of species in the genus *Nitrosomonas*. Multiple *Nitrosomonas* species may become established in a single aquarium filter (Burrell *et al.*, 2001). *Nitrosomonas* species are Gram-negative bacteria that belong to the Phylum *Pseudomonadota*. They occur widely in fresh water, soils and sediments (He *et al.*, 2018). In addition to the ammonia-oxidising bacteria, there are also ammonia-oxidising archaea. *Nitrosopumilus maritimus* (the Maritime Ammonia-bug) was the first ammonia-oxidising archaeon to be isolated in culture. The Maritime Ammonia-bug was discovered in a seawater aquarium in Seattle, Washington (Könneke *et al.*, 2005). Ammonia-oxidising archaea have since been detected in soil and fresh water as well as in geothermal springs in Yellowstone National Park and on the Kamchatka Peninsula in Russia (Sauder *et al.*, 2011).

The second step in the nitrification process, the conversion of nitrite to nitrate is performed by nitrite-oxidising bacteria. Nitrite-oxidising bacteria include members of the genera *Nitrobacter* and *Nitrospira*. *Nitrobacter* species can be found in soil, fresh water and seawater. *Nitrobacter* species are members of the Phylum *Pseudomonadota*. *Nitrospira* bacteria, however, belong to the Phylum *Nitrospirota*, a taxonomic group that also includes some iron-oxidising and thermophilic sulphate-reducing bacteria. *Nitrospira* species occur in soils, fresh water and seawater and play an important role in nitrite oxidation in wastewater treatment plants. The cells of *Nitrospira* bacteria have a lumpy, vaguely spiral shape. The first species to be formally described was *Nitrospira marina* (The Marine Nitrite-bug) in 1986.

The classic nitrification process, consisting of two stages performed separately by ammonia-oxidising and nitrite-oxidising bacteria was described by Winogradsky in the late nineteenth century. The received wisdom that the microbial breakdown of nitrogenous waster involved two different groups of bacteria persisted for more than a century. In 2015, two papers published in the journal *Nature* recorded the discovery of bacteria that perform a complete nitrification process (Daims *et al.*, 2015; van Kessel *et al.*, 2015). Bacteria capable of complete nitrification were detected in biofilm samples from aquarium filters in Vienna, Austria and also from aquarium filters in the Netherlands at Radboud University, Nijmegen. These novel bacteria, which are capable of complete nitrification belong to the genus *Nitrospira*, a genus that contains some well-known nitrite-oxidising bacteria. *Nitrospira* species capable of complete nitrification were soon detected in tap water, soil and sediments (Koch *et al.*, 2019).

Denitrification

One of the facets of modern agricultural practice involves the application of large volumes of nitrate fertiliser to arable fields. Nitrates act as a source of nitrogen to growing plants, increase crop yields and obviate the need to let fields lay fallow. A drawback of the widespread application of nitrate fertilisers is that nitrate-rich agricultural runoff may reach the water supplies and contaminate the soils of sensitive conservation areas, resulting in the rapid growth of rank vegetation and the disappearance of plants that are adapted to nutrient-poor conditions. Whereas the conversion of ammonia to nitrate is known as nitrification, the conversion of nitrate to gaseous nitrogen is called denitrification. Denitrification is a microbial process performed by denitrifying bacteria, archaea and fungi. At Cothill National Nature Reserve special clean-up zones were constructed in order to counter the threat posed by high nitrate levels to the local fen flora. The clean-up zones consist of log dams filled with heaps of rotting reed stems harvested from the fen (Fig. 3.40). The rotting stems provide an ideal habitat for denitrifying bacteria which convert nitrate to gaseous nitrogen, thereby preventing the ingress of excess levels of nutrients to more sensitive areas of the fen.

Methanogenic archaea

Oxygen is a highly reactive substance. In oxygenated environments, aerobic organisms such as animals, fungi and aerobic bacteria manage to extract copious amounts of energy from organic matter by oxidising it completely to yield carbon dioxide as a waste product. Anaerobic microbes that live in anoxic environments, such as the sediments beneath a stagnant pond make do with chemical reactions that yield relatively small amounts of energy and which result in the release of incompletely oxidised waste products such as hydrogen, acetate and methanol. These simple molecules act as sources of food for sulphate-reducing bacteria and for methanogenic archaea. Sulphate-reducing bacteria depend on the availability of sulphate, which they use as a weak oxidising agent. Enormous numbers of sulphate-reducing bacteria can be found in sulphate-rich habitats, such as estuarine marshes and marine sediments. In sulphate-depleted habitats such as peat bogs, the degradation of simple organic compounds is executed predominantly by methanogenic archaea.

Fig. 3.40. Rotting reed stems colonised by denitrifying bacteria are used to reduce nitrate levels at Cothill National Nature Reserve, UK (Judy Webb).

Methanogens utilise a variety of simple substrates as sources of food and energy. Some use acetate or methanol as food sources. Many of them, such as *Methanococcus maripaludis* (Fig. 3.41), an aquatic methanogen which was originally cultured from salt marsh sediment, can subsist on a simple diet of hydrogen and carbon dioxide. *Methanococcus maripaludis* grows rapidly under laboratory conditions and is often used as a model organism for the study of metabolic processes in archaea (*Ding et al.*, 2016). Methanogenic archaea release methane as a waste product. Methane (CH_4) is the simplest of all organic compounds.

Although methanogens are themselves invisible, indirect evidence of their existence can easily be found during the course of a routine field trip to a freshwater pond. Bubbles that rise to the surface of a pond when sediment is disturbed by a fishing net or a wellington boot are composed largely of methane produced by methanogens (Clymo, 2015). Casual observers will be familiar with the presence of bubbles in pond water but may not be aware of the existence of methanogenic archaea. The bubbles of methane are a bit like the shadows in Plato's allegorical tale about the inmates of a cave who can see the shadows of figures on a wall but cannot see the figures themselves. The methanogens as a taxonomic group can be traced back to a time when the earth was not an oxygenated planet. In the Pilbara region of Western Australia, mineral deposits exist that contain 3.5 billion year old methane bubbles. Analysis of the carbon isotope ratios of these ancient bubbles indicates that the methane trapped within them was produced by methanogenic microbes (Ueno *et al.*, 2006).

Oxygen's high reactivity makes it toxic to some organisms. Some microbes known as strict anaerobes cannot survive in the presence of oxygen. Methanogenic archaea occur in environments where oxygen is absent or present only in small amounts. In addition to pond sediments, methanogens can be found in many other oxygen-deficient environments, including salt marshes, hot springs, paddy fields, landfill sites, anaerobic digesters and the intestinal tracts of animals. Methanogens found in soil crusts and a species of marine methanogen, *Methanosarcina acetivorans* (Fig. 3.42) are distinguished by their ability to

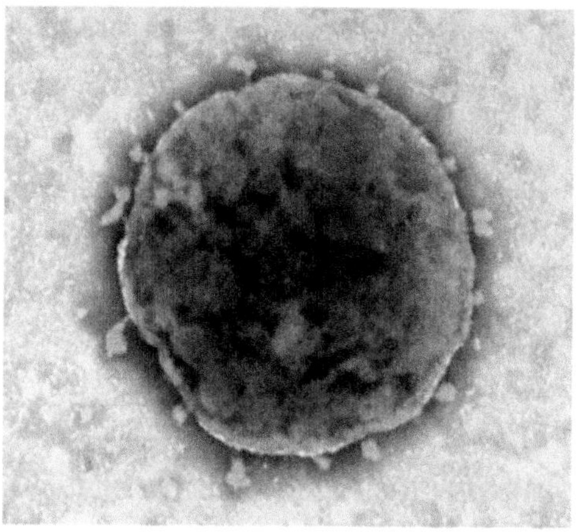

Fig. 3.41. *Methanococcus maripaludis*, a methanogen associated with salt marsh habitat. (Image cropped from figure 5 from Ding *et al.*, 2016). Creative Commons Attribution Licence.

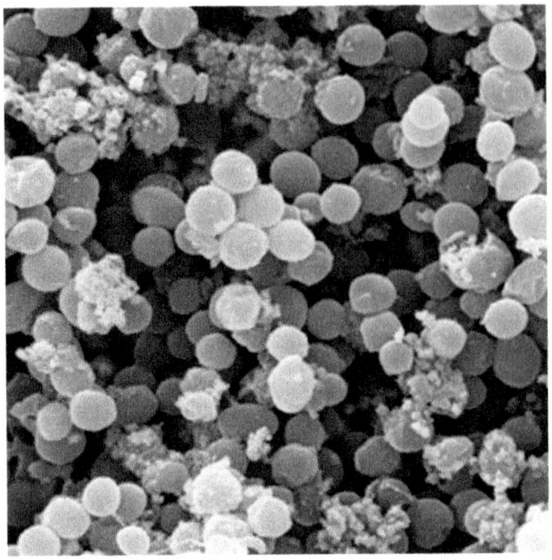

Fig. 3.42. *Methanosarcina acetivorans*, an oxygen-tolerant methanogen. (Image cropped from figure 5 from Jasso-Chávez *et al.*, 2015). Creative Commons Attribution Licence.

tolerate exposure to higher concentrations of oxygen than other methanogens (Jasso-Chávez *et al.*, 2015).

Different kinds of methanogen are associated with different kinds of habitat (Table 3.1). Almost all of the known methanogens belong to the Phylum *Methanobacteriota* (previously known as the "*Euryarchaeota*"), a large taxonomic grouping composed of diverse microorganisms, including many non-methanogens. Recently, a thermophilic methanogen which belongs to another phylum (Phylum *Thermoproteota*) was discovered in a Chinese oil field, whilst two novel phyla have been proposed to accommodate some distinctive novel methanogens detected in anaerobic digesters and coal bed methane wells.

In the intestinal tracts of animals, methanogens work in tandem with fermentative microorganisms to degrade plant material. Fermentative microbes yield carbon dioxide and hydrogen as waste products, which are used as food sources by methanogenic archaea. *Methanobrevibacter smithii* (Smith's Methanogen) is a species of methanogen associated with the human intestinal flora (Samuel and Gordon, 2006). You may be among the 50–85% of humans whose intestines are colonised by methanogenic archaea. Smith's Methanogen helps us to extract energy from food. During the digestive process, gut bacteria break down carbohydrates in food and release hydrogen as a waste product. High concentrations of hydrogen inhibit the fermentation reactions that bacteria use to break down carbohydrate. By converting hydrogen to methane, Smith's Methanogen facilitates fermentation and could therefore be considered to aid the digestion.

Methane-oxidising bacteria (methanotrophs)

Methanotrophs are methane-oxidising bacteria that feed on methane. Some of them, the obligate methanotrophs, depend on methane as their sole source of

Table 3.1. Methanogenic archaea: the major taxonomic groups and their habitats.

Taxonomic group	Habitat
Phylum *Methanobacteriota* (previously "*Euryarchaeota*")	
Class *Methanobacteria*	
Order *Methanobacteriales*	Marine and estuarine sediments, animal intestines, hot springs, soda lakes, peat bogs
Class *Methanosarcinia*	
Order *Methanosarcinales*	Freshwater, hypersaline, estuarine and marine sediments, peat bogs, landfill sites, animal intestines
Order *Methanocellales*	Paddy fields, peat bogs
Class *Methanococci*	
Order *Methanococcales*	Marine and estuarine sediments, hydrothermal vents
Class *Thermoplasmata*	
Order *Methanomassiliicoccales*	Mammal intestines, insect intestines, wetlands
Class "*Methanomicrobia*"	
Order *Methanomicrobiales*	Ponds, estuarine and marine sediments, soda lakes, paddy fields, peat bogs, landfill sites, wastewater treatment plants, animal intestines
Class *Methanonatronarchaei*	
Order *Methanonatronarchaeales*	Hypersaline lakes
Class *Methanopyri*	
Order *Methanopyrales*	Black smoker hydrothermal vent on seafloor
Class "*Candidatus* Methanofastidiosa"	Fresh water, marine sediments, bioreactors
Phylum "*Candidatus* Bathyarchaeota"	
BA1, BA2	Cold bed methane wells
Phylum "*Candidatus* Verstraetearchaeota"	
"*Candidatus* Methanomethylicuia"	Anaerobic digesters, tailings ponds
"*Candidatus* Methanosuratus"	Cold bed methane wells

food. Others are more versatile and can utilise ethanol, propane and acetate as alternatives. Methanotrophs inhabit oxygenated environments that are close to sources of methane. They frequent the same habitats as methanogens, including fresh water, hypersaline lakes, soda lakes, seawater, sediments, peat bogs, upland and forest soils, paddy fields, landfill sites, sewage and the intestines of mammals and termites. Methanotrophs can be found in oxygenated pockets within what are otherwise waterlogged anoxic soils and sediments, inhabited by methanogens. Methane-oxidising bacteria occur in association with the roots of vascular plants that grow in wetland soils. The porous architecture of plant stems allows oxygen to reach the root layers of the soil and creates an oxygenated microhabitat in close proximity to sources of methane. Some of the methane produced in wetland soils and sediments escapes to the atmosphere through plant stems (Fig. 3.43). Methane contributes to the buoyancy of water plants (Dacey and Klug, 1979).

Numerous genera and species of methane-oxidising bacteria have been described, including *Methylocystis heyeri* (the Marsh Gas Guzzler), *Methylomonas methanica* (the Methane Gas Guzzler) and *Methylosarcina lacus* (the Lacustrine Gas Guzzler), all of which are members of the Phylum *Pseudomonadota*. Acidic geothermal habitats are frequented by a phylogenetically unrelated group of methane-oxidising bacteria that belong to the Phylum *Verrucomicrobia*. Examples include "*Candidatus* Methylacidiphilum infernorum" (the Hell's Gate Gas Guzzler), "*Candidatus* Methylacidiphilum fumariolicum" (the Neapolitan Gas Guzzler) and *Methylacidiphilum kamchatkense* (the Kamchatka Gas Guzzler), all of which thrive at temperatures of around 60°C. As a general rule, *Methyl-* is used to form the generic names of methanotrophs and *Methano-* is used to form the generic names of methanogens.

Bacteria in bogs and fens

Peat consists of the compressed remains of sphagnum moss and other kinds of vegetation and includes the chitinous remains of beetles and other organisms. Peat deposits in the British Isles have been accumulating since the end of the last Ice Age and peatlands now account for approximately 12% of land cover in the UK (Yu *et al.*, 2011; Evans *et al.*, 2017a; Evans *et al.*, 2017b). A characteristic vegetation develops in association with peat bogs, including sphagnum mosses that colonise peatland pools (Fig. 3.44). Peat bogs are characterised by

Fig. 3.43. Methane contributes to the buoyancy of water plants. Water lilies. Hitchcopse Pit, near Oxford, UK.

Fig. 3.44. Sphagnum bog, Abernethy Forest, UK.

acidity (pH 3–5), paucity of minerals and waterlogged anoxic conditions, which hinder the decomposition of dead plant matter. In order to degrade woody materials bacteria and other microbes require access to oxygen. Wooden objects composed of lignin are resistant to decay in anoxic environments. The stumps of ancient trees emerging from layers of eroded peat are a familiar sight in the boggier parts of the Scottish Highlands. Similarly, bog oak and 'bog deal' (pine) are features of peat bogs in Ireland (Fig. 3.45). The presence of these ancient wooden remnants indicates that the peat deposits from which they emerged are anoxic habitats.

Waterlogged anoxic deposits of sphagnum moss and peat are environments in which organic matter cannot be oxidised completely. The anaerobic bacteria that dwell in peatlands release hydrogen and simple organic compounds as waste products, which are used as substrates by communities of methanogenic archaea. Diverse assemblages of methanogenic archaea occur in peatlands (Carson *et al.*, 2019). The methanogens produce methane, which may dissolve or accumulate as gaseous bubbles within the layers of peat (Clymo, 2015). In addition to supporting methanogens, peat deposits that contain quantities of sulphate may also support sulphate-reducing bacteria, which compete with the methanogens for hydrogen and low-molecular weight organic substrates. Peat cores extracted from a bog at the Moor House National Nature Reserve in northern England, are marked at a depth of 10–18 cm by a black zone of black iron sulphide, which indicates the presence of sulphate-reducing bacteria. Rates of methane production are much lower in the black sulphide zone than in

Fig. 13.45. Trunk of ancient 'bog deal' - a negative marker of microbe-mediated decay. Brackagh Moss, County Armagh, Northern Ireland.

other peat layers (McDonald *et al.*, 1999). In nutrient-poor bogs methanogens and sulphate-reducing bacteria primarily use hydrogen and carbon dioxide as substrates. Fens differ from bogs in being alkaline rather than acidic and being richer in nutrients. Fens support diverse communities of deep-rooted vascular plants. In fenland, methanogens and sulphate-reducing bacteria make use of acetate released into the rhizosphere from plant roots as a food source (Galand *et al.*, 2005).

Because methane is a greenhouse gas, there is interest in the factors that affect its production by methanogenic archaea and its degradation by methane-oxidising bacteria in peatlands. It has been shown that rates of peat decomposition and methane production increase with temperature. By incubating samples of peat slurry obtained from moorland near Lake Vyrnwy in Wales, scientists based at the University of Aberdeen showed that methane production was low at 4°C but increased at higher temperatures, peaking at a temperature of 25°C (Freitag and Prosser, 2009). The oxidation of methane by methane-oxidising bacteria (methanotrophs) is also affected by temperature. Although methanotrophs present in some habitats are most active at temperatures above 20°C, the methanotrophs at Moor House National Nature Reserve were found to be attuned to lower temperatures. At Moor House 100% of available methane was oxidised at temperatures below 15°C whereas at 25°C only 50% of the available methane was oxidised (van Winden *et al.*, 2012).

In peat bogs, methanotrophs inhabit oxygenated layers of peat and sphagnum in moorland pools. Methanotrophs detected at Moor House National Nature Reserve include bacteria belonging to the genera *Methylococcus* and *Methylosinus* as well as members of an as yet uncultured group of methanotrophs designated informally as the Moor House Peat Cluster (McDonald *et al.*, 1996; Knief, 2015). Methanotrophs are relatively undemanding organisms. Cultures can be maintained in the laboratory for years on a diet of little more than water, methane and oxygen (Dedysh *et al.*, 1998).

Soil bacteria

A gram of soil may contain as many as 10 billion bacteria (Raynaud and Nunan, 2014). Bacteria and archaea share the soil with fungi, protozoa, animals and the roots of plants. Bacteria enrich soils by fixing carbon and nitrogen. Bacteria synthesise a range of organic molecules that are utilised as nutrients by other organisms. Bacteria mediate the degradation of organic compounds, including cellulose $(C_6H_{10}O_5)_n$, a structural component of plants, as well as chitin $(C_8H_{13}O_5N)_n$, a structural component of invertebrate animals. Bacteria eliminate the nitrogenous waste products of animals. Nitrifying bacteria oxidise ammonia, to yield nitrite and nitrate. Denitrifying bacteria subsequently reduce nitrate to yield nitrogen gas. Some soil bacteria can tolerate ionising radiation, while others can tolerate toxic metals. There are also soil bacteria which can degrade toxins including pesticides.

Soil bacteria (Figs. 3.46–3.47) are diverse but most of the species that can be detected using molecular methods have not been cultured. Nevertheless, genetic sequence data analysis makes it possible to allocate individual strains to phylogenetic groups, determine the functions of their genes and predict their roles in the soil ecosystem. Sequence data can be used to obtain measures of diversity within and between different locations. Such an approach was used to investigate 1010 soil samples obtained during the Countryside Survey in 2007 (Griffiths *et al.*, 2011). Study samples were chosen that represented a full range of soil types from across Great Britain. Soils with the lowest

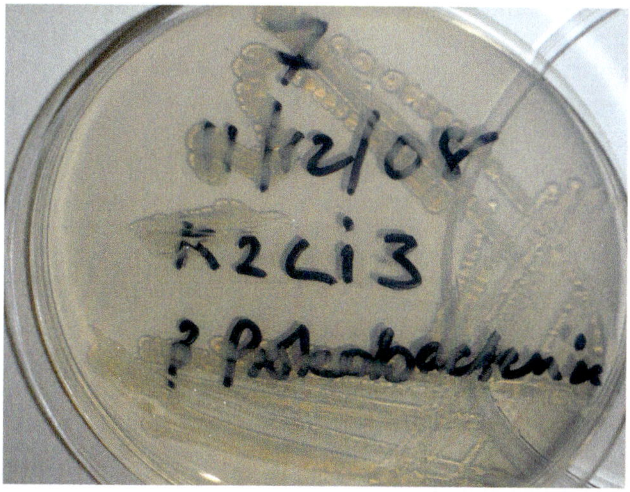

Fig. 3.46. *Janthinobacterium*, a proteobacterium cultured from soil. (Isolate kindly provided by Minna Männistö.)

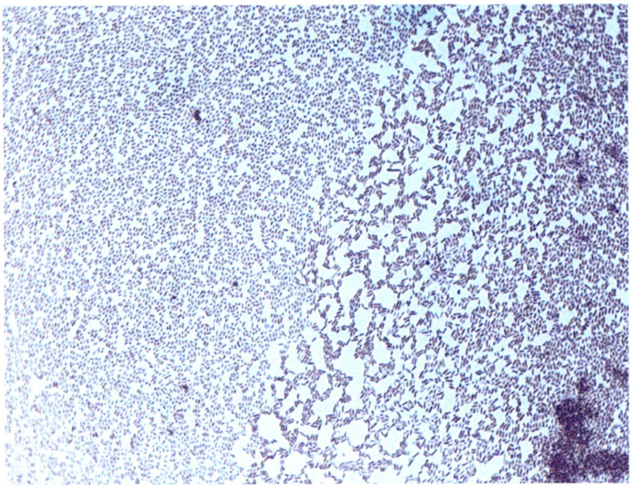

Fig. 3.47. Microscopy of *Janthinobacterium*, a proteobacterium cultured from soil. (Isolate kindly provided by Minna Männistö.)

within-sample microbial diversity were found to be acidic soils from upland locations. Similarly, in the Americas, acidic soils were shown to be the least microbially diverse (Fierer and Jackson, 2006). In Great Britain higher species totals were detected in agricultural soil samples with higher pH values from lowland sites than in the more acidic upland soil samples. Lowland soil microbial communities, however, were found to be relatively homogeneous across the extent of their range, whereas upland soil microbial communities varied in composition between sites. These differences may reflect the varied ecological conditions which exist across upland Britain. The composition of communities of soil bacteria has been shown to vary in accordance with the plant communities present at different locations. For example, sites with fast-growing plants and an abundance of nitrogenous leaf litter host large numbers of soil bacteria including those involved in the process of denitrification (Orwin *et al.*, 2010). In addition to pH and vegetation cover, soil microbes are also affected by rainfall. A study of soil microbes at 12 distinctive grassland sites in England showed that numbers of bacteria increase with increasing rainfall (de Vries *et al.*, 2012). Studies conducted in the Americas show that the microbial biomass of soils increases with rainfall, although the most microbially diverse soils in the Americas are desert soils (Fierer and Jackson, 2006; de Vries *et al.*, 2012).

In addition to pH, moisture and vegetation cover, soil bacteria community structure is also influenced by salinity, metal toxicity and the availability of nutrients (Gans *et al.*, 2005; Lozupone and Knight, 2007; Langenheder and Prosser, 2008). In upland soils, the the microbial community structure is influenced by the duration of snow cover. In cold climates bacteria are sensitive to small changes in temperature. Under thick snow soil stays at a temperature of around 0°C, which allows bacteria to remain metabolically active (Zinger *et al.*, 2009). In northern Finland, different species of acidobacteria are associated with exposed windswept sites compared with locations that are subject to deep snow cover in winter (Männistö *et al.*, 2013).

Soil acidity and acidobacteria

Molecular studies indicate that members of the Phylum *Acidobacteria* form an important component of soil microbial communities. Soil surveys conducted in northern Finland showed that acidobacteria (Fig. 3.48) were predominant members of the soil flora and an altitudinal transect study of Peruvian soils showed that acidobacteria accounted for a significant proportion of the microflora at higher altitudes (Männistö *et al.*, 2013; Nottingham *et al.*, 2018). However, the acidobacteria are poorly understood in comparison with other phylogenetic groups of bacteria that predominate in soils. Acidobacteria are recalcitrant to culture and isolation using traditional culture methods. The first acidobacterium to be isolated in pure culture was *Acidobacterium capsulatum* (the Capsulated Acid-bug), which was formally described as recently as 1991 on the basis of strains that were cultured from acidic mining waste in Japan in the 1980s (Kishimoto and Tano, 1987). Additional species have since been described from a variety of habitats, including soils. Acidobacteria secrete extracellular polymeric substances that may enhance their resilience in hostile and variable soil environments (Kielak *et al.*, 2016). Molecular investigations of environmental samples show acidobacteria to be as abundant and diverse as the better known groups of bacteria and that acidic soils show a relative abundance of acidobacteria (Lauber *et al.*, 2009). For practical purposes acidobacteria are classified according to a scheme of numeric subgroups.

Multiple subgroups of acidobacteria have been recognised, which respond to different environmental conditions. For example, acidic upland soils in the

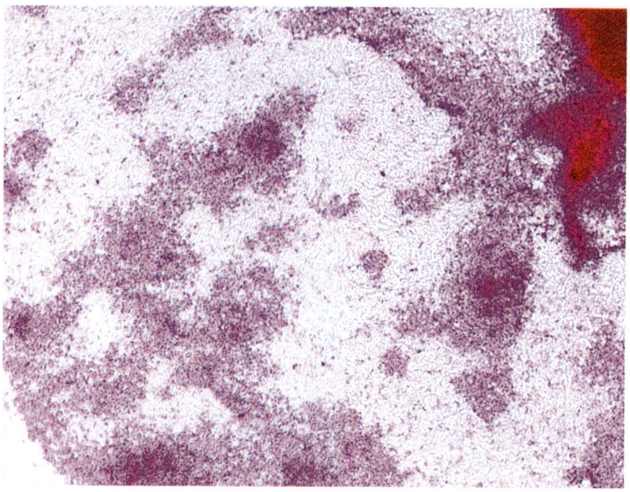

Fig. 3.48. *Terriglobus saanensis*, an acidobacterium cultured from tundra soil in northern Finland. (Isolate kindly provided by Minna Männistö.)

British Isles are characterised by high proportions of *Acidobacteria* subgroup 1, whereas neutral and slightly alkaline soils are characterised by higher proportions of *Acidobacteria* subgroup 6 (Griffiths *et al.*, 2011). Soil acidity can be seen to influence the composition of microbial communities even over short distances, such as within the bounds of a single field (Baker *et al.*, 2009). At the Rothamsted Research Station in Hertfordshire, there is an experimental strip of arable land where the soil acidity changes over a distance of less than 200 m from pH 4.0 to pH 8.5. The acid gradient strip was established as a result of a mid-nineteenth century trial that involved treatment of the ground with chalk. The strip has never been fertilised. Following its creation, the strip was sown annually with spring barley. More recently it has been sown with winter wheat. Soil conditions hardly vary along the length of the strip apart from the change in pH. At the more acidic end of the strip the crops that are planted regularly fail and the organic carbon content of the soil is accordingly lower than at the more alkaline end of the strip.

Whilst populations of fungi remain fairly constant along the strip, dramatic shifts can be observed in the communities of bacteria. *Acidobacteria* subgroup 1 are most abundant at the acidic end of the strip, as are *Acidobacteria* subgroups 2 and 3, which are also favoured by acidic conditions. *Acidobacteria* subgroup 4 (as well as members of the *Pseudomonadota*) are more abundant at the alkaline end of the strip (Rousk *et al.*, 2010). A similar acid soil gradient exists at the Craibstone Experimental Farm, near Aberdeen, where there are soil plots with pH values that vary between 4.5–7.5. The acidities of the different plots have been maintained for 50 years through the application of aluminium sulphate and calcium carbonate. Bacteria are more diverse at the near-neutral end of the acidity gradient, whereas the more acidic soil contains a larger proportion of acidobacteria that belong to subgroups 1, 2 and 3. The acid soil also favours other bacteria such as *Dyella* and *Rhodanobacter* (Bartram *et al.*, 2014). The genera *Dyella* and *Rhodanobacter* include species such as *Dyella acidisoli* (the Acid Soil Bacterium) and *Rhodanobacter aciditrophus* (the Acidic Wastewater Bacterium) which grow optimally in mildly acidic conditions. *Dyella* and *Rhodanobacter* are members of the Phylum *Pseudomonadota*.

The influence of land use on soil bacteria

Changes in land use affect soil microbes. Improvement of upland soils by the application of lime and sewage sludge stimulates the activity of nitrifying bacteria and increases soil pH which correlates with an increase in bacterial diversity (Gray *et al.*, 2003). Mining and industrial practices can cause long-term changes to the composition of soils by affecting patterns of drainage, increasing the acidity and by enriching soils with toxic metals.

In a 1970s television film entitled *The Stone Tape* (1972), a group of scientists discover to their horror that the stonework of their research centre resonates with an ancient remembrance of past events. In similar vein, soil microbiologists in Germany have ventured to suggest that soil bacteria may record an 'ecological memory' of former land use and that the microbial composition of a soil may reflect its ancient history as well as its current environmental conditions (Köhler *et al.*, 2020). A microbial soil survey was made of 10 forested prehistoric earthworks. Although now forested, it is believed that the ancient enclosures were once cattle pounds. As expected, soil bacteria communities were shown to correlate with soil acidity and to reflect the underlying geology. But more subtle differences were also detected: bacteria were detected inside the enclosures that did not occur on the outside, including the methane-oxidising genera *Methylcapsa* and *Methylocella* as well as members of the Class *Ktedonobacteria*. *Ktedonobacteria* are filamentous soil bacteria that occur in diverse environments. *Ktedonobacteria* have the capacity to degrade cellulose and the numbers of *Ktedonobacteria* increase in soils subject to grazing by cattle (Ding *et al.*, 2023).

Cyanobacteria (blue-green algae)

Cyanobacteria are pioneering organisms which can colonise bare rocks, dunes and deserts (Fig. 3.49). Resistance to desiccation allows them to proliferate in environmental conditions that are too dry for plants and because cyanobacteria can fix nitrogen they do not have to rely on extraneous sources of nitrogen compounds. Cyanobacteria secrete sticky extracellular polymeric substances that bind sand grains together. Cyanobacteria form crusts that protect desert soils against erosion. Mats of cyanobacteria stabilise sandy intertidal flats. In montane and polar regions cyanobacteria colonise the surfaces of

Fig. 3.49. Dune crust formed by cyanobacteria and various other microbes, Brancaster, Norfolk, UK.

glaciers. The visual appeal of geothermal sites is enhanced by colourful mats and streamers made by thermophilic cyanobacteria, which thrive at high temperatures (Fig. 3.50). Cyanobacteria are often noticeable even in habitats that are dominated by plants and animals. Some species form colonies that are visible to the naked eye, whilst strikingly pigmented blooms of planktonic cyanobacteria can occur in both freshwater and saline water bodies. Because cyanobacteria are often visible to the naked eye as films, mats or streamers and because the different genera can be identified according to their characteristic morphological features, the cyanobacteria constitute a group of bacteria that is unusually accessible to study by amateur microscopists.

Cyanobacteria are photosynthetic organisms that contain chlorophyll. They use carbon dioxide and water as substrates for the synthesis of organic carbon compounds, as do plants. Similarly, they release oxygen as a waste product. In addition to green chlorophyll, which captures red and blue light, cyanobacteria use accessory pigments that capture light of other wavelengths. Some use a blue pigment, phycocyanin, which captures orange light. The combination of green chlorophyll and blue phycocyanin imparts a blue-green colour which gives cyanobacteria their name. Some cyanobacteria make use of an orange pigment, phycoerythrin which absorbs blue-green light. Cyanobacteria also use orange carotenoid pigments that protect against oxidative stress and yellow-brown scytonemin and red gloeocapsin pigments that protect them against ultraviolet radiation. The combination of pigments displayed by cyanobacteria varies with species, habitat and environmental conditions. Cyanobacteria can make do with relatively low intensity light exposure and can colonise deep water, cave entrances and the undersurfaces of rocks (Fig. 3.51). A subterranean

Fig. 3.50. Thermophilic cyanobacteria at La Duke Spring, Montana, USA.

Fig. 3.51. Cyanobacteria growing on the underside of a stone. Whitbarrow Scar, Cumbria, UK.

microbial ecosystem in Spain was found to be dominated by a cyanobacterium which lives entirely in the dark. The novel cyanobacterium was detected in rock samples retrieved from a 600-m deep borehole. It appears to derive its energy by using an as yet undetermined oxidising agent to oxidise hydrogen (Puente-Sánchez *et al.*, 2018).

Compared with the average bacterium, cyanobacteria have relatively large cells (Fig 3.52). Cyanobacteria contain chlorophyll and can easily be confused with eukaryotic algae. From an ecological perspective, cyanobacteria and eukaryotic algae share a number of common features. Cyanobacteria were studied initially by botanists who called them blue-green algae, or cyanophytes. Many of them were described on the basis of morphology and named in accordance with the International Code of Botanical Nomenclature. This means that numerous species of cyanobacteria have been described according to botanical practice on the basis of quite subtle morphological features, which they may share with closely related taxa (Wright *et al.*, 2001). A formal proposal to place the nomenclature of cyanobacteria under the rules of the International Code of Nomenclature of Bacteria was not published until the 1970s (Stanier *et al.*, 1978).

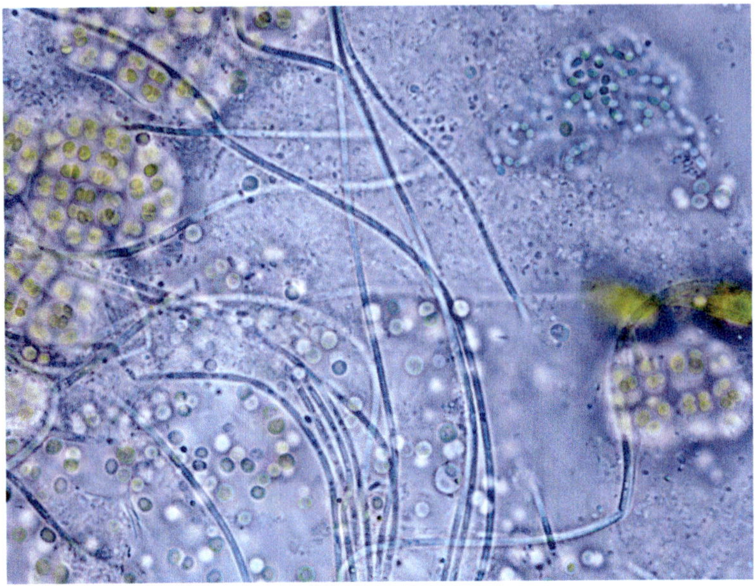

Fig. 3.52. Globular and filamentous cyanobacteria from a rock scraping, Alum Bay, Isle of Wight, UK.

Fig. 3.53. Coastal rocks coated with bright green eukaryotic algae and darkly pigmented cyanobacteria, Alum Bay, Isle of Wight, UK.

Cyanobacteria compared with eukaryotic algae

The term alga was originally applied to seaweeds. Later the use of the term was extended to include a number of groups of single-celled eukaryotic organisms, including the green algae, diatoms, photosynthetic flagellates and dinoflagellates. All of these rather diverse microscopic organisms, as well as seaweeds, are considered by some authors to belong to the Kingdom Protoctista. Many eukaryotic algae are bright yellow-green in colour, whereas cyanobacteria are often blue-green in colour, although some are dark brown and some are bright orange. Cyanobacteria share a wide range of habitats with eukaryotic algae (Fig. 3.53). Water bodies are often tinted green by planktonic cyanobacteria as well as by eukaryotic green algae. Cyanobacteria and eukaryotic algae can tolerate desiccation and shade and both groups of organisms can often be found growing together on walls, under stones and in cave entrances. There are even eukaryotic algae that grow in acidic hot springs. Although cyanobacteria can tolerate high temperatures and can be found in hot springs they do not tolerate conditions of high acidity. The eukaryotic algae that belong to the Order Cyanidiales are exceptional among photosynthetic microbes in being able to tolerate acidic (pH 2) hot water. At Yellowstone National Park, eukaryotic algae impart a vibrant green tinge to the run-off channels of acidic hot springs, a habitat which they share with acidophilic thermophilic bacteria but not with cyanobacteria. *Chlamydomonas acidophila* is an acidophilic eukaryotic alga that tolerates high concentrations of copper and other metals (Fig. 3.54). It can be seen in drainage channels at Parys Mountain on Anglesey (Dean *et al.*, 2019). Another true alga, *Dunaliella salina* thrives in conditions of high salinity and is an important primary producer in salt lakes. *Dunaliella salina*, as well as many of the halophilic bacteria and archaea that live in hypersaline

Fig. 3.54. *Chlamydomonas acidophila* an acidophilic eukaryotic alga. Parys Mountain, Anglesey, UK.

water bodies have cells that contain red carotenoid pigments (Dunal, 1837). Together they account for the pink colouration of hypersaline lakes (Fig. 3.55). In many different habitats, cyanobacteria can often be found living in association with eukaryotic algae. Many ecologists and biologists who study photosynthetic microorganisms take an interest in cyanobacteria as well as in eukaryotic algae. The standard identification guide to British freshwater algae includes the cyanobacteria (John *et al*., 2002).

Algal blooms

Proliferations of planktonic cyanobacteria, as well as those of eukaryotic algae, are commonly referred to as algal blooms (Figs. 3.56–3.59). Cyanobacterial blooms are triggered by increases in temperature, light intensity and concentrations of nutrients. Blooms are accordingly most often apparent during the summer months. Dense concentrations of cells turn water bodies bright green or blue-green in colour. A scum consisting of cellular debris may accumulate on the shoreline. Blooms may deprive other organisms of sunlight and oxygen. Cyanobacterial blooms occur sporadically throughout the British Isles (Turner *et al*., 2018). The steady supply of phosphate and nitrogen to a water body can create conditions that are conducive to the regular occurrence of algal blooms. An excessive build-up of nutrients in a water body is referred to as eutrophication.

Fig. 3.55. Hutt Lagoon, Western Australia, tinted by the pink pigments of prokaryotic halophiles and the eukaryotic alga *Dunaliella*.

Fig. 3.56. Cyanobacterial bloom warning sign. Loch Garten, UK.

Lough Neagh, Ireland's largest lake, is relatively shallow. Ingress of agricultural run-off and treated sewage promote the growth of cyanobacteria and make the lake susceptible to algal blooms (Skvortsov *et al.*, 2016). Nevertheless, Lough Neagh is of great interest to the naturalist (Fig. 3.60). Its aquatic life includes a species of whitefish, the pollan, *Coregonus pollan*, an Ice Age relic. Lough Neagh's sandy shores appeal to coleopterists on account of the existence of a number of scarce beetles, including the ground beetles *Dyschirius obscurus* and *Pelophila borealis*. Despite such riches, Lough Neagh has been described as one of the most eutrophic lakes in the British Isles. Sediment cores show that the eutrophication process began in about 1955. Eutrophication has been linked to phosphate derived from various sources around the lake margin (Bunting *et al.*, 2007). Since 1984, sewage treatment plants situated near the lake have adopted the practice of extracting phosphate before discharging treated sewage into Lough Neagh.

Even Lake Victoria (Nyanza), the largest lake in Africa is susceptible to eutrophication (Fig. 3.61). Lake Victoria is almost as big as Ireland. Run-off from agricultural communities situated around its shoreline, coupled with industrial effluent and a tropical climate have created conditions that favour the proliferation of cyanobacteria in the lake's shallow bays and gulfs, where the planktonic microflora is dominated by cyanobacteria that belong to the genus *Microcystis* (Sitoki *et al.*, 2012). *Microcystis* and other genera of cyanobacteria produce toxins known as microcystins. Microcystins have been

Fig. 3.57. Cyanobacterial toxin warning sign in Welsh. Freshwater East, Pembrokeshire, UK.

detected in samples taken from Lake Victoria (Miles *et al.*, 2013). Exposure to microcystins can lead to eye irritation, respiratory tract irritation, headache, fever, abdominal pain, nausea and vomiting. Microcystins accumulate in the liver and cause liver toxicity. In 1996 seventy-six patients in Brazil died as a result of acute liver failure, following exposure to untreated reservoir water while receiving haemodialysis (Carmichael *et al.*, 2001). Recreational exposure to a harmful algal bloom at a beach resort in Uruguay in 2015 led to a case of hepatic failure in a 20-month-old child that necessitated treatment by liver transplantation (Vidal *et al.*, 2017). Blooms of toxic cyanobacteria may cause the mass die-off of populations of fish and water birds. During the course of a visit to the Salton Sea, an inland lake in California, I became aware of a

Fig. 3.58. A stream at Freshwater East Pembrokeshire, UK, affected by a cyanobacterial bloom.

strange tactile sensation while tramping along the shoreline. The cause was a thick layer of bleached bones that crunched underfoot. The Salton Sea is situated 85 m below sea level, is subject to high summer temperatures and is supplied with nutrients from agricultural run-off. The lake's nutrient-rich waters are concentrated by a high rate of evaporation, which creates ideal conditions for cyanobacteria. In 1992, approximately 150,000 birds died on the Salton Sea. Most of them were Eared Grebes, *Podiceps nigricollis*. Cyanobacterial toxins were found in tissue samples (Carmichael and Li, 2006). Incidents of microcystin poisoning have occasionally been reported in the UK. In 1989 it was reported that 20 sheep and 15 dogs had died as a result of microcystin toxicity after coming into contact with contaminated water and scum deposits at Rutland Water. Also in 1989, two army recruits were admitted to hospital with pneumonia, after taking part in canoe exercises on Rudyard Lake in Staffordshire (Pearson, 1990; Turner *et al.*, 1990; Turner *et al.*, 2018).

Cyanobacteria engage in symbiotic relationships with fungi and plants. Plants such as cycads and *Gunnera* species that host nitrogen-fixing cyanobacteria derive benefit from access to nitrogen compounds provided by their

Fig. 3.59. Cyanobacterial bloom. Dungeness, Kent, UK.

Fig. 3.60. Loch Neagh, a water body prone to algal blooms. Northern Ireland.

cyanobacterial symbionts. A lichen is a symbiotic consortium composed of a fungus and a photosynthetic partner. The partner may be either a green alga or a cyanobacterium. Some lichens include two partners; a true alga whose primary role is photosynthetic as well as a nitrogen-fixing cyanobacterium. Many lichen-associated cyanobacteria belong to the genus *Nostoc*. Lichen-associated

Fig. 3.61. Cyanobacterial bloom seen from the air. Lake Victoria (Nyanza), Kenya.

symbiotic strains are sometimes referred to as *Nostoc punctiforme*. Numerous free-living species of nostoc have been described on the basis of morphological differences and habitat preferences. *Nostoc commune* is a terrestrial species found in calcareous habitats. *Nostoc commune* is generally unobtrusive but jelly-like colonies of nostoc may become visible on grassland, stony surfaces and garden paths during periods of wet weather (Figs. 1.21–1.23, 3.62–3.64). *Nostoc pruniforme*, also known as Mare's Eggs, is an aquatic species, colonies of which may grow into large gelatinous spheroids. *Nostoc verrucosum* is a freshwater species that occurs on rocks in streams and waterfalls (Fig. 2.4).

Tufa

Tufa is a soft porous limestone that develops in response to the precipitation of calcium carbonate in the vicinity of springs, streams, seepages and waterfalls. Tufa deposition creates a special kind of habitat, populated by characteristic communities of bacteria, algae, mosses and other plants. In carbonate-rich waters, bacteria, algae and bryophytes induce mineral precipitation and mediate patterns of carbonate deposition. Tufa formation has been linked in particular with the calcification of biofilms composed of cyanobacteria (Zippel and Neu, 2011). Cells of cyanobacteria and fragments of plants, such as stems and leaves become coated in calcium carbonate and become incorporated within the tufa matrix. In marl lakes and calcareous fens algae belonging to the genus *Chara* also act as important substrates for carbonate precipitation. Over the course

Fig. 3.62. Slimy flagstones, coated with microscopic cyanobacteria and fungi. Upper Beeding, Sussex, UK.

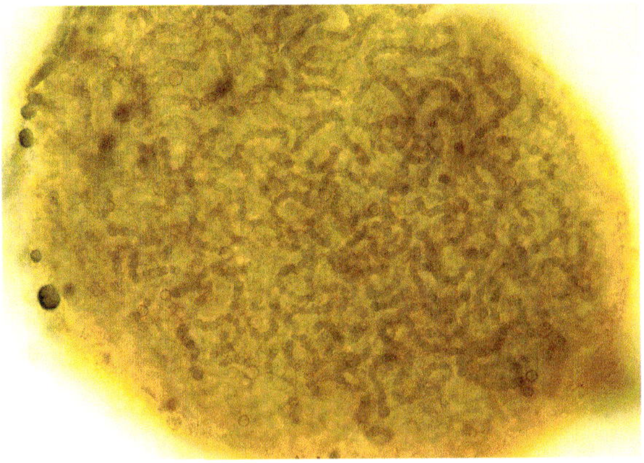

Fig. 3.63. Microscopic ball of nostoc seen in scraping from the patio shown in Fig. 3.62.

of time, organic matter entombed in tufa decomposes to create a complex network of rot holes of various sizes. The final product is a light, sponge-like limestone. Tufa formation depends on a supply of water that is rich in carbonate, derived from dissolved limestone. Limestone is susceptible to dissolution by carbonic acid, which originates in soils when rain water absorbs carbon dioxide from the atmosphere. The erosion of limestone is further accelerated by bacteria and fungi which form biofilms on rock surfaces, penetrate the rocks along lines of weakness and release organic acids as waste products of metabolism. Water that percolates through limestone becomes saturated with dissolved

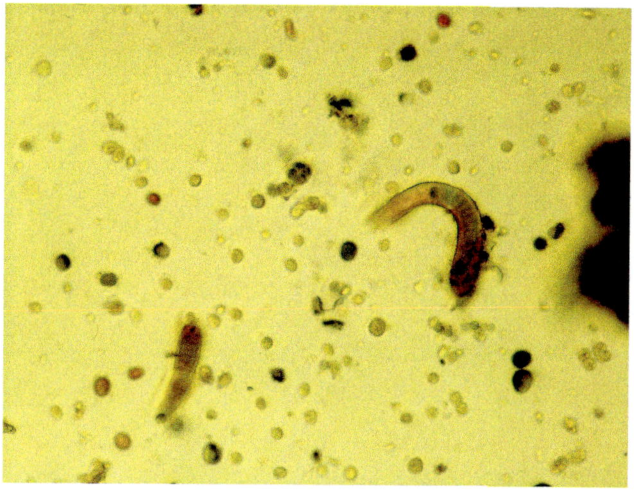

Fig. 3.64. Cyanobacteria and other microbes from the patio shown in Fig. 3.62.

calcium carbonate. When carbonate-saturated water re-emerges at the surface, carbon dioxide is lost to the atmosphere, which results in the precipitation of calcium carbonate. The rate of mineral precipitation is accelerated by the uptake of carbon dioxide by photosynthetic cyanobacteria, algae and plants (Zhu and Dittrich, 2016). Furthermore, the cells of cyanobacteria, algae and plants act as templates for calcium carbonate precipitation.

Signs of carbonate precipitation can be seen in limestone districts throughout the British Isles (Fig. 3.65). Areas that are known for tufa formation include the Craven District of Yorkshire, parts of Wales, Derbyshire, Gloucestershire, Somerset, Lismore Island and calcareous areas of Ireland such as the Burren in County Clare (Pentecost, 1981; Baker and Simms, 1998; Farr *et al.*, 2014; Lyons and Kelly, 2016; Farr and Graham, 2017; Faulkner and Crae, 2018). In the area around Malham Cove, Yorkshire, numerous and varied tufa features occur in streams, around waterfalls and on lake margins (Fig. 3.66). Fast-flowing turbulent streams increase contact between carbonate-rich water and the air leading to calcium carbonate precipitation. Streams with tufa formation include Gordale Beck, Waterfall Beck and Cote Gill. As well as tufa hummocks in streams, there are tufa dams and around the margins of Malham Tarn the surfaces of stones are coated with thin spongy tufa crusts. Spongy microbialite crusts are a feature of the marl lakes of the limestone districts of Ireland (Doddy *et al.*, 2019). During the summer months at Lough Gealain in the Burren, thick orange and green sponge-like cyanobacterial mats accumulate along the shoreline (Fig. 3.67). Crusted stones represent another manifestation of microbial activity which can be seen around lough margins in the Burren (Fig. 3.68). In the German language such stones are referred to as *Krustensteine*. Ecologists use

Fig. 3.65. Tufa dam in a stream at Hinksey Heights near Oxford, UK.

the word krustenstein as a technical term for marl stones, or crust stones. The surfaces of these crust stones are intricately marked: they are continuously reworked by microbes and aquatic invertebrates. The microbial community associated with krustenstein includes cyanobacteria belonging to the genus *Rivularia*. Colonies of these filamentous cyanobacteria induce the growth of little pimples of calcium carbonate that speckle rock surfaces and stones in streams and on lake margins. The endolithic cyanobacterium *Schizothrix calicola* (the Limestone Hair-splitter) grows in abundance at sites of active tufa formation. Filaments of *Schizothrix calicola* grow within gelatinous sheaths beneath the surface of the developing tufa. Cyanobacteria produce carotenoid pigments which protect them against damage from exposure to ultraviolet radiation (Llewellyn *et al.*, 2020). Pale orange cyanobacterial pigmentation makes sites of active tufa formation stand out from the surrounding rocks and vegetation.

My first experience of tufa as a habitat occurred by chance when during the course of a summer holiday I cycled to Aust Cliff to look for

Fig. 3.66. Tufa screen at Gordale Scar, Yorkshire, UK.

fossils. Before reaching the fossiliferous section of the cliff my gaze was drawn to a multicoloured dripping mass of overhanging vegetation, strands of which appeared to be turning to stone (Fig. 3.69). I ended up spending more time picking my way through the reeds and brambles to investigate this mysterious tufaceous backwater than in looking for fossils. The tufa cascade at Aust is still much as it was when I first saw it. It is one of a number of sites that feature in a recent survey of 'petrifying springs with tufa formation' in Gloucestershire (Farr and Graham, 2017). The tufaceous seepages at Aust Cliff present the observer with a vertical patchwork of moist yellowing calcium carbonate deposits and luxuriant mosses, among which are the Whorled Tufa-moss, *Eucladium verticillatum* and Olive Beard-moss, *Didymodon tophaceus* (Farr and Graham, 2017). In Gloucestershire, which is particularly rich in tufa springs, a number of different limestone formations converge to act as sources of carbonate (Figs. 3.70–3.71). At Aust, the local source of calcium carbonate is a layer of Liassic limestone, whilst not so far away there are also some exposures of Carboniferous limestone. Further inland are the oolitic limestones of the

Fig. 3.67. Spongy microbial crust, Lough Gealain, County Clare, Ireland.

Fig. 3.68. Crust stone at Lough Bunny, County Clare, Ireland.

Cotswold Hills. In the Cotswolds there are numerous steep-sided valleys that encourage the formation of seepages, tufa springs, tufa streams and tufa dams.

The archaeologist and publisher John Henry Parker (1806–1884) noted that: 'Some of the streams in Gloucestershire deposit tufa, and it is there known by the name of puff-stone' (Parker, 1845). Puff-stone was quarried at

Fig. 3.69. Tufa mound. Aust, Gloucestershire, UK.

Fig. 3.70. Tufa with mosses and liverworts. Lower Woods, Gloucestershire, UK.

Fig. 3.71. Tufa mound. Chedworth, Gloucestershire, UK.

Dursley and used in the construction of Berkeley Castle, Gloucester Cathedral and churches at Dursley and Cam (Lewis, 1848). Tufa is strong but also porous and less dense than other building stones, making it suitable for vaulting. Blocks of tufa can also be seen incorporated into the walls of buildings, such as at the church of St Andrew and St Cuthman in Steyning, West Sussex (Fig. 3.72). Blocks of tufa feature in the walls of the second century Roman lighthouse at Dover in Kent.

Tufa and travertine

Confusingly, some authors make no distinction between tufa and travertine. In a glossary of architectural terms, John Henry Parker described tufa as 'a porous stone deposited by calcareous waters' and added that 'when compact it is called Travertine' (Parker, 1845). Authors continue to use the terms tufa, to describe soft porous rocks, and travertine to describe denser forms, especially those that are more suitable for use as building stone (Pentecost and Viles, 1994). Others reserve the word travertine to refer to carbonate deposits associated with hot springs, such as those at Tivoli, near Rome. The original *lapis tiburtinus* (travertine) of the Romans was quarried at Tibur near Rome, a settlement that later became Tivoli. Some deposits referred to as travertine represent a recrystallised form of tufa (Pentecost, 1981). The term travertine is also used by some authors to refer to deposits of calcium carbonate that have been precipitated in cave systems.

Fig. 3.72. Tufa block used as a building material. Steyning Church, West Sussex, UK.

Plants and animals found in association with tufa

Curled Hook-moss, *Palustriella commutata* (previously known as *Cratoneuron commutatum*) is the predominant plant species at many tufa sites. The close correlation of Curled Hook-moss with tufa springs was recognised in the drafting of the Habitats Directive of the European Union, which includes this species among the defining features of a type of habitat designated as 'H7220 Petrifying springs with tufa formation (*Cratoneurion*)'. Other plants found in association with tufa include Pendulous Sedge, *Carex pendula*, Greater Horsetail, *Equisetum telmateia*, Hart's-tongue fern, *Asplenium scolopendrium* and Hemp-agrimony, *Eupatorium cannabinum* (Farr and Graham, 2017). In calcareous fenland and in upland areas of the British Isles, Black Bog-rush, *Schoenus nigricans* and Common Butterwort, *Pinguicula vulgaris* occur in association with tufa springs (Fig. 3.73). In Ireland, Large-flowered Butterwort, *Pinguicula grandiflora* may be present (Lyons and Kelly, 2016). In northern England the Bird's-eye Primrose, *Primula farinosa* is a characteristic member of the flora that can be found in association with tufa seepages.

Seepages associated with tufa-formation provide important refuges for molluscs, water beetles, crane flies and soldier flies. In Ireland, the water beetle, *Ochthebius nilssoni* ekes out an existence by foraging on the surfaces of microbial biofilms which encrust the margins of marl lakes in the Burren and at Lough Carra in County Mayo. *Ochthebius nilssoni* is otherwise known from Scandinavia and has not been recorded from Britain. The larva of the rare

Fig. 3.73. Butterwort, *Pinguicula vulgaris*, Lismore Island, UK.

soldier fly, *Stratiomys chamaeleon* inhabits tufa-rich seepages in calcareous fens (Drake, 2017). *Stratiomys chamaeleon* is known currently only from a few widely separated sites in Anglesey, Black Isle and from Cothill near Oxford. The larva is a detritivore whose diet includes bacteria. Cothill Fen also harbours specialised plant species, including Grass of Parnassus, *Parnassia palustris* and Marsh Helleborine, *Epipactis palustris*.

Living tufa has its own particular aesthetic, which is contingent upon its sponge-like texture, the orange and yellow carotenoid pigmentation supplied by its resident cyanobacteria, the green mosses that colonise its glistening surface, the sound of trickling water and the scenery that one associates with limestone country.

References

Baker, A. and Simms, M.J. (1998) Active deposition of calcareous tufa in Wessex, UK, and its implications for the "late-Holocene tufa decline". *The Holocene*, 8, 359–365.

Baker, K.L., Langenheder, S., Nicol, G.W., Ricketts, D., Killham, K. *et al.* (2009) Environmental and spatial characterisation of bacterial community composition in soil to inform sampling strategies. *Soil Biology and Biochemistry*, 41, 2292–2298.

Bartram, A.K., Jiang, X., Lynch, M.D.J., Masella, A.P., Nicol, G.W. *et al.* (2014) Exploring links between pH and bacterial community composition in soils from the Craibstone Experimental Farm. *FEMS Microbiology Ecology*, 87, 403–415.

Bellini, S. (2009) On a unique behaviour of freshwater bacteria. *Chinese Journal of Oceanology and Limnology*, 27, 3–5.

Blakemore, R. (1975) Magnetotactic bacteria. *Science*, 190, 377–379.

Blackwell, N., Perkins, W., Palumbo-Roe, B., Bearcock, J., Lloyd, J.R. *et al.* (2019) Seasonal blooms of neutrophilic betaproteobacterial Fe(II) oxidisers and Chlorobi in iron-rich coal mine drainage sediments. *FEMS Microbiology Ecology*, 95 (10), fiz140.

Bunting, L., Leavitt, P.R., Gibson, C.E., McGee, E.J. and Hall, V. (2007) Degradation of water quality in Lough Neagh, Northern Ireland, by diffuse nitrogen flux from a phosphorus-rich catchment. *Limnology and Oceanography*, 52, 354–369.

Burrell, P.C., Phalen, C.M. and Hovanec, T.A. (2001) Identification of bacteria responsible for ammonia oxidation in freshwater aquaria. *Applied and Environmental Microbiology*, 67, 5791–5800.

Carmichael, W.W., Azevedo, S.M.F.O., An, J.S., Molica, R.J.R., Jochimsen, E.M. *et al.* (2001) Human fatalities from cyanobacteria: chemical and biological evidence for cyanotoxins. *Environmental Health Perspectives*, 109, 663–668.

Carmichael, W.W. and Li, R. (2006) Cyanobacteria toxins in the Salton Sea. *Saline Systems*, 2, 5.

Carson, M.A., Bräuer, S. and Basiliko, N. (2019) Enrichment of peat yields novel methanogens: approaches for obtaining uncultured organisms in the age of rapid sequencing. *FEMS Microbiology Ecology*, 95, 2.

Clymo, R.S. (2015) Why are there few gas bubbles in deep peat in British raised and blanket peat bogs? *Mires and Peat*, 16, 1–20.

Corstjens, P.L., de Vrind, J.P., Westbroek, P. and de Vrind-de Jong, E.W. (1992) Enzymatic iron oxidation by *Leptothrix discophora*: identification of an iron-oxidizing protein. *Applied and Environmental Microbiology*, 58, 450–454.

Dacey, J.W. and Klug, M.J. (1979) Methane efflux from lake sediments through water lilies. *Science*, 203, 1253–1255.

Daims, H., Lebedeva, E.V., Pjevac, P., Han, P., Herbold, C. *et al.* (2015) Complete nitrification by *Nitrospira* bacteria. *Nature*, 528, 504–509.

de Vries, F.T., Manning, P., Tallowin, J.R.B., Mortimer, S.R., Pilgrim, E.S. *et al.* (2012) Abiotic drivers and plant traits explain landscape-scale patterns in soil microbial communities. *Ecology Letters*, 15, 1230–1239.

Dean, A.P., Hartley, A., McIntosh, O.A., Smith, A., Feord, H.K. *et al.* (2019) Metabolic adaptation of a *Chlamydomonas acidophila* strain isolated from acid mine drainage ponds with low eukaryotic diversity. *Science of the Total Environment*, 647, 75–87.

Dedysh, S.N., Panikov, N.S., Liesack, W., Großkopf, R., Zhou, J. *et al.* (1998) Isolation of acidophilic methane-oxidizing bacteria from northern peat wetlands. *Science*, 282, 281–284.

Demory, F., Oberhänsli, H., Nowaczyk, N. R., Gottschalk, M., Wirth, R. *et al.* (2005) Detrital input and early diagenesis in sediments from Lake Baikal revealed by rock magnetism. *Global and Planetary Change*, 46, 1–4.

Ding, L., Tian, L., Li, J., Zhang, Y., Wang, M. and Wang, P. (2023) Grazing lowers soil multifunctionality but boosts soil microbial network complexity and stability in a subtropical grassland of China. *Frontiers in Microbiology*, 13, 1027097.

Ding, Y., Vrionis, H.A., Schneider, J., Berezuk, A., Khursigara, C.M. *et al.* (2016) Complementation of an aglB mutant of *Methanococcus maripaludis* with heterologous oligosaccharyltransferases. *PLoS ONE*, 11 (12), e0167611.

Doddy, P., Roden, C.M. and Gammell, M.P. (2019) Microbialite crusts in Irish limestone lakes reflect lake nutrient status. *Biology and Environment: Proceedings of the Royal Irish Academy*, 119B. 1, 1–11.

Drake, C.M. (2017) A review of the status of larger Brachycera flies of Great Britain – Species Status No. 29. Natural England Commissioned Reports, Number192.

Dunal, M.F. (1837) Note sur les algues qui colourent en rouge certaines eaux des marais salants méditerranéens. *Comptes Rendus de l'Académie des Sciences, Séries III, Sciences de la Vie*, 15, 585–587.

Dyer, B.D. (2003) A field guide to bacteria. Cornell, Ithaca, New York.

Evans, C., Artz, R., Moxley, J., Smyth, M-A., Taylor, E. *et al.* (2017) Implementation of an emission inventory for UK peatlands. Report to the Department for Business, Energy and Industrial Strategy. Centre for Ecology and Hydrology, Bangor, UK.

Evans, C., Morrison, R., Burden, A., Williamson, J., Baird, A. *et al.* (2017) Interim report on project SP1210: Lowland peatland systems in England and Wales – evaluating greenhouse gas fluxes and carbon balances. Report to Defra, Centre for Ecology and Hydrology, Bangor, UK.

Farr, G., Graham, J. and Stratford, C. (2014) Survey, characterisation and condition assessment of *Palustriella* dominated springs 'H7220 Petrifying springs with tufa formation (*Cratoneurion*)' in Wales. Centre for Ecology and Hydrology and the British Geological Survey (NERC).

Farr, G. and Graham, J. (2017) Survey, characterisation and condition assessment of *Palustriella* dominated springs 'H7220 Petrifying springs with tufa formation (*Cratoneurion*)' in Gloucestershire, England. *British Geological Survey*, OR/17/020.

Faulkner, T. and Crae, J. (2018) Distribution of tufa and speleothem deposits on the island of Lismore, Argyll, Scotland. *Cave and Karst Science*, 45, 101–110.

Fierer, N. and Jackson, R.B. (2006) The diversity and biogeography of soil bacterial communities. *Proceedings of the National Academy of Sciences of the United States of America*, 103, 626–631.

Fleming, E.J., Woyke, T., Donatello, R.A., Kuypers, M.M.M., Sczyrba, A. *et al.* (2018) Insights into the fundamental physiology of the uncultured Fe-oxidizing bacterium *Leptothrix ochracea. Applied and Environmental Microbiology*, 84, e02239-17.

Frankel, R.B. (2009) The discovery of magnetotactic/magnetosensitive bacteria. *Chinese Journal of Oceanology and Limnology*, 27, 1–2.

Freitag, T.E. and Prosser, J.I. (2009) Correlation of methane production and functional gene transcriptional activity in a peat soil. *Applied and Environmental Microbiology*, 75, 6679–6687.

Galand, P.E., Fritze, H., Conrad, R. and Yrjälä, K. (2005) Pathways for methanogenesis and diversity of methanogenic archaea in three boreal peatland ecosystems. *Applied and Environmental Microbiology*, 71, 2195–2198.

Gans, J., Woilinsky, M. and Dunbar, J. (2005) Computational improvements reveal great bacterial diversity and high metal toxicity in soil. *Science*, 309, 1387–1390.

Goddard, V.J., Baker, A.C., Davy, J.E., Adams, D.G., De Ville, M.M. *et al.* (2005). Temporal distribution of viruses, bacteria and phytoplankton throughout the water column in a freshwater hypereutrophic lake. *Aquatic Microbial Ecology*, 39, 211–223.

Grau, R.R., Limansky, A.S., Ricci, J.C.D. and de Mendoza, D. (1991) The Winogradsky column: a simple and inexpensive approach to teach environmental biochemistry. *Biochemical Education*, 19, 143–147.

Gray, N., Hastings, R., Sheppard, S., Loughnane, P., Lloyd, D. *et al.* (2003) Effects of soil improvement treatments on bacterial community structure and soil processes in an upland grassland soil. *FEMS Microbiology Ecology*, 46, 11–22.

Griffiths, R.I., Thomson, B.C., James, P., Bell, T., Bailey, M.J. *et al.* (2011) The bacterial biogeography of British soils. *Environmental Microbiology*, 13, 1642–1654.

Hanzlik, M., Winklhofer, M. and Petersen, N. (1996) Spatial arrangement of chains of magnetosomes in magnetotactic bacteria. *Earth and Planetary Science Letters*, 145, 125–134.

He, H., Zhen, Y., Mi, T., Fu, L. *et al.* (2018) Ammonia-oxidizing archaea and bacteria differentially contribute to ammonia oxidation in sediments from adjacent waters of Rushan Bay, China. *Frontiers in Microbiology*, 9, 116.

Herbert, R.A. (1985) Development of mass blooms of photosynthetic bacteria on sheltered beaches in Scapa Flow, Orkney Islands. *Proceedings of the Royal Society Edinburgh, Section B, Biological Sciences*, 87, 15–25.

Hubas, C., Jesus B., Ruivo M., Meziane T. and Thiney N. (2013) Proliferation of purple sulphur bacteria at the sediment surface affects intertidal mat diversity and functionality. *PLoS ONE*, 8, e82329.

Hubas, C., Boeuf, D., Jesus, B., Thiney, N., Bozec, Y. *et al.* (2017) A nanoscale study of carbon and nitrogen fluxes in mats of purple sulfur bacteria: implications for carbon cycling at the surface of coastal sediments. *Frontiers in Microbiology*, 8, 1995.

Jasso-Chávez, R., Santiago-Martínez, M.G., Lira-Silva, E., Pineda, E., Zepeda-Rodríguez, A. *et al.* (2015). Air-adapted *Methanosarcina acetivorans* shows high methane production and develops resistance against oxygen stress, *PLoS ONE*, 10 (2), e0117331.

John, D.M., Whitton, B.A. and Brook, A.J. (eds.) (2002) The freshwater algal flora of the British Isles: An identification guide to freshwater and terrestrial algae. Cambridge University Press, Cambridge, UK.

Kielak, A.M., Cipriano, M.A. and Kuramae, E.E. (2016) Acidobacteria strains from subdivision 1 act as plant growth-promoting bacteria. *Archives of Microbiology*, 198, 987–993.

Kiene, R.P., Linn, L.J. and Bruton, J.A. (2000) New and important roles for DMSP in marine microbial communities. *Journal of Sea Research*, 43, 209–224.

Kishimoto, N. and Tano, T. (1987) Acidophilic heterotrophic bacteria isolated from acidic mine drainage, sewage, and soils. *Journal of General and Applied Microbiology*, 33, 11–25.

Knief, C. (2015) Diversity and habitat preferences of cultivated and uncultivated aerobic methanotrophic bacteria evaluated based on *pmoA* as molecular marker. *Frontiers in Microbiology*, 6, 1346.

Koch, H., van Kessel, M.A.H.J. and Lücker, S. (2019) Complete nitrification: insights into the ecophysiology of comammox *Nitrospira*. *Applied Microbiology and Biotechnology*, 103, 177–189.

Köhler, J.M., Beetz, N., Günther, P.M., Möller, F., Schüler, T. *et al.* (2020) Microbial community types and signature-like soil bacterial patterns from fortified prehistoric hills of Thuringia (Germany). *Community Ecology*, https://doi.org/10.1007/s42974-020-00017-4.

Kondo, R., Purdy, K.J., Silva, S.D.Q. and Nedwell, D.B. (2007) Spatial dynamics of sulphate-reducing bacterial compositions in sediment along a salinity gradient in a UK estuary. *Microbes and Environments*, 22, 11–19.

Könneke, M., Bernhard, A.E., de la Torre, J.R., Walker, C.B., Waterbury, J.B. *et al.* (2005) Isolation of an autotrophic ammonia-oxidizing marine archaeon. *Nature*, 437, 543–546.

Langenheder, S. and Prosser, J.I. (2008) Resource availability influences the diversity of a functional group of heterotrophic soil bacteria. *Environmental Microbiology*, 10, 2245–2256.

Lankester, E. (1842) An account of Askern and its mineral springs. John Churchill, London, UK.

Lankester, R. (1873) On a peach-coloured bacterium – *Bacterium rubescens* n.s. *Quarterly Journal of Microscopic Science*, 13, 408–425.

Larsen, K. and Olsen, S. (1950) Ochre suffocation of fish in the river Tim Aa. *Report of the Danish Biological Station*, 6, 3–27.

Lauber, C.L., Hamady, M., Knight, R. and Fierer, N. (2009) Pyrosequencing-based assessment of soil pH as a predictor of soil bacterial community structure at the continental scale. *Applied and Environmental Microbiology*, 75, 5111–5120.

Le Nen, P. (2009) En Bretagne, après la mort d'un cheval, les marées vertes sont montrées du doigt. *Le Monde*, 8 August 2009.

Lefèvre, C.T. and Bazylinski, D.A. (2013) Ecology, diversity, and evolution of magnetotactic bacteria. *Microbiology and Molecular Biology Reviews*, 77, 497–526.

Lewis, S. (1848) A Topographical Dictionary of England. Samuel Lewis, London, UK.

Llewellyn, C.A., Airs, R.L., Farnham, G. and Greig, C. (2020) Synthesis, regulation and degradation of carotenoids under low level UV-B radiation in the filamentous cyanobacterium *Chlorogloeopsis fritschii* PCC 6912. *Frontiers in Microbiology*, 11, 163.

Lowenstam, H.A. (1962) Magnetite in denticle capping in recent chitons (Polyplacophera). *Geological Society of America Bulletin*, 73, 435.

Lozupone, C.A. and Knight, R. (2007) Global patterns in bacterial diversity. *Proceedings of the National Academy of Sciences of the United States of America*, 104, 11436–11440.

Lyons, M.D. and Kelly, D.L. (2016) Monitoring guidelines for the assessment of petrifying springs in Ireland. Irish Wildlife Manuals, No. 94. National Parks and Wildlife Service, Department of Arts,. Heritage, Regional, Rural and Gaeltacht Affairs, Ireland.

Männistö, M.K., Kurhela, E., Tiirola, M. and Häggblom, M.K. (2013) *Acidobacteria* dominate the active bacterial communities of Arctic tundra with widely divergent winter-time snow accumulation and soil temperatures, *FEMS Microbiology Ecology*, 84, 47–59.

McDonald, I.R., Hall, G.H., Pickup, R.W. and Murrell, J.C. (1996) Methane oxidation potential and preliminary analysis of methanotrophs in blanket bog peat using molecular ecology techniques, *FEMS Microbiology Ecology*, 21, 197–211.

McDonald, I., Upton, M., Hall, G., Pickup, R.W., Edwards, C. *et al.* (1999). Molecular ecological analysis of methanogens and methanotrophs in blanket bog peat. *Microbial Ecology*, 38, 225–233.

Montgomery, P. (1994) The magnetostratigraphy of Coniacian-Late Campanian chalk sequences of Southern England. University of Southampton, Oceanography, Doctoral Thesis.

Montgomery, P., Hailwood, E.A., Gale, A.S. and Burnett, J.A. (1998) The magnetostratigraphy of Coniacian late Campanian chalk sequences in southern England. *Earth Planetary Science Letters*, 156, 209–224.

Nedwell, D.B., Embley, T.M. and Purdey, K.J. (2004) Sulphate reduction, methanogenesis and phylogenetics of the sulphate reducing bacterial communities along an estuarine gradient. *Aquatic Microbial Ecology*, 37, 209–217.

Nickless, E.F.P., Booth, S.J. and Mosley, P.N. (1976) The celestite resources of the area north-east of Bristol. Mineral Assessment Report 25. Institute of Geological Sciences. HMSO, London.

Nottingham, A.T., Fierer, N.,Turner, B.L.,Whitaker, J., Ostle, N.J. *et al.* (2018). Microbes follow Humboldt: Temperature drives plant and soil microbial diversity patterns from the Amazon to the Andes. *Ecology*, 99, 2455–2466.

Orwin, K.H., Buckland, S.M., Johnson, D., Turner, B.L., Smart, S. *et al.* (2010), Linkages of plant traits to soil properties and the functioning of temperate grassland. *Journal of Ecology*, 98, 1074–1083.

Parker, J.H. (1845) A glossary of terms used in Grecian, Roman, Italian, and Gothic architecture (4th ed. enlarged). John Henry Parker, Oxford, UK.

Parkes, R.J., Gibson, G.R., Mueller-Harvey, I., Buckingham W.J. and Herbert R. (1989) Determination of the substrates for sulphate-reducing bacteria within marine and estuarine sediments with different rates of sulphate reduction. *Journal of General Microbiology*, 135, 175–187.

Pearson, M.J., Ferguson, A.J.D., Codd, G.A., Reynolds, C.S, Hamilton, R.M. *et al.* (1990) Toxic blue-green algae. a report by the UK National Rivers Authority. Water Quality Series No. 2.

Peck, J.A. and King, J.W. (1996) Magnetofossils in the sediment of Lake Baikal, Siberia. *Earth and Planetary Science Letters*, 140, 159–172.

Pentecost, A. (1981) The tufa deposits of the Malham district, North Yorkshire. *Field Studies*, 5, 365–387.

Pentecost, A. and Viles, H. (1994) A review and reassessment of travertine classification. *Géographie physique et Quaternaire*, 48, 305–314.

Puente-Sánchez, F., Arce-Rodríguez, A., Oggerin, M., García-Villadangos, M., Moreno-Paz, M. *et al.* (2018). Viable cyanobacteria in the deep continental subsurface. *Proceedings of the National Academy of Sciences of the United States of America*, 115, 10702–10707.

Raynaud, X. and Nunan, N. (2014) Spatial ecology of bacteria at the microscale in soil. *PLoS ONE*, 9, e87217.

Roth, A.W. (1797) Catalecta botanica quibus plantae novae et minus cognitae describuntur atque illustrantur. I.G. Müller, Leipzig, Germany.

Rousk, J., Brookes, P.C. and Bååth, E. (2010) The microbial PLFA composition as affected by pH in an arable soil. *Soil Biology and Biochemistry*, 42, 516–520.
Samuel, B.S. and Gordon, J.I. (2006) A humanized gnotobiotic mouse model of host-archaeal-bacterial mutualism. *Proceedings of the National Academy of Sciences of the United States of America*, 103, 10011–10016.
Sauder, L.A., Engel, K., Stearns, J.C., Masella, A.P., Pawliszyn, R. *et al.* (2011) Aquarium nitrification revisited: *Thaumarchaeota* are the dominant ammonia oxidizers in freshwater aquarium biofilters. *PLoS ONE*, 6, e23281.
Scott, P.W., Ealey, P.J. and Rollinson, G.K. (2011) Bog iron ore from Lowland Point, St Keverne, Lizard, Cornwall. *Geoscience in South-West England*, 12, 9.
Sitoki, L., Kurmayer, R. and Rott, E. (2012). Spatial variation of phytoplankton composition, biovolume, and resulting microcystin concentrations in the Nyanza Gulf (Lake Victoria, Kenya). *Hydrobiologia*, 691, 109–122.
Skvortsov, T., de Leeuwe, C., Quinn, J.P., McGrath, J.W., Allen, C.C.R. *et al.* (2016) Metagenomic characterisation of the viral community of Lough Neagh, the largest freshwater lake in Ireland. *PLoS ONE*, 11, e0150361.
Stanier, R.Y, Sistrom, W.R., Hansen, T.A., Whitton, B.A., Castenholz, R.W. *et al.* (1978) Proposal to place the nomenclature of the cyanobacteria (Blue-Green Algae) under the rules of the International Code of Nomenclature of Bacteria. *International Journal of Systematic Bacteriology*, 28, 335–336.
Stevenson, R.C.R. and Masson, J. (2016).*Cladopodiella fluitans* and iron deposition on Dersingham Bog National Nature Reserve. *Field Bryology*, 116, 22–27.
Tank, M., Blümel, M. and Imhoff, J.F. (2011) Communities of purple sulfur bacteria in a Baltic Sea coastal lagoon analyzed by *pufLM* gene libraries and the impact of temperature and NaCl concentration in experimental enrichment cultures. *FEMS Microbiology Ecology*, 78, 428–438.
Trimmer, M., Purdy, K.J. and Nedwell, D.B. (1997) Process measurement and phylogenetic analysis of the sulfate reducing bacterial communities of two contrasting benthic sites in the upper estuary of the Great Ouse, Norfolk, UK. *FEMS Microbiology Ecology*, 24, 333–342.
Turner, P.C., Gammie, A.J., Hollinrake, K. and Codd, G.A. (1990) Pneumonia associated with cyanobacteria. *British Medical Journal*, 300, 1440–1441.
Turner, A.D., Dhanji-Rapkova, M., O'Neill, A., Coates, L., Lewis, A. *et al.* (2018). Analysis of microcystins in cyanobacterial blooms from freshwater bodies in England. *Toxins*, 10, 39.
Ueno, Y., Yamada, K., Yoshida, N., Maruyama, S. and Isozaki, Y. (2006) Evidence from fluid inclusions for microbial methanogenesis in the early Archaean era. *Nature*, 440, 516–519.
van Gemerden, H., De Wit, R., Tughan, C.S. and Herbert, R.A. (1989) Development of mass blooms of *Thiocapsa roseopersicina* on sheltered beaches on the Orkney Islands. *FEMS Microbiology Ecology*, 62, 111–118.
van Kessel, M.A.H.I., Speth, D.R., Albertsen, M., Nielsen, P.H., Op den Camp, H.J.M. *et al.* (2015) Complete nitrification by a single microorganism. *Nature*, 528, 555–559.
van Winden, J.F., Reichart, G.-J., McNamara, N.P., Benthien, A. and Damsté, J.S.S. (2012) Temperature-induced increase in methane release from peat bogs: a mesocosm experiment. *PLoS ONE*, 7, e39614.
Vidal, F., Sedan, D., D'Agostino, D., Cavalieri, M.L., Mullen, E. *et al.* (2017) Recreational exposure during algal bloom in Carrasco Beach, Uruguay: a liver failure case report. *Toxins*, 9, 267.
Webster, G., Rinna, J., Roussel, E.G., Fry, J.C., Weightman, A.J. *et al.* (2010) Prokaryotic functional diversity in different biogeochemical depth zones in tidal sediments of the Severn Estuary, UK, revealed by stable-isotope probing. *FEMS Microbiology Ecology*, 72, 179–197.
Wellsbury, P., Herbert, R.A. and Parkes, R.J. (1996) Bacterial activity and production in near-surface estuarine and freshwater sediments, *FEMS Microbiology Ecology*, 19, 203–214.

Whetton, J.T. and Myers, J.O. (1951) Geophysical survey of magnetite deposits in Strath, Island of Skye. *Transactions of the Geological Society of Glasgow*, 21, 263–277.
Winogradsky, S. (1949) Microbiologie du sol: problèmes et méthodes: cinquante ans de recherches. Masson et Cie, Paris, France.
Wright, D., Prickett, T., Helm, R.F. and Potts, M. (2001) Form species *Nostoc commune* (Cyanobacteria). *International Journal of Systematic and Evolutionary Microbiology*, 51, 1839–1852.
Yu, Z., Beilman, D. W., Frolking, S., MacDonald, G. M., Roulet, N. T. *et al.* (2011), Peatlands and their role in the global carbon cycle. *Eos Transactions AGU*, 92, 97.
Zhu, T. and Dittrich, M. (2016) Carbonate precipitation through microbial activities in natural environment, and their potential in biotechnology: A review. *Frontiers in Bioengineering and Biotechnology*, 4, 4.
Zinger, L., Shahnavaz, B., Baptist, F., Geremia, R. and Choler, P. (2009) Microbial diversity in alpine tundra soils correlates with snow cover dynamics. *ISME Journal*, 3, 850–859.
Zippel, B. and Neu, T.R. (2011) Characterization of glycoconjugates of extracellular polymeric substances in tufa-associated biofilms by using fluorescence lectin-binding analysis. *Applied and Environmental Microbiology*, 77, 505–516.

4 Extreme bacteria

> Whether lakes of brine, or those subterranean ones hidden beneath volcanic mountains—warm mineral springs—the wide expanse and depths of the ocean—the upper parts of the atmosphere, and even the surface of the perpetual snow—all support organic beings.
>
> Charles Darwin (1839). *The Voyage of the Beagle.*

Bacteria and archaea occupy spaces left vacant by other organisms. Most plants and animals thrive only in physical conditions that we would usually describe as being moderate and while many species of bacteria also flourish in conditions that can be regarded as moderate, there are also numerous species of bacteria and archaea that grow optimally under conditions that we might consider to be extreme. They are known as extremophiles. Extremophiles include psychrophiles, which thrive at low temperatures, thermophiles and hyperthermophiles that live at high and at very high temperatures, acidophiles that flourish in conditions of extreme acidity, halophiles that live in unusually salty habitats and alkaliphiles that require conditions of high alkalinity. Bacteria able to resist high pressures, exposure to ionising radiation, to desiccation and to metal toxicity and microbes that obtain food and energy independently of sources derived from sunlight extend the limits of the known biosphere deeper and higher than those of the zone that is inhabited by plants and animals. Extreme microbial habitats include hot springs, hydrothermal vents, deep oceanic sediments, petroleum reservoirs, cliff faces, landfill sites, salt lakes, soda lakes, frozen rocks and derelict copper mines. Bacteria and archaea inhabit all of them with aplomb. Uncluttered by plants and animals, these seemingly empty quarters perhaps provide a clearer introduction to microbial ecology and to the things that bacteria do in nature, than do more familiar habitats such as soils, grasslands and forests.

DOI: 10.1079/9781836993933.0004

Although we may not regard the natural environments of the British Isles as being particularly extreme, these islands do in fact provide numerous opportunities for the naturalist to engage with habitats that are frequented by extremophiles and to experience bacteria that find space to live in places that are not occupied by plants and animals. Even habitats that cannot be experienced at first hand, such as petroleum reservoirs, can be investigated indirectly by visiting natural oil seepages and commercial extraction sites. Show caves offer glimpses of subsurface environments. Acidophiles can be seen in the streams that drain heaps of abandoned mining waste. Halophiles dependent on high salt concentrations inhabit salt mines. Furthermore, an awareness that crystals of rock salt may contain living halophiles allows one's imagination to transform the tedium of an icy winter's drive along a gritted highway into a spectacle of microbial theatre.

In fact Britain's subterranean rock salt deposits represent the desiccated remnants of massive hypersaline water bodies which would once have shimmered with the bright photosynthetic pigments of innumerable halophiles. The halophiles are still with us except that now their presence is more subdued. Britain's geological record incorporates traces of other extreme environments. At Parys Mountain on Anglesey, sulphide deposits rich in metal ores represent the remaining vestiges of hydrothermal vents that once rose from an ancient seabed. The unusual geology at Parys Mountain now provides a home for acidophiles. Beneath a grassy field at Rhynie in Aberdeenshire are rocks that bear witness to the former existence of a steaming hot spring. It would have looked resplendent when surrounded by microbial mats composed of multicoloured thermophiles. When we visit these sites, we can imagine what these places were once like and consider how they might compare with some of the extreme microbial habitats that still exist elsewhere in the world today. Knowledge of the specialised microbes that occupy these spaces allows for a more complete understanding of microbial ecology.

Parys Mountain, Anglesey: past and present

Black smokers are towering structures that rise from the depths of the seafloor. Black smokers belch out superheated solutions of iron sulphide from hydrothermal vents. Sulphide-oxidising bacteria use chemical energy to fix carbon dioxide in order to build organic carbon compounds. In the abyssal depths where there is no sunlight, sulphide-oxidising bacteria are the primary producers. Their productivity supports complex ecosystems, populated by worms, molluscs and crustacea. A similar scene would have existed about 450 million years ago during the late Ordovician Period on a volcanic section of a seafloor that was later uplifted to become Parys Mountain (Mynydd Parys) on the island of Anglesey (Ynys Môn). Thick deposits of iron sulphide accumulate around black smokers. Such accumulations are referred to by geologists as volcanogenic massive sulphide deposits. They are rich in iron, copper, lead, zinc and other metals and form the basis of economically important ore deposits. As yet the metal ores at Parys Mountain are the only volcanogenic massive sulphide deposits to have been recognised as such in Britain's geological record. In Ireland, similar deposits occur at Avoca, County Wicklow (Platt, 1977; Younger and Potter, 2012; Hollis *et al.*, 2014). Parys Mountain was mined for copper during the bronze age and later during the eighteenth century it became the largest copper mine in the world (Johnson, 2012). Eventually mining operations came to a halt, although they may yet resume at some time in the future. The residues of pyrite (iron sulphide, $FeSu_2$) that remain in the flooded abandoned mine workings have made them a paradise for acidophiles. The metal toxicity and extremely acid conditions at Parys Mountain make it unsuitable for most plants and animals. There are some rare metal-tolerant lichens and scattered clumps of heather, *Calluna vulgaris*, where the occasional Mottled Grasshopper, *Myrmeleotettix maculatus* can be seen hopping about but there is not much else to see in the way of plants and animals amongst the heaps of strangely pigmented rocks etched bare by acid (Fig. 4.1). The pools that lie dotted about in this weirdly tinted landscape seem strangely inert and look as though they might be devoid of life but a drop of water placed under the microscope can be seen to be teeming with acidophilic bacteria (Fig. 1.14).

The acidophiles of Parys Mountain

Organisms that grow optimally at acidities of less than pH 3 can be described as extreme acidophiles and organisms that grow at acidities that range from pH 3–5 can be regarded as moderate acidophiles (Johnson, 2012). The acidity

Fig. 4.1. Parys Mountain, home of acidophiles. Anglesey, UK.

of water and sediments at Parys Mountain varies from between about pH 2–3. The flooded mine workings and generally low temperatures (8–9°C) provide an ideal habitat for *Acidithiobacillus ferrivorans* (the Cool Acid-bug) as well as for *Acidithiobacillus ferrooxidans* (the Iron Acid-bug). Unlike its congener, the Cool Acid-bug can grow at lower temperatures (as low to 4°C). In the presence of oxygen, the Cool Acid-bug and Iron Acid-bug obtain energy by oxidising iron. The Cool Acid-bug and Iron Acid-bug can oxidise sulphide as well as iron, both of which are abundantly available as pyrite (iron sulphide) at Parys Mountain. Ferric sulphate and sulphuric acid are released as waste products. In this way *Acidithiobacillus* bacteria contribute to the acidification of their own favoured habitat. Ferric sulphate and sulphuric acid also have the combined effect of leaching copper from copper ore which emerges as soluble copper sulphate in floodwater. When mining operations became quiescent at Parys Mountain, copper extraction continued for a while by exploiting a process known as *in situ* bioleaching. This involved allowing the abandoned mine workings to flood at periodic intervals. Flooding promoted the production of ferric sulphate and sulphuric acid by *Acidithiobacillus* species, resulting in the liberation of soluble copper sulphate into the floodwater. From time to time the floodwater was drained into pools in order to harvest the copper. When this practice eventually ceased at some time in the early 1950s the neglected mine workings gradually became transformed into a voluminous subterranean acid lake. When it was realised that this posed a hazard to neighbouring communities, the lake's contents were pumped out and allowed to discharge into the Irish Sea (Coupland and Johnson, 2004). A colourful acidic stream continues to discharge metal-contaminated water from the former mine workings.

Despite its desolate appearance, Parys Mountain has a rich flora of microscopic bacteria and archaea. Iron-oxidising bacteria such as *Acidithiobacillus* species are abundant in surface waters where they can readily access oxygen. In addition to *Acidithiobacillus ferrooxidans* (the Iron Acid-bug) and its congener *Acidithiobacillus ferrivorans* (the Cool Acid-bug), which are members of the Phylum *Pseudomonadota*, bacteria from several other phyla have found a home on Parys Mountain. Many of them are iron-oxidising acidophiles (Table 4.1). One of them is *Ferromicrobium acidiphilum* (the Acid Iron Microbe), a Gram-positive bacterium and member of the Phylum *Actinomycetota* (Johnson *et al.*, 2009). The Acid Iron Microbe grows as a mat of tangled filaments. In the presence of oxygen, *Ferromicrobium acidiphilum* gains energy by oxidising iron. In anoxic conditions it can switch to iron reduction. It is tolerant of high concentrations of copper. *Leptospirillum ferrooxidans* (the Coiled Iron-bug) is another iron-oxidising acidophile and a member of the Phylum *Nitrospirota*. The Coiled Iron-bug has a distinctively spiral morphology.

The Afon Goch (red river) is a brightly coloured stream of highly acidic (pH 2.5) effluent, which emerges from a mine adit portal at the foot of Parys Mountain. The Afon Goch derives its water supply from the flooded subterranean mine workings of the mountain. The Afon Goch contains a toxic cocktail of copper and other metal ions. The hostile conditions of the Afon Goch exclude the bulk of animal and plant species that one might usually associate with freshwater streams, making it an ideal location for the observation of

Table 4.1. Some acidophilic bacteria and archaea and their habitats.

Taxonomic group and species	Habitat
Domain *Bacteria*	
Phylum *Pseudomonadota*	
Acidithiobacillus ferrooxidans	Landslips, old mine workings
Acidithiobacillus ferrivorans	Acid mine drainage
Acidithiobacillus thiooxidans	Sulphur-rich soils, caves, sewers
Acidithiobacillus caldus	Coal spoil heaps
Phylum *Nitrospirota*	
Leptospirillum ferrooxidans	Old mine workings
Phylum *Acidobacteria*	
Acidobacterium capsulatum	Old mine workings
Phylum *Actinomycetota*	
Ferromicrobium acidiphilum	Old mine workings
Ferrithrix thermotolerans	Hot springs
Phylum *Bacillota*	
Sulfobacillus acidophilus	Coal spoil heaps, hydrothermal vent
Domain *Archaea*	
Class *Thermoprotei*	
Sulfolobus acidocaldarius	Volcanic habitats
Saccharolobus solfataricus	Volcanic habitats
Class *Thermoplasmata*	
'E-plasma'	Old mine workings
Thermoplasma acidophilum	Coal spoil heaps
Thermoplasma volcanium	Volcanic habitats
Class *Halobacteria*	
Halarchaeum acidiphilum	Acidic salt deposits

bacteria en masse. Acidophiles flourish in the highly acidic conditions. Luxuriant apricot-coloured mats of bacteria carpet the bed and banks of the channel and jelly-like streamers, composed of dense strings of innumerable microbes, sway like a swarm of elvers from side to side in the gentle current (Fig. 4.2). These acid streamer communities are composed largely of a small number of species of specialised acidophilic iron-oxidising bacteria, including *Acidithiobacillus ferrivorans* (the Cool Acid-bug), "*Ferrovum myxofasciens*" (the Slimy Acid-bug) and *Acidithrix ferrooxidans* (the Acid Streamer-bug) (Kay *et al*., 2013). The acidic water bodies of Parys Mountain also provide a home for acid-tolerant eukaryotic green algae. Their vivid lime green filaments complement the more subtle apricot tones provided by the bacterial streamers.

Acid mine drainage occurs where water flows through rocks that have a high sulphur content and is a feature of abandoned coal mines as well as of metal extraction sites. Mine drainage from former metal extraction sites poses a pollution hazard because of toxic concentrations of cadmium, copper, lead, nickel and other metals (Gozzard *et al*., 2011; Jones *et al*., 2018). Toxic metal ores are associated with communities of specialised metal-tolerant bacteria. *Sulfuriferula plumbiphila* (the Galena Acid-bug) is a moderate acidophile

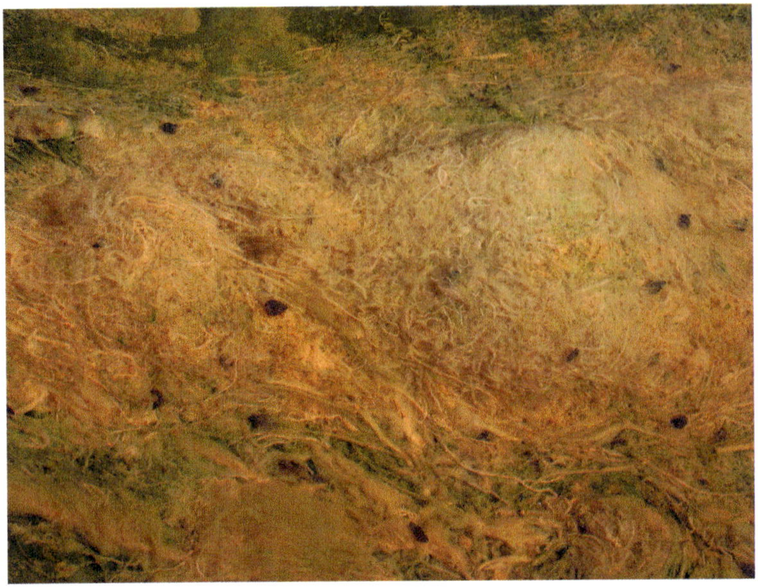

Fig. 4.2. Streamers composed of acidophilic bacteria. Afon Goch, Anglesey, UK.

that was discovered in a uranium mine in Germany. The Galena Acid-bug oxidises galena (lead sulphide, PbS) present in lead ore deposits. The iron-oxidising acidophile *Acidithiobacillus ferrooxidans* (the Iron Acid-bug) is also capable of oxidising lead sulphide. Britain's industrial heritage means that large areas of the country are dotted with abandoned mines. Lead has been mined in Britain since Roman times. Former lead mines are associated with toxic spoil heaps and polluted soils. Such sites are of botanical interest because of the presence of specialist metallophytes such as the nationally scarce Alpine Penny-cress, *Noccaea caerulescens*, which can be found growing on lead-contaminated soils in Northumberland and Somerset. The grasslands that develop on soils with high metal concentrations are known as calaminarian grasslands and because of their special plants are considered to be of conservation value (Richards and Waite, 2017; Hodd and Hodgetts, 2018). The heaps of mining rubble at Parys Mountain are considered to be of conservation value because of their rare metal-tolerant lichens. Metal-tolerant microbes are of equal interest to microbiologists. Perhaps they too should receive attention when assessing the conservation value of former industrial sites.

Acidophiles in natural habitats

Naturally acidic conditions occur in the British Isles in places where coastal erosion exposes pyrite (iron sulphide) deposits to the elements, allowing them to be oxidised by bacteria that yield acidic waste products. Such situations occur along the slumping cliffs of Dorset, Hampshire and the Isle of Wight. Amongst the jumbled rocks of the undercliff, where the scent of hydrogen sulphide hangs in the air, the visiting naturalist should expect to encounter freshly exposed mineral deposits which are tinted with shades of orange and yellow by the oxidation products of iron sulphide (Fig. 4.3). Acidic pyrite exposures provide a natural habitat for *Acidithiobacillus ferrooxidans* (the Iron Acid-bug), an organism which mediates the oxidation of pyrite. One of the oxidation products of pyrite is jarosite ($KFe_3(SO_4)_2(OH)_6$), an eye-catching crystalline yellow mineral that superficially resembles elemental sulphur (Fig. 4.4). The bright yellow hue of jarosite helps to define the special character of coastal landslips in this region of southern England. In some places ancient deposits of jarosite form part of the geological record. At Alum Bay on the Isle of Wight, quartz grains that are finely coated with jarosite crystals endow the Eocene sand deposits of the Barton Clay Formation with a gaudy yellow appearance (Wright and Wilson, 1970). One of the byproducts of jarosite precipitation by bacteria is sulphuric acid (Grishin *et al.*, 1988). Sulphuric acid is also one of the breakdown products of degraded jarosite. Among the bacteria capable of degrading jarosite are *Geobacter metallireducens* (the Metal Reducer) and *Shewenella oneidensis* (the Oneida Lake-bug) (Jones *et al.*, 2006; Ouyang *et al.*, 2014).

Acidithiobacillus thiooxidans (the Sulphur Acid-bug) is closely related to *Acidithiobacillus ferrooxidans* (the Iron Acid-bug). The Sulphur Acid-bug oxidises sulphur and produces sulphuric acid as a waste product. Unlike the Iron Acid-bug it cannot oxidise iron. The Sulphur Acid-bug is an inhabitant of soils. Famously it also contributes to the growth of pendulous viscous highly acidic biofilms in caves, known as snottites. The Sulphur Acid-bug is known to colonise sewer pipes and septic tanks where it can cause corrosion. The Sulphur

Fig. 4.3. Coastal landslip with sulphur minerals: a habitat for acidophiles. Isle of Wight, UK.

Fig. 4.4. Bacteria are involved in the precipitation and degradation of the sulphur mineral jarosite. Dorset, UK.

Acid-bug was originally described (as *Thiobacillus thiooxidans*) from an isolate cultured from soil at Rutgers University in New Jersey, USA (Waksman and Joffe, 1922). The original type culture was lost. When in 2000 the Sulphur Acid-bug was reassigned to the genus *Acidithiobacillus*, a replacement type culture was selected from the archive of the former National Chemical Laboratory in Teddington, UK. It was a strain recorded as having been isolated in 1930 from 'Kimmeridge clay' (presumably from coastal exposures at Kimmeridge in Dorset) (Kelly and Wood, 2000).

Coal spoil as a habitat for thermoacidophiles

Thermoplasma acidophilum (the Acid Plasma-bug) is a strange amorphous organism that lacks a cell wall. It is a member of the Domain *Archaea* and belongs to the Order *Thermoplasmatales*. The Acid Plasma-bug was first detected amongst smoking piles of coal in Indiana and was then found to be widespread in smouldering coal refuse heaps in the neighbouring state of Pennsylvania (Brock, 1978). It is a thermophilic acidophile that grows under acidic conditions (pH 2) at temperatures of about 60°C. Early attempts to locate a more natural home for the Acid Plasma-bug at geothermal sites in Yellowstone National Park drew a blank but eventually, populations of *Thermoplasma acidophilum* were found living in the wild in volcanic areas of Italy, Japan and Indonesia. The search for *Thermoplasma acidophilum* led to the discovery of a related species at Isola Vulcano in Italy, which was named *Thermoplasma volcanium* (the Vulcan Plasma-bug) (Segerer *et al.*, 1988).

Distant relatives of *Thermoplasma acidophilum* have been discovered at Parys Mountain. Novel species of archaea were detected by analysing sequence data obtained from DNA extracted from acidic mine drainage sediment samples. Sequence analysis indicates that they belong to the same taxonomic group as *Thermoplasma*, the Order *Thermoplasmatales*, but unlike *Thermoplasma* which requires high temperatures, the novel archaea at Parys Mountain reside in cool acidic sediments. As yet these novel organisms have not been cultured in the laboratory. The most abundant lineage of these organisms at Parys Mountain is known informally as 'E-plasma' (Korzhenkov *et al.*, 2019). Both *Thermoplasma* and E-plasma appear to eke out a living as molecular scavengers by foraging for organic molecules derived from other microorganisms.

Coal spoil bacteria in Britain

It is amazing how quickly habitats can change or even disappear. In the relatively recent past, gigantic heaps of coal spoil were a once familiar site in many parts of Britain. Coal spoil creates a productive habitat, which is favourable to certain pioneering organisms, including specialised bacteria and some scarce beetles (Eyre *et al.*, 2002). Following the collapse of the coal industry in Britain during the late twentieth century, so-called slag heaps are on their way to becoming a thing of the past. Coal spoil has the potential to become hot and acidic. Surveys of the Warwickshire Coalfield during the late 1970s and 1980s resulted in the discovery of thermoacidophiles that were new to science. These interesting organisms were sampled from what has become a lost world.

Like its close relative *Acidithiobacillus thiooxidans* (the Sulphur Acid-bug), *Acidithiobacillus caldus* (the Hot Sulphur Acid-bug) is an acidophile that obtains energy by oxidising sulphur, except that unlike its congener, the Hot Sulphur Acid-bug is a moderate thermophile (Marsh and Norris, 1983; Hallberg and Lindström, 1994). Its optimum temperature for growth is 45°C. The Hot Sulphur Acid-bug was found in hot coal spoil at the site of the former Kingsbury Colliery, where mining operations had already ceased some years earlier in 1968. The Hot Sulphur Acid-bug has since been found to occur widely in mining environments. By oxidising sulphur it facilitates the dissolution of metals and is now used for the bioleaching of commercially valuable metal ores.

Another novel sulphur-oxidising thermoacidophile *Sulfobacillus acidophilus* (the Acid Sulphur Bacillus) was retrieved from coal spoil at Alvecote at the site of a former colliery, which had already closed in 1965 (Norris *et al.*, 1996). The residual coal spoil was later incorporated into Pooley Country Park (Fig. 4.5). In 2011, Gold Leaf, a 12-m high sculpture by the Scottish artists Matthew Dalziel and Louise Scullion was erected on top of the remaining spoil heap. Gold Leaf is a stack of gilded aluminium birch leaves, which draws attention to the pioneering capacity of birch trees to colonise skeletal soils. *Sulfobacillus acidophilus* (the Acid Sulphur Bacillus) has since been recovered from a hydrothermal vent at a depth of 3 km in the Pacific Ocean (Li *et al.*, 2011). A related thermoacidophile, *Sulfobacillus thermosulfidooxidans* (the

Fig. 4.5. Residual spoil heap in Pooley Country Park, Warwickshire, UK, once the home of thermoacidophilic bacteria.

Hot Sulphur Bacillus) was isolated from coal spoil at the Birch Coppice Colliery (Fig. 4.6). The Hot Sulphur Bacillus grows at temperatures of 45–55°C. It is globally distributed in hot springs and in acid mine drainage.

Neutral and alkaline hot springs

Geothermal springs in Britain are mostly warm rather than hot. A few of them are hot but not very hot. The thermal springs at Bath reach outflow temperatures of 46°C, making them suitable for organisms which may be categorised as moderate thermophiles. In 1985, scientists at the University of Warwick described *Thiobacillus tepidarius* (the Tepid Bath Bacterium) as a new species (Wood and Kelly, 1985) (Fig. 4.7). The Tepid Bath Bacterium is a moderate thermophile that was isolated from the inflow of the Great Bath at the Roman Baths in Bath. The Tepid Bath Bacterium grows optimally at temperatures of 43–45°C and can tolerate temperatures that range between 20–52°C. The

Fig. 4.6. The now vegetated Birch Coppice Colliery spoil heap, Warwickshire, UK, a former site for thermoacidophiles.

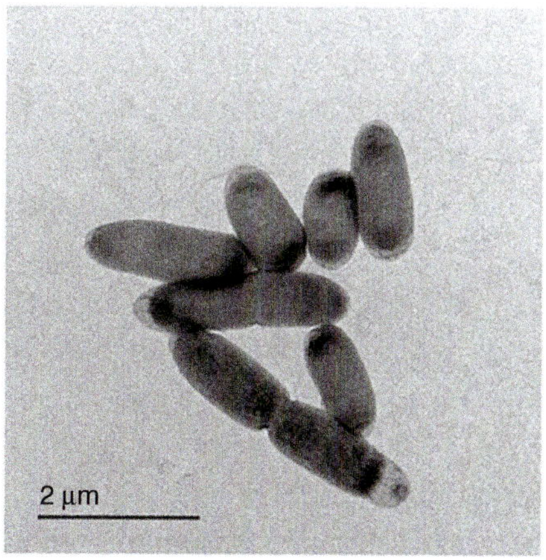

Fig. 4.7. *Thermithiobacillus tepidarius*, a thermophile discovered in the thermal waters at Bath, UK (Boden *et al.*, 2016). Creative Commons.

Tepid Bath Bacterium obtains energy for growth by oxidising thiosulphate (S_2O_3), sulphide and other reduced forms of sulphur. The Tepid Bath Bacterium was later reclassified as *Thermithiobacillus tepidarius* (Boden *et al.*, 2016). Another thiosulphate-oxidising bacterium from the Roman Baths at Bath was described in 1988 as *Thiobacillus aquaesulis* (Ann Wood's Bacterium). Ann

Wood's Bacterium is a moderate thermophile that uses carbon dioxide as a source of carbon and ammonia as a source of nitrogen (Wood and Kelly, 1988). In 2017, *Thiobacillus aquaesulis* was reclassified as *Annwoodia aquaesulis*. The new genus was named in honour of the lead author of the original description of *Thiobacillus aquaesulis*.

Thermophiles are at their most exuberant at geothermal sites whose waters are extravagantly hot and where diverse physicochemical conditions allow microbes to exercise their self-expression to the maximum. In 1969, two microbiologists based at Indiana University described *Thermus aquaticus* (the Hot Water Bacterium), a new species of bacterium that they had isolated by culturing samples obtained from Yellowstone National Park (Figs. 4.8–4.9) and from Pacheteau's Calistoga Hot Springs in California. The new organism grew in nutrient media at a temperature of 70°C (Brock and Freeze, 1969). *Thermus aquaticus* grows as part of a microbial mat community in hot spring water where it lives as a scavenger or saprophyte, feeding on the organic remains of dead microbes. At the time of its discovery, *Thermus aquaticus* set a new upper limit for the known temperature at which living organisms were known to thrive. *Thermus aquaticus* stimulated interest and further research into life at high temperatures (Brock, 1978). Numerous other species of thermophilic bacteria and archaea were soon discovered, including *Thermus scotoductus* (the Hot Water Pipe-bug), a close relative of *Thermus aquaticus*. The Hot Water Pipe-bug, which like its congener grows at 70°C, is capable of colonising manmade habitats and has been cultured from a hot water tap in a London hospital. Yellowstone National Park is awash with exotic geothermal features, including neutral and alkaline hot springs, solfataric fields, clay pools, acidic channels, iron springs and travertine deposits. A commentary on Yellowstone's thermal habitats and their associated microbes is provided in Chapter 9, *Bacteria tourism*.

Fig. 4.8. Streamers of *Thermus aquaticus*. Yellowstone National Park, USA. (Image kindly provided by Thomas Brock, with permission.)

Fig. 4.9. Collection site for *Thermus aquaticus*. Yellowstone National Park, USA. (Image kindly provided by Thomas Brock, with permission.)

The deep biosphere

Below the seafloor and below the surface of the land there is a habitable zone which has been referred to as the deep biosphere (Edwards *et al.*, 2012). The idea of a deep biosphere was popularised by the scientist Thomas Gold (1920–2004) in his book, *The Deep Hot Biosphere* (Gold, 1999). Samples from deep drilling sites and from deep mines can be examined in order to obtain information about microbial life in the deep biosphere. Beneath the seafloor exist two distinctive substrates: compacted oceanic sediments and igneous crustal rocks formed by volcanic action. Both are inhabited by bacteria. Suboceanic crustal rocks are porous and are permeated by aquifers through which the water of the world's oceans circulates. Crustal aquifers operate as convection systems, powered by heat supplied by an underlying layer of magma. Seawater that is drawn into porous seafloor rocks proceeds to percolate through crustal aquifers until eventually it is expelled through hydrothermal vents to rejoin the ocean's watery mass. The constant flow of seawater ferries a supply of nutrients to the microbial communities that reside within the porous suboceanic crust. The bacteria that reside in the upper layers of the earth's crust and in deep suboceanic sediments have been estimated to account for a large proportion of the world's biomass (Bar-On *et al.*, 2018).

Ordovician Britain: views of an ancient suboceanic deep biosphere

The presence of pyrite (iron sulphide) in the suboceanic crust demonstrates that the porous igneous rocks of which the crust is composed are inhabited by sulphate-reducing bacteria. Sulphate reducers release hydrogen sulphide as a waste product, which reacts with iron ions to generate deposits of iron sulphide. The activity of sulphate-reducing bacteria within the suboceanic crust is limited by the availability of sulphate and by the availability of organic carbon compounds. The suboceanic crustal world is not easily accessible to microbiological study, although samples can be obtained by drilling boreholes into the ocean floor. An alternative approach is to study rocks derived from the suboceanic crust which have been uplifted and preserved as part of our terrestrial geological record. The early Ordovician Period, approximately 450 million years ago, was a time both of organic sedimentation, and also of intense volcanic activity, leading to the creation of igneous crustal rocks on the seafloor. The compacted Ordovician sediments eventually became black shale deposits. Large exposures of Ordovician shale can be seen in the English Lake District, the Isle of Man, southern Scotland, Ireland and Wales.

Suboceanic habitats like those that existed during the early Ordovician Period can still be seen in the world today, especially in the vicinity of tectonic plate boundaries. During the Ordovician Period, much of the land that eventually became the British Isles was situated in the southern hemisphere where it formed part of a microcontinent called Avalonia. Avalonia was steadily drifting northwards, causing the oceanic crust of a neighbouring tectonic plate to disappear under the encroaching land mass. The subduction zone was marked by volcanism and the eruption of basaltic pillow lavas on the seafloor. The Ordovician sea was thick with accumulated organic sediments which served as a source of organic carbon for microbes that became established within the basalt. Ordovician basalt exposures can be seen at a number of sites in the British Isles in places where they occur in close proximity to Ordovician shales. Researchers based in Aberdeen and Glasgow measured sulphur isotope ratios in basalt samples in order to gain insights into the activities of sulphate-reducing bacteria during the Ordovician Period. Because bacteria preferentially reduce sulphate which contains a particular sulphur isotope (sulphur-32), sulphur isotope ratios in ancient crustal rocks provide a measure of the productivity of the sulphate-reducing bacteria which once inhabited them. The most productive locations, Llanwrtyd Wells in Wales and Helen's Bay (Fig. 4.10) in Northern Ireland feature basaltic rocks with unusually high carbon contents. At other locations the activity of sulphate-reducing bacteria was limited by the lower availability of organic carbon (Parnell *et al.*, 2015).

Suboceanic sediment as a habitat

In contrast to crustal rocks, suboceanic sediments are relatively impermeable and are populated by bacteria that become incorporated within the sediments as they gradually accumulate on the seafloor. Once incorporated within sediments, access to supplies of nutrients becomes severely restricted. The novelist Franz Kafka (1883–1924) wrote a short story entitled *A Hunger Artist* about a man who made it his profession to give public performances that entailed extended periods of fasting (Kafka, 1922). As time passed people lost interest

Fig. 4.10. Carbon-rich basalt, formerly a productive habitat for sulphate-reducing bacteria during the Ordovician Period, Helen's Bay, County Down, Northern Ireland.

in fasting as a form of entertainment. The poor hunger artist was ignored and eventually he wasted away completely. Sadly, this is the fate that befalls many of the bacteria that become entombed within sediments. Drill core samples from beneath the Pacific Ocean show that during the first million years after sedimentation, the numbers of viable bacterial cells rapidly decline (D'Hondt *et al.*, 2015). After about 5 million years the situation begins to stabilise. The small numbers of bacteria that survive do so by entering into almost inactive states in which individual cells use as little energy as possible. Viable bacteria can still be recovered from the deepest sediments that have been buried for more than 70 million years. The survivors at such depths though are few and far between, amounting to fewer than 1000 cells per cubic centimetre of sediment. Bacteria in deep sediments are hardly active at all. It is estimated that such bacteria may replicate as infrequently as once every several thousand years (Bradley *et al.*, 2019).

The terrestrial deep biosphere: the bacteria beneath our feet

The thermophilic microbes at the Roman Baths in Bath provide a rare glimpse of the terrestrial deep biosphere. The thermal waters at Bath travel upwards through fractures in the rocks from a depth of approximately 2.5 km (about 1.5 miles), bringing with them exotic bacteria which inhabit subsurface environments. Not much is known about Britain's subsurface microflora. With the exception of petroleum reservoirs and salt mines, deep subsurface environments are seldom sampled by microbiologists. To gain insights into the lives of the microbes that exist beneath our feet it is necessary to piece together different strands of evidence acquired as a result of studies of the bacteria found in subsurface environments from around the world.

"*Candidatus* Desulforudis audaxviator", the Bold Traveller

Ecosystems generally exist as extremely complex networks of interdependent organisms. The simplest known ecosystem is that of the Mponeng gold mine in South Africa, whose only natural inhabitants belong to a single species of bacterium, "*Ca.* Desulforudis audaxviator" (the Bold Traveller), a bacterium which was named in reference to a quotation from Jules Verne's novel, *A Journey to the Centre of the Earth* (Verne, 1864). The Bold Traveller thrives at a depth of almost 3 km, far beyond the reach of sunlight and independently of the support of other organisms. In the absence of sunlight, the Bold Traveller relies on supplies of sulphate generated indirectly by the radioactive decay of uranium and other radioactive elements in the mine as a source of energy. The Bold Traveller uses sulphate as a source of chemical energy to manufacture organic carbon compounds from carbon dioxide. "*Ca.* Desulforudis audaxviator" can also fix nitrogen which it uses to synthesise amino acids and proteins. The Bold Traveller is the most frequently detected organism at depths of more than 1.5 km throughout the Witwatersrand Basin in South Africa (Lin *et al.*, 2006; Chivian *et al.*, 2008). Following the discovery of "*Ca.* Desulforudis audaxviator" in South Africa, closely related bacteria were detected in samples retrieved from a wide range of subsurface locations, including a 1500-m deep borehole in Finland, a geothermal well in Germany, lake sediment in Turkey and radioactive water from a nuclear device test cavity in Nevada (Alawi *et al.*, 2011; Itävaara *et al.*, 2011; Balci *et al.*, 2012; Moser, 2012).

"*Ca.* Desulforudis audaxviator" has been shown to be widely distributed in subsurface environments. The original discovery and characterisation of "*Ca.* Desulforudis audaxviator" was achieved by sequencing bacterial DNA fragments from water samples from a fracture zone deep in a South African gold mine (Figs. 4.11–4.12). DNA sequence data were used to assemble the organism's genetic code and determine its metabolic processes. The Bold Traveller's genome includes some archaea-like genes which it seems to have acquired via lateral gene transfer from an archaeal donor organism at some time during its evolutionary history. Although it is a remarkably self-sufficient organism, at some of its known localities "*Ca.* Desulforudis audaxviator" lives alongside other species of bacteria and archaea.

Eventually, the Bold Traveller was cultured in the laboratory by a team of scientists based at the Tomsk State University in Siberia. "*Ca.* Desulforudis audaxviator" was isolated from water sampled from a 2-km deep aquifer in the Tomsk region (Karnachuk *et al.*, 2019). Comparison of sequence data indicate that European, Siberian and South African strains of the organism are genetically closely related, which suggests that the Bold Traveller may be capable of dispersal over long distances. Although intolerant of oxygen, "*Ca.* Desulforudis audaxviator" has the capacity to produce resilient spores. An alternative school

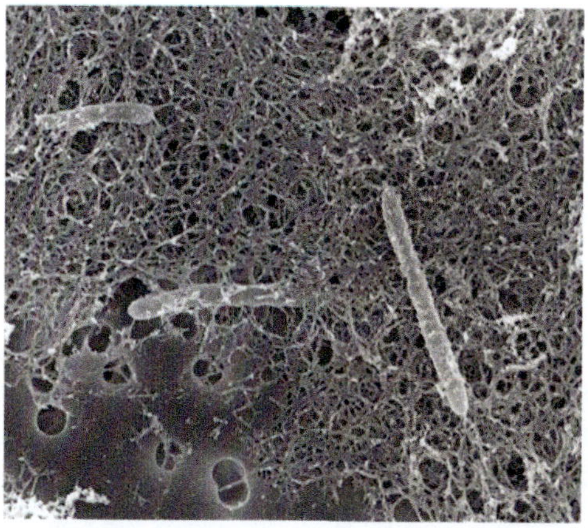

Fig. 4.11. "*Candidatus* Desulforudis audaxviator" discovered in the Mponeng gold mine near Johannesburg, South Africa. (NASA) Public Domain.

Fig. 4.12. Subterranean habitat of "*Candidatus* Desulforudis audaxviator": a fault zone in a South African gold mine. (Gordon Southam, with permission.)

of thought favours a view that populations of the bacterium found in different continents have been separated since the break-up of the supercontinent Pangaea.

Petroleum reservoirs

Near the village of Kimmeridge in Dorset, a solitary beam pump pumps oil from the Kimmeridge Oil Field (Fig. 4.13). The site is marked on the Ordnance Survey map as an oil well. The beam pump is clearly visible from the adjacent

Fig. 4.13. Beam pump used to extract petroleum. Kimmeridge, Dorset, UK.

coast path. The workings of oil extraction sites in Britain are not generally readily accessible to the public gaze but at Kimmeridge there is even an interpretative board, explaining that the well was drilled in 1959 and that the oil originates from a depth of 520 m below sea level. The Kimmeridge beam pump presents the observer with a point of contact between the familiar terrestrial environment and the deep subsurface habitat of a petroleum reservoir.

Petroleum consists of a blend of hydrocarbons sequestered within rock strata. The hydrocarbons that constitute petroleum vary in density from tar, or bitumen, through oil to natural gas. Petroleum is a fossil fuel, derived from the decomposed remains of dead organisms, such as algae, cyanobacteria, marine plankton and plants, that have been buried in sediments and altered by the effects of heat and pressure. The molecular composition of a petroleum reserve is determined by its source material, the subsurface temperatures to which the maturing hydrocarbons have been exposed and by the extent to which they have subsequently been degraded by colonising bacteria. The hydrocarbons of which petroleum is composed vary in size and shape from methane, with its single carbon atom, to more complex substances, including chain-like molecules consisting of many carbon atoms known as hopenoids. Hopenoids are produced by bacteria and are constituents of bacterial cells (Welander *et al.*, 2010). When detected in the geological record, hopenoids can be regarded as molecular fossils. They can be used as biomarkers that provide clues about the origins of organic deposits.

Petroleum begins with the burial of organic matter under anoxic conditions, such as may occur in stagnant sediments, where partial but incomplete degradation by bacteria can take place. Organic matter that becomes incorporated in sedimentary rocks is known as kerogen. Deep below the surface, the exposure of kerogen-bearing sediments to high temperatures (50–150°C) causes the breakdown of kerogen, which results in the release of oil and gas (Planckaert, 2005). The resulting buoyant hydrocarbons migrate upwards through porous rocks unless progress is halted by a layer of impervious cap rock that traps the hydrocarbons and prevents their escape to the surface. Even when trapped by cap rock, some hydrocarbons may continue to make their way to the surface. Naturally occurring oil seeps act as markers of deeper petroleum reservoirs. In the British Isles, terrestrial petroleum reservoirs and associated oil seepages occur in several regions, including the Orcadian basin and Midland Valley in Scotland, areas adjacent to the Pennines, the Weald and the Wessex Basin (Selley, 1992). Petroleum reservoirs contain oil, gas and water, or hypersaline brine, which are stratified according to density.

The world's largest known oil reserves are clustered around the Persian Gulf. The Gulf today is known for shallow marginal mudflats known as sabkhas that are colonised by bacterial mats. The Gulf is also known for whiting events caused by the precipitation of calcium carbonate, triggered by blooms of cyanobacteria and unicellular eukaryotes. Whiting events result in the deposition of lime sediments that eventually consolidate to become limestone. At the same time, tonnes of organic matter derived from the dead cells of microorganisms that have the potential to become petroleum become incorporated in the sediments. The shallow coastal margins and arid climate in the Persian Gulf also create ideal conditions for the deposition of evaporite minerals such as gypsum and salt. Evaporite deposits have the capacity to trap and contain petroleum reservoirs. The geological history of the Gulf suggests that conditions suitable for the generation and containment of petroleum have occurred repeatedly in the region at least since the Triassic Period, which may account for the existence of its enormous oil reserves (Kendall *et al.*, 2007).

Oil seepage bacteria

Natural oil seepages occur at several sites along the Dorset coast (Lees and Cox, 1937). Samples taken from an acidic (pH 3.4–3.6) oil seep near Lulworth Cove were found to be dominated by acidophilic bacteria (Röling *et al.*, 2006). Among them were *Acidithiobacillus ferrooxidans* and *Acidithiobacillus thiooxidans*. The oil seepages on the Dorset coast make their way to the surface through beds containing iron pyrite (FeS_2) that have been disturbed by coastal erosion. Their microbial communities have much in common with those found in acidic landslip habitats elsewhere along the south coast of England. *Acidithiobacillus ferrooxidans* and *Acidithiobacillus thiooxidans* occur together at Lulworth in the company of members of the genera *Acidiphilium* and *Acidocella* as part of a wider acidophilic microbial community. One of the bacteria found at the Lulworth oil seepage was *Methylorubrum extorquens* (the Twisted Meths Drinker), a soil bacterium which is not especially associated with highly acidic conditions. The Twisted Meths Drinker is unusual in being one of a fairly limited number of bacteria which are capable of utilising methanol as a food source. *Methylorubrum* species typically grow in culture as distinctive pink-pigmented colonies.

Sulphate-reducing bacteria and archaea

During the 1920s, the American geologist Edson S. Bastin (1878–1953) initiated a series of pioneering investigations into the microbiology of American oil fields. Bastin repeatedly cultured sulphate-reducing bacteria from oil samples and reasoned that the bacteria might have originated from the brine layer in the petroleum reservoirs. Sulphate-reducing bacteria have since been shown to form part of a characteristic oil field flora. Sulphate reducers release hydrogen sulphide as a waste product which acidifies the reservoir and reduces the quality of the oil that is harvested. They are also associated with the corrosion of steel pipes and structures. Therefore, they have become a focus for research by petroleum microbiologists. The sulphate-reducing bacteria that inhabit oil fields are much more diverse than the sulphate reducers found in more moderate habitats. Most of the sulphate-reducing bacteria that flourish in the black sediments in coastal salt marshes belong to a single phylum, the *Pseudomonadota*. Oil field sulphate-reducing prokaryotes include heat tolerant and thermophilic members of the *Pseudomonadota* and also thermophilic bacteria from other phyla as well as hyperthermophilic *Archaea*. Thermophilic sulphate-reducing archaea occur in hot springs as well as in oil reservoirs. Very distantly related sulphate-reducing bacteria and archaea share similar genes that encode an essential enzyme involved in sulphate reduction. It has been argued that thermophilic sulphate-reducing bacteria and archaea might have acquired sulphate-reduction genes from a sulphate-reducing donor (a member of the *Pseudomonadota*) during the course of their evolutionary history via lateral gene transfer (Klein *et al.*, 2001). This suggests that in hot sulphate-rich environments a strong selective advantage acts in favour of organisms that are capable of reducing sulphate.

The British oil field microflora

Petroleum reservoirs are hot, anoxic environments. Microbial diversity and abundance varies according to depth, temperature and the geological history of the reservoir. Earth temperatures increase with increasing depth. Beyond a certain depth it becomes too hot for bacteria and archaea. Even hyperthermophiles do not flourish at temperatures of much greater than about 120°C (Takai *et al.*, 2008). Over the course of geological time oil deposits move up and down in synchrony with movements of the earth. Although the oil fields of the Wessex Basin at Wytch Farm and Kimmeridge Bay in Dorset are relatively cool, with exit temperatures of around 65°C, the oil extracted from them does not show strong signs of degradation by microbial action. However, these reserves have been uplifted by earth movements over the course of geological time. Once they were situated 2 km deeper and were exposed to very high temperatures which would have exceeded the tolerances of any colonising microbes (Wilhelms *et al.*, 2001). In effect they have undergone a process akin to pasteurisation.

At temperatures of about 100°C, some of the oil fields situated under the North Sea are hot but fortunately for the microbes, these reservoirs are currently situated at their maximum depth. They have never been much hotter and they abound with hyperthermophiles. Oil fields can be recognised as a distinctive kind of habitat that is populated by a characteristic microbial flora. Closely related strains of bacteria and archaea have been recovered from distantly located petroleum reservoirs, suggesting that they have some capacity for dispersal. There is some overlap between the microbial floras of petroleum reservoirs and the microbes associated with deep sea hydrothermal vents. Because of the economic interest in petroleum, oil fields are among the most intensively studied of subsurface habitats. Oil prospecting and extraction operations that involve drilling down through layers of rock enable the collection of samples for microbiological study. Interpretation of the results of microbiological investigations of oil fields must take into account that samples may become contaminated with bacteria derived from surface environments. The most reliable samples are considered to be those collected at the well heads of newly opened reservoirs. Samples collected from reservoirs that have been flooded with water to aid the extraction of petroleum are likely to have been contaminated with exogenous microbes.

Several species of sulphate-reducing archaea, including *Archaeoglobus fulgidus* (the Glossy Globe-bug) and *Archaeoglobus profundus* (the Abysmal Globe-bug), have been cultured from the Thistle field in the East Shetland basin of the North Sea where reservoir temperatures may reach 110°C. The optimal temperatures for growth for *Archaeoglobus fulgidus* and *Archaeoglobus profundus* were found to be 94°C and 90°C, respectively (Stetter *et al.*, 1993; Stetter and Huber, 1999). The Thistle field has yielded other hyperthermophilic archaea, including *Thermococcus celer* (the Boiling Oil-bug) and an as yet undescribed member of the genus *Pyrococcus* (the Innominate Fireball). These microbes are not sulphate reducers but sulphur reducers: they use elemental sulphur as an oxidising agent. Members of both genera have been cultured from hydrothermal vents in various parts of the world. They appear to be widespread in deep, hot habitats. *Thermococcus sibiricus* (the Siberian Boiling Oil-bug), for example, is a *Thermococcus* species described from a high-temperature oil reservoir in Siberia. Oil reservoirs also provide a home for methanogenic archaea. Thermophilic methanogens cultured from the Thistle field in the North Sea include *Methanobacterium thermoautotrophicum* (the Self-sufficient Methanogen) and *Methanothermococcus thermolithotrophicus* (the Hot Rock Methanogen). In addition to thermophilic archaea, North Sea petroleum reservoirs are populated by thermophilic bacteria. One of them, *Petrotoga mobilis* (the Mobile Petrol Head) is a sulphur-reducing bacterium whose optimal temperature for growth is 60°C (Lien *et al.*, 1998). Members of the genus *Petrotoga* belong to a group of thermophilic bacteria whose cells are enclosed by a sheath-like structure or 'toga'.

Caves

Cave environments may appear to be almost devoid of life but even the deepest recesses of caves provide homes for communities of resourceful bacteria that mediate the precipitation of cave mineral deposits. Chemoautotrophic bacteria that can use chemical energy to fix carbon dioxide and synthesise organic compounds help to support the more complex organisms that live in subterranean spaces. Cave entrances lead the visitor towards an interior twilight zone that provides shelter from the elements (Fig. 4.14). Cave entrances are shadowy places that encourage the growth of ferns, mosses, liverworts and green algae. As light intensities diminish further, cyanobacteria become the predominant photosynthetic organisms (Figs. 4.15–4.16). Many cave cyanobacteria are species that also grow on shaded cliffs and in other dimly lit habitats (Pentecost and Whitton, 2000; Popović *et al.*, 2017). Among them are *Gloeocapsa sanguinea* (the Gory Gumball), *Gloeocapsa alpina* (the Alpine Gumball) and *Chroococcus turgidus* (the Puffed-up Paintball). Two species of cyanobacteria, *Scytonema julianum* (the Julian Cave-bug) and *Geitleria calcarea* (Geitler's Cave-bug) are cave specialists that induce the precipitation

Fig. 4.14. Cave entrance. Smoo cave, Sutherland, UK.

Fig. 4.15. Microbial crusts. Smoo cave, Sutherland, UK.

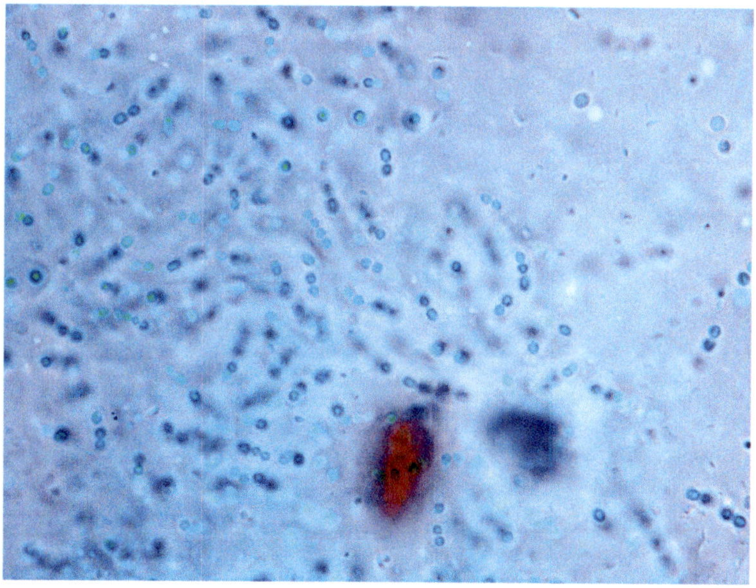

Fig. 4.16. Cyanobacteria seen on microscopy. Smoo cave, Sutherland, UK.

of calcium carbonate and become clothed by microscopic crystals of calcite. Geitler's Cave-bug was discovered in a cave in Israel and has since been reported from caves in France and Spain but as yet has not been recorded from caves in the British Isles.

In the deeper recesses of the cave, where there is no natural light at all, photosynthesis becomes impossible. As a consequence, nutrient availability becomes severely restricted. In the sunless depths, the only primary producers are bacteria that obtain energy for growth by oxidising chemical substances obtained from rocks. Bacteria create oxide residues that stain cave walls (Northup and Lavoie, 2001). Additional sources of nutrients are supplied by waters that percolate through cave systems. Temporary visitors, such as hibernating mammals, may also act as sources of nutrients inside the cave environment. Cave dwelling animals are sparse and specialised. Invertebrates recorded from British caves include the ground beetle, *Trechoblemus micros*, which is also associated with mammal burrows, the millipede, *Blaniulus guttulatus*, which is also found in soil, the cave isopod, *Proasellus cavaticus* and the cave shrimp, *Niphargus fontanus*. Surprisingly, despite limited supplies of food, underground watercourses can even support populations of fish. Caves in South Wales are known for their blanched specimens of brown trout, *Salmo trutta* (Ford, 1989; Chapman, 1993).

Cave systems are dynamic environments where minerals are dissolved and precipitated and where bacteria participate in remodelling the subterranean landscape. Britain's most extensive cave systems came into being as a result of the dissolution of limestone. Limestone caves occur in northern Scotland, the Pennines, South Wales, the Mendips, Devon and in other areas. In Ireland, extensive cave systems occur in County Fermanagh and County Clare. In coastal locations there are also sea caves that have been modelled by wave action. The distinctive rock formations composed of mineral deposits that develop inside caves are known technically as speleothems (Figs. 4.17–4.18).

Fig. 4.17. Speleothems at Smoo cave, Sutherland, UK.

Fig. 4.18. Speleothems exposed to the outside world by quarrying at Portland, Dorset, UK.

Speleothems are major contributors to the beauty of caves and are populated by diverse communities of bacteria and fungi. Speleothems include stalactites, stalagmites, flowstones that coat cave walls and paste-like deposits known as moonmilk. Speleothems develop at sites where carbonate-rich waters emerge into underground cavities. Loss of carbon dioxide to the air triggers the precipitation of calcium carbonate. Colonising microbes mediate the precipitation of carbonates while others bore into calcite crystals and dissolve them. Numerous species of bacteria that have been recovered from cave samples can be shown to induce carbonate precipitation and calcite crystals grown in the laboratory can be shown to contain microscopically visible bacterial filaments. Bacteria-shaped impressions can be seen in samples taken from stalactites and stalagmites. Bore-holes and extracellular polymeric substances secreted by bacteria can be seen on microscopic examination of carbonate crystals. Studies based on the measurement of carbon isotope ratios in carbonate samples suggest that bacteria are important inducers of mineral deposition during the early stages of speleothem generation but that abiotic chemical processes become more important as the structures mature (Cacchio *et al.*, 2012; Sanchez-Moral *et al.*, 2012; Dhami *et al.*, 2018).

Many of the bacteria found in association with speleothems belong to the Phylum *Pseudomonadota* and the Phylum *Actinomycetota*. Supplies of nutrients are severely limited in cave environments. Accordingly, among the cave-dwelling *Pseudomonadota* are microbes that can fix carbon dioxide, bacteria that degrade a wide range of organic compounds and bacteria that can fix nitrogen. The Phylum *Actinomycetota* includes species of bacteria that are associated with carbonate mineralisation. Moonmilk is an unusual paste-like substance found on the floors, walls and ceilings of caves (Fig. 4.19). It has a complex microstructure that consists of needle-shaped and plate-like crystals of calcite. Clusters of bacterial and fungal cells can be seen in spaces between carbonate crystals in samples of moonmilk. Moonmilk is associated with a diverse assortment of archaea, bacteria and fungi although their roles in the creation of moonmilk deposits are uncertain (Rooney *et al.*, 2010; Maciejewska *et al.*, 2017; Dhami *et al.*, 2018).

Rock surfaces in caves and in old mine workings are often coated with orange, brown and black crusts of metal oxides, especially oxides of iron and manganese known as ferromanganese deposits (Fig. 4.20). Microscopic examination of such crusts reveals the presence of clumps of bacteria. Ferromanganese deposits resemble desert varnish in that they consist of mixtures of clay and iron and manganese oxides. Iron oxide deposition is promoted by the presence of sticky extracellular polymeric substances secreted by bacteria. Crystalline ferromanganese deposits can be produced under laboratory conditions from cultures of bacteria grown in the presence of rocks that contain iron and manganese (Spilde *et al.*, 2005; Pacton *et al.*, 2013).

Humans have a long history as explorers of caves. Cave tourism modifies the subterranean environment by causing changes in temperature, humidity and air flow and by bringing in minerals and nutrients from outside. The installation of electric lighting in show caves creates dimly illuminated conditions similar to those present inside cave entrances. A characteristic lamp flora develops in the vicinity of electric lights, known in French as *la maladie verte* and in German as *Lampenflora* (Fig. 4.21). Photosynthetic cyanobacteria and green algae colonise the illuminated rock surfaces. Cyanobacteria and algae fix carbon dioxide, synthesise organic molecules and become the primary producers in what would

Fig. 4.19. Moonmilk, Ogof Ffynnon Ddu, South Wales. (Andrew Lewington, with permission.)

Fig. 4.20. Ferromanganese deposits in a derelict mine in Kakamega Forest, Kenya.

Fig. 4.21. Lamp flora with cyanobacteria. Ailwee Cave, County Clare, Ireland.

otherwise be a nutrient-poor environment. Complex biofilms emerge that are populated by different species of bacteria, fungi and algae. Bacteria secrete slimy extracellular polymeric substances that protect biofilms from desiccation. Eventually little subterranean gardens of mosses and ferns become established.

Numerous species of cyanobacteria have been identified from illuminated show caves, including *Gloeocapsa sanguínea* (the Gory Gumball), *Gloeocapsa aeruginosa* (the Verdigris Gumball), *Chroococcus minor* (the Lesser Paintball), *Chroococcus helveticus* (the Swiss Paintball) and *Oscillatoria rupicola* (the Stoned Swinger).

Cave tourism can be detrimental to prehistoric art works. Since the discovery in 1940 of an outstanding collection of Palaeolithic paintings at Lascaux near the village of Montignac in France, the Lascaux Cave has endured a series of microbial infestations, which have threatened the art works and have necessitated closure of the cave to visitors (Bastian *et al.*, 2010). Following the installation of electric lighting and modifications to the cave to allow the admittance of tourists into the cave, Lascaux became colonised with a green alga, *Bracteacoccus minor*, which formed biofilms on the walls, leading to the closure of the cave in 1963. In 2001 fluffy white fungi identified as *Fusarium solani* appeared on the walls, which were found growing in association with a bacterium identified as *Pseudomonas fluorescens*. In 2006 black stains appeared on the ceilings and walls caused by fungi, including a species described as new to science, *Ochroconis lascauxensis*. Each of the microbial outbreaks was treated aggressively with antimicrobial substances including quicklime and formaldehyde. Remedial action even included the topical application of the antibiotics penicillin, streptomycin and polymixin to combat bacteria. Despite restricted access to visitors, later surveys have shown Lascaux to be the home of diverse bacteria, including *Microbacterium murale* (the Mouldy Wall-bug), originally described from a mouldy wall in Germany and *Brachybacterium fresconis* (the Black Fresco-bug), a bacterium first isolated from a blackened wall painting at the Castle Herberstein in Austria (Bastian *et al.*, 2010; Martin-Sanchez *et al.*, 2014). Other show caves have been adversely affected by photosynthetic microbes, fungi and the growth of biofilms. The Altamira and Tito Bustillo caves in Spain were both closed in order to protect their Palaeolithic art works from degradation by colonising microbes (Estévez *et al.*, 2019).

Halophiles

Salt-seared roadsides dotted with the purplish-white flowers of Danish scurvy grass, *Cochlearia danica* have become so commonplace that nowadays they go almost unnoticed. The colonisation of British roadsides by salt-tolerant maritime plants was first noted in the 1970s. Since the time of those initial observations, halophytes, including Danish scurvy grass have become an established feature of heavily salted road margins (Matthews and Davison, 1976; Scott and Davison, 1982; James, 2010). Less well known is the fact that the rock salt applied to winter roads comes complete with halophilic microbes and that salt-treated road margins are suitable habitats for salt-tolerant and salt-dependant bacteria and archaea as well as for salt-tolerant plants. The rock salt applied to British roads in winter is sourced from salt mines in Cheshire and North Yorkshire. Both sources are known to be inhabited by halophilic microbes. Therefore it seems not unlikely that if efforts were made to look for them, halophilic microbes would be detected sooner or later on some salty road

Fig. 4.22. Roadside grit bunker. Arnside, Cumbria, UK.

verge somewhere in Britain (Fig. 4.22). When such a study was conducted in the USA, soil samples from salted roadsides and salt samples from a salt storage facility in Baltimore, Maryland were found to be colonised by halotolerant cyanobacteria and by bacteria belonging to the moderately halophilic genera *Thalassobacillus* and *Halobacillus*. Halophilic archaea belonging to the genera *Natrinema* and *Haloterrigena* were also detected (Pecher *et al.*, 2019).

Halophiles (Table 4.2) are organisms that thrive in places where the concentration of salt (sodium chloride, NaCl) significantly exceeds that of seawater. The salinity of seawater expressed in terms of mass per volume is about 3.5%, which corresponds to a salt concentration of 35 grams per litre. Diverse communities of halophilic bacteria and archaea can be found in water bodies whose salt concentrations reach 100 grams per litre or more. Halophiles can thrive even in solutions that are almost saturated with salt at concentrations of around 300 grams per litre (Oren, 2002). Some halophiles actually require high salinity for their survival and will burst if dropped into anything as dilute as seawater. In contrast to true halophiles, halotolerant organisms are ones which although able to tolerate high salinity, will grow better at lower salt concentrations. Hypersaline environments bring microbes into prominence by excluding the presence of unspecialised plants and animals. These are places that may seem dead to uninitiated observers and accordingly have attracted place names such as Death Valley, Dead Sea and Aigues Mortes. Many species of halophilic bacteria and archaea produce red, green, orange or purple pigments. Salt lakes that are densely populated by brightly coloured microbes are among the world's great natural spectacles. The presence of halophiles is more understated in the British Isles where they can be found living unobtrusively in salt mines, coastal sediments and in the decayed stonework of old buildings.

Table 4.2. Some halophilic microbes and their habitats.

Taxonomic group and species	Habitat
Domain *Bacteria*	
Phylum *Cyanobacteriota*	
Aphanothece halophytica	Seawater and brines
Phylum *Pseudomonadota*	
Halorhodospira halophila	Crystalliser ponds
Halovibrio variabilis	Hypersaline lakes and tamarisk leaves
Halomonas utahensis	Hypersaline lakes
Chromohalobacter israelensis	Hypersaline lakes
Phylum *Rhodothermota*	
Salinibacter ruber	Crystalliser ponds
Phylum *Actinomycetota*	
Actinopolyspora mortivallis	Saline soil
Phylum *Bacillota*	
Halanaerobium praevalens	Saline lake and fermented herrings
Halobacillus sp.	Gritted roadsides
Orenia marismortui	Hypersaline lakes
Thalassobacillus sp.	Gritted roadsides
Virgibacillus marismortui	Hypersaline lakes
Domain *Archaea*	
Class *Halobacteria*	
Haloarcula japonica	Hypersaline lakes and salterns
Haloarcula marismortui	Hypersaline lakes
Haloarcula vallismortis	Saline soils and lakes
Halobacterium salinarum	Salt lakes and salted fish
Halococcus dombrowskii	Salt mines and shearwater salt glands
Halococcus salifodinae	Salt mines
Haloferax mediterranei	Salterns
Haloquadratum walsbyi	Crystalliser ponds
Halorubrum lacusprofundi	Antarctic saline lake
Natrialba magadii	Soda lakes
Class *Methanosarcinia*	
Methanohalophilus halophilus	Cyanobacteria mat, saline bay
Phylum "*Ca.* Nanohalarchaeota"	
"*Ca.* Nanohalarchaeum antarcticus"	Antarctic saline lake
"*Ca.* Nanopetramus"	Desert halite crystals
"*Ca.* Nanosalinarum"	Hypersaline lake
Domain Eukarya (Eukaryota)	
Phylum (Division) Chlorophyta	
Dunaliella salina	Salt lakes and salterns

Halophiles in the British Isles

Habitats that exist in the British Isles today do not afford halophiles the opportunity to express themselves as fully as do the extremely salty environments that exist elsewhere in the world in places where the climate is more arid and where halophiles occur in such abundance that whole lakes may be painted pink by their photosynthetic pigments. Halophiles really come into their own in such places as salt lakes situated in desert landscapes and in the evaporation ponds of solar salterns used for the commercial production of sea salt. A commentary on examples of some hypersaline sites is provided in Chapter 9, *Bacteria tourism*. The halophiles that occur in Britain live unostentatiously in places where they are outnumbered by other microbes and in subterranean salt deposits where they lie hidden from view.

In order to flourish most halophiles require conditions of high salinity. Some halophiles, however, can tolerate conditions of low salinity and can survive exposure to seawater (Purdy *et al.*, 2004). One of the locations in Britain where halophiles have been detected is Colne Point in Essex (Fig. 4.23). Colne Point is a mosaic of estuarine creeks, salt marsh, sand and shingle situated at the mouth of the Colne Estuary in Essex. It happens that the Colne Estuary is a place that has been subjected to microbiological scrutiny for many decades (Nedwell *et al.*, 2016). The salt marsh sediments at Colne Point are not much more salty than seawater and are regularly inundated by the sea. Nevertheless,

Fig. 4.23. Salt marsh habitat for halophiles at Colne Point, Essex, UK.

scientists based at Essex University managed to detect halophilic archaea at Colne Point by analysing DNA sequence data obtained from salt marsh sediment samples (Munson *et al.*, 1997). Colne Point was shown to host halophilic archaea related to *Haloferax mediterranei* (the Mediterranean Salt-bug), a species which occurs in solar salterns around the Mediterranean and *Natrialba magadii* (the Magadi Soda-bug), a microbe known from Lake Magadi, a hypersaline alkaline lake in Kenya (Fig. 4.24). Colne Point is not unlike a number of other estuarine locations in Britain and it would seem more than likely that halophilic microbes could be found at coastal sites elsewhere around the British Isles.

Salt mines

Redolent of the legend that King Arthur will reemerge one day from his resting place are halophilic archaea (haloarchaea) that lie entombed within halite crystals in ancient salt deposits. Halophiles become incorporated into liquid inclusions of concentrated brine inside crystals of halite during the final stages of evaporation and crystallisation of brines in salterns and evaporating saline pools. Halophiles that have been experimentally sealed inside salt crystals in the laboratory can be revived years later. This observation suggests that these organisms have become adapted to periods of dormancy inside salt crystals as a means of coping with the long-term cycles of drying and wetting that occur in hypersaline environments. It may even be the case that halophiles use salt crystals as a dispersal mechanism. Haloarchaea that become trapped in fluid

Fig. 4.24. Halophile habitat. Magadi, Kenya. (Bill Grant, with permission.)

inclusions inside halite crystals do not form spores as do some bacteria but instead become almost inactive and assume a spherical morphology. Their long-term survival is enhanced by the fact that haloarchaea are polyploid: within each cell there are multiple copies of the genome. If one copy is damaged, for example by ionising radiation, other copies provide templates for repair. The excess DNA itself serves as a phosphate store (Jaakkola *et al.*, 2014; Stan-Lotter and Fendrihan, 2015). The survival of haloarchaea in salt crystals explains how these organisms are able to colonise salted foods and hides and act as spoilage agents. Solar salt harvested months previously from solar salterns still contains viable organisms when it is applied as a curing agent to organic material such as meat or fish. Proteolytic halophiles thrive on salted fish or hides where they cause red discolouration of the salted product.

Numerous reports document the revival of halophiles from intact halite crystals extracted from salt mines or from drill core samples. Ancient evaporite deposits occur in many parts of the world. In the British Isles significant sources of rock salt include 200 million year old Triassic rock salt deposits in Cheshire and in County Antrim and 230 million year old Permian deposits in North Yorkshire. Halite deposits, at the Winsford mines in Cheshire are situated at a depth of 170 m. In Cheshire and in a few places in Staffordshire and Worcestershire Triassic salt deposits also occur close to the surface and are associated with brine springs and inland salt marshes. Many of these locations have been subjected to salt mining and other forms of land use and inland salt marshes are an endangered habitat. Salt marsh vegetation can be seen at Upton Warren, a site previously worked for brine extraction in Worcestershire but the only surviving example of a natural inland salt marsh may be Pasturefields Salt Marsh near Stafford. Sea plantain, *Plantago maritima* and sea milkwort, *Lysimachia maritima* are among the halophilic species that have been recorded at Pasturefields (Edees, 1972). The Cheshire salt deposits have been exploited for centuries as a source of salt. Much of what is excavated is spread onto icy roads in winter. The Permian deposits in Yorkshire are situated more than a kilometre below the surface, where they are harvested at Boulby Mine on the edge of the North York Moors (Fig. 4.25). The focus of commercial interest at Boulby is an evaporite deposit containing potassium chloride or 'potash' used to manufacture fertiliser. Halite is mined as a consequence of potash extraction and used for gritting roads in winter (Fig. 4.26).

The Permian rocks of Yorkshire and Durham were laid down at the margins of the ancient Zechstein Sea, a huge inland sea that once covered large areas of what is now northern Europe. The Zechstein Sea reached as far as Poland and the Alps. Arid conditions prevailed during the Permian Period, leading to accumulation of thick deposits of rock salt and other evaporite minerals. Desert conditions continued into the ensuing Triassic Period when incursions of seawater and its subsequent evaporation led to the deposition of the rock salt deposits of Cheshire and Northern Ireland.

Haloarchaea have been cultured from intact halite crystals retrieved from deep inside the salt mines of Cheshire and North Yorkshire (Norton *et al.*, 1993) (Fig. 4.27). Similar isolates have been recovered from mines and drill cores in Austria, Germany and Poland. *Halobacterium noricense* (the Noric

Fig. 4.25. Boulby mine, Yorkshire, UK.

Fig. 4.26. Rock salt (halite).

Salt-bug) was described on the basis of an isolate cultured from a drill core obtained at Altaussee in Austria. Strains of haloarchaea identified as the Noric Salt-bug have also been recovered from salt mines at Bad Ischl in Austria, from Winsford in Cheshire and from Wieliczka in Poland (Gruber *et al.*, 2004).

Fig. 4.27. Saline agar slope used for the culture of halophiles.

A strain related to the Noric Salt-bug was also isolated from a more recent 40 million year old salt deposit at the Yunying in China (Jaakkola *et al.*, 2014). The salt mine at Yunying yielded another species, *Halococcus salfodinae* (the Salt Mine Salt-bug) which has been cultured from halite samples from salt mines at Bad Ischl in Austria and Lostock in Cheshire and from a drill core sample from Berchtesgaden in Germany (Denner *et al.*, 1994; Stan-Lotter *et al.*, 2000; Jaakkola *et al.*, 2014). *Halococcus dombrowskii* (Dombrowski's Salt-bug) is another member of the genus *Halococcus* that was described as a new species on the basis of an isolate cultured from an Austrian salt mine (Stan-Lotter *et al.*, 2002). Although attempts to isolate halophiles from intact halite crystals from a salt mine at Kilroot in County Antrim were not successful, culture of brine samples from the mine yielded representatives of six archaeal genera and 18 genera of bacteria (Megaw *et al.*, 2019). The archaeal isolates include members of the genera *Haloarcula*, *Halobacterium*, *Halococcus* and *Haloferax*.

Halophilic bacteria

Halophilic and halotolerant bacteria are phylogenetically diverse (Table 4.2). Moderately salty water bodies host salt-tolerant cyanobacteria which are important primary producers in such habitats. Some specialised cyanobacteria can tolerate very high salt concentrations (Oren, 2015). In addition to cyanobacteria, hypersaline environments provide a home for Gram-positive halophilic bacteria in the Phylum *Bacillota* and in the Phylum *Actinomycetota*. Among the Gram-negative bacteria, halophiles are well represented by members of the Phylum *Pseudomonadota*. One particular species of halophilic bacterium stands out as being so unusual that a new phylum was designated for it and for its relatives, the Phylum *Rhodothermota* (Munoz *et al.*, 2016). *Salinibacter ruber* (the Rechargeable Salt-bug) was first isolated from brine samples obtained from the salt ponds at Santa Pola near Alicante and from Majorca where it was found to account for 5–25% of the prokaryotes present in the evaporation ponds (Antón *et al.*, 2002). It has since been recorded from many hypersaline water bodies from around the world (Oren, 2013). The Rechargeable Salt-bug requires a salt concentration of 150 grams per litre for growth and grows optimally at salt concentrations of 200–300 grams per litre. To maintain osmotic balance in hypersaline solutions *Salinibacter ruber* accumulates potassium chloride (potassium salt) inside its cells. The accumulation of potassium chloride as an osmoprotectant is an archaeal trait. Although a bacterium, the Rechargeable Salt-bug is biologically similar to the halophilic archaea. In common with the haloarchaea, the Rechargeable Salt-bug has acidic amino acids which allow its enzymes to function in the presence of high concentrations of potassium chloride. High concentrations of potassium chloride are toxic to most organisms. Cells of the Rechargeable Salt-bug contain a number of reddish pigments, including salinixanthin, which absorbs light energy and xanthorhodopsin which pumps protons across the cell membrane, as part of a process something akin to charging a battery. *Salinibacter ruber* uses the resulting proton gradient as an energy source.

Halophilic archaea

The Domain *Archaea* includes numerous halophilic species (Fig. 4.28). Many of them belong to one particular group, the Class *Halobacteria*; a potentially confusing name which pre-dates the recognition of archaea and bacteria as two separate groups. In the more recent literature the informal term 'haloarchaea' is generally used for them. The type species that gives the Class *Halobacteria* its name is *Halobacterium salinarum* (the Salina Salt-bug). All members of the Class *Halobacteria* are halophiles. They are specialist microbes. Depending on the particular species they grow optimally at salt concentrations of between 100–250 grams per litre (Oren, 2002).

Triangular and square-shaped haloarchaea

The morphologies of some halophilic archaea conform to the usual spherical and rod shapes that are typical of prokaryotes. Others, such as members of the genus *Haloferax* are round and flat in shape and resemble dried apricots. *Haloarcula japonica* (the Salt Triangle) is flat and triangular in shape. The Salt Triangle was described as a new species in Japan and was subsequently identified among strains of halophilic archaea recovered from a coastal hypersaline lake at Sahline in Tunisia and from the Great Salt Lake in Utah (Nicolaus *et al.*, 1999; Kemp *et al.*, 2018) (Figs. 4.29–4.30). Archaea related to *Haloarcula japonica* have been detected in a salt mine at Kilroot in County Antrim. *Haloquadratum walsbyi* is flat and square or rectangular in shape (Fig. 1.38). It was discovered in a hypersaline pool in Sinai by Anthony Walsby, a microbiology professor at the University of Bristol (Walsby, 1980). The square flat cells might have been mistaken for salt crystals were it not for the

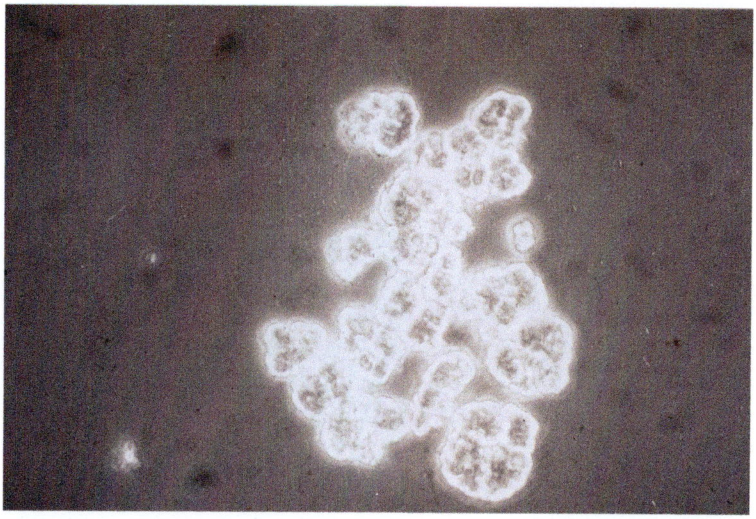

Fig. 4.28. Electron micrograph of haloarchaea. (Bill Grant, with permission.)

Fig. 4.29. Sahline, Tunisia, a locality for *Haloquadratum walsbyi*, *Haloarcula japonica* and other halophiles.

Fig. 4.30. The Great Salt Lake, Utah, USA.

fact that they contain visible gas vesicles that are thought to allow the organism to maintain its position in the water column. Initially this strange microbe became known as Walsby's square bacterium and then following the recognition of the *Archaea* as a separate domain it became known as Walsby's Square Archaeon. Although first reported in 1980, it took another 25 years until the organism was isolated in pure culture, by two separate groups of researchers in Europe and Australia (Bolhuis *et al.*, 2004; Burns *et al.*, 2004). It was formally described as *Haloquadratum walsbyi* in 2007 (Burns *et al.*, 2007). At the time of writing it remains the only member of the genus *Haloquadratum*. It grows best in nearly saturated brine.

This unusual organism is flat, square-shaped and transparent. Walsby's Square Archaeon functions like a solar panel. It uses the protein bacteriorhodopsin to capture light energy in order to pump protons across its cell membrane to generate a protein gradient that it can use as a source of energy. Walsby's Square Archaeon replicates by division along both axes of its rectangular cell. Cells remain attached for many divisions leading to an appearance that resembles a sheet of postage stamps. Since Walsby's discovery in Sinai, this organism has been found to occur abundantly in hypersaline pools in other parts of the world, including Spain, Australia, Peru, Tunisia, Turkey and the USA (Stan-Lotter and Fendrihan, 2015). Distantly separated populations of *Haloquadratum walsbyi* are remarkably similar. Genetic variation within and between populations has been measured as being less than 2% and isolates of this organism from opposite ends of the earth have been shown to be almost genetically identical (Dyall-Smith *et al.*, 2011). *Haloquadratum walsbyi* reminds me of a scene from the film *Superman* (1978) in which three miscreants imprisoned inside a sheet of glass are jettisoned into space.

Initially regarded as a curiosity Walsby's Square Archaeon was subsequently found to be the dominant species of microbe in salt-saturated solar salterns and salt lakes where population densities may reach 100 million cells per millilitre of brine (Bolhuis *et al.*, 2017). As many as four-fifths of the microbes in a crystalliser pond may be *Haloquadratum walsbyi*. In such environments there can be little competition from other organisms. A disadvantage of living in crowded surroundings, however, is that they are conducive to the spread and maintenance of viral infections. Enormous numbers of lytic viruses share the same crystalliser pools as *Haloquadratum walsbyi*. Viral infection must be a major cause of mortality.

Dispersal of halophiles by migrating birds

Saline water bodies are regularly visited by migrating birds. It has been suggested that birds might act as vectors of halophilic microbes and assist in the dispersal of these organisms from one site to another. A halophilic bacterium was cultured from the nasal salt gland of a Scopoli's shearwater, *Calonectris diomedea* in Spain (Brito-Echeverría *et al.*, 2009). Shearwaters, petrels, fulmars and albatrosses (Order: Procellariiformes, tube-nosed seabirds) are pelagic birds that drink seawater. The nasal salt gland of tube-nosed seabirds serves as a mechanism to excrete excess salt. Accordingly their salt gland secretions are hypersaline. Although Scopoli's shearwater is primarily a western Mediterranean species, in 2004 a Scopoli's shearwater was sighted off the Isles of Scilly (Fisher and Flood, 2010). In addition to Scopoli's shearwater it is possible that other tube-nosed seabirds might accommodate halophilic microbes in their nasal salt glands. The halophilic bacterium cultured from the Scopoli's shearwater in Spain was identified as *Halococcus dombrowskii* (Dombrowski's Salt-ball), a species originally described from a strain isolated from a salt mine in Austria.

The nasal salt glands of shearwaters may serve as long-term reservoirs for halophilic microorganisms and facilitate their dispersal. The plumage of migrating water birds might also act as a vehicle for the dispersal of halophiles. Halophilic archaea belonging to the genus *Haloarcula* were cultured from salt crystals harvested from the feathers of migrating pelicans at the Great Salt Lake in Utah (Fig. 4.30). Strains cultured from the pelican feathers were genetically closely related to strains isolated from other hypersaline locations (Kemp *et al.*, 2018). Sequence data for small ribosomal RNA genes showed near-identical matches between Great Salt Lake isolates and isolates from disparate sites around the globe. A strain of *Haloarcula hispanica* (the Spanish Salt-bug) showed 99% similarity to a strain from Santa Pola in Spain. A strain of *Haloarcula japonica* (the Salt Triangle) showed 99% similarity to a strain from Japan. A strain of *Haloarcula marismortui* (the Dead Sea Salt-bug) showed 98% similarity to a strain from the Dead Sea. However, all of these places are located far from the migration routes of the pelicans which visit the Great Salt Lake.

The discovery of closely related haloarchaea from different continents and from geological halite deposits – localities that are widely separated by time and space – poses something of a puzzle. It is oddly reminiscent of the phenomenon of particle entanglement in quantum physics, where two distantly located particles appear to act as one. Despite their exacting environmental requirements, haloarchaea are able to disperse over great distances and survive in extreme conditions for long periods. A complementary additional possibility is that the genomes of these organisms may be extremely stable and may hardly change over the course of time.

Landfill sites

Humans are making the geology of the future. In 2018, landfill sites in the UK received 7.2 million tonnes of biodegradable municipal waste, consisting of organic materials such as food and paper (Anon., 2020). The anoxic conditions of a landfill site create an environment in which degraded waste can be converted into methane by methanogenic archaea. Fermentative and acetogenic bacteria convert various forms of carbohydrate into acetate, hydrogen and carbon dioxide, which are then converted into methane by methanogens. Microbiological surveys of sites in Essex and Somerset demonstrated that rubbish dumps are colonised by diverse communities of methanogenic archaea (Luton *et al.*, 2002). Within a few years of its inception, a new landfill site will emit significant amounts of methane and continue to emit the gas for decades after the dumping of rubbish has ceased. It was estimated that in 1995, landfill sites in the UK emitted 887,000 metric tonnes of methane (Milton *et al.*, 1997). Global methane emissions from landfill sites during the 1980s and 1990s were estimated to have amounted to something in the region of 16 million to 57 million metric tonnes per year (Bognor and Matthews, 2003). At some landfill sites methane is piped to flaming stacks where it is burnt off or piped to power plants where the gas is exploited as a fuel source.

Plant material, paper and clothes made of cellulose account for much of the rubbish submitted to landfill sites. Microbial degradation of cellulose is the first step in the conversion of rubbish to methane. Cotton lures, incubated in fluids that have percolated through the rubbish can be used to investigate cellulose-degrading bacteria (Ransom-Jones *et al.*, 2017). Within weeks the cotton becomes colonised by a biofilm composed of organisms that steadily degrade the cotton fibres. Samples taken from waste facilities in North Wales and the Wirral harboured cellulose-degrading microbes from multiple phyla. The composition of the microflora showed similarities to the stomach flora of ruminants. Members of the Phylum *Fibrobacteriota* were detected, a small phylum of distinctive bacteria that form part of the gut flora of herbivorous mammals. Anaerobic decomposition of organic matter, especially cellulose, yields large volumes of methane. Lignin, a major constituent of woody plant tissues is resistant to decay in anoxic conditions. Wooden items that have been consigned to burial in landfill sites may fail to decompose. In this and in other respects, rubbish dumps are similar to peat bogs and waterlogged sediments.

Methane-oxidising bacteria

Methane that reaches oxygenated pockets of soil or sediment can be oxidised by methanotrophic bacteria to yield carbon dioxide and water as waste products. For this reason landfill sites are often covered with a layer of soil to encourage oxidation. Earthworm activity increases the rate of methane oxidation in landfill cover soils (Héry *et al.*, 2008). The rate of methane oxidation is temperature dependent. During the summer months much of the methane produced in a landfill is oxidised by methanotrophs. Methane emissions peak in winter as rates of oxidation decrease (Borjesson and Svensson, 1997). Core temperatures inside landfill sites may be maintained by the heat generated by the decomposition of organic matter. In contrast cover soils may become so cold in winter as to impede the activity of methane-oxidising bacteria.

The methane-oxidising bacteria found in rubbish dumps belong to numerous genera and species (Cébron *et al.*, 2007). All of them though are members of a single phylum, the *Pseudomonadota*. The methanotrophic bacteria that live in rubbish dumps are essentially the same bacteria as the ones that occur more widely in other habitats (see Chapter 3). Nevertheless, because landfill sites have been subjected to detailed microbiological investigation, methanotrophs new to science have been discovered amongst the heaps of rotting rubbish. Two novel species of methanotroph were described on the basis of isolates cultures from a landfill site near Athens, Georgia. The new species were named *Methylosarcina fibrata* (The Garland Gas-guzzler) and *Methylosarcina quisquiliarum* (the Garbage Gas-guzzler) (Wise *et al.*, 2001).

In addition to the escape of methane and other gases, landfill sites are a potential source of pollutants that may be leached from the site by rain water that percolates through the rubbish. New landfill sites can be lined with clay and polyethylene to preserve their hydrological integrity. Landfill sites are microbiologically interesting. They provide a model for the study of the degradation of organic matter by microbes in the wider world. It is tempting to suggest that some landfill sites could be designated as nature reserves in recognition of the distinctive communities of microorganisms that populate them.

Landfill sites constitute a significant microbial habitat. Landfill sites attract birdwatchers and they make eminently suitable destinations for the adventurous bacteria tourist. During the course of my quest to obtain suitable photographs for *Bacteria in the Natural World* I made a detour to reconnoitre a landfill facility on Horsea Island on the outskirts of Portsmouth. While driving along the adjacent motorway I had often noticed bulldozers moving earth around on the dome-shaped expanse of compacted rubbish. Following the recent closure of the landfill, part of the site had already been converted into a marina, which was attached to a shopping complex with a cinema and luxury housing. Glittering steel methane gas vents embellish the flower beds that surround the public carpark (Fig. 4.31). Their tops rotate in the briny air like kinetic sculptures. It is almost as though the methanogenic microbes mean to send us signals of their existence from their underground hideaway. I strolled through the shrubbery and across the turf to the perimeter fence. The former dump had already become colonised with wild flowers. On examination of

Fig. 4.31. A forest of methane vents in the vicinity of a landfill site at Port Solent, Hampshire, UK.

some seed heads of wild carrot I found Lesne's Earwig, *Forficula lesnei* and an attractive blue ground beetle, *Ophonus ardosiacus*. A few years later I learned that Horsea Island had become colonised by a newly established species of chafer, *Oxythyrea funesta*. More recently I read of plans to turn the area into a country park.

Air

The sudden blighting of an entire field of spring wheat in Montana by the Leaf Blight Bacterium *Pseudomonas syringae* showed that when it rains, it rains bacteria. The wheat field was being monitored as a study site for leaf blight. The soil had previously tested negative for strains of *Pseudomonas syringae* that cause leaf blight and the wheat seed had been treated with a proven antimicrobial agent prior to planting. *Pseudomonas syringae* was subsequently cultured from raindrops and hailstones sampled from the airspace above the field. These observations suggested that the atmospheric dispersal of *Pseudomonas syringae* might constitute a normal part of the life cycle of this widely distributed bacterium. Thermal columns can transport plant debris, dust particles and bacterial cells high into the atmosphere (Walker and Patel, 1964; Schnell and Vali, 1972). In the relatively warm lower reaches of the atmosphere, aerial suspensions of living bacteria act as important catalysts of snow and ice formation. In the atmosphere, suspended droplets of supercooled moisture may remain in the form of liquid water even at temperatures as low as −15°C. However, it has been shown that living cells of *Pseudomonas syringae* can trigger the freezing of supercooled water at temperatures of −1.8 to −3.8°C (Maki *et al.*, 1974). In this way, intact cells of *Pseudomonas syringae* act as templates or nuclei for the formation of hailstones. Particles of inanimate matter, including dead bacteria do not act as effective ice nucleation catalysts. Raindrops and hailstones act as re-entry vehicles that bring aerial suspensions of *Pseudomonas syringae* back down to earth where the bacteria may find new opportunities to infect susceptible plants (Sands *et al.*, 1982).

The troposphere is an aerial habitat that extends from the ground upwards for about 13 km (about 8 miles, or roughly 40,000 feet above sea level). The troposphere is thinner at the poles and thicker at the equator. It is a turbulent layer of the atmosphere that interacts with the ground and provides bacteria with a temporary home and dispersal system. Aerial sampling over the USA with an Aircraft Bioaerosol Collector system demonstrated a homogeneous distribution of bacteria up to an altitude of about 12 km (Smith *et al.*, 2018).

Aerial concentrations of bacteria are higher in dusty atmospheric conditions, such as occur downwind of arid environments. Airborne bacteria counts are generally higher in urban than in rural areas and higher at inland sites than at coastal locations. In a study conducted in Sweden, high urban counts were attributed to disturbance by humans and motor traffic leading to high densities of suspended dust. High counts in rural areas could be associated with local disturbance of the terrestrial environment (Bovallius *et al.*, 1978a). High counts of airborne bacteria sampled in the Botanical Gardens at Birmingham University occurred when the wind was blowing from the direction of the city centre. In contrast, counts of airborne bacteria at a sparsely populated coastal site on the Lizard Peninsula in Cornwall did not vary with wind direction (Harrison *et al.*, 2005). Counts for airborne bacteria are generally higher in summer than in winter and the numbers of bacteria generally increase with increasing temperature.

Once airborne, bacteria attached to dust particles can travel for many thousands of miles before coming down to earth. Counts of aerial bacteria in the Caribbean increased significantly with the arrival of an African dust storm, while high air counts for bacteria at a sampling site in Sweden could be linked to the arrival of windblown dust that originated near the Black Sea (Bovallius *et al.*, 1978b; Griffin *et al.*, 2001). Rain and snow samples collected at a montane sampling station in the Pyrenees showed that precipitation in summer was associated with the arrival of desert bacteria associated with dust from the Sahara Desert and with marine bacteria from the Mediterranean Sea. In winter, when the prevailing winds blow in from the Atlantic, forest microbes were detected that had possibly originated in North American boreal forests (Cáliz *et al.*, 2018). Surveillance of the aerial microflora at a montane observatory in Oregon recorded bacteria that arrived in association with dust plumes that originated in Asia (Smith *et al.*, 2013). Counts of airborne bacteria taken over the ocean are lower than those obtained in coastal areas. One-third to two-thirds of the aerial microflora that exist over the oceans is estimated to be of marine origin (Mayol *et al.*, 2017).

Unlike other continents, Antarctica exists in a meteorological state of relative isolation on account of the effects of the Antarctic circumpolar current.

Nevertheless, globally distributed bacteria have been detected in Antarctic air samples. Air sampling in the McMurdo Dry Valleys, which are among the coldest and driest places on earth, yielded thermophilic bacteria, including *Thermoaerobacter subterraneus* (the Hot Air-bug) and *Anoxybacillus tepidamans* (the Lukewarm-bug). The Hot Air-bug was described on the basis of an isolate cultured from the outflow water of an artesian borehole in Australia. Its optimal temperature for growth is 70°C. The Lukewarm-bug was originally found at a sugar beet processing plant in Austria and also in Yellowstone National Park. Its optimal temperature for growth is 55°C. The origin of the Antarctic isolates is something of a mystery, although it is possible that these thermophilic organisms could have originated from volcanic soils in the Antarctic interior (Bottos *et al.*, 2014).

Cloud water is a microbial habitat. Cloud water contains simple organic compounds and may even be more nutritious than lake water. A cloud is, nonetheless, an impoverished habitat. Cloud droplets sampled as snow at an observatory in the Austrian Alps contained as few as 1500 bacterial cells per millilitre. When incubated at 0°C the cells were shown to be metabolically active and replicated every 3.6–19.5 days (Sattler *et al.*, 2001). At higher temperatures, bacteria replicate more frequently. In a simulated humid atmosphere experiment, cells of *Serratia marcescens* (Serrati's Bacterium) doubled every 1–2 hours in a rotating drum that was maintained at a temperature of 30°C (Dimmick *et al.*, 1979).

Temperatures in the lowest layer of the earth's atmosphere, the troposphere, decrease with altitude. Above the troposphere, temperatures in the next highest layer, the stratosphere, increase with altitude to reach a maximum of approximately −15°C at the top of the stratosphere. The stratosphere lies above

Fig. 4.32. High altitude balloon used by the Indian Space Research Agency to collect bacteria from the stratosphere. (Courtesy: ISRO.)

the earth's weather and is where the ozone layer exists. In order for bacteria to survive in the stratosphere they must be resistant to desiccation, ultraviolet radiation and gamma radiation. *Deinococcus aetherius* (the Ethereal Bacterium) is such a bacterium, which was collected from the stratosphere over Japan (Yang *et al.*, 2010). In 2005 the Indian Space Research Agency (ISRO) launched a high altitude balloon, which carried sterilised steel sampling probes soaked in liquid neon (Fig. 4.32). Air samples collected in the stratosphere at altitudes between 27–41 km (17–25 miles) yielded three novel species of bacteria: *Janibacter hoylei* (Sir Fred Hoyle's Bacterium), *Solibacillus isronensis* (the ISRO Bacterium) and *Bacillus aryabhattai* (Aryabhatta's Bacterium) (Shivaji *et al.*, 2009). An isolate of Sir Fred Hoyle's Bacterium was subsequently isolated from a blood culture taken from a febrile 8-week-old child in Korea (Lim *et al.*, 2017). Aryabhatta's Bacterium has since been cultured from soil and water samples.

References

Alawi, M., Lerm, S., Vetter, A., Wolfgramm, M., Seibt, A. *et al.* (2011) Diversity of sulfate-reducing bacteria in a plant using deep geothermal energy. *Grundwasser*, 16, 105.

Anon. (2020) UK Statistics on waste: March 2020 update. Department of Environment Food and Rural Affairs: Government Statistical Service.

Antón, J., Oren, A., Benlloch, S., Rodríguez-Valera, F., Amann, R. *et al.* (2002) *Salinibacter ruber* gen. nov., sp. nov., a novel, extremely halophilic member of the *Bacteria* from saltern crystallizer ponds. *International Journal of Systematic and Evolutionary Microbiology*, 52, 485–491.

Balci, N., Vardar, N., Yelboga, E. and Karaguler, N. (2012) Bacterial community composition of sediments from artificial Lake Maslak, Istanbul, Turkey. *Environmental Monitoring and Assessment*, 184, 5641–5650.

Bar-On, Y. M., Phillips, R. and Milo, R. (2018) The biomass distribution on Earth. *Proceedings of the National Academy of Sciences of the United States of America*, 115, 6506–6511.

Bastian, F., Jurado, V., Novakova, A., Alabouvette, C. and Saiz-Jimenez C. (2010) The microbiology of the Lascaux Cave. *Microbiology*, 156, 644–652.

Boden, R., Hutt, L.P., Huntemann, M., Clum, A., Pillay, M. *et al.* (2016) Permanent draft genome of *Thermithiobacillus tepidarius* DSM 3134T, a moderately thermophilic, obligately chemolithoautotrophic member of the *Acidithiobacillia*. *Standards in Genomic Science*, 11, 74.

Bogner, J. and Matthews, E. (2003) Global methane emissions from landfills: new methodology and annual estimates 1980–1996. *Global Biogeochemisty Cycles*, 17, 1065.

Bolhuis, H., Poele, E.M. and Rodriguez-Valera, F. (2004) Isolation and cultivation of Walsby's square archaeon. *Environmental Microbiology*, 6, 1287–1291.

Bolhuis, H., Martín-Cuadrado, A.B., Rosselli, R., Pašić, L. and Valera, F.R. (2017) Transcriptome analysis of *Haloquadratum walsbyi*: vanity is but the surface. *BMC Genomics,* 18, 510.

Borjesson, G. and Svensson, B.H. (1997) Seasonal and diurnal methane emissions from a landfill and their regulation by methane oxidation. *Waste Management Research*, 15, 33–54.

Bottos, EM., Laughlin, D.C., Herbold, C.W., Lee, C.K., McDonald, I.R. *et al.* (2020) Abiotic factors influence patterns of bacterial diversity and community composition in the Dry Valleys of Antarctica. *FEMS Microbiology Ecology,* 96, Issue 5.

Bovallius, A., Bucht, B., Roffey, R. and Anäs, P. (1978a) Three-year investigation of the natural airborne bacterial flora at four localities in Sweden. *Applied and Environmental Microbiology*, 35, 847–852.

Bovallius, A., Bucht, B., Roffey, R. and Anäs, P. (1978b) Long-range air transmission of bacteria, *Applied and Environmental Microbiology*, 35, 1231–1232.

Bradley, J.A., Amend., J.P. and LaRowe, DE. (2019) Survival of the fewest: microbial dormancy and maintenance in marine sediments through deep time. *Geobiology*, 17, 43–59.

Brito-Echeverría, J., López-López, A., Yarza, P., Antón, J. and Rosselló-Móra, R. (2009) Occurrence of *Halococcus* spp. in the nostrils salt glands of the seabird *Calonectris diomedea*. *Extremophiles*, 13, 557–565.

Brock, T.D. (1978) Thermophilic microorganisms and life at high temperatures. Springer-Verlag, New York.

Brock, T.D. and Freeze, H. (1969) *Thermus aquaticus* gen. n. and sp. n., a nonsporulating extreme thermophile. *Journal of Bacteriology*, 98, 289–297.

Burns, D.G., Camakaris, H.M., Janssen, P.H. and Dyall-Smith, M.L. (2004) Cultivation of Walsby's square haloarchaeon. *FEMS Microbiol Letters*, 238, 469–473.

Burns, D.G., Janssen, P.H., Itoh, T., Kamekura, M., Li, Z., Jensen, G. *et al.* (2007) *Haloquadratum walsbyi* gen. nov., sp. nov., the square haloarchaeon of Walsby, isolated from saltern crystallizers in Australia and Spain. *International Journal of Systematic and Evolutionary Microbiology*, 57, 387–392.

Cacchio, P., Ercole, C., Contento, R., Cappuccio, G., Preite Martinez M. *et al.* (2012). Involvement of bacteria in the origin of a newly described speleothem in the gypsum cave of Grave Grubbo (Crotone, Italy). *Journal of Cave and Karst Studies*, 74, 7–18.

Cáliz, J., Triadó-margarit, X., Camarero, L. and Casamayor, E.O. (2018) A long-term survey unveils strong seasonal patterns in the airborne microbiome coupled to general and regional atmospheric circulations. *Proceedings of the National Academy of Sciences of the United States of America*, 115, 1–6.

Cébron, A., Bodrossy, L., Chen, Y., Singer, A.C., Thompson, I.P. *et al.* (2007) Identity of active methanotrophs in landfill cover soil as revealed by DNA-stable isotope probing. *FEMS Microbiology and Ecology*, 62, 12–23.

Chapman, P. (1993) Caves and cave life. New Naturalist. Collins, London, UK.

Chivian, D., Brodie, E.L., Alm, E.J., Culley, D.E., Dehal, P.S. *et al.* (2008) Environmental genomics reveals a single-species ecosystem deep within earth. *Science*, 322, 275–278.

Coupland, K. and Johnson, D.B. (2004) Geochemistry and microbiology of an impounded subterranean acidic water body at Mynydd Parys, Anglesey, Wales. *Geobiology*, 2, 77–86.

Denner, E.B.M., McGenity, T.J., Busse, H.-J., Wanner, G., Grant, W.D. *et al.* (1994) *Halococcus salifodinae* sp. nov., an archaeal isolate from an Austrian salt mine. *International Journal of Systematic Bacteriology*, 44, 774–780.

Dhami, N.K., Mukherjee, A. and Watkin, E. (2018) Microbial diversity and mineralogical-mechanical properties of calcitic cave speleothems in natural and *in vitro* biomineralization conditions. *Frontiers in Microbiology*, 9, 40.

D'Hondt, S., Inagaki, F., Zarikian, C.A., Abrams, L.J., Dubois, N. *et al.* (2015) Presence of oxygen and aerobic communities from sea floor to basement in deep-sea sediments. *Nature Geoscience*, 8, 299–304.

Dimmick, R.L., Wolochow, H. and Chatigny, M.A. (1979) Evidence for more than one division of bacteria within airborne particles. *Applied and Environmental Microbiology*, 38, 642–643.

Dyall-Smith, M.L., Pfeiffer, F., Klee, K., Palm, P., Gross, K. *et al.* (2011) *Haloquadratum walsbyi*: limited diversity in a global pond. *PLoS ONE*, 6, e20968.

Edees, E.S. (1972) Flora of Staffordshire. David and Charles, Newton Abbot, UK.

Edwards, K.J., Becker, K. and Colwell, F. (2012) The deep, dark energy biosphere: intraterrestrial life on earth. *Annual Review of Earth and Planetary Sciences*, 40, 551–568.

Estévez, C.B., Moreno-Merino, L., Román, A. and Durán Valsero, J. (2019) The Lampenflora in show caves and its treatment: an emerging ecological problem. *International Journal of Speleology*, 48, 249–277.

Eyre, M.D., Luff, M.L. and Woodward, J.C. (2002) Rare and notable Coleoptera from post-industrial and urban sites in England. *Coleopterist*, 11, 91–101.

Fisher, A. and Flood, R. (2010) Scopoli's shearwater off Scilly: new to Britain. *British Birds*, 103, 712–715.

Ford, T.D. (ed.) (1989) Limestones and caves of Wales. Cambridge University Press, Cambridge, UK.

Gold, T. (1999). The Deep hot biosphere. Springer, New York.

Gozzard, E., Mayes, W.M., Potter, H.A.B. and Jarvis, A.P. (2011) Seasonal and spatial variation of diffuse (non-point) source zinc pollution in a historically metal mined river catchment, UK. *Environmental Pollution*, 159, 3113–3122.

Griffin, D., Garrison, V., Herman, J. and Shinn, E. (2001) African desert dust in the Caribbean atmosphere: microbiology and public health. *Aerobiologia*, 17, 203–213.

Grishin, S., Bigham, J. and Tuovinen, O. (1989) Characterization of jarosite formed upon bacterial oxidation of ferrous sulfate in packed-bed reactor. *Applied and Environmental Microbiology*, 54, 3101–3106.

Gruber, C., Legat, A., Pfaffenhuemer, M., Radax, C., Weidler, G. *et al.* (2004) *Halobacterium noricense* sp. nov., an archaeal isolate from a bore core of an alpine Permo-Triassic salt deposit, classification of *Halobacterium* sp. NRC-1 as a strain of *Halobacterium salinarum* and emended description of *Halobacterium salinarum*. *Extremophiles*, 8, 431–439.

Hallberg, K.B. and Lindström, E.B. (1994) Characterization of *Thiobacillus caldus* sp. nov., a moderately thermophilic acidophile. *Microbiology*, 140, 3451–3456.

Harrison, R.M., Jones, A.M., Biggins, P.D.E., Pomeroy, N., Cox, C.S., *et al.* (2005) Climate factors influencing bacterial count in background air samples. *International Journal of Biometeorology*, 49, 167–178.

Héry, M., Singer, A., Kumaresan, D., Bodrossy, L., Stralis-Pavese, N. *et al.* (2007) Effect of earthworms on the community structure of active methanotrophic bacteria in a landfill cover soil. *ISME Journal*, 2, 92–104.

Hodd, R.L. and Hodgetts, N.G. (2018) Results of a survey to monitor the EU Annex I habitat Calaminarian grassland, 2018. Irish Wildlife Manuals, No. 105. National Parks and Wildlife Service, Department of Culture, Heritage and the Gaeltacht, Ireland.

Hollis, S.P., Roberts, S., Earls, G., Herrington, R., Cooper, M.R. *et al.* (2014) Petrochemistry and hydrothermal alteration within the Tyrone Igneous Complex, Northern Ireland: implications for VMS mineralization in the British and Irish Caledonides. *Mineralium Deposita*, 49, 575–593.

Itävaara, M., Nyyssönen, M., Kapanen, A., Nousiainen, A., Ahonen, L. *et al.* (2011) Characterization of bacterial diversity to a depth of 1500 m in the Outokumpu deep borehole, Fennoscandian Shield. *FEMS Microbiology Ecology*, 77, 295–309.

Jaakkola, S.T., Zerulla, K., Guo, Q., Liu, Y., Ma, H. *et al.* (2014) Halophilic archaea cultivated from surface sterilized middle-late Eocene rock salt are polyploid. *PLoS ONE*, 9, e110533.

James, T.J. (2010) Roadside halophytes, including *Parapholis incurva* and *P. strigosa*, on trunk roads in Herts. (v.c.20) and Cambs. (v.c.29) *BSBI News*, 115, 45–46.

Johnson, D.B., Bacelar-Nicolau, P., Okibe, N., Thomas, A. and Hallberg, K.B. (2009) *Ferrimicrobium acidiphilum* gen. nov., sp. nov. and *Ferrithrix thermotolerans* gen. nov., sp. nov.: heterotrophic, iron-oxidizing, extremely acidophilic actinobacteria. *International Journal of Systematic Evolutionary Microbiology*, 59, 1082–1089.

Johnson, D.B. (2012) Geomicrobiology of extremely acidic subsurface environments, *FEMS Microbiology Ecology*, 81, 2–12.

Jones, E.J.P., Nadeau, T.-L., Voytek, M.A. and Landa, E.R. (2006) Role of microbial iron reduction in the dissolution of iron hydroxysulfate minerals. *Journal of Geophysical Research*, 111, G01012.

Jones, J.I., Murphy, J.F., Collins, A.L., Spencer, K.L., Rainbow, P.S. *et al.* (2018). The impact of metal-rich sediments derived from mining on freshwater stream life. In: Reviews of Environmental Contamination and Toxicology (Continuation of Residue Reviews), 1–79. Springer, New York.

Kafka, F. (1922) Ein Hungerkünstler. *Die Neue Rundschau*, 33, 983–992. S. Fischer Verlag, Berlin and Leipzig, Germany.

Kay, C.M., Rowe, O.F., Rocchetti, L., Coupland, K., Hallberg, K.B. *et al.* (2013) Evolution of microbial "streamer" growths in an acidic, metal-contaminated stream draining an abandoned underground copper mine. *Life*, 3, 189–210.

Kelly, D.P. and Wood, A.P. (2000) Reclassification of some species of *Thiobacillus* to the newly designated genera *Acidithiobacillus* gen. nov., *Halothiobacillus* gen. nov. and *Thermithiobacillus* gen. nov. *International Journal of Systematic and Evolutionary Microbiology*, 50, 511–516.

Kemp, B.L., Tabish, E.M., Wolford, A.J., Jones, D.L., Butler, J.K. *et al.* (2018) The biogeography of Great Salt Lake halophilic Archaea: testing the hypothesis of avian mechanical carriers. *Diversity*, 10, 124.

Kendall, C.G.St.C., Shinn, G. and Janson, X. (2007) Holocene cyanobacterial mats and lime muds: links to Middle East carbonate source rock potential. Abstract in: The American Association Petroleum Geologists Search and Discovery. Article #900063, Houston, Texas.

Klein, M., Friedrich, M., Roger, A.J., Hugenholtz, P., Fishbain, S. *et al.* (2001). Multiple lateral transfers of dissimilatory sulfite reductase genes between major lineages of sulfate-reducing prokaryotes. *Journal of Bacteriology*, 183, 6028–6035.

Korzhenkov, A., Toshchakov, S., Bargiela, R., Gibbard, H., Ferrer, M. *et al.* (2019) Archaea dominate the microbial community in an ecosystem with low-to-moderate temperature and extreme acidity. *Microbiome*, 7, 11.

Lees, G.M. and Cox, P.T. (1937) The geological basis of the present search for oil in Great Britain by the D'Arcy Exploration Company, Limited. *Quarterly Journal of the Geological Society*, 93, 156–194.

Li, B., Chen, Y., Liu, Q., Hu, S. and Chen, X. (2011) Complete genome analysis of *Sulfobacillus acidophilus* strain TPY, isolated from a hydrothermal vent in the Pacific Ocean. *Journal of Bacteriology*, 193, 5555–5556.

Lien, T., Madsen, M., Rainey, F.A. and Birkeland, N.-K. (1998) *Petrotoga mobilis* sp. nov., from a North Sea oil-production well. *International Journal of Systematic Bacteriology*, 48, 1007–1013.

Lim, Y.K., Kweon, O.J., Kim, H.R., Kim, T.H. and Lee, M.K. (2017) First case of bacteremia caused by *Janibacter hoylei*. *APMIS*, 125, 665–668.

Lin, L., Wang, P., Rumble, D., Lippmann-Pipke, J., Boice, E. *et al.* (2006) Long-term sustainability of a high-energy, low-diversity crustal biome. *Science*, 314, 479–482.

Luton, P., Wayne, J., Sharp, R. and Riley, P. (2002) The *mcrA* gene as an alternative to 16S rRNA in the phylogenetic analysis of methanogen populations in landfill. *Microbiology*, 148, 3521–3530.

Maciejewska, M., Adam, D., Naômé, A., Loïc, M., Elodie, T. *et al.* (2017) Assessment of the potential role of *Streptomyces* in cave moonmilk formation. *Frontiers in Microbiology*, 8, 1181.

Maki, L.R., Galyan, E.L., Chang-Chien, M.M. and Caldwell, D.R. (1974) Ice nucleation induced by *Pseudomonas syringae*. *Applied Microbiology*, 28, 456–459.

Marsh, R.M. and Norris, P.R. (1983) The isolation of some thermophilic, autotrophic, iron- and sulfur-oxidizing bacteria. *FEMS Microbiology Letters*, 17, 311–315.

Martin-Sanchez, P., Jurado, V., Porca, E., Bastian, F., Lacanette, D. *et al.* (2014) Airborne microorganisms in Lascaux Cave (France). *International Journal of Speleology*, 43, 295–303.

Matthews, P. and Davison, A.W. (1976) Maritime species on roadside verges. *Watsonia*, 11, 146–147.

Mayol, E., Arrieta, J.M., Jiménez, M.A., Martínez-Asensio, A., Garcias-Bonet, N. *et al.* (2017) Long-range transport of airborne microbes over the global tropical and subtropical ocean. *Nature Communications*, 8, 201.

Megaw, J., Kelly, S.A., Thompson, T.P., Skvortsov, T. and Gilmore, B.F. (2019) Profiling the microbial community of a Triassic halite deposit in Northern Ireland: an environment with significant potential for biodiscovery. *FEMS Microbiology Letters*, 366, 22:fnz242.

Milton, M.J.T., Partridge, R.H., Goody, B. and Andrews, A.S. (1997) An estimate of methane emissions from UK landfill sites based on direct flux measurements at representative sites. National Physical Laboratory Report QM 134. HMSO, London.

Moser, D.P. (2012) Deep microbial ecosystems in the U.S. Great Basin: a second home for *Desulforudis audaxviator*? Abstract B41F-08 presented at 2012 Fall Meeting, AGU, San Francisco, California.

Munoz, R., Rosselló-Móra, R. and Amann R. (2016) Revised phylogeny of *Bacteroidetes* and proposal of sixteen new taxa and two new combinations including *Rhodothermaeota* phyl. nov. *Systematic and Applied Microbiology*, 39, 281–296.

Munson, M.A., Nedwell, D.B. and Embley, T.M. (1997) Phylogenetic diversity of Archaea in sediment samples from a coastal salt marsh. *Applied and Environmental Microbiology*, 63, 4729–4733.

Nedwell, D.B., Underwood, G.J.C., McGenity, T.J., Whitby, C. and Dumbrell, A.J. (2016) The Colne Estuary: a long-term microbial ecology observatory. *Advances in Ecological Research*, 55, 227–281.

Nicolaus, B., Lama, L., Esposito, E., Manca, M.C., Improta, R. *et al.* (1999) *Haloarcula* spp. able to biosynthesize exo- and endopolymers. *Journal of Industrial Microbiology and Biotechnology*, 23, 489–496.

Norris, P.R., Clark, D.A., Owen, J.P. and Waterhouse, S. (1996) Characteristics of *Sulfobacillus acidophilus* sp. nov. and other moderately thermophilic mineral-sulphide-oxidizing bacteria. *Microbiology*, 142, 775–783.

Northup, D.E. and Lavoie, K. (2001) Geomicrobiology of caves: a review. *Geomicrobiology Journal*, 18, 199–222.

Norton, C.F., McGenity, T.J. and Grant, W.D. (1993) Archaeal halophiles (halobacteria) from two British salt mines. *Journal of General Microbiology*, 139, 1077–1081.

Oren, A. (2002) Halophilic microorganisms and their environments. Kluwer Academic, Dordrecht, Netherlands.

Oren, A. (2013) *Salinibacter*: an extremely halophilic bacterium with archaeal properties. *FEMS Microbiol Letters*, 342, 1–9.

Oren, A. (2015) Cyanobacteria in hypersaline environments: biodiversity and physiological properties. *Biodiversity and Conservation*, 24, 781–798.

Ouyang, B., Lu, X., Huan, L., Juan, L., Zhu, T. *et al.* (2014) Reduction of jarosite by *Shewanella oneidensis* MR-1 and secondary mineralization. *Geochimica et Cosmochimica Acta*, 124, 54–71.

Pacton, M., Breitenbach, S.F.M., Lechleitner, F.A., Vaks, A., Rollion-Bard, C. *et al.* (2013) The role of microorganisms in the formation of a stalactite in Botovskaya Cave, Siberia – paleoenvironmental implications. *Biogeosciences*, 10, 6115–6130.

Parnell, J., Hole, M. and Boyce, A. (2015) Evidence for microbial activity in British and Irish Ordovician pillow lavas. *Geological Journal*, 50, 497–508.

Pecher, W.T., Al Madadha, M.E., DasSarma, P., Ekulona, F., Schott, E.J. *et al.* (2019) Effects of road salt on microbial communities: Halophiles as biomarkers of road salt pollution. *PLoS ONE*, 14, e0221355.

Pentecost, A. and Whitton, B.A. (2000) Limestone. In: *The ecology of cyanobacteria*. Eds. Whitton, B.A. and Potts, M. Kluwer, Dordrecht, Netherlands.

Planckaert M. (2005) Oil reservoirs and oil production, 3–19. In Ollivier, B. and Magot, M. (ed), Petroleum microbiology. ASM Press, Washington, DC.

Platt, J.W. (1977) Volcanogenic mineralization at Avoca, Co. Wicklow, Ireland and its regional implications. In: Volcanic processes in ore genesis. Geological Society, London, Special Publications, 7, 163-170.

Popović, S., Simić, G.V.S., Stupar, M., Unković, N., Krunić, O. *et al.* (2017) Cave biofilms: characterization of phototrophic cyanobacteria and algae and chemotrophic fungi from three caves in Serbia. *Journal of Cave and Karst Studies*, 79, 10–23.

Purdy, K.J., Cresswell-Maynard, T.D., Nedwell, D.B., McGenity, T.J., Grant, W.D. *et al.* (2004). Isolation of haloarchaea that grow at low salinities. *Environmental Microbiology*, 6, 591–595.

Ransom-Jones, E., McCarthy, A.J., Haldenby, S., Doonan, J. and McDonald, J.E. (2017) Lignocellulose-degrading microbial communities in landfill sites represent a repository of unexplored biomass-degrading diversity. *mSphere*, 2, e00300-17.

Richards, J. and Waite, N. (2017) Calaminarian grassland management guide. Northumberland Wildlife Trust, UK.

Röling, W.F., Ortega-Lucach, S., Larter, S.R. and Head, I.M. (2006) Acidophilic microbial communities associated with a natural, biodegraded hydrocarbon seepage. *Journal of Applied Microbiology*, 101, 290–299.

Rooney, D.C., Hutchens, E., Clipson, N., Baldini, J. and McDermott, F. (2010) Microbial community diversity of moonmilk deposits at Ballynamintra Cave, Co. Waterford, Ireland. *Microbial Ecology*, 60, 753–761.

Sanchez-Moral, S., Portillo, M.C., Janices, I., Cuezva, S., Fernández-Cortés, A. *et al.* (2012) The role of microorganisms in the formation of calcitic moonmilk deposits and speleothems in Altamira Cave. *Geomorphology*, 139–140, 285–292.

Sands, D.C., Langhans, V.E., Scharen, A.L. and de Smet, G. (1982) The association between bacteria and rain and possible resultant meteorological implications. *Idojaras*, 86, 148–152.

Schnell, R.C. and Vali, G. (1972) Atmospheric ice nuclei from decomposing vegetation. *Nature*, 236, 163–165.

Scott, N.E. and Davison, A.W. (1985) The distribution and ecology of coastal species on roadsides. In: Beeftink, W.G., Rozema, J. and Huiskes, A.H.L. (eds.) Ecology of coastal vegetation. *Advances in vegetation science*, vol. 6. Springer, Dordrecht, Netherlands.

Segerer, A., Langworthy, T.A. and Stetter, K.O. (1988) *Thermoplasma acidophilum* and *Thermoplasma volcanium* sp. nov. from solfatara fields. *Systematic and Applied Microbiology*, 10, 161–171.

Selley, R.C. 1992 Petroleum seepages and impregnations in Great Britain. *Marine and Petroleum Geology,* 9, 226–244.

Shivaji, S., Chaturvedi, P., Begum, Z., Pindi, P. K., Manorama, R. *et al.* (2009). *Janibacter hoylei* sp. nov., *Bacillus isronensis* sp. nov. and *Bacillus aryabhattai* sp. nov., isolated from cryotubes used for collecting air from the upper atmosphere. *International Journal of Systematic and Evolutionary Microbiology*, 59, 2977–2986.

Smith, D.J., Timonen, H.J., Jaffe, D.A., Griffin, D.W., Birmele, M.N. *et al.* (2013) Intercontinental dispersal of bacteria and archaea by transpacific winds. *Applied and Environmental Microbiology*, 79, 1134–1139.

Smith, D.J., Ravichandar, J.D., Jain, S., Griffin, D.W., Hongbin, Y. *et al.* (2018) Airborne bacteria in earth's lower stratosphere resemble taxa detected in the troposphere: results from a new NASA aircraft bioaerosol collector (ABC). *Frontiers in Microbiology*, 9, 1752.

Spilde, M., Northup, D., Boston, P., Schelble, R., Dano, K. *et al.* (2005) Geomicrobiology of cave ferromanganese deposits: a field and laboratory investigation. *Geomicrobiology Journal*, 22, 99–116.

Stan-Lotter, H., Radax, C., Gruber, C., McGenity, T.J., Legat, A. *et al.* (2000) The distribution of viable microorganisms in Permo-Triassic rock salt. In: Geertman, R.M. (ed.). SALT 2000. 8th World Salt Symposium. Elsevier Science BV, Amsterdam, Netherlands.

Stan-Lotter, H., Pfaffenhuemer, M., Legat, A., Busse, H.-J., Radax, C. *et al.* (2002) *Halococcus dombrowskii* sp. nov., an archaeal isolate from a Permo-Triassic alpine salt deposit. *International Journal of Systematic and Evolutionary Microbiology*, 52, 1807–1814.

Stan-Lotter, H. and Fendrihan, S. (2015) Halophilic archaea: life with desiccation, radiation and oligotrophy over geological times. *Life*, 5, 1487–1496.

Stetter, K., Huber, R., Blöchl, E., Kurr, M., Eden, R.D. *et al.* (1993) Hyperthermophilic archaea are thriving in deep North Sea and Alaskan oil reservoirs. *Nature*, 365, 743–745.

Stetter, K.O. and Hubber R. (1999) The role of hyperthermophilic prokaryotes in oil fields. Microbial ecology of oil fields? Proceedings of the 8th International Symposium on Microbial Ecology. Bell, C.R., Brylinsky, M., Johnson-Green, P. (ed.) Atlantic Canada Society for Microbial Ecology, Halifax.

Takai, K., Nakamura, K., Toki, T., Tsunogai, U., Miyazaki, M. *et al.* (2008). Cell proliferation at 122°C and isotopically heavy CH4 production by a hyperthermophilic methanogen under high-pressure cultivation. *Proceedings of the National Academy of Sciences of the United States of America*, 105, 10949–10954.

Verne, J. (1864) Voyage au centre de la terre. Pierre-Jules Hetzel, Paris, France.

Waksman, S.A. and Joffe, J.S. (1922) Microorganisms concerned in the oxidation of sulfur in the soil. II. *Thiobacillus thiooxidans*, a new sulfur-oxidizing organism isolated from the soil. *Journal of Bacteriology*, 7, 239–256.

Walker, J.C. and Patel, P.N. (1964) Splash dispersal and wind as factors in epidemiology of halo blight of bean. *Phytopathology*, 54, 140–141.

Walsby, A.E. (1980) A square bacterium, *Nature*, 283, 69–71.

Welander, P.V., Coleman, M.L., Sessions, A.L., Summons, R.E. and Newman, D.K. (2010) Identification of a methylase required for 2-methylhopanoid production and implications for the interpretation of sedimentary hopanes. *Proceedings of the National Academy of Sciences of the United States of America*, 107, 8537–8542.

Wilhelms, A., Larter S.R., Head I., Farrimond P., di-Primio R. *et al.* (2001) Biodegradation of oil in uplifted basins prevented by deep-burial sterilization. *Nature*, 411, 1034–1037.

Wise, M.G., McArthur, J.V. and Shimkets, L.J. (2001) *Methylosarcina fibrata* gen. nov., sp. nov. and *Methylosarcina quisquiliarum* sp. nov., novel type 1 methanotrophs. *International Journal of Systematic and Evolutionary Microbiology*, 51, 611–621.

Wood, A.P. and Kelly, D.P. (1985) Physiological characteristics of a new thermophilic, obligately chemolithotrophic *Thiobacillus* species, *Thiobacillus tepidarius* sp. nov. *International Journal of Systematic Bacteriology*, 35, 91–94.

Wood, A.P. and Kelly, D.P. (1988) Isolation and physiological characterisation of *Thiobacillus aquaesulis* sp. nov., a novel facultatively autotrophic moderate thermophile. *Archives in Microbiology*, 149, 339–343.

Yang, Y., Itoh, T., Yokobori, S.I., Shimada, H., Itahashi, S. *et al.* (2010) *Deinococcus aetherius* sp. nov., isolated from the stratosphere. *International Journal of Systematic and Evolutionary Microbiology*, 60, 776–779.

Younger, P. and Potter, H.A.B. (2012) Parys in springtime: hazard management and steps towards remediation of the UK's most polluted acidic mine discharge. In: 9th International Conference on Acid Rock Drainage (ICARD), 20th–26th May 2012, Ottawa, Canada.

5 Animals

if a female, she has killed the male.

David Lindsay (1920). *A Voyage to Arcturus*.

Large numbers of species of insects are prone to infection with strains of the bacterium, *Wolbachia*. It has been estimated that *Wolbachia* infects a significant proportion of the world's butterfly species (Ahmed *et al*., 2015). Several British butterfly species are known to be susceptible to *Wolbachia* infection, including some of conservation concern. One of them, the Marsh Fritillary butterfly, *Euphydryas aurinia* (Fig. 5.1) was investigated for the presence of *Wolbachia* at a number of its remaining British localities. The Marsh Fritillary is a declining species in Europe. In the British Isles it is now restricted to a relatively small number of sites, mainly in the western parts of Britain and in Ireland. A study of Marsh Fritillary colonies, mainly in Southwest England but also in Wales and Scotland revealed the presence of a novel strain of *Wolbachia*. All of the butterflies sampled at a total of 19 sites were found to be infected (Smee, 2011). As yet, the consequences of *Wolbachia* infection in the Marsh Fritillary are unknown. *Wolbachia* is also known to infect the Large Blue butterfly, *Phengaris arion* (Fig. 5.2). The Large Blue was successfully reintroduced to Britain using stock from Sweden following its extinction as a British species in 1979. Its parasitic caterpillars are taken into the nests of the ant *Myrmica sabuleti*, where it feasts on the ant larvae. The Large Blue is itself parasitised by *Wolbachia*. All butterflies sampled in a study of the Large Blue in Italy and Poland tested positive for *Wolbachia* (Patricelli *et al*., 2013).

Wolbachia: a male-killing bacterium

The intracellular bacterium *Wolbachia* (Phylum: *Pseudomonadota*, Class: *Alphaproteobacteria*) (Fig. 5.3), has been detected in many insect species,

DOI: 10.1079/9781836993933.0005

Fig. 5.1. Marsh Fritillary, *Euphydryas aurinia*, a butterfly prone to *Wolbachia* infection. Murlough National Nature Reserve, County Down, Northern Ireland.

Fig. 5.2. Large Blue, *Phengaris arion*, a butterfly prone to *Wolbachia* infection. Muskmyr, Sweden.

in some crustaceans and also in the bodies of filarial nematode worms. The diverse *Wolbachia* strains that infect these organisms are classified according to a system of supergroups (A to U), which are currently treated as members of a single species, *Wolbachia pipientis*. Within each supergroup different strains can be differentiated by sequence typing. *Wolbachia* is an example of an infectious agent than can be transmitted vertically, meaning that the agent can be passed matrilineally from an infected female to her offspring. When vertically transmitted, *Wolbachia* is passed to the next generation via the cytoplasm of the ovum (Werren *et al.*, 1995). Wolbachia can also be transmitted horizontally from one organism to another. The invasive bacteria become incorporated inside host cells, where they exercise a degree of control over the cell biology of the infected organism. There is even evidence that *Wolbachia*

DNA can become integrated with the DNA of the host genome within the host cell nucleus. It has been shown, for example, that the genome of the Glanville Fritillary butterfly, *Melitaea cinxia* (Fig. 5.4), includes a 350 base pair sequence of DNA which almost exactly matches a DNA sequence found in the genome of a known *Wolbachia* strain (Ahmed, 2016).

Relationships between *Wolbachia* symbionts and their hosts vary. Some relationships are mutualistic. Mutualistic symbionts confer a fitness advantage on their host. Mutualistic symbionts, for example, may increase the host's access to nutrients, or they may enhance the host's defences against predation or infection. Some fruit flies, *Drosophila melanogaster*, have been shown to be infected with mutualistic strains of *Wolbachia* that render the flies resistant to viral infections.

Many insect species, however, are infected with strains of *Wolbachia* which act to the detriment of their host. *Wolbachia* and a number of other insect-associated symbiotic bacteria (Table 5.1) are reproductive parasites that employ

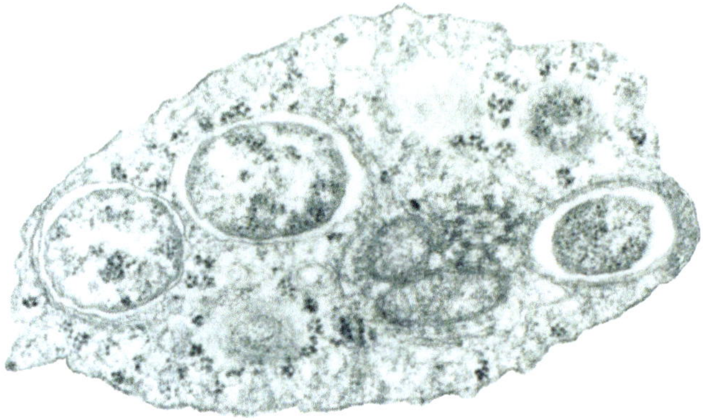

Fig. 5.3. Transmission electron micrograph of *Wolbachia* within an insect cell (Scott O'Neill, 2004; *PLoS Biol.* 2(3): e76). CC BY 2.5.

Fig. 5.4. Glanville Fritillary, Alum Bay, Isle of Wight, UK. The genome of the Glanville Fritillary includes a DNA sequence apparently derived from the bacterium *Wolbachia*.

Table 5.1. Some bacterial symbionts and their insect hosts.

Symbiotic bacterium	Insect host	Consequence of symbiosis
Class *Alphaproteobacteria*		
Wolbachia	many insect species	female-biased sex ratio, enhanced nutrient access, resistance to infection
Rickettsia	many insect species	female-biased sex ratio, resistance to heat stress, resistance to infection
"*Ca.* Tisiphia"	Bed bugs, flies	reduced fecundity in bed bugs
Class *Gammaproteobacteria*		
"*Ca.* Williamhamiltonia defensa"	aphids	resistance to parasitoids
"*Ca.* Regiella insecticola"	aphids	resistance to parasitoids
Serratia symbiotica	aphids	resistance to heat stress
Buchnera aphidicola	aphids	enhanced nutrient access
"*Ca.* Stammera capleta"	a tortoise beetle	enhanced nutrient access
Symbiopectobacterium purcellii	a leafhopper	enhanced nutrient access
"*Ca.* Symbiopectobacterium" strain SyEv	a leafhopper	enhanced nutrient access
"*Ca.* Symbiopectobacterium" strain SyCt	bullrush bug	enhanced nutrient access
"*Ca.* Symbiopectobacterium" strain SyCl	bed bugs	enhanced nutrient access
"*Ca.* Riesia pediculicola"	body and head lice	enhanced nutrient access
"*Ca.* Riesia pthiripubis"	pubic lice	enhanced nutrient access
"*Ca.* Riesia pediculischaeffi"	chimpanzee lice	enhanced nutrient access
"*Ca.* Curculioniphilus buchneri"	weevils, *Curculio* spp.	enhanced nutrient access
Sitophilus oryzae principal endocytobiote	a rice weevil	enhanced nutrient access
Pseudomonas-like bacterium	*Paederus* beetles	toxin production
"*Ca.* Macropleicola appendiculatae"	reed beetles	cocoon water resistance
"*Ca.* Macropleicola muticae"	reed beetles	cocoon water resistance
Sodalis glossinidius	tsetse fly	enhanced nutrient access
Phylum *Bacteroidota*		
Blattabacterium	cockroaches, termites	enhanced nutrient access
Blattabacterium-like bacterium	ladybirds	female-biased sex ratio
"*Ca.* Cardinium hertigii"	*Encarsia* wasps	female-biased sex ratio
Phylum *Actinomycetota*		
"*Ca.* Streptomyces philanthi"	beewolf larvae	resistance to fungi
Actinomyces sp. strain Lu9419	rose chafer larvae	cetoniacytone production
Coriobacterium glomerans	firebugs	enhanced nutrient access
Phylum *Bacillota*		
Lactobacillus kunkeei	honey bees	honey production
Phylum *Mycoplasmatota*		
Spiroplasma	ladybirds	female-biased sex ratio

a number of different strategies to increase the proportion of the host's offspring that are infected with *Wolbachia*. They do so by interfering in some way with the reproductive processes of the host insect. *Wolbachia* can induce parthenogenesis, can convert male insects into females and can induce the death of males during early development. These mechanisms increase the proportion of female descendants, all of which have the potential to transmit *Wolbachia* in turn to their offspring. *Wolbachia* can also induce a phenomenon known as cytoplasmic incompatibility, which causes males infected with *Wolbachia* to be unable to produce viable offspring unless they mate with a female infected with the same strain of *Wolbachia*. Cytoplasmic incompatibility selects against uninfected females. By manipulating the reproductive biology of its host in various ways, *Wolbachia* can modify the population structure of an insect species. *Wolbachia* and other reproductive parasites are often referred to as male-killing bacteria or as male killers, even though male-killing is only one of the ways in which these bacteria manipulate the reproductive biology of their host.

Hybrid zones between subspecies of the Meadow Grasshopper and their strains of *Wolbachia*

Grasshopper species found in Britain that are known to be susceptible to *Wolbachia* infection include the Stripe-winged Grasshopper, *Stenobothrus lineatus* (Fig. 5.5), the Common Green Grasshopper, *Omocestus viridulus*, the Rufous Grasshopper, *Gomphocerripus rufus* and the Lesser Marsh Grasshopper, *Chorthippus albomarginatus*, all of which are widely distributed in mainland Europe and in temperate Asia (Duron *et al.*, 2008; Ilinsky *et al.*, 2022). The

Fig. 5.5. Stripe-winged Grasshopper, *Stenobothrus lineatus*, a species prone to *Wolbachia* infection. Anchor Bottom, West Sussex, UK.

biological significance of *Wolbachia* infection in these insects is as yet largely unknown. Much more is known about the relationship between *Wolbachia* and the Meadow Grasshopper, *Pseudochorthippus parallelus* (Fig. 5.6), a species found in grassland habitats in Britain and Western Europe. Over much of its range, specimens of the Meadow Grasshopper look much the same. The general body colouration is usually green or brown, or occasionally pink or purple. The hind tibiae are yellowish-brown in colour. As one ventures southwards into the Pyrenees, one begins to encounter grasshoppers with red hind tibiae. Grasshoppers from the Iberian Peninsula and areas in the French Pyrenees belong to a separate subspecies, *Pseudochorthippus parallelus erythropus* (Butlin and Hewitt, 1985). The calling songs of the males are slightly different and there are other differences (Ragge and Reynolds, 1998). The two subspecies of the Meadow Grasshopper are considered to represent the descendants of populations of grasshoppers which became separated during the Ice Age. When the ice retreated the two populations met up again in the Pyrenees, where a hybrid zone now exists between the two subspecies (Fig. 5.7). Laboratory-reared male hybrids between the two subspecies are healthy but sterile. Nevertheless, prolonged contact between the two subspecies in the wild over many millennia has led to the creation of a hybrid zone, populated by fertile hybrid grasshoppers. The hybrid zone has been subjected to decades of study.

Areas to the north and south of the Pyrenees and the hybrid zone itself are associated with distinctive patterns of *Wolbachia* infection. Grasshoppers found to the north of the Pyrenees, including Britain, are infected mainly by Supergroup B strains and to a lesser extent by Supergroup F strains of *Wolbachia*. Grasshoppers in the Iberian Peninsula are infected mainly by Supergroup F strains of *Wolbachia*. Supergroup F strains found in the Iberian Peninsula differ in sequence type from Supergroup F strains found to the north

Fig. 5.6. Meadow Grasshopper, *Pseudochorthippus parallelus*, a species prone to *Wolbachia* infection. Anchor Bottom, West Sussex, UK.

Fig. 5.7. A valley in the Pyrenees populated by hybrid Meadow Grasshoppers (subspecies *parallelus* x *erythropus*), infected with recombinant *Wolbachia* strains.

of the Pyrenees, suggesting that during the time when the grasshoppers were busy diverging into separate subspecies, the bacteria themselves were mutating into different sequence types. It seems likely that Supergroup B strains were acquired relatively recently by the Meadow Grasshopper as a result of horizontal transmission from an as yet undetermined donor insect species. The detection of indistinguishable Supergroup B strains in other species of grasshopper, including the Heath Grasshopper, *Chorthippus vagans* (Fig. 5.8) shows that these strains are readily transmissible between different species of grasshopper (Martínez-Rodríguez, 2013).

There is not only a grasshopper hybrid zone: there is also a bacteria hybrid zone. Within the grasshopper hybrid zone complex patterns of *Wolbachia* infection exist, involving Supergroup B and F *Wolbachia* strains as well as recombinant strains derived from different sequence types. It is likely that *Wolbachia* helps to maintain the grasshopper hybrid zone by inducing cytoplasmic incompatibility between grasshoppers infected with different strains of *Wolbachia*. Meanwhile, the discovery of novel recombinant strains demonstrates that *Wolbachia* itself continues to evolve (Martínez-Rodríguez and Bella, 2018).

Wolbachia infection in an expanding population of Blue-tailed Damselflies

The distribution patterns of insects continue to change in response to changes in climate. When the climate gets warmer, dragonflies and damselflies move

Fig. 5.8. Heath Grasshopper, *Chorthippus vagans*, a species prone to *Wolbachia* infection. New Forest, UK.

further north. During recent decades the range of the Blue-tailed Damselfly, *Ischnura elegans*, has expanded northwards into new areas of Scandinavia and Scotland (Dudaniec *et al.*, 2018; Wood *et al.*, 2019). Blue-tailed Damselflies as well as a related species, the Scarce Blue-tailed Damselfly, *Ischnura pumilio*, are prone to infection with *Wolbachia*. Blue-tailed Damselflies sampled from multiple sites in Europe and Japan were found to be infected with a total of three different *Wolbachia* strains. Only one strain was detected in northwestern Europe. In Sweden and Scotland, almost 100% of damselflies were infected, whereas some populations of the damselfly in southern Europe yielded no evidence of infection. It has been estimated that the *Wolbachia* strain found in northwestern Europe began to emerge around the time that the ice retreated at the end of the Ice Age and at a time when populations of damselflies began to expand and spread in response to the warming climate (Deng *et al.*, 2021). Insect populations which are expanding at the edge of their range may be more prone to infection with *Wolbachia*, an organism that manipulates the reproductive biology of its host to increase the numbers of *Wolbachia*-infected offspring.

The Blue-tailed Damselfly is a common species in Britain. It can tolerate a wide range of environmental conditions. The Scarce Blue-tailed Damselfly is more fussy about where it lives. The larvae favour shallow, sparsely vegetated waters. It can be found breeding in runnels around the edges of bogs and also at brownfield sites, such as old quarries and derelict mining land. A strain of *Wolbachia* found in Scarce Blue-tailed Damselflies sampled in Sweden was shown to differ from any of the strains that are known to infect the more common damselfly species.

The Common Blue butterfly: conversion of males to females by *Wolbachia*

Common Blue butterflies are sexually dimorphic. In males the wings are blue on the upper side and spotted on the underside (Fig. 5.9). Female upper sides have orange marginal lunules and a scattering of blue scales against a darker background (Fig. 5.10). The Common Blue, *Polyommatus icarus*, occurs widely in the British Isles from Orkney to the Isles of Scilly. Populations of

Fig. 5.9. Common Blue butterfly, *Polyommatus icarus*, male from Uig Sands, Isle of Lewis, UK. Hebridean Common Blues show high rates of *Wolbachia* infection.

Fig. 5.10. Common Blue butterfly, *Polyommatus icarus*, female from Uig Sands, Isle of Lewis, UK, a locality where phenotypical females exhibit male genetic markers.

the butterfly from different areas of its range of distribution show differences in size, colour and biology. The Common Blue is double-brooded in southern England and single-brooded in northern Britain. Scottish specimens are often larger than those found further south and the blue colouration of Scottish males tends to be brighter than that of English males. Females from northwestern Scotland tend to have larger areas of blue colouration than English females. In this respect they resemble subspecies *mariscolore*, described from Ireland. Butterflies from northwestern Scotland are active at lower temperatures than those from England (Howe *et al.*, 2007). A study of Common Blue butterflies sampled from multiple localities in the British Isles revealed that butterflies from northwestern Scotland are genetically distinguishable from butterflies from other areas of Britain. The study also detected high rates of *Wolbachia* infection in butterflies from northwestern Scotland. All of ten females and six out of eight males from Uig Sands (Tràigh Uige) on the Isle of Lewis were infected with *Wolbachia* (Arif *et al.*, 2021). Butterflies from central and eastern Scotland also show significant rates of *Wolbachia* infection but they are infected with a different strain of the bacterium. By comparison, only one *Wolbachia* infection was detected among English samples (from a butterfly collected in County Durham). Butterflies from northwestern Scotland share a distinctive mitochondrial sequence type with populations of the Common Blue found in parts of Ireland, Scandinavia, Belle-Île, Spain and Provence. These butterflies situated on the western fringes of Europe may represent the relicts of an ancient lineage of the Common Blue that colonised western Europe soon after the ice retreated roughly 12,000 years ago. Other lineages of the Common Blue found in Britain, Ireland and mainland Europe presumably arrived at a later date.

Whereas sex in mammals is determined by a system of XY (male) and XX (female) sex chromosomes, in the Common Blue butterfly, sex is determined by a system of ZZ (male) and ZW (female) sex chromosomes. *Wolbachia* is known to induce the feminisation of males of some species of Lepidoptera. Feminised genetic males function reproductively as females. Feminisation is more likely to occur when a developing insect is infected with large numbers of *Wolbachia* cells. Butterflies sampled in the Western Isles showed high loads as well as high rates of *Wolbachia* infection. Phenotypical females, as determined by anatomy and wing colouration, from Lewis and Berneray in the Western Isles were found to exhibit male instead of female genetic markers. These insects may actually be genetic males that have been feminised by *Wolbachia* (Arif *et al.*, 2021).

Coinheritance of host mitochondria with *Wolbachia* in Brown Argus butterflies

Mitochondria are intracellular organelles that supply the cells of eukaryotic organisms, including animals and plants with energy by oxidising organic compounds such as sugars (see Chapter 8). The genetic code of mitochondria is maintained separately from that of the host cell in the form of a small circular chromosome of DNA. Because mitochondrial genomes are much smaller than those of their host cells, mitochondrial DNA sequence analysis, or mitochondrial sequence typing, is widely used as a convenient and reasonably economical way to investigate the phylogeny and population structure of insects and other eukaryotes. Mitochondria are inherited matrilineally via the ovum in the same way as *Wolbachia*. Hence, when a particular strain of *Wolbachia* becomes prevalent in a population of insects, the mitochondrial sequence type co-inherited with the successful *Wolbachia* strain will also become prevalent. The expansion of a successful *Wolbachia* strain in tandem with its co-inherited mitochondrial sequence-type may result in the replacement of other mitochondrial sequence types; a phenomenon known as a selective sweep. Accordingly, insect populations with high rates of *Wolbachia* infection tend to exhibit a low diversity of mitochondrial sequence types. Such is the case with the Northern Brown Argus, *Aricia artaxerxes*, a butterfly found in Scotland, northern England, in northern Europe and in montane areas further south. Scottish specimens, which belong to the nominotypical subspecies *Aricia artaxerxes artaxerxes*, can easily be identified by a prominent white spot in the centre of each forewing (Fig. 5.11). In examples from northern England, which belong to the subspecies *Aricia artaxerxes salmacis*, the white spots are less distinct or may be absent. The type locality for subspecies *salmacis* is Castle Eden Dene in County Durham, where the butterfly may still be seen in small numbers on coastal slopes close to where the Castle Eden Burn reaches the sea. The northern English subspecies more closely resembles the Brown Argus, *Arica agestis*, a butterfly that is widely distributed across southern Britain (Fig. 5.12). The situation is complicated by evidence indicating that populations of butterflies at some sites in North Wales and in northern England have been affected at some time in the past by hybridisation events involving both species (Mallet *et al.*, 2011).

Fig. 5.11. Northern Brown Argus, *Aricia artaxerxes artaxerxes*, a butterfly prone to *Wolbachia* infection. Glen Tilt, Perthshire, UK.

Fig. 5.12. Brown Argus, *Arica agestis*, a butterfly prone to *Wolbachia* infection. Westonbirt, Gloucestershire, UK.

Continental populations of the Northern Brown Argus also lack white spots on the forewings and may be difficult to distinguish from examples of the Brown Argus simply by looking at them. However, an accurate identification can usually be made by analysing the DNA sequences of mitochondrial genes (Sucháčková Bartoňová *et al.*, 2021). A team of European researchers combined mitochondrial sequence analysis with *Wolbachia* sequence analysis to investigate populations of Northern Brown Argus butterflies and Brown Argus butterflies from a wide range of localities in Europe. Both species were shown to be infected by a small number of distinctive *Wolbachia* strains; 93% of Northern Brown Argus butterflies were found to be infected with *Wolbachia*, compared with only 22% of Brown Argus butterflies. The range of mitochondrial sequence types associated with the Northern Brown Argus was much narrower than the range of types associated with the Brown Argus. As mitochondria and *Wolbachia* are co-inherited, it seems likely that the low diversity of mitochondrial sequence types seen in the Northern Brown Argus can be attributed to the predominance of butterflies that are infected with successful strains of *Wolbachia*.

Horizontal transmission of *Wolbachia* between butterfly species

In most areas of Europe, Northern Brown Argus and Brown Argus butterflies do not share the same strains of *Wolbachia*. In central Europe, however, some neighbouring populations of Northern Brown Argus and Brown Argus butterflies were found to be infected with indistinguishable strains of *Wolbachia*. This suggests that at some point *Wolbachia* has been transmitted horizontally between the two species of butterfly. As well as being inherited maternally, *Wolbachia* can be transmitted horizontally between different species of insects.

It has been speculated that transmission events might occur via contaminated food plants or nectar sources that are shared by more than one insect species, or perhaps as a result of exposure to parasitic mites or parasitic wasps. A global survey of *Wolbachia* infection in butterflies and other arthropods revealed numerous examples of similar *Wolbachia* strains that were shared by unrelated butterfly species (Ahmed, 2016). It has been estimated that butterflies first became infected with *Wolbachia* as recently as about 23 million years ago, whereas the Lepidoptera themselves and related insect orders began to diversify approximately 350 million years ago (Ahmed, 2016).

Horizontal transmission of *Wolbachia* between different wood ant species

Wood ants are large bicoloured brown and red ants that build easily recognisable thatched nest mounds, usually in woodland. British species include *Formica rufa*, in southern Britain, *Formica aquilonia*, in Scotland and Northern Ireland, *Formica lugubris*, in northern Britain and Ireland and *Formica pratensis*. *Formica pratensis* is thought to be extinct on the British mainland but still occurs in Jersey and Guernsey. Other species occur in mainland Europe, including *Formica polyctena*. When compared with continental specimens of *Formica rufa* and *Formica polyctena*, British ants identified as *Formica rufa* show intermediate morphological features, suggesting that the British ants may actually represent a hybrid population (Seifert, 2021). The wood ants constitute a subgroup of closely related species within the genus *Formica*. It has been estimated that the wood ants diverged from a common ancestor into separate species as recently as approximately half a million years ago (Viljakainen *et al*., 2008).

A study of wood ants sampled from multiple sites in Britain, mainland Europe and Siberia showed high rates of *Wolbachia* infection. Wood ants appear to be unusually susceptible to infection. Most ants across the whole of the study area were infected with two different strains of *Wolbachia*. DNA sequencing of *Wolbachia* genes revealed that disparate lineages of ants were infected with identical *Wolbachia* sequence types, indicating that horizontal transmission of *Wolbachia* has occurred between different wood ant species. Gene sequences from *Wolbachia* found in wood ants were an exact match also for those reported from *Wolbachia* infections in a more distantly related ant species, the Narrow-headed ant, *Formica exsecta* (Viljakainen *et al*., 2008). The biological significance of *Wolbachia* infection in these ants is currently unknown.

Parasitic symbionts confer a fitness cost to their host species. Phylogenetic trees for deleterious symbionts do not tend to map neatly onto the family trees of host species. Instead, lineages of susceptible hosts are more likely to demonstrate repeated patterns of elimination followed by reinfection with novel parasites. The detection of horizontal transmission events between distantly related species of butterflies, ants and other arthropods suggests that the overall effects of *Wolbachia* infection on these organisms are likely to be deleterious, or at best neutral.

Coevolution of mutualistic *Wolbachia* symbionts with their hosts

Because mutualistic relationships favour the host species as well as the symbiont they tend to be long-lasting. Beneficial symbionts coevolve with their hosts, a phenomenon known as co-cladogenesis. Accordingly, the family trees of mutualistic symbionts map onto the phylogenetic trees of their hosts. Such is the case with the *Wolbachia* symbionts of filarial nematode worms. Filarial nematode worms cause a variety of tropical diseases, including onchocerciasis (river blindness). Filarial worms depend on symbiotic *Wolbachia* for their reproductive fitness. Treatment of worms with anti-*Wolbachia* antibiotics interrupts the development of female worms. Accordingly, there is interest in the use of antibiotics as antifilarial agents.

Bed bugs, *Cimex lectularius* (Order Hemiptera, Family Cimicidae) and related species, including the Tropical Bed Bug, *Cimex hemipterus*, the Bat Bug, *Cimex pipistrelli* and the Swallow Bug, *Cimex hirudinis,* exhibit mutualistic relationships with their *Wolbachia* symbionts, which supply their hosts with the essential nutrients riboflavin (vitamin B2) and biotin (vitamin B7). The phylogenetic trees for bed bugs and their *Wolbachia* symbionts are roughly concordant, showing that bed bugs have coevolved with their hosts (Balvín *et al.*, 2018). Bed bugs, *Cimex lectularius*, are inconspicuous lentil-like true bugs, which venture out at night to feed on the blood of their sleeping victims (Fig. 5.13). A day or so later, mysterious raised itchy spots start to appear. During the daylight hours bed bugs secrete themselves into narrow spaces within the fabric and furniture of the room. Until around the time of the millennium, bed bugs had all but become a thing of the past (Paul and Bates, 2000). Recent decades, however, have seen a worldwide resurgence of bed bugs.

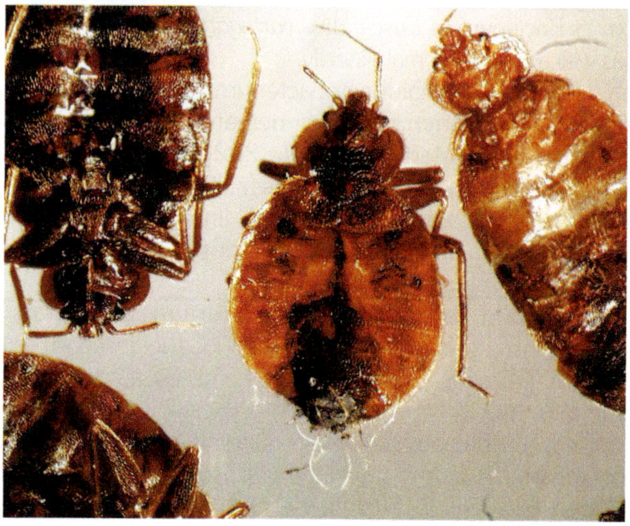

Fig. 5.13. Bed bugs, *Cimex lectularius.*

Other symbionts of true bugs (Hemiptera)

Collectively known as true bugs, the Order Hemiptera includes bed bugs, shield bugs, assassin bugs, cicadas, leafhoppers and aphids. True bugs subsist on restricted, nutrient-deficient diets. Many species utilise a single food source throughout their developmental stages. True bugs harbour symbiotic bacteria which supplement their hosts with essential nutrients. True bugs have sucking mouthparts. Some species are sap-suckers; others, such as bed bugs, are blood-suckers.

In addition to *Wolbachia*, bed bugs regularly host at least two other bacterial symbionts. One of them, "*Candidatus* Tisiphia" is a member of the Class *Alphaproteobacteria*. Bed bugs infected with "*Candidatus* Tisiphia" produce fewer eggs than uninfected females (Thongprem *et al.*, 2020). Another bed bug symbiont, "*Candidatus* Symbiopectobacterium" strain SyCl is a member of the Class *Gammaproteobacteria* (Cagatay *et al.*, 2025).

The name "*Candidatus* Symbiopectobacterium" refers to *Symbiopectobacterium purcellii*, a closely related symbiotic bacterium which was cultured from a species of leafhopper, *Empoasca decipiens* (Order Hemiptera, Family Cicadellidae). The type strain of *Symbiopectobacterium purcellii* was cultured from a leafhopper collected in Princes Park, Liverpool (Nadal-Jimenez, 2022). The genome of *Symbiopectobacterium purcellii* encodes synthetic pathways for a variety of vitamins and amino acids and it is assumed that *Symbiopectobacterium purcellii* and related symbionts benefit their hosts by providing enhanced access to nutrients. Related strains of symbiotic bacteria have been cultured from the Bulrush Bug, *Chilacis typhae* (Order Hemiptera, Family Lygaeidae) and from another species of leafhopper, *Euscelidius variegatus* (Purcell *et al.*, 1986; Kuechler *et al.*, 2011). They are unusual among insect-associated symbiotic bacteria in that they can easily be isolated in culture as free-living bacteria. It is known that the strain associated with *Euscelidius variegatus* ("*Candidatus* Symbiopectobacterium" strain SyEv, also known as Bacterium of *Euscelidius variegatus* or BEV) can be transmitted horizontally as well as inherited matrilineally.

"*Candidatus* Symbiopectobacterium" strain SyHa is an obligate symbiont of *Howardula aoronymphium*, a nematode worm that parasitises flies in the genus *Drosophila* (Martinson *et al.*, 2020). Symbiont-free worms which have been treated with antibacterial agents fail to thrive (Martinson *et al.*, 2020). The "*Candidatus* Symbiopectobacterium" lineage is related to *Pectobacterium carotovorum*, a plant pathogen that causes soft rot in carrots, cabbages, lettuces and other plant species.

A symbiotic bacterium, *Coriobacterium glomerans* (Phylum: *Actinomycetota*) forms part of the gut flora of the Firebug, *Pyrrhocoris apterus* (Fig. 5.14). Firebugs feed on mallows and are a common sight in continental Europe and the Channel Islands. A long established colony is known from the Ore Stone, an islet off the Devon coast. In the 1990s a breeding colony was discovered on the British mainland at Epsom in Surrey and more recently there have been sightings in Sussex, Kent, Essex and Suffolk. They are distinctive black and red insects and not easily overlooked, although they can be confused with some similarly coloured bugs. Newly hatched nymphs obtain symbiotic

Fig. 5.14. Firebug, *Pyrrhocoris apterus*. Epsom, Surrey, UK.

bacteria from the outside of their eggshells and can also acquire them from exposure to the faeces of other firebugs (Kaltenporth *et al.*, 2009).

Buchnera aphidicola (Class *Gammaproteobacteria*) is a symbiotic bacterium which provisions aphids with a supply of essential amino acids. Some aphids, such as greenfly and blackfly are well-known garden pests. Aphids are sap-feeding bugs. Plant sap is deficient in essential amino acids. The relationship between *Buchnera aphidicola* (Fig. 5.15) and its aphid host is a mutualistic symbiosis. The bacteria reside in vacuoles in specialised cells called bacteriocytes. They are transmitted matrilineally through the egg stage to the next generation of aphids. The phylogenetic trees for aphids and their symbiotic bacteria are congruent and it has been estimated that aphids and their symbionts have been coevolving for more than 150 million years. Other bacteria associated with aphids include *Serratia symbiotica*, "*Candidatus* Williamhamiltonia defensa" and "*Candidatus* Regiella insecticola" (Vorburger *et al.*, 2010). *Serratia symbiotica* (Class *Gammaproteobacteria*) protects conifer aphids, *Cinara* spp. against heat stress. The aphids accommodate the bacteria in bacteriocytes (Lamelas *et al.*, 2011). One particular species of conifer aphid, *Cinara cupressi*, is responsible for the destruction of leylandii hedges in Britain in recent summers. Two other aphid symbionts, "*Candidatus* Hamiltonella defensa" and "*Candidatus* Regiella insecticola" protect aphids and whiteflies against parasitoid wasps (Vorburger *et al.*, 2010).

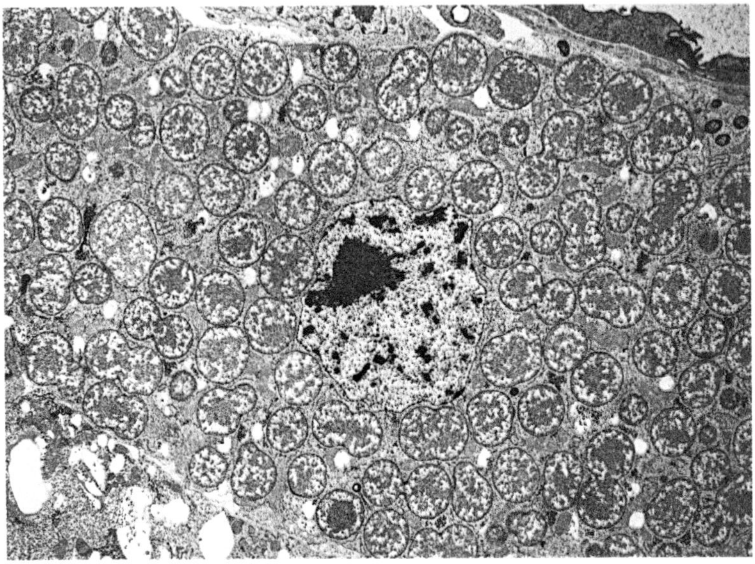

Fig. 5.15. *Buchnera aphidicola* within a bacteriocyte of a pea aphid (*Acyrthosiphon pisum*). The central object is the host nucleus; *Buchnera* cells are round and packed into the cytoplasm (J. White & N. Moran, University of Arizona; *PLOS Biology*). Creative Commons Attribution 2.5 Generic licence.

Blattabacterium: mutualistic symbionts of cockroaches

Mutualistic symbionts coevolve and diversify alongside their hosts. Cockroaches (Order Dictyoptera) harbour symbiotic bacteria of the genus *Blattabacterium*, a Gram-negative bacterium and member of the Phylum *Bacteroidota*. *Blattabacterium* lives in the fat bodies of the cockroach and supplies its host with amino acids and vitamins and allows cockroaches to survive on nitrogen-poor diets (Sabree *et al.*, 2009). Phylogenetic analysis based on rRNA genes of bacteria and cockroach mitochondrial genes shows that *Blattabacterium* and cockroaches have coevolved: the family tree for cockroaches mirrors the family tree of the symbiotic bacteria. All cockroaches which have been examined, apart from the divergent cave-dwelling genus *Nocticola*, harbour *Blattabacterium* in specialised cells (Lo *et al.*, 2013). It has been estimated that cockroaches have been accommodating *Blattabacterium* as a symbiont for at least 200 million years. *Blattabacterium* is also a symbiont of the Giant Northern Termite, *Mastotermes darwiniensis,* a primitive species of termite found in northern Australia. Termites began to evolve as a distinctive infraorder of insects, after diverging from a cockroach-like ancestor roughly 170 million years ago (Kinjo *et al.*, 2022). Today *Blattabacterium* persists only in the most primitive of termites, *Mastotermes darwiniensis*.

Cockroaches are omnivorous scavengers, best known for a small number of cosmopolitan pest species, which in temperate countries infest permanently heated buildings, such as hospitals, hotels, bakeries and botanical hot-houses.

The discovery of cockroach egg cases, or oothecae, from the remains of a communal bakery, dated to 65–80 AD and excavated during the construction of the Bloomberg London Building, showed that the Common Cockroach, *Blatta orientalis* (Fig. 5.16) was already present in Britain during Roman times (Smith *et al.*, 2023). Other indoor pest species found in Britain include the German cockroach, *Blattella germanica*, the American Cockroach, *Periplaneta americana*, and the Australian Cockroach, *Periplaneta australasiae*, all of which probably originated in Africa. Britain's native Cockroach species include the Lesser Cockroach, *Ectobius panzeri*, the Dusky Cockroach, *Ectobius lapponicus* and the Tawny Cockroach, *Ectobius pallidus*. They are dainty inoffensive creatures. All of them are locally distributed across southern Britain and associated with natural habitats, including sand dunes, chalk grassland, heathland and woodland clearings. Our native Cockroaches seldom occur indoors and they are not pests. The Swedish botanist Linnaeus (1707–1778) did, however, note that the Dusky Cockroach, *Ectobius lapponicus*, was known to consume dried fish.

Nymphs of the German Cockroach, *Blattella germanica*, are attracted to volatile substances produced by particular kinds of enteric bacteria present in the digestive tracts of other cockroaches and in cockroach faeces. Accordingly

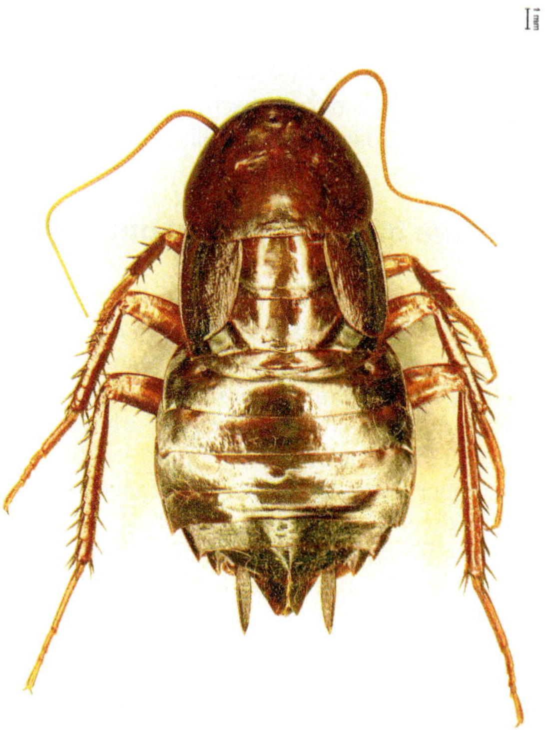

Fig. 5.16. Common Cockroach, *Blatta orientalis*.

they are attracted to cockroach faeces and to other cockroaches and typically cluster together in groups (Wada-Katsumata *et al.*, 2015).

Because cockroaches will feed on faeces as well as on human foodstuffs there is a long tradition of research on cockroaches as potential reservoirs and vectors of pathogenic bacteria (Roth and Willis, 1957). Many species of medically important bacteria have been isolated from wild cockroaches or have been shown to persist in the digestive tract of cockroaches which have been fed with bacteria. Nevertheless, epidemiological evidence to support a role for cockroaches in the spread of disease is scanty and the significance of cockroaches as disease vectors is uncertain. One of the more persuasive studies to have implicated cockroaches as disease vectors concerns an outbreak of gastroenteritis caused by *Salmonella* on a paediatric ward in a Belgian hospital, which was infested with *Blattella germanica*. *Salmonella* was isolated from one of the cockroaches. Conventional infection control measures had no effect on the outbreak, which was brought to a close only when the cockroaches were exterminated (Graffar and Mertens, 1950). In recent decades increased attention to hygiene means that cockroaches have become a rare sight in British hospitals. As recently as the 1980s it was not particularly unusual to encounter examples of *Blatta orientalis* or *Blattella germanica*, especially when working at night and while walking down dimly lit hospital corridors. A study of the gut flora of cockroaches collected from hospitals and other institutions in London in the 1970s resulted in the discovery of a bacterium new to science, which was named *Escherichia blattae* and later renamed *Shimwellia blattae* (Burgess *et al.*, 1973). Non-native cockroaches are more easily found nowadays in botanical hot-houses. The Australian Cockroach, for example, can be seen in the rainforest biome at the Eden Project in Cornwall.

Male-killing bacteria of ladybirds and other beetles

An as yet unnamed bacterium, allied to *Blattabacterium*, induces male-killing in the Adonis Ladybird, *Hippodamia variegata*, and Cream-streaked Ladybird, *Harmonia quadripunctata*, as well as in an American ladybird, the Spotted Lady Beetle, *Coleomegilla maculata* (Hurst *et al.*, 1999; Majerus, 2006). The Adonis and Cream-streaked Ladybird are two of Britain's 20-odd 'typical' ladybird species. In addition to the typical ladybirds the Family Coccinellidae includes a number of mostly smaller beetle species that are not so obviously recognisable as ladybirds. A large proportion of ladybird species are known to be susceptible to infection with male-killing bacteria and female-biased sex ratios have been recorded in populations of numerous well-known ladybird species, including the Two-spot Ladybird, *Adalia pipunctata*, the Ten-spot Ladybird, *Adalia decempunctata* and the invasive Harlequin Ladybird, *Harmonia axyridis* (Majerus, 1994; Weinert *et al.*, 2007). Bacteria detected in British ladybirds include members of the genera *Wolbachia*, *Rickettsia* and *Spiroplasma*. *Spiroplasma* is a member of the *Mycoplasmatota*, a phylum composed of unusual bacteria which lack cell walls. *Wolbachia* and *Rickettsia* belong to the Class *Alphaproteobacteria*. The male-killing bacterium found in association with Adonis and Cream-streaked Ladybird belongs to the Phylum *Bacteroidota*. The association of male-killing traits with such diverse taxonomic groups of bacteria suggests that male-killing mechanisms have evolved independently in several different lineages of bacteria.

Of 300 European beetle species sampled for potential reproductive parasites almost a third of them were found to be infected with *Wolbachia* (Kolasa *et al*., 2018). Other intracellular bacteria – *Rickettsia*, *Spiroplasma* and "*Candidatus* Cardinium" – were also detected but were less prevalent than *Wolbachia*. *Wolbachia* is known to induce male-killing in the flour beetle, *Tribolium madens* and there is some evidence that *Wolbachia* may be responsible for female-biased sex ratios in a species of flea beetle, *Altica lythri* (Fialho and Stevens, 2000; Jäckel *et al*., 2013). Coleopterists sometimes share anecdotes about the scarcity of males in other beetle species and it seems plausible that infection with male-killing bacteria could account for at least some of these observations. But for most species of beetles, such as the ground beetle, *Carabus glabratus* (Fig. 5.17), any possible effects caused by *Wolbachia* and other symbiotic bacteria on their reproductive biology have yet to be investigated.

Some beetles have been shown to be infected with multiple strains of *Wolbachia* or with multiple species of male-killing bacteria. *Polydrusus inustus*, a species of European weevil was shown to be infected with *Rickettsia* and *Spiroplasma* as well as *Wolbachia*. Numerous weevil species are known to be parthenogenetic, including *Polydrusus inustus*. Weevils can easily be translocated to new sites in tandem with their host plants. Because only a single female is needed to found a new population, several parthenogenetic weevil species have become widespread pests of agricultural and garden plants. The parthenogenetic Vine Weevil, *Otiorhynchus sulcatus*, a common pest of gardens in Britain and elsewhere, is a well-known example. Treatment of weevils with anti-*Wolbachia* antibiotics interferes with egg hatching, suggesting that

Fig. 5.17. *Carabus glabratus*, a ground beetle. One of many beetle species prone to infection with *Wolbachia*. Cairngorm National Park, UK.

fertility in *Otiorhynchus sulcatus* depends in some way on the presence of *Wolbachia* (Son *et al.*, 2008). However, the significance of *Wolbachia* and other bactéria as potential inducers of parthenogenesis in weevils has not been ascertained. Most parthenogenetic weevil species are thought to have arisen as a result of hybridisation (Mazur *et al.*, 2016).

"*Candidatus* Cardinium"

"*Candidatus* Cardinium" (Phylum *Bacteroidota*) is the proposed name of a genus of male-killing bacteria originally described as a symbiont of *Encarsia hispida*, a species of parasitic wasp. "*Candidatus* Cardinium" (Fig. 5.18) has also been detected as an intracellular symbiont in spiders, mites, planthoppers and beetles, including the flea beetle *Altica oleracea*, the tortoise beetle *Cassida viridis* and the weevil *Tanymecus palliatus*. *Encarsia* wasps are tiny parasitoids of whitefly. One species, *Encarsia formosa*, is used commercially as a biological control agent of whitefly in greenhouses. "*Candidatus* Cardinium" induces parthenogenesis in several species of *Encarsia* wasp (Zchori-Fein *et al.*, 2001). Infection with "*Candidatus* Cardinium" can also cause cytoplasmic incompatibility, resulting in high rates of mortality among the offspring of matings between infected males with uninfected female wasps. "*Candidatus* Cardinium" spreads rapidly among captive colonies of *Encarsia* wasps following exposure to an infected female (Harris *et al.*, 2010). Electron micrographs of infected insect cells reveal arrays of fibril-like structures inside the bacteria. It has been suggested that the fibril-like structures might operate as protein secretion systems. A likely mechanism by which the bacteria might be able to manipulate the reproductive biology of the host insect is by introducing effector proteins into the host cell (Penz *et al.*, 2012).

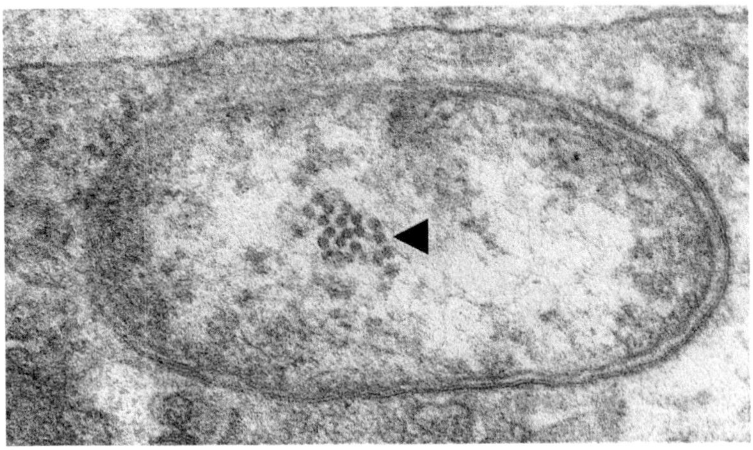

Fig. 5.18. Electron micrograph showing *Cardinium* in *Encarsia pergandiella* ovary follicle cell. Arrow points to fibril structures representing the putative secretion system for translocation of effector proteins into the host cells (Image cropped from figure 3 in Penz *et al.*, 2012). Creative Commons Attribution Licence.

Bacterial symbionts of bees

The British fauna includes about 20 species of bumblebee (*Bombus* species) and more than 200 different solitary bees. Bumblebees, and the honey bee, *Apis mellifera* are eusocial insects that nest in colonies and take care of their young. The colonial lifestyle of bumblebees and honey bees lends itself to the establishment and maintenance of a distinctive and beneficial socially transmitted gut flora that sets them apart from solitary bees. The European honey bee, *Apis mellifera* now occurs worldwide, largely as a domesticated and feral insect. Most of the seven species in the genus *Apis* occur in Asia and it seems possible that the genus might have originated in Asia. Bumblebees occur naturally in the Americas, Europe and Asia. Several species have been introduced to New Zealand. Honey bees and bumblebees collect pollen to feed their larvae and forage for nectar as a source of food, some of which they store as honey.

The honey bee, *Apis mellifera*, collects nectar from flowers and transports it back to the hive in its crop, a specialised chamber of the bee's foregut. The crop contains a complex community of lactic acid bacteria. Lactic acid bacteria are fermentative bacteria which are involved in the conversion of nectar to honey (Vásquez *et al.*, 2012). Humans use lactic acid bacteria to produce a range of fermented foodstuffs and it could be argued that honey is a fermented food produced by bees. The most abundant species of lactic acid bacterium in the honey bee crop flora is *Lactobacillus kunkeei*, a bacterium which was originally described in the context of the wine industry as a detrimental bacterium that inhibits the growth of yeast. As a member of the honey bee crop flora, it seems likely that the bacterium plays a role as a growth suppressant of environmental yeasts and other fungi.

Social bees appear to have coevolved with their symbiotic bacteria. Bumblebees, which also produce honey, are associated with their own particular strains of lactic acid bacteria, which differ from those found in honey bees (McFrederick *et al.*, 2013). Whereas the symbiotic bacteria associated with some insects are transmitted via the ovum, bee symbionts are transmitted socially. It has been shown that the gut bacteria of adult bumblebees are acquired after emergence by exposure to the faeces of other bees in the nest. Most of the bacteria that populate the midgut of bumblebees belong to a select coterie of specialised bee symbionts, including members of the genera *Snodgrassella* (Class *Betaproteobacteria*), *Gilliamella* (Class *Gammaproteobacteria*), "*Candidatus* Schmidhempelia" (*Gammaproteobacteria*), *Bombiscardovia* (Phylum *Actinomycetota*) and *Bombilactobacillus* (Phylum *Bacillota*). Many of these genera can also be found in the midgut of worker honey bees (Hammer *et al.*, 2021).

Lactic acid bacteria help to protect honey bees against *Melissococcus plutonius*, a pathogenic bacterium that causes European foulbrood and *Paenibacillus larvae*, the causative agent of American foulbrood. *Paenibacillus larvae* is a spore-forming bacterium that occurs worldwide. Infected bee larvae die sealed within their cells. The larval remains remain contaminated with infectious spores which can survive for years. Researchers in Switzerland have shown that the normal gut flora of the Buff-tailed Bumblebee, *Bombus terrestris*, a common species in Britain and Europe, protects bees from infection with a parasitic protozoan, *Crithidia bombi* (Koch and Schmid-Hempel, 2011).

Queen honey bees host a distinctive gut microflora characterised by large numbers of acetic acid bacteria. Worker bees also host acetic acid bacteria but not in such large numbers as queens. Acetic acid bacteria convert sugars to acetic acid. *Bombella apis* (*Alphaproteobacteria*) is an acetic acid bacterium that forms part of the gut flora of queen honey bees. It can also be found in royal jelly, a secretion produced by worker bees that is fed to the larvae of queens as well as workers (Smith *et al.*, 2021a). *Commensalibacter melissae* is another species of acetic acid bacterium which has been cultured from honey bees. Other members of the genus have been cultured from fruit flies and butterflies. *Commensalibacter papalotli* is an acetic acid bacterium cultured from the gut contents of an overwintering Monarch butterfly, *Danaus plexippus*, in Mexico (Servín-Garcidueñas, 2014). Monarch butterflies are rare visitors to the British Isles (Fig. 5.19). Most sightings are made during late summer and early autumn, a time of year that corresponds with the annual migration of North American Monarchs to their overwintering grounds in Mexico.

Reed beetle symbionts

Reed beetles (subfamily Donaciinae) host symbiotic bacteria that enable reed beetle larvae to survive on a nutritionally poor diet of plant sap and reed beetle adults to digest pectin-rich plant tissue (Reis *et al.*, 2020). Reed beetles are conspicuous insects. Some are metallic green, while others are bronze, purple or multicoloured. Reed beetles are represented in the British Isles by three genera: *Macroplea* (two species), *Donacia* (15 species) and *Plateumaris* (four species). Some are widespread and common while others such as the multicoloured Zircon

Fig. 5.19. Monarch butterfly found on the seafront at Brighton, Sussex, UK in the wake of Tropical Storm Humberto, 9 October 2001.

reed beetle, *Donacia aquatica*, are local and rare. The adult beetles can be found during the warmer months in wetland habitats, often on their food plants, which depending on the species of beetle include water-lilies, broad-leaved pondweed, *Potamogeton natans* and members of the grass family (Poaceae), such as bur-reeds, *Sparganium* and common reed, *Phragmites australis*. Unlike the terrestrial adults, the aquatic larvae feed on the nutrient-deficient sap of submerged vegetation. The foregut of the larva contains special diverticula or pouches that house symbiotic bacteria. The symbionts provide reed beetle larvae with essential amino acids and vitamins. Reed beetle larvae construct submerged air-filled cocoons in which the larvae can pupate and safely overwinter. Reed beetle larvae line their cocoons with a water-repellent proteinaceous substance. Construction of the cocoons is impeded in symbiont-free larvae treated with antibiotics. The symbiotic bacteria persist in adult beetles in the Malpighian tubules, which are specialised branches from the hindgut. Female beetles inoculate the outer surfaces of their eggs with bacteria, allowing the transmission of symbionts to the next generation of larvae (Kleinschmidt and Kölsch, 2011).

The symbiotic relationship between reed beetles and bacteria was documented in the 1930s by a German biologist who noted the presence of bacteria inside specialised cells or bacteriocytes seen on microscopic examination of the larvae (Stammer, 1935). As is often the case with specialised symbionts, reed beetle bacteria have not been cultured in the laboratory, although genetic analysis shows that they are members of the Class *Gammaproteobacteria* and are more closely related to other known symbionts than to free-living bacteria. The symbionts associated with two species of reed beetle in the genus *Macroplea*, both of which are rare beetles in Britain, have been described under the designation *Candidatus* as "*Candidatus* Macropleicola appendiculatae" and "*Candidatus* Macropleicola muticae". The symbionts associated with other reed beetle genera have not been formally described but enough is known about them at a molecular level to allow the construction of a phylogenetic tree to illustrate their genetic relatedness. The phylogenetic tree for reed beetle symbionts maps closely onto the phylogenetic tree of their beetle hosts, showing that the bacteria and beetles have coevolved. Reed beetle lineages associated with pectin-rich plants, such as water lilies and pond weeds, host symbionts that produce pectinase enzymes. However, the capacity to produce pectinases has been lost by the symbionts of beetle lineages that feed on pectin-deficient reeds, sedges and rushes (Poaceae). The reed beetles are estimated to have emerged as a group somewhere between 75 and 100 million years ago (Kölsch *et al.*, 2008). Reed beetles are classified as a distinctive subfamily, Donaciinae within the leaf beetle family, Chrysomelidae. Most chrysomelids are essentially terrestrial although members of another subfamily, Sagrinae which occur in Asia, also build impervious cocoons and are colonised with symbionts that are related to reed beetle symbionts (Kölsch and Synefiaridou, 2012).

Antibiotic-producing symbionts of invertebrates

Many of the antibiotics developed by the pharmaceutical industry to treat bacterial and fungal infections have been derived from naturally occurring substances produced by soil bacteria. Similarly, some invertebrates derive protective benefit from naturally occurring antibiotics of bacterial origin. Leaf-cutter ants which occur in tropical and subtropical parts of the Americas harvest leaves that they use as the raw material for the cultivation of fungus gardens in their nests. The ants themselves are colonised with Gram-positive actinomycete bacteria belonging to the genus *Pseudonocardia* which appear to play a role in protecting the ants' fungus gardens from attack by a parasitic fungus called *Escovopsis* (Mueller *et al.*, 2010). There is also some evidence to suggest that the symbiotic bacteria offer some protection to the ants from direct attack by pathogenic fungi.

Originally from Asia, the Oriental Shrimp, *Palaemon macrodactylus*, is a globally invasive crustacean that has spread to North and South America, Australia and Europe. In 2002 the Oriental Shrimp was found in the rivers Stour and Orwell in Essex and Suffolk and in 2006 it was the most frequently collected palaemonid shrimp in the water intake of Tilbury Power Station in the Thames Estuary (Worsfold and Ashelby, 2008). Embryos of *Palaemon macrodactylus* are unusually resistant to attack by a fungus called *Lagenidium callinectes*, an otherwise lethal pathogen of related marine crustaceans. Protection from fungal infection is conferred by the presence of a symbiotic member of the genus *Alteromonas*, which produces isatin (2,3-indolinedione), a compound with antifungal properties (Gil-Turnes *et al.*, 1989). Isatin is the same substance as that derived from woad, *Isatis tinctoria*, used as a basis for the production of the well-known natural blue dye.

Antifungal symbiont of beewolves

Beewolves (Crabronidae: Philanthinae) are big attractive solitary wasps (Fig. 5.20). Beewolves prey on bees, with whose bodies they stock their burrows as food for their larvae. The species of beewolf found in Britain is *Philanthus triangulum*, which is represented by several different subspecies in mainland Europe, Asia and Africa. Other species of beewolf occur in Europe, Asia and North America. Until recently the beewolf was a rare insect in Britain and hymenopterists wishing to see one would have to make a special trip to the Isle of Wight. The Isle of Wight is still an excellent place to see beewolves. In summer it is easy to find aggregations of beewolf nests on the island in dry sandy soils at such locations as Redcliff and Headon Warren. Since the 1980s the beewolf has been spreading northwards. The species is now widespread in southern England and there are scattered records also from the north. A nesting colony was found on Battersea Bridge roundabout in London (Edwards, 1997). Adult beewolves are on the wing from about July to September. The female stocks each brood cell with one or more bees. Beewolf larvae spin cocoons in which they overwinter. During the course of construction of each brood cell the female secretes large amounts of a white substance (Fig. 5.21) from special antennal glands (Kaltenpoth *et al.*, 2005). The white secretion appears to protect the larva from attack by fungi during its long sojourn underground in its humid cell. Experiments with captive larvae have shown that the white secretion exerts

Fig. 5.20. Beewolf, *Philanthus triangulum*. Fishersgate, West Sussex, UK.

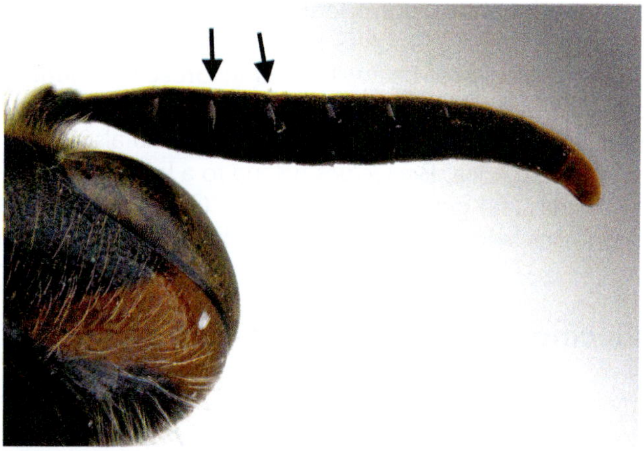

Fig. 5.21. Beewolf antenna showing white secretion which harbours bacteria with antifungal properties.

a protective effect. A Gram-positive bacterium (Fig. 5.22) has been detected in the antennal gland secretions of the female beewolf and described as a new species, '*Candidatus* Streptomyces philanthi' (Phylum *Actinomycetota*) (Kaltenpoth *et al*., 2006; Nechitaylo *et al*., 2014). Using molecular methods it was shown that the beewolf bacterium belonged to the genus *Streptomyces* and that it differed sufficiently from known species to be considered a new species, although because it was not possible to culture the new species in the laboratory it was described under the designation of *Candidatus*. Streptomycetes are slow-growing filamentous soil bacteria that are well-known as producers of antimicrobial substances. Many of the antibiotics that are used to treat infections

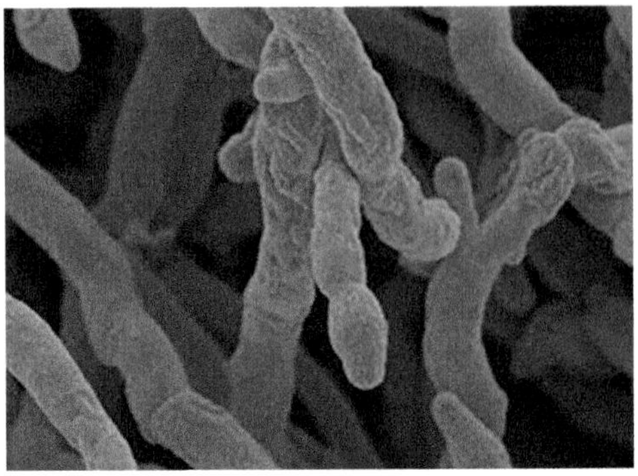

Fig. 5.22. Electron micrograph of the beewolf symbiont "*Candidatus* Streptomyces philanthi" (Image cropped from figure 1 in Nechitaylo *et al.*, 2014). Creative Commons Public Domain.

are obtained from streptomycetes. At some time in evolutionary history, a symbiosis developed between the ancestor of modern beewolves and a streptomycete with antifungal properties. '*Candidatus* Streptomyces philanthi' has been detected in all beewolves tested for the bacterium, including *Philanthus venustus*, a species of beewolf found in Greece, *Philanthus coarctus*, a beewolf species from Oman, *Philanthus triangulum diadema*, which is the South African subspecies of the European beewolf and also many species of beewolf from North America. Beewolves belong to a large family of solitary wasps with many species and genera. A search for '*Candidatus* Streptomyces philanthi' among wasps belonging to other related genera drew a blank.

Toxin-producing symbionts of rove beetles

Dermatitis linearis is a medical term used to describe blistering of the skin caused by exposure to pederin, a toxin present in the body fluids of beetles of two rove beetle genera, *Paederus* and *Paederidus* (Family Staphylinidae). The condition is also known as pederin dermatitis. Pederin, the vesicant toxin present in the haemolymph of rove beetles is produced by an as yet unnamed symbiotic bacterium allied to *Pseudomonas* (Class *Gammaproteobacteria*). Pederin dermatitis is similar in appearance to that caused by the better known and taxonomically and toxicologically unrelated Spanish fly and blister beetles of the families Meloidae and Oedomeridae. *Paederus* species are contrastingly coloured and in common with other rove beetles have short wing cases and rather elongated bodies. The genus has a worldwide distribution. Four species occur in the British Isles. All of them have contrasting red and black markings and metallic blue-black wing cases. *Paederus littoralis* is generally common in southern Britain in both dry and wet habitats. *Paederus fuscipes* and *Paederus riparius* occur more locally in marshes and on river banks. *Paederus caligatus* is very local and recorded mainly from bogs in the New Forest. Cases of pederin dermatitis has been recorded from France, Italy, Turkey, Iran, Iraq,

Pakistan, India, China, Malaysia, Australia and large areas of Africa and South America (Bong *et al*., 2015). Several *Paederus* species have been implicated, including two of the British species, *Paederus fuscipes* and *Paederus riparius*. Pederin dermatitis is a frequently reported dermatological condition in northern Iran, where it has been linked to exposure to *Paederus fuscipes*. Rove beetles sometimes take to the wing in large numbers and it is not uncommon for rove beetles to fly into eyes or to be accidentally rubbed or crushed against the skin. In East Africa the rove beetles *Paederus eximius* and *Paederus sabaeus* are known locally as Nairobi fly, or as Nairobi eye (Fig. 5.23). One might imagine that entomologists in Britain who regularly handle beetles in the field would be at risk of pederin dermatitis. However, an appeal circulated to the Beetles of the British Isles online discussion forum generated only one response from a coleopterist who described a burning reaction after handling one of the beetles.

Adult female beetles are either pederin producers or non-producers and males and larvae that contain pederin have either acquired the toxin maternally as eggs from a pederin-positive female or have obtained the toxin by eating other pederin-positive beetles. The capability to produce pederin is inherited from pederin-producing females but can also be acquired by pederin-negative females that eat pederin-positive eggs (Kellner, 2000; Piel, 2002). Molecular

Fig. 5.23. *Paederus sabaeus*, a species of rove beetle known in East Africa as Nairobi eye. Nairobi, Kenya.

studies show that pederin-producing beetles harbour an as yet uncultured bacterium related to *Pseudomonas aeruginosa*. The gene cluster for pederin synthesis appears to be part of a genomic island of transferable DNA that was acquired most probably from some other kind of organism. Curiously, marine sponges are also known to produce pederin-like compounds.

The book of Exodus describes a series of plagues which occurred in Egypt, most probably at some time around 1450 BC, or possibly at a later date around 1225 BC. The first plague, when the water turned to blood, has been interpreted as a toxic algal bloom, otherwise known as a red tide. Such an event would have triggered a die off of the aquatic fauna. Decaying organic matter on the banks of the Nile would have generated ideal conditions for a massive expansion of detritivorous invertebrates as well as predatory rove beetles, leading to the third and fourth plagues when swarms of insects invaded the Egyptian dwellings. An epidemic of blistering pederin dermatitis caused by exposure to swarms of rove beetles is a possible explanation for the sixth plague of boils (Norton and Lyons, 2002).

Tetrodotoxin-producing bacteria

Many species of bacteria produce tetrodotoxin, a neurotoxin which inhibits muscle contraction. Creatures as diverse as octopuses, puffer fish, newts, toads, crabs, arrow worms and ribbon worms accumulate tetrodotoxin in their tissues. Tetrodotoxin provides them with protection against predation. Some of them utilise tetrodotoxin to suppress their prey. Blue-ringed octopuses, *Hapalochlaena* spp., which occur in the Pacific and Indian Oceans from Australia to Japan, concentrate tetrodotoxin in the posterior salivary gland (Fig. 5.24). The sea creatures on which the octopus feeds are envenomed by its bite. Several different species of tetrodotoxin-producing bacteria have been cultured from the posterior salivary gland and intestinal flora of the octopus.

Arrow worms (Chaetognatha) are small planktonic marine animals that grow to a maximum length of about 4 cm. They occur throughout the world's oceans where they feed on other forms of planktonic life. Species recorded from British waters include *Flaccisagitta lyra* and *Eukrohnia hamata*. A tetrodotoxin-producing bacterium, *Vibrio alginolyticus*, was cultured from the heads of arrow worms – including the globally distributed species *Flaccisagitta lyra* and *Eukrohnia hamata* – captured by researchers working off the coast of Japan (Thuesen *et al.*, 1988). It is thought that arrow worms may use tetrodotoxin to subdue their prey.

Ribbon worms (Nemertea) occur worldwide and can grow to an enormous size. According to the *Guinness Book of Records*, the longest worm ever recorded was an example of a species of ribbon worm known as the Bootlace worm, *Lineus longissimus*, washed up at St Andrews in 1864, which measured 55 m (180 feet) in length. Specimens of *Lineus longissimus* and six other species of ribbon worm collected from the intertidal zone of the North Wales coast were positive for tetrodotoxin and yielded toxin-producing bacteria (Carroll *et al.*, 2003). Ribbon worms are known to use a variety of toxins

Fig. 5.24. Poster warning of blue-ringed octopus, Western Australia.

for their own defence as well as for subduing their prey. When handled the Bootlace worm produces copious quantities of toxic mucus. Ribbon worms form part of the diet of puffer fish.

Puffer fish (family Tetraodontidae) occur globally in the oceanic waters of the tropics and subtropics. Occasional specimens are reported from Cornwall and in 2014 one was found washed up on Chesil Beach in Dorset. Puffer fish feed mainly on invertebrates, including ribbon worms. Puffer fish contain high concentrations of tetrodotoxin, especially in the liver, ovaries and intestines. Consumption of a few milligrams of toxin may lead to paralysis and death. Although puffer fish are potentially lethal unless correctly prepared, they are considered to be a delicacy in Japan, China and Korea. A variety of different toxin-producing bacteria have been cultured from different kinds of puffer fish. It is thought that puffer fish might obtain toxin-producing bacteria, as well as supplies of tetrodotoxin itself, from their diet of toxic invertebrates. Puffer fish also contain the neurotoxin saxitoxin, which is produced by cyanobacteria and dinoflagellates.

Ticks and bacteria

Ticks (Class Arachnida) are parasitic arthropods that feed on the blood of mammals, birds and reptiles. About 20 species occur in the British Isles. Some of them have narrow host ranges and limited geographical distributions. The Sand Martin Tick, *Ixodes lividus*, for example, feeds on sand martins whilst the Marsh Tick, *Ixodes apronophorus*, is recorded mainly from East Anglia. The Sheep Tick, *Ixodes ricinus* and the Hedgehog Tick, *Ixodes hexagonus* (Fig. 5.25) are common species that feed on multiple host species and are the ticks most often found attached to humans. Removing ticks is a routine activity for field naturalists who like to spend time grovelling in the undergrowth. It is well known that ticks can transmit pathogenic bacteria, such as *Borrelia burgdorferi*, the agent of Lyme disease, to humans and other animals (Saylor *et al.*, 2022). Ticks also host symbiotic bacteria which are transmitted to the next generation of ticks matrilineally.

"*Candidatus* Midichloria mitochondrii" (Class *Alphaproteobacteria*) is an uncultured bacterium found in the ovaries of female ticks. Some members of the *Alphaproteobacteria* share a propensity to engage in intimate relationships with eukaryotic cells, living either as mutualistic intracellular symbionts or as pathogenic agents. "*Candidatus* Midichloria mitochondrii" is apparently unique among intracellular symbionts in that it resides and replicates inside mitochondria. The ovaries of infected ticks continue to function and the bacterium is transmitted vertically to the next generation of ticks (Sacchi *et al.*, 2004). The first published report of these strange bacteria dates from the 1970s, when microorganisms were seen inside mitochondria in electron micrographs of the ovaries of captive Sheep Ticks in England (Lewis, 1979). Similar organisms were detected in ticks collected in Switzerland and Italy.

Fig. 5.25. Hedgehog tick, *Ixodes hexagonus*, East Sussex, UK.

A prevalence study later showed that '*Candidatus* Midichloria mitochondrii' was present in the ovaries of all female Sheep Ticks examined from 12 different countries. Similar bacteria have been found in association with various other tick species.

Other tick-associated intracellular symbionts include *Anaplasma, Rickettsia* and *Spiroplasma*. Anaplasmatosis is a disease of livestock, domestic animals, deer, foxes, hedgehogs, birds, rodents and humans, caused by *Anaplasma phagocytophilum* (Class *Alphaproteobacteria*). Human cases of the disease are rarely diagnosed. Symptoms include fever, headache, muscle pain and nausea. The bacterium shows a propensity to invade white blood cells. The bacterium can survive in chilled blood. On rare occasions disease transmission has occurred via blood transfusion. The majority of infections are acquired from tick bites, although only a minority of ticks are infected. A survey involving the testing of almost 3000 tick specimens detected a prevalence of infection of less than 5% in ticks in the UK (Keyte *et al.*, 2021).

The genus *Rickettsia* is named in honour of Howard Taylor Ricketts (1871–1910) an American pathologist who made the discovery that Rocky Mountain spotted fever, a disease caused by *Rickettsia rickettsiae*, was transmitted by ticks. Ricketts died after contracting endemic typhus, a louse-borne infection caused by the bacterium *Rickettsia prowazekii* during the course of a disease outbreak investigation in Mexico City. Stanislaus von Prowazek (1875–1915) was a Czech zoologist who died after contracting endemic typhus while conducting laboratory experiments. A multiplicity of tick-borne rickettsial diseases, have been described from around the world. The bacteria that cause them are known collectively as the spotted fever group of rickettsias. Rickettsial spotted fevers, which vary in their severity, usually present with a blackish scab, or eschar (Fig. 5.26) at the bite site, headache, fever and rash. Rickettsial infections can be treated with antibiotics. Although cases of imported rickettsial infections are diagnosed in the UK on a fairly regular basis, the presence of spotted fever group rickettsias in British ticks is a recent finding (Tijsse-Klasen *et al.*, 2013). As a result of surveys in Wales and southern England, *Rickettsia helvetica* was detected in sheep ticks, *Rickettsia raoultii*

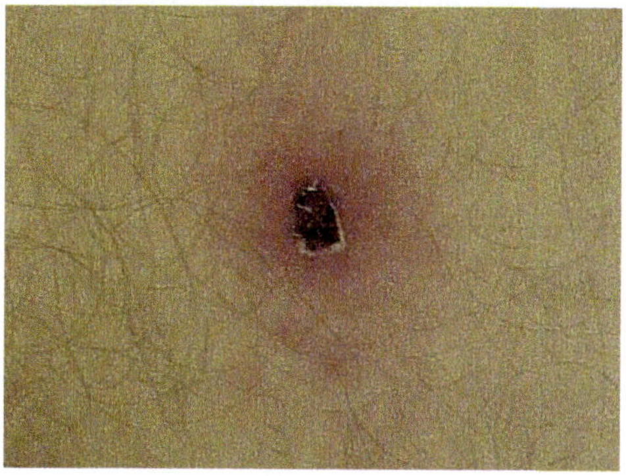

Fig. 5.26. Blackish eschar at a tick bite site of a patient with a rickettsial spotted fever.

was found in association with Ornate Cow Ticks, *Dermacentor reticulatus*, and a bacterium related to *Rickettsia massiliae* was detected in Red Sheep Ticks, *Haemaphysalis punctata*. *Rickettsia felis* is a matrilineally inherited symbiont of cat fleas, *Ctenocephalides felis* (Fig. 5.27) and the agent of flea-borne spotted fever, also known as cat-flea typhus, a disease with a worldwide distribution, which is particularly common in sub-Saharan Africa.

Lyme disease is a tick-borne disease caused by *Borrelia burgdorferi* (Phylum *Spirochaetota*) and other related species including *Borrelia afzelii* and *Borrelia garinii*. Early symptoms and signs include headache, fever, fatigue and in many cases a gradually spreading rash known as erythema chronicum migrans or simply erythema migrans (Fig. 5.28). The rash, which does not start to appear until at least several days after being bitten, slowly expands to reach a size of several inches in diameter. Sometimes the centre of the lesion assumes a bluish appearance. In some cases the rash develops into a ring-shaped or target-like lesion. Erythema migrans is not usually itchy or painful. Erythema migrans usually looks and feels quite unlike the raised itchy bumps that normally develop in reaction to tick bites and to insect bites in general. The late complications of untreated Lyme disease include facial palsy, arthritis, swollen joints and severe headaches. Transmission of Lyme disease can be prevented by removing ticks, ideally within the first day or two after being bitten. Adult ticks and nymphs are more easily spotted than the much smaller larvae. Lyme disease can be treated with antibiotics.

Borrelias are spirochaetes, motile spiral-shaped organisms, able to corkscrew their way through viscous body fluids and tissues (Fig. 5.29). Lyme borreliosis is maintained in nature through networks of transmission between ticks and wild mammals. The tick life cycle involves four stages, egg, nymph, larva and adult. Larvae and nymphs tend to feed on the blood of small rodents, whilst adults tend to feed on larger mammals such deer. Ticks acquire

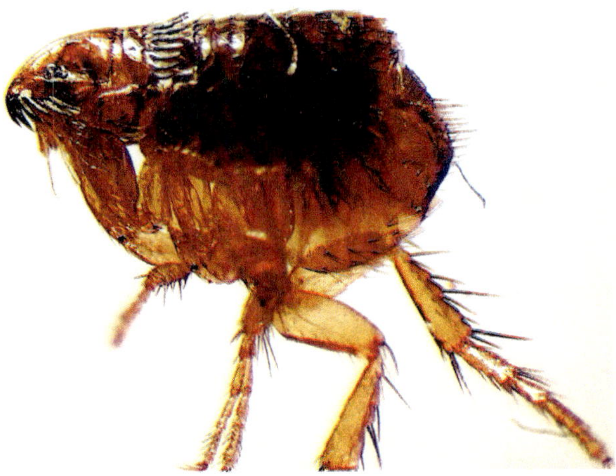

Fig. 5.27. Cat flea, *Ctenocephalides felis*, a host of *Bartonella henselae* and *Rickettsia felis*.

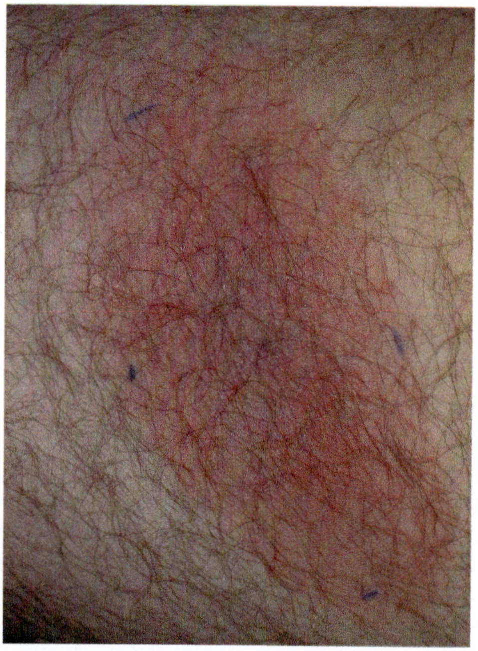

Fig. 5.28. Erythema migrans, a sign of Lyme disease.

Fig. 5.29. *Borrelia burgdorferi*, the Lyme disease spirochaete (Image cropped from figure 4 in Saylor *et al.*, 2022; *PLoS ONE* 17(9): e0274125). Creative Commons Attribution Licence.

spirochaetes by feeding on an infected mammal and remain infectious through successive moults. Humans and domestic animals do not form part of the natural cycle of Lyme borreliosis and act as dead-end hosts for unfortunate spirochaetes. Hotspots for Lyme disease in England are concentrated around the Hampshire Basin, and include the New Forest, the Isle of Purbeck and parts of the intervening Bournemouth and Poole hinterland (Tulloch *et al.*, 2019). The Hampshire Basin is richly endowed with areas of scrubby heathland suitable for ticks and the mammalian hosts of *Borrelia* species. The so-called sheep tick, *Ixodes ricinus*, is the principal vector of Lyme disease in Europe. The Lyme disease agent *Borrelia afzelii* has been detected in Hedgehog Ticks, *Ixodes hexagonus* in Belgium (Jahfari *et al.*, 2017).

Deer ked dermatitis

The bacterium *Bartonella schoenbuchensis* (Class *Alphaproteobacteria*) is a suspected agent of deer ked dermatitis, a dermatological condition associated with deer ked bites (Dehio *et al.*, 2004). Deer keds, *Lipoptena cervi* (Order Diptera, family Hippoboscidae) are blood-sucking flies that feed on members of the deer family, including red deer, roe deer and fallow deer (Fig. 5.30). After reaching a suitable host, deer keds shed their wings and assume a louse-like existence. *Bartonella schoenbuchensis* is one of a number of *Bartonella* species known to infect ruminants and other mammals. *Bartonella schoenbuchensis* has been isolated from deer ked adults as well as larvae, indicating that the bacterium can be passed on vertically to the next generation of flies and that deer keds probably act as vectors of the bacterium (Regier *et al.*, 2018). The term deer ked dermatitis refers to raised pigmented lumpy lesions that develop at deer ked bite sites (Fig. 5.31). The lumpy lesions may persist for months and share features with similar cutaneous lesions that are associated

Fig. 5.30. A deer ked, *Lipoptena cervi*, that bit the author while investigating a deer carcass, Hurston Warren, West Sussex, UK.

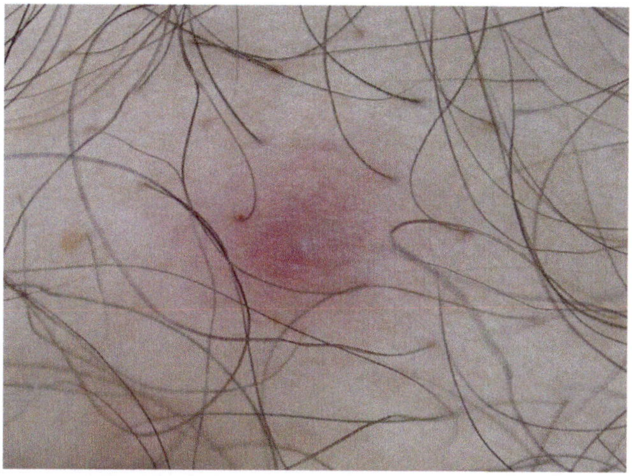

Fig. 5.31. Deer ked dermatitis.

with a number of diseases caused by *Bartonella* species. Cases of cat scratch fever, for example, a disease caused by *Bartonella henselae*, present with fever, swollen lymph nodes and papular skin lesions which resemble those of deer ked dermatitis. *Bartonella henselae* is a gut symbiont of cat fleas, *Ctenocephalides felis*. Viable bacteria persist in flea faeces and it is thought that humans acquire the infection following the inoculation of flea faeces into scratch wounds made by contaminated cats' claws (Billeter *et al*., 2008). Cats infected with *Bartonella henselae* usually tolerate the infection without showing symptoms.

Symbionts of blood-sucking leeches

When medicinal leeches (Phylum Annelida) feed they more than double their weight by ingesting a blood meal. Medicinal leeches subsist on a diet of blood (Fig. 5.32). Ingested blood is stored in a crop which forms the anterior part of the leech's gut. The leech's ability to store intact red blood cells in a crop allows it to survive for months without feeding. The crop flora is dominated by two symbionts, an *Aeromonas* species (Class *Gammaproteobacteria*) and an anaerobic bacterium, *Mucinivorans hirudinis* (Phylum *Bacteroidota*). The two symbionts also occur in the leech's hind gut where they reside in the company of additional species of bacteria. The two principal symbionts use proteins from ingested blood as a food source. *Mucinivorans hirudinis* also feeds on mucus secreted by cells lining the leech's gut and releases acetic acid as a fermentation product (Marden *et al*., 2016). Blood is deficient in B vitamins. The genome of *Mucinivorans hirudinis* encodes synthetic pathways involved in the production of vitamin B12.

Aeromonas species are associated with aquatic environments. *Aeromonas* species have the capacity to produce a haemolytic toxin and are potential

Fig. 5.32. *Three leeches in the role of physicians attend a grasshopper in the role of the patient and propose diet and bloodletting*. Coloured lithograph after J.J. Grandville, c. 1832 (Wellcome Collection). Public Domain Mark.

pathogens. Wound infection is a recognised complication of fish bites, reptile bites and leech bites. Nevertheless, medicinal leeches are still used in some areas of clinical practice, for example to manage congested tissues after certain kinds of surgery. Only one species of medicinal leech, *Hirudo medicinalis*, is native to the British Isles. A second species of medicinal leech, *Hirudo verbana*, occurs in mainland Europe. Both species, as well as additional species from other parts of the world have been marketed as medicinal leeches. *Hirudo medicinalis* has a limited geographical distribution in Britain and the medicinal leech has long been regarded as a rarity. Recent records indicate the presence of scattered populations in Scotland, Wales, the New Forest, Cumbria and Kent. Writing in 1915, the zoologist Arthur Shipley documented its decline, stating that '*this species is now almost extinct in England, although I know of a naturalist who can still find them in the New Forest, but he will not tell me where*'. A residual hotspot for leeches today is the Romney Marsh and Dungeness area of Kent. Medicinal leeches are attracted to disturbance and it is not particularly unusual to come across the occasional specimen when investigating the margins of disused gravel pits at Dungeness. Of the medicinal leech Shipley noted: '*It is a delicious harmony of reddish-browns and greens and blacks and yellows, a beautiful soft symphony of velvety orange and olive black, the markings being repeated on each segment, but not to the extent of a tedious repetition. So beautiful are they that the fastidious ladies who adorned the* salons *at the height of the leech mania, during the beginning of the eighteenth century, used to deck their dresses with embroidered leeches* ...' (Shipley, 1915).

Vertebrates

Animals are habitats that provide food and shelter for specialised bacteria. The bodies of vertebrate animals are a composite of microbial sub-habitats that consist of internal and external surfaces and cavities. In order to understand vertebrate animals as habitats for bacteria, it is helpful to distinguish between normally sterile and normally non-sterile body sites. Normally sterile sites include the bloodstream, bones, muscles, brain, cerebrospinal fluid, liver, kidneys and various other organs and tissues. These are normally microbe-free zones that are kept sterile by the integrity of the skin, by other physical barriers and by the immune system. The presence of bacteria in such normally sterile sites is not normal in healthy vertebrate animals and is a signal of invasive bacterial disease. In contrast to vertebrate animals, many invertebrate species accommodate symbiotic bacteria in their body fluids and inside specialised cells within their body tissues.

Most of our cells are bacteria. A human body consists of about 10 trillion (10^{13}) human cells and harbours about 100 trillion (10^{14}) bacterial cells. Bacteria though are much smaller than the eukaryotic cells that make up the bodily tissues of mammals and much of the human bacterial flora resides within the confines of the gastrointestinal tract. The bacterial floras, or microbiomes, of different animal species are broadly similar, although particular species of bacteria may be host-specific, or limited to a narrow range of host species. The genus *Staphylococcus*, for example, includes numerous species of bacteria that live on the skin and mucosal surfaces of animals. Cattle, goats, horses, bats, otters, pigs, dogs, voles and humans are all associated with particular species of staphylococci. Similarly, rats, mice, cats, hamsters, ferrets, otters, Canada geese and humans each have their own distinctive species of *Helicobacter*. *Helicobacter* species are specialised bacteria that reside in the stomach linings of birds and mammals. Much of the world's population is infected with *Helicobacter*. Some people suffer from gastritis as a result of infection but most remain without symptoms. *Escherichia coli* is more of a generalist. *Escherichia coli* is a genetically variable and versatile bacterium which can be cultured from many different kinds of animal.

The normal human flora is composed of bacteria that are often referred to as commensal bacteria. Commensals are usually harmless or at least not especially harmful. Commensal bacteria may benefit the host by inhibiting colonisation by pathogenic organisms. Lactobacilli colonise the female genital tract where they convert glycogen to lactic acid. Acidity, as well as antimicrobial substances produced by lactobacilli inhibit the growth of potential pathogens such as the yeast *Candida albicans*. Similarly, the gut flora appears to protect against colonisation by pathogenic bacteria. Disruption of the normal gut flora by treatment with broad spectrum antibiotics creates conditions that are favourable for the growth of toxigenic strains of *Clostridioides difficile* (formerly known as *Clostridium difficile*).

Our commensal bacteria are, like us, heterotrophs that depend on an external supply of organic carbon compounds as a source of food. Gut bacteria feed on the food that we ingest. Other commensals feed on substances present in secretions on mucosal surfaces or the skin. At sites such as the teeth, bacteria live as biofilms and secrete an extracellular polymeric substance to create a favourable bacterial microhabitat that we call dental plaque. In 1683 Leeuwenhoek reported to the Royal Society that he had discovered animalcules by looking at scrapings from his teeth through one of his homemade microscopes. This is considered by many to represent the earliest recorded unequivocal observation of bacteria. The lumen of the gut is a special habitat for bacteria. Animal guts contain vast numbers of bacteria that outnumber the number of cells in their own bodies. Much of the gut is a relatively anoxic environment. In the absence of oxygen, gut bacteria are not able to respire aerobically. Many gut bacteria obtain energy from carbohydrates, derived from the food eaten by their host organism, via anaerobic respiration (fermentation). Many gut bacteria are strict anaerobes such as members of the genus *Bacteroides* (Phylum *Bacteroidota*) and *Fusobacterium* (Phylum *Fusobacteriota*). Others, such as *Escherichia coli*, can switch between aerobic and anaerobic respiration depending on circumstances. Fermentative break down of carbohydrates releases compounds of value to the animal host as well as to the bacteria themselves.

Breakdown products of carbohydrates also sustain methanogenic archaea that live in the gut. Hydrogen at high concentration inhibits fermentative processes. By converting hydrogen into methane, methanogens facilitate fermentation and increase the availability of small carbohydrate molecules both to the gut flora and to the host organism. Herbivorous animals that are able to process cellulose release large quantities of methane as a result of the activity of their gut methanogens. Anaerobic bacteria also produce volatile fatty acids as waste products. Different anaerobes produce different waste products and at one time there was a vogue in diagnostic microbiology laboratories for using liquid-gas chromatography to characterise volatile carbohydrate profiles as a means of identifying anaerobic bacteria. Gnotobiotic mice that are delivered by caesarean section and raised in the laboratory to be free of bacteria have been noted to produce odourless faeces. Gnotobiotic mice allow assessment of the roles of particular species of bacteria in mammalian health. It was shown that a gut bacterium, *Bacteroides thetaiotaomicron*, shared by humans and mice, plays a role in the development of a healthy mammalian immune system. Relationships with bacteria from which the host derives benefit are described as mutualistic symbioses. Much of what is known about the normal flora of animals relates to studies of humans and laboratory animals such as mice and domestic animals of commercial importance.

Commensals and pathogens of vertebrate animals

Most of the commensal bacteria that normally colonise our bodies are harmless or almost harmless. Nevertheless, an otherwise harmless bacterium in the wrong place may become a pathogen. Normally harmless bacteria from the normal oral flora, for example, are the main causes of bacterial endocarditis, a potentially life-threatening condition in which bacteria proliferate on the valves of the heart. Bacteria associated with endocarditis, such as oral streptococci, specialise in binding to tissue surfaces. If introduced to the bloodstream they have a propensity to bind to the surfaces of damaged heart valves. Some of the bacteria that reside as usually harmless commensals in the upper airways, such as *Streptococcus pneumoniae* (the pneumococcus, a common agent of pneumonia), *Neisseria meningitidis* (the meningococcus, a cause of bacterial meningitis) and *Haemophilus influenzae* (a cause of meningitis and other infections) sometimes behave as pathogens. Different strains of these bacteria vary in virulence and their human hosts vary in susceptibility to invasive disease. Bacteria that usually live harmlessly on body surfaces, such as *Staphylococcus aureus*, can cause soft tissue infection and bloodstream infection if allowed to gain entry through penetrating injuries of the skin.

The bacterial diseases of animals are similar to those of man. Some pathogenic bacteria have narrow host ranges, while others such as *Mycobacterium tuberculosis*, act as agents of infection in a number of different animal species. Infections acquired from animals are called zoonoses. Infections acquired from humans by animals are called anthroponoses. Because birds have higher body temperatures than mammals, bird-associated bacteria may grow optimally at slightly higher temperatures than the bacteria associated with mammals.

The normal flora of the skin

The mammalian skin is composed of sub-habitats, which include: relatively dry potentially sun-exposed superficial surfaces; partially enclosed microhabitats such as hair follicles and sebaceous glands; as well as anatomically distinctive sub-habitats, such as the spaces between digits. Bacteria that normally live on the skin may benefit the host by providing protection against colonisation by pathogenic microbes. Volatile substances produced by skin bacteria contribute to body odour. Skin bacteria may act as pathogens by causing skin infections, such as impetigo, by causing wound infection, or by causing infections associated with invasive medical devices. Bacteria which can easily be cultured from skin samples include a variety of Gram-positive cocci such as *Staphylococcus epidermidis* (Phylum *Bacillota*) and Gram-positive bacilli such as *Propionibacterium acnes* (Phylum *Actinomycetota*). *Staphylococcus epidermidis* produces isovaleric acid, a volatile substance that gives body odour a characteristic smell. Gram-negative bacteria also make up a large proportion of the skin flora. Investigations of the skin flora using molecular techniques indicate that many species of skin bacteria exist which have not yet been cultured and characterised. The skin flora amounts to something in the region of a thousand species (Grice *et al.*, 2009). Comparison by molecular methods of human skin sampled from the antecubital fossa (i.e. the inner elbow) with mouse skin sampled from the ear indicates that mice and humans share broadly similar skin floras. In addition to readily culturable Gram-positive bacteria, molecular methods indicate that Gram-negative bacilli, such as *Pseudomonas* and *Janthinobacter*, are significant members of the skin flora of humans as well as of mice (Grice *et al.*, 2008).

Staphylococcus aureus

The name staphylococcus (Fig. 5.33) was first used by Sir Alexander Ogston (1844–1929). While working as a surgeon in Aberdeen Ogston became curious about the nature of post-surgical wound infection and sepsis. He used a grant awarded by the British Medical Association to buy a microscope and he built a small laboratory on a plot of land behind his home. By inoculating hens' eggs with samples obtained from infected patients Ogston succeeded in isolating bacteria, which on microscopy resembled bunches of grapes when stained with a violet dye. At the suggestion of W.D. Geddes (1828–1900), professor of Greek in Aberdeen, Ogston called them staphylococci, 'from σταφυλή [*staphyle*], a bunch of grapes …' (Ogston, 1882). Ogston showed that his staphylococcus acted as a pathogen when inoculated experimentally into mice and claimed that it was a cause of sepsis in humans. Initially Ogston's findings were not accepted by the British medical establishment. After Ogston presented an account of his research in Berlin, German researchers successfully duplicated his experiments. The German microbiologist Friedrich Rosenbach (1842–1923) used agar media to culture strains of staphylococci and formally described a strain that produced golden yellow pigment as *Staphylococcus pyogenes aureus* (Rosenbach, 1884). The golden yellow pigment produced by the bacterium now known as *Staphylococcus aureus* is a carotenoid that appears to protect the organism from oxidative stress when attacked by white cells of the host immune system (Liu *et al.*, 2005). Some strains of *Staphylococcus aureus* do not produce golden pigment.

Staphylococcus aureus (a Gram-positive coccus and member of the Phylum *Bacillota*) is a member of the normal skin flora. About a third of

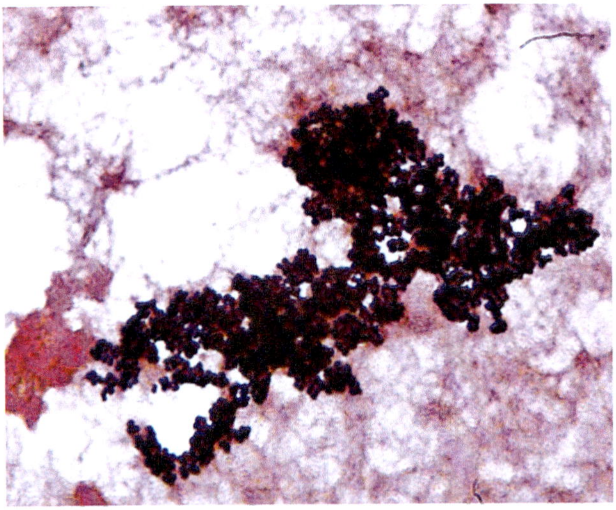

Fig. 5.33. Gram-positive clusters of staphylococci. Microscopy, Gram's stain.

humans are colonised with *Staphylococcus aureus*. Common carriage sites include the groin, the armpits (axillae), the inside of the nose (anterior nares) and the throat. As a member of the normal skin flora *Staphylococcus aureus* is unusual in that it bristles with virulence factors and has the capacity to cause skin infections, such as folliculitis and even to cause potentially life-threatening invasive disease. The virulence profile of *Staphylococcus aureus* varies from strain to strain. Coagulase, an enzyme which causes plasma to clot, is a virulence factor produced by all strains of the organism. Methods used to identify *Staphylococcus aureus* in diagnostic laboratories use tests based on the detection of coagulase. Other staphylococci associated with humans do not generally produce coagulase. *Staphylococcus argenteus*, a close relative of *Staphylococcus aureus* from which it is not readily distinguishable, is a rare exception. Coagulase-positive staphylococci associated with other animals include: *Staphylococcus delphini*, with badgers, ferrets, mink and horses; *Staphylococcus hyicus*, with pigs; *Staphylococcus intermedius*, with dogs and horses; and *Staphylococcus schweitzeri*, which was isolated from a red-tailed monkey, *Cercopithecus ascanius*, in Gabon. *Staphylococcus aureus* is an important pathogen of animals as well as humans. It causes mastitis in ruminants and rabbits, skin infections in horses and foot infections (bumble foot) in poultry.

Staphylococcus aureus is an important cause of hospital-acquired infection. Surgical incisions and invasive medical devices which allow the ingress of the skin flora provide staphylococci with an opportunity to reach the bloodstream. A study in which strains cultured from superficial sites were compared with bacteria cultured from the blood or deep tissues of patients suffering from invasive staphylococcal disease showed that invasive strains rapidly adapt to the hostile internal environment of the body. Bacteraemic staphylococcal disease is, however, something of a tragic event for the bacterium as well as for the unfortunate patient because for *Staphylococcus aureus* bloodstream infection is likely to prove a dead end that lacks much scope for onward transmission to a new host (Young *et al.*, 2017).

The population structure of *Staphylococcus aureus* is clonal. Successful clones composed of related organisms expand and become dominant before being superseded by new strains of the bacterium. The 1990s and 2000s saw the spread and rapid clonal expansion of strains resistant to methicillin and related antibiotics in UK hospitals (Figs. 5.34–5.36). Almost all methicillin-resistant strains detected in UK hospitals belonged to two clonal complexes, CC22 and CC30. At the same time, a separate clonal expansion involving methicillin-resistant CC1 strains was detected among injecting drug users (Atkinson *et al.*, 2009). But within a decade or so, strains of methicillin-resistant *Staphylococcus aureus* (MRSA) were already in relative decline, possibly as a result of the restricted use of antibiotics or because of the implementation of enhanced infection control measures. Additionally, antibiotic-resistant strains themselves began to lose their resistance genes (Ledda *et al.*, 2017).

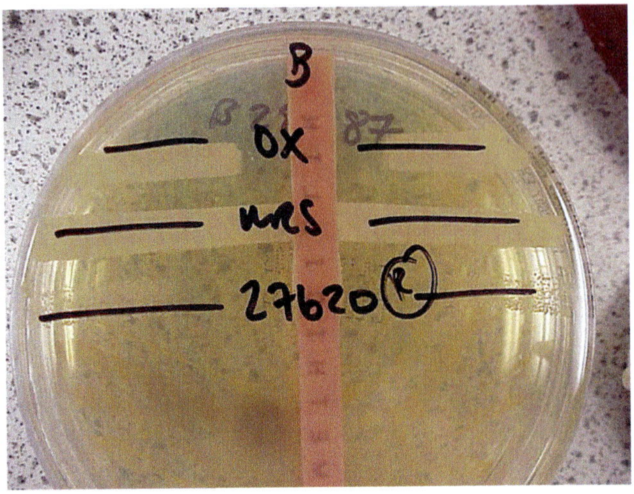

Fig. 5.34. Susceptibility test plate showing methicillin-sensitive (labelled: 'OX') and methicillin-resistant strains of *Staphylococcus aureus* (labelled: 'MRS' and '27620'), both growing up to the central antibiotic-impregnated paper strip.

Fig. 5.35. Detail from the *MRSA Quilt*, an artwork by Anna Dumitriu; cotton squares dyed blue by chromogenic medium from MRSA testing plates. (Anna Dumitriu.)

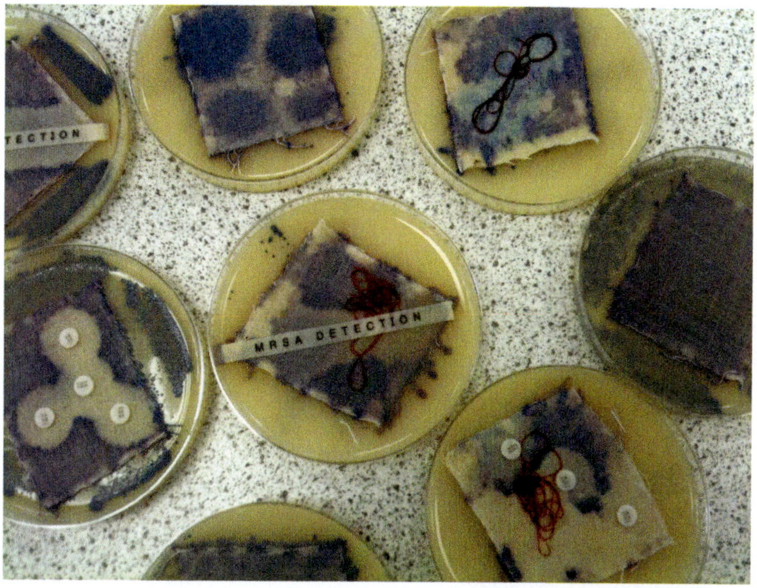

Fig. 5.36. Cotton squares embedded in chromogenic agar MRSA testing plates. The squares were used to make the *MRSA Quilt*. (Anna Dumitriu.)

Normal oral flora

Over 700 species of bacteria have been cultured or detected by molecular methods in the human mouth, although only about 140 species account for the predominant members of the normal oral flora in healthy subjects (Aas *et al.*, 2005). The mouth contains aerobic and anaerobic sub-habitats, which vary in texture from the hard surfaces of the teeth to the soft tissues of the tongue and other structures. Bacteria in the mouth are supplied by nutrients from saliva, ingested food and the products of other oral bacteria. Oral microbes live together as microbial communities in biofilms on oral tissue surfaces (Table 5.2). Many oral bacteria are Gram-positive cocci belonging to the genus *Streptococcus* and related bacteria in the genera *Gemella*, *Abiotrophia* and *Granulicatella*. *Streptococcus mitis* is a bacterium that occurs throughout the mouth. Others such as *Streptococcus sanguinis* specialise in living in biofilms on the tooth surface. Besides Gram-positive cocci, other tooth surface inhabitants include rod-shaped Gram-positive bacteria, including *Rothia dentocariosa*, which despite its name is one of the members of the normal tooth surface flora in healthy subjects.

The lactic acid-producing bacterium *Streptococcus mutans* is a member of the oral flora which shows a strong association with tooth decay. Other lactic acid-producing oral bacteria include *Actinomyces gerencseriae*, a bacterium which is thought to play a role in the initiation of tooth decay and members of the genus *Bifidobacterium* that colonise deep cavities (Becker *et al.*, 2002). Lactic acid, which inhibits the growth of bacteria at high concentrations is

Table 5.2. Some oral bacteria and their microhabitats in healthy humans.

Taxonomic group and species	Distribution in mouth
Phylum *Bacillota*	
Streptococcus mitis	Throughout the mouth
Streptococcus intermedius	Subgingival plaque
Streptococcus salivarius	Tongue
Streptococcus sanguinis	Tooth surface
Streptococcus gordonii	Tooth surface
Abiotrophia defectiva	Tooth surface
Granulicatella adiacens	Throughout the mouth
Gemella sanguinis	Throughout, except tooth surface
Phylum *Actinomycetota*	
Actinomyces naeslundii	Tooth surface
Rothia dentocariosa	Tooth surface
Rothia mucilaginosa	Tongue
Corynebacterium matruchotii	Subgingival plaque
Phylum *Pseudomonadota*	
Neisseria subflava	Throughout, except subgingival
Campylobacter concisus	Throughout, except tooth surface
Phylum *Fusobacteria*	
Fusobacterium nucleatum	Tooth surface, subgingival plaque
Phylum *Spirochaetota*	
Treponema lecithinolyticum	Subgingival plaque

Source: Table based on data from Paster *et al.*, 2001; Kazor *et al.*, 2003; Aas *et al.*, 2005.

converted to propionic and acetic acid by members of the genus *Veillonella*, including *Veillonella denticariosi*, a species found in association with dental caries (Table 5.3). *Veillonella* is a genus of anaerobic bacteria which form part of the oral and intestinal flora of man and other mammals. The genus was named after the French microbiologist Adrien Veillon (1864–1931). *Veillonella* species stain pink with Gram's stain. Technically they are Gram-negative cocci. However, despite their staining characteristics *Veillonella* species are classified as aberrant members of the Gram-positive phylum *Bacillota*. The conversion of lactic acid to propionic and acetic acid by *Veillonella* moderates the acidity in deep tooth decay lesions and enables the continued growth of *Streptococcus mutans* and other lactic acid-producing bacteria.

The subgingival crevice where the tooth disappears from view into the gums provides a habitat for diverse bacteria, including anaerobes such as *Fusobacterium nucleatum* and microaerophiles such as *Streptococcus intermedius* that thrive in environments with low concentrations of oxygen. Comparison of subgingival plaque samples from healthy subjects and patients with periodontal disease suggests that a suite of about 30 species of bacteria may act as pathogens (Paster *et al.*, 2001). Bacteria associated with periodontal disease include the anaerobes *Porphyromonas gingivalis* and *Tannerella forsythia*, as well as uncultured members of the little known phylum *Deferribacterota* (Kumar *et al.*, 2003).

Table 5.3. Some bacteria implicated in oral disease in humans.

Taxonomic group and species	Associated complaint
Phylum *Bacillota*	
Streptococcus mutans	Dental caries
Veillonella denticariosi	Dental caries
Streptococcus constellatus	Periodontal disease
Filifactor alocis	Periodontal disease
Catonella morbi	Periodontal disease
Dialister pneumosintes	Periodontal disease
Gemella haemolysans	Periodontal disease
Solobacterium moorei	Halitosis
Eubacterium sulci	Halitosis
Phylum *Actinomycetota*	
Actinomyces gerencseriae	Dental caries
Atopobium parvulum	Periodontal disease, halitosis
Atopobium rimae	Periodontal disease
Phylum *Pseudomonadota*	
Campylobacter rectus	Periodontal disease
Campylobacter gracilis	Periodontal disease
Phylum *Bacteroidota*	
Bacteroides forsythus	Periodontal disease
Prevotella tannerae	Periodontal disease
Porphyromonas gingivalis	Periodontal disease, halitosis
Phylum *Fusobacteria*	
Fusobacterium animalis	Periodontal disease
Fusobacterium periodonticum	Halitosis
Phylum *Spirochaetota*	
Treponema socranskii	Periodontal disease
Treponema denticola	Halitosis

Source: Based on data from Paster, 2001; Becker, 2002; Kazor, 2003.

The upper surface of the tongue (Fig. 5.37) with its complex texture of taste buds provides a special habitat for bacteria. Proteolytic bacteria that generate malodorous sulphur compounds and that live in the coating that covers the dorsal surface of the tongue are thought to account for halitosis or bad breath in some individuals. Implicated bacteria include *Eubacterium sulci* and *Solobacterium moorei*.

The bacteria that make up the commensal oral flora sometimes reach other sites in the body where they may act as agents of disease. Microaerophilic streptococci such as *Streptococcus intermedius* can cause dental abscesses. Microaerophilic streptococci are also implicated in the formation of lung abscesses (following inhalation of stomach contents containing oral flora) or may be disseminated by the bloodstream to the liver or brain. The microaerophilic oral streptococci that are associated with abscess formation are sometimes referred to as the 'milleri' group of streptococci, in honour of Willoughby D. Miller (1853–1907) an American dentist, pioneer of oral microbiology research and author of '*The micro-organisms of the human mouth: the*

Fig. 5.37. Bacteria on tongue, scanning electron micrograph (David Gregory and Debbie Marshall. Attribution 4.0 International). CC BY 4.0.

local and general diseases which are caused by them' (Miller, 1890). The Sindecuse Museum of Dentistry, University of Michigan, Ann Arbor, preserves Miller's laboratory apparatus and writings.

Other oral streptococci such as *Streptococcus sanguinis* are known collectively as the 'viridans' streptococci because of green colouration produced by colonies when grown on blood agar. The green colour is caused by conversion of red haemoglobin in blood cells to green methaemoglobin by hydrogen peroxide released by the streptococci, a phenomenon described by microbiologists as alpha-haemolysis, although the term is a misnomer as the blood cells are not actually lysed. (Barnard and Stinson, 1996). The green colour is due to the oxidation of iron in the haemoglobin molecule. Hydrogen peroxide produced by viridans streptococci inhibits competition by other bacteria and possibly acts as a virulence factor through its toxic effects on human cells. The viridans streptococci use a variety of mechanisms, including adhesion proteins and secretion of extracellular polymeric substance to bind to tissues in the mouth to form layers or plaque. If they enter the bloodstream they can bind to damaged heart valves and cause infective endocarditis, a potentially fatal condition unless treated with antibiotics and which in severe cases may necessitate surgical valve replacement. Besides viridans streptococci, other oral bacteria which are similarly capable of causing endocarditis and other forms of invasive disease include *Abiotrophia* and *Granulicatella*. Members of both genera require pyridoxal (vitamin B6) for growth. They were formerly known as 'nutritionally variant streptococci'.

Otherwise harmless oral commensal bacteria can cause disease when implanted deep into soft tissues via bite injuries. The oral flora of humans has been well characterised but less is known about the oral microbiome of animals. The bacteriology of animal bites provides clues about the oral bacteria found in association with animals (Table 5.4). Snake bites are associated

Table 5.4. Some bacteria associated with wound infection following animal bites.

Taxonomic group and species	Animal species
Phylum *Bacillota*	
Streptococcus canis	dogs
Phylum *Mycoplasmatota*	
Mycoplasma phocacerebrale	seals
Phylum *Pseudomonadota*	
Aeromonas hydrophila	snakes, leeches
Eikenella corrodens	*Homo sapiens*
Pasteurella aerogenes	pigs
Pasteurella multocida	cats, dogs
Pasteurella caballi	horses
Pasteurella canis	dogs, cats
Actinobacillus lignieresii	horses, sheep
Actinobacillus suis	pigs, horses
Neisseria weaveri	dogs, cats
Neisseria canis	dogs, cats
Neisseria animaloris	dogs, cats, grey seals
"*Spirillum minus*"	rats
Phylum *Bacteroidota*	
Bergeyella zoohelkum	dogs, cats
Capnocytophaga canimorsus	dogs
Capnocytophaga cynodegmi	dogs
Phylum *Fusobacteria*	
Streptobacillus moniliformis	rats

with wound infections caused by *Aeromonas hydrophila*, an environmental bacterium, otherwise found in aquatic habitats and also as a symbiont of blood-sucking leeches. *Aeromonas hydrophila* can be cultured from oral samples obtained from snakes. The canine oral flora is dominated by members of the *Pasteurellaceae*, *Bergeyella*, *Conchiformibius* and *Porphyromonas cangingivalis*. Other bacteria associated with the oral flora of dogs include *Neisseria weaveri* and *Capnocytophaga* species (Ruparell *et al*., 2020). *Bergeyella zoohelkum*, *Pasteurella multocida* (Fig. 5.38), *Neisseria weaveri* and *Capnocytophaga canimorsus* are regularly cultured from infected dog bite wounds. *Capnocytophaga* species (Phylum *Bacteroidota*) are odd bacteria that characteristically demonstrate gliding motility when cultured on solid media. The bacteria are capable of hauling their own cells en masse across solid surfaces such as glass slides to which they attach themselves by secreting mobile adhesin molecules in order to gain traction (Shrivastava *et al*., 2018).

When visualised microscopically, *Neissera* species (Class *Betaproteobacteria*) typically appear as pairs of adpressed cells, described by microbiologists as kidney-shaped Gram-negative diplococci (Fig. 5.39). Species cultured from animals include: *Neisseria iguanae*, from lizards; *Neisseria musculi*, from mice; *Neisseria animaloris*, *Neisseria canis* and *Neisseria weaveri* from cats and dogs; and *Neisseria arctica* from the eggs of White-fronted geese. *Neisseria animaloris*, has been cultured from dead and moribund porpoises,

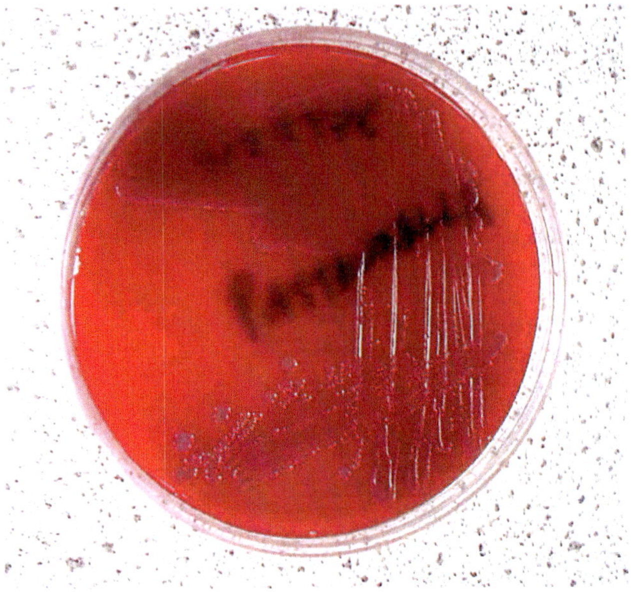

Fig. 5.38. Colonies of *Pasteurella multocida* growing on blood agar.

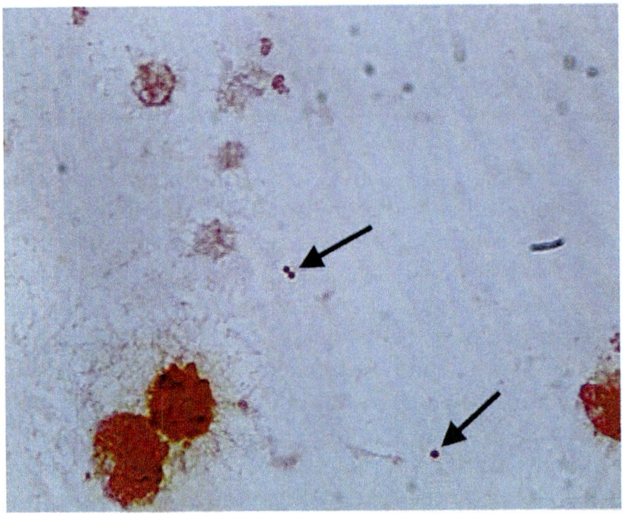

Fig. 5.39. Gram-negative diplococci, *Neisseria meningitidis*. Microscopy, Gram's stain.

found stranded at various locations on the shores of the North Sea. Affected animals presented with multiple abscesses and all of them showed signs of seal bite injuries. Having initially survived the bite injuries, the porpoises appear to have succumbed to disseminated bacterial infection (Foster *et al.*, 2019).

Neisseria species associated with the oral and upper respiratory tract of humans include *Neisseria cinerea*, *Neisseria elongata*, *Neisseria flava*, *Neisseria flavescens* and *Neisseria meningitidis*. Usually a harmless upper airways commensal the meningococcus *Neisseria meningitidis* has the capacity to behave as a potentially life-threatening invasive pathogen. The closely related gonococcus *Neisseria gonorrhoeae* is the causative agent of gonorrhoea, a sexually transmitted infection, known since antiquity. *Neisseria gonorrhoeae* attaches to the epithelial cells of mucosal surfaces, usually of the male urethra and female cervix. It can also colonise the upper airways and is a rare cause of pharyngitis. The gonococcus is a naturally transformable organism that readily acquires chunks of DNA from other gonococci and related species of bacteria, resulting in a multiplicity of distinguishable strains of *Neisseria gonorrhoeae*. Nevertheless, investigations based on the comparison of conserved parts of the gonococcal genome make it possible to plot networks of transmission. A study of gonorrhoeal infections in Brighton demonstrated the continual arrival of new strains in the city and also revealed close links with cases of infection in London, with other parts of the UK as well as the USA (De Silva *et al.*, 2016).

Helicobacter

When food is swallowed and enters the acidic environment of the human stomach it reaches a zone of low microbial diversity. Some bacteria, including potential pathogens may survive the journey through the stomach and reach the intestine. *Helicobacter pylori* is unusual in being a bacterium that normally resides in the stomach lining. *Helicobacter pylori* is a microaerophile, that grows in conditions of low oxygen tension. It is a Gram-negative spiral-shaped member of the Class *Epsilonproteobacteria* and is distantly related to *Campylobacter*. Helicobacters release urease, an enzyme that splits urea secreted by cells in the lining of the stomach to yield alkaline ammonia. This leads to the neutralisation of stomach acid. Much of the world's population is infected with *Helicobacter pylori*, although rates of infection are declining in developed countries. Most people tolerate helicobacter infection and exhibit no symptoms. A minority of those infected suffer from gastritis.

The persistence of *Helicobacter pylori* over the life of an individual and the fact that transmission often occurs in families make it a suitable surrogate marker for the study of human migration. *Helicobacter pylori* strains from East Asia differ from European strains. Amerindians living in remote areas of South America are infected with strains that resemble those found in East Asia, whereas city dwellers in South America are infected with European strains. Whereas urban populations in the Americas are infected with strains that arrived with European colonists over the last 500 years, Amerindian populations living in remote locations appear to be infected with strains derived from people who migrated to the Americas from Asia, more than 11,000 years ago (Ghose *et al.*, 2002).

Different helicobacter species reside in the stomachs of various mammals, including the cheetah, dog, cat, ferret and pig (Table 5.5). Some animals may be symptomatic. Some captive cheetahs, for example, have been reported to suffer from gastritis. The helicobacter associated with cheetahs is called

Table 5.5. *Helicobacter* species and their mammalian hosts.

Helicobacter species	Host animal species	Gastric or enterohepatic
Helicobacter acinonychis	cheetah	gastric
Helicobacter anseris	Canada goose	enterohepatic
Helicobacter aurati	Syrian hamster	gastric
Helicobacter baculiformis	cat	gastric
Helicobacter bizzozeronii	dog	gastric
Helicobacter brantae	Canada goose	enterohepatic
Helicobacter canadensis	birds	enterohepatic
Helicobacter canis	dog	enterohepatic
Helicobacter cetorum	cetaceans	gastric
Helicobacter cinaedi	Syrian hamster	enterohepatic
Helicobacter cynogastricus	dog	gastric
Helicobacter equorum	horse	enterohepatic
Helicobacter felis	cat, dog	gastric
Helicobacter fennelliae	man	enterohepatic
Helicobacter ganmani	mouse	enterohepatic
Helicobacter hepaticus	mouse	enterohepatic
Helicobacter marmotae	woodchuck, cat	enterohepatic
Helicobacter mastomyrinus	rodents	enterohepatic
Helicobacter mesocricetorum	Syrian hamster	enterohepatic
Helicobacter muridarum	rat	enterohepatic
Helicobacter mustelae	ferret	gastric
Helicobacter pametensis	birds	enterohepatic
Helicobacter pullorum	birds	enterohepatic
Helicobacter pylori	man	gastric
Helicobacter rodentium	mouse	enterohepatic
Helicobacter salomonis	dog	gastric
Helicobacter suis	pig	gastric
Helicobacter trogontum	rat	enterohepatic
Helicobacter typhlonius	mouse	enterohepatic

Helicobacter acinonychis. *Helicobacter cetorum*, a species associated with whales and dolphins has been isolated from post-mortem samples taken from stranded dolphins in Cornwall (Davison, 2014). *Helicobacter cetorum* was originally described on the basis of strains isolated from captive and wild dolphins and a captive beluga whale (Harper *et al*., 2002). In addition to gastric species, others reside elsewhere in the gastrointestinal tract and are known as the enterohepatic helicobacters (Solnick and Schauer, 2001). Some of these have been found in animals diagnosed with inflammatory bowel disease and some animal-associated helicobacters have been implicated as occasional causes of diarrhoea or invasive disease in humans.

Breakdown and absorption of nutrients by vertebrates

The gastrointestinal tracts of different vertebrate species follow much the same overall pattern from mouth to anus. Many animals process food initially by biting, chewing and exposing it to salivary enzymes before swallowing it. Most vertebrates have a recognisable stomach where food is processed further by exposure to hydrochloric acid and the enzyme pepsin that breaks down proteins. In many vertebrate species a part of the gastrointestinal tract is modified to form a kind of holding area where food is retained and fermented by bacteria. The anatomical position of this specialised fermentation area varies with species. Some animals have modified stomachs whereas others have enlarged areas in parts of the midgut or hindgut (Stevens and Hume, 1998). The fermentation products are utilised by the host animal as well as the bacteria. In humans the hindgut or colon is enlarged and segmented by narrow bands into a series of pouches called haustrations. The gut microbiome is the richest of all bacterial habitats of the human body both in absolute numbers of cells and probably also in diversity of species. In many species, including man, rodents, horses and reptiles the hindgut functions as a retaining area for the processing of its contents by bacteria. Bacteria in the hindgut of humans may reach densities of 1 trillion (10^{12}) cells per millilitre. A survey of the gut flora of 124 human subjects recorded a total of more than 1000 species of bacteria (Qin, *et al.*, 2010). Many of them were members of the Phylum *Bacteroidota* and the Phylum *Bacillota*.

Unlike humans, some herbivorous animals have complex stomachs that act as fermentation tanks. Plant matter ingested by such animals is retained in the stomach allowing the enzymatic degradation of polysaccharides to sugars and the fermentation of complex carbohydrates, such as cellulose to yield short chain fatty acids, acetic acid, propionic acid and butyric acid. These simpler substances can be absorbed through the gut lining. The rumen also supports methanogenic archaea and large volumes of methane are belched into the atmosphere by cattle and other ruminants. Ruminants such as sheep have a four-chambered stomach, including a rumen that acts as a fermentation tank. To make the most of their food, ruminants regurgitate partially digested food and chew it a second time. Rumen bacteria may reach densities of 10 billion (10^{10}) to 100 billion (10^{11}) cells per millilitre. Large numbers of eukaryotic protists also populate the rumen. Ruminants are able to thrive on a diet that consists mostly of plant fibre. Plant fibre is not a particularly nutritious diet. Animals that eat mostly plant fibre tend to be large and spend a lot of time eating. The gastrointestinal tracts of herbivorous dinosaurs were probably populated by fermentative bacteria. Rumen bacteria consist mostly of Gram-positive bacteria and members of the Phylum *Bacteroidota*. Whales that feed on krill have rumen-like fore-stomachs which are densely populated by bacteria capable of degrading the chitinous exoskeletons of crustacea (Olsen *et al.*, 2000). Chitin-degrading bacteria also feature in the midgut flora of some fish.

Fish that frequent the murky sediments of estuarine waters consume large numbers of environmental bacteria. Bacteria were shown to account for 15–30% of the stomach contents of grey mullet, *Mugil cephalus* (Moriarty, 1976). Surgeonfish (Family Acanthuridae), which graze on algae and cyanobacteria, have long digestive tracts populated by huge (600 µm) symbiotic gut bacteria that are just about visible to the naked eye. "*Epulopiscium fischelsoni*" (Fischelson's Fish-feaster) was described as a symbiont of a surgeonfish found in the Red Sea. Other species of surgeonfish harbour large symbiotic bacteria. Fischelson's Fish-feaster was at first mistaken for a protist on account of its enormity. Its huge cells harbour thousands of copies of its genome. The availability of multiple genomes may allow a single cell to interact with its micro-environment on a local level. Multiple genome copies are passed on to daughter cells when the bacteria replicate, giving this species the potential to inherit multiple versions of its genome (Mendell *et al.*, 2008).

Campylobacter

Campylobacters (Class *Epsilonproteobacteria*) are Gram-negative curved rod-shaped bacteria. Campylobacters are microaerophilic bacteria that require low concentrations of oxygen for growth. Two closely related species, *Campylobacter jejuni* and *Campylobacter coli* commonly cause diarrhoeal illnesses in humans. Campylobacters grow well at temperatures of 37°C and 42°C, showing that they are well adapted to life in the intestines of both mammals and birds. A wide range of animals and birds can harbour the two species most commonly isolated from humans. Campylobacters are naturally transformable organisms that can acquire genetic material from other strains through horizontal gene transfer. Genetic variation also occurs as a result of mutation. *Campylobacter jejuni* and *Campylobacter coli* are closely related organisms. *Campylobacter jejuni* is closely associated with humans but *Campylobacter coli* is more closely associated with pigs. It has been estimated that *Campylobacter jejuni* might have diverged as a separate species from an ancestral population of *Campylobacter coli* as recently as approximately 6500 years ago (Wilson *et al.*, 2009).

Campylobacter species which are more usually associated with animals include: *Campylobacter upsaliensis* and *Campylobacter helveticus*, which are associated with cats and dogs; *Campylobacter lari*, which has been isolated from gulls and marine bivalve molluscs; *Campylobacter fetus* subspecies *fetus* and *Campylobacter fetus* subspecies *venerealis*, which cause abortion in sheep and cattle; and *Campylobacter insulaenigrae*, which was described on the basis of strains isolated from marine mammals in Scotland (Foster *et al.*, 2004). None of these species is associated strictly with a single host and all have been isolated at some time from humans.

Salmonella

Salmonellas are Gram-negative rod-shaped bacteria (Phylum *Pseudomonadota*, Class *Gammaproteobacteria*) that colonise a wide range of mammals, birds and reptiles. Salmonellas are a common cause of diarrhoeal illness in humans. Asymptomatic carriage may also occur. Most infections in humans are zoonoses acquired from animals via contaminated food products. According to the current scheme of bacterial nomenclature, there are two salmonella species, *Salmonella bongori* and *Salmonella enterica*. *Salmonella bongori* was first isolated from a lizard and is seldom associated with mammals, although sporadic cases of infections have been reported in humans from Sicily. *Salmonella enterica* includes several thousand distinguishable serotypes or serovars, grouped into six subspecies of which *Salmonella enterica* subspecies *enterica* is the subspecies most closely associated with man and mammals. One of its serovars, *Salmonella* serovar Typhi infects only humans and primates. *Salmonella* Typhi is the agent of typhoid fever, a potentially life-threatening infectious disease. For many years it was classified as a separate species, *Salmonella typhi*. *Salmonella* Typhi strains collected from around the world show little genetic diversity. *Salmonella* Typhi has been estimated to have emerged as a distinct human pathogen roughly 50,000 years ago (Kidgell *et al.*, 2002).

Pathogenic mycobacteria

Members of the genus *Mycobacterium* have a propensity to invade and survive inside eukaryotic cells, including the cells of unicellular organisms, the cells of humans and the cells of other vertebrates. Some mycobacteria, such as *Mycobacterium tuberculosis* (Fig. 5.40) are specialised pathogens. *Mycobacterium tuberculosis* is the agent of tuberculosis in humans and other mammals. Until recently, animal-associated strains were classified as separate species within what was known as the *Mycobacterium tuberculosis* complex of bacteria. In 2018 some of the animal-adapted strains were reclassified as *Mycobacterium tuberculosis* in recognition of the close genetic relatedness (more than 99.9% similarity) of animal-adapted and human-adapted strains (Riojas *et al.*, 2018). Cattle-adapted strains of the bacterium (previously known as *Mycobacterium bovis*) cause bovine tuberculosis in cattle, badgers

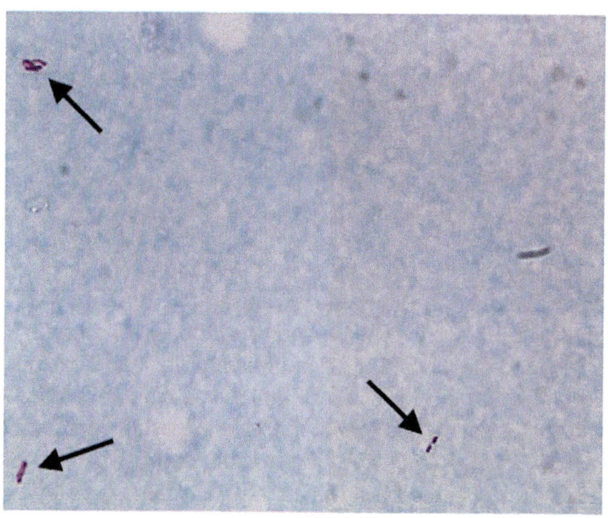

Fig. 5.40. *Mycobacterium tuberculosis*. Microscopy, Ziehl–Neelsen stain.

and red deer. Rodent-adapted strains (previously known as *Mycobacterium microti*) occasionally infect other mammals, including wild boar, ferrets, cats and dogs but rarely infect humans (Brites *et al.*, 2018; van de Weg *et al.*, 2020). Similarly, infections caused by seal-adapted strains (previously known as *Mycobacterium pinnipedii*) are seldom detected in humans. Humans can contract bovine tuberculosis by drinking unpasteurised milk, although bovine strains of the bacterium are not particularly transmissible between humans.

Infected hosts vary in their susceptibility to tuberculosis and different strains of the bacterium vary in virulence. The principal route of tuberculosis transmission is via the inhalation of aerosolised droplets containing bacteria. The majority of people infected with tuberculosis manage to contain the spread of the disease following primary infection. Initially, they remain clinically well and some may remain asymptomatic indefinitely. However, they may fall ill later after a period of latent infection. Patients may present with pulmonary tuberculosis or with disseminated tuberculosis as manifested for example by tuberculosis of a lymph gland. People with pulmonary tuberculosis are a source of infection to others. The clinical features of tuberculosis seen in other mammals resemble those seen in humans.

It seems probable that *Mycobacterium tuberculosis* walked out of Africa with modern humans (Comas *et al.*, 2013). Comparison of mycobacterial genomic data with human mitochondrial sequence data shows congruence between phylogenetic trees for the bacterium and for humans, indicating that humans and *Mycobacterium tuberculosis* coevolved. It has been estimated that *Mycobacterium tuberculosis* emerged as a pathogen somewhere in Africa sometime around 70,000 years ago. When human populations were small and dispersed *Mycobacterium tuberculosis* would probably have behaved as a relatively benign low-grade pathogen. Virulent life-threatening strains of *Mycobacterium tuberculosis* may not have become prevalent until after the Neolithic revolution when the establishment of settled communities created conditions capable of supporting the on-going transmission of more lethal strains (Comas *et al.*, 2013). Animal-adapted strains appear to be derived from ancestral human strains and appear to have arisen on multiple separate occasions (Brites *et al.*, 2018).

Leprosy

Leprosy has been recognised since antiquity as a disease of humans, *Homo sapiens*. Leprosy in humans is caused by two related species of bacteria, *Mycobacterium leprae* and *Mycobacterium lepromatosis*. More recently leprosy has been recognised as a disease of armadillos and red squirrels (see Chapter 9, *Bacteria tourism*). Leprosy is also known as Hansen's disease. In 1873, Gerhard Henrik Armauer Hansen (1841–1912) visualised leprosy bacilli in tissue sections taken from Norwegian leprosy patients. Leprosy bacilli are slow-growing bacteria, which replicate about once every 14 days. Untreated leprosy is a disease that progresses slowly. The leprosy bacillus attacks nerves and deprives sufferers of sensation. Sensory deficit leads to injury and disfigurement. Leprosy bacilli

(Fig. 5.41) can be visualised in stained tissue sections and leprosy bacilli can be grown in the footpads of immunosuppressed mice but leprosy bacteria cannot be isolated in pure culture using artificial culture media. Compared with free-living mycobacteria, leprosy bacilli exhibit reduced complements of genes involved in the processing of nutrients. Leprosy bacilli have become dependent on the cells that host them to supply their needs. Investigations into the natural history of leprosy were impeded for decades by an inability to culture and identify leprosy bacilli in a routine laboratory. More recently it has become possible to conduct leprosy research using molecular methods based on the detection and sequencing of bacterial DNA. The development of such methods has made it possible to detect *Mycobacterium leprae* and *Mycobacterium lepromatosis* in environmental samples and in animals.

It is not certain exactly how leprosy is transmitted. It it generally assumed that close contact with an infected individual may account for most cases of infection but it has been hard to study transmission of the disease because it is not possible to culture leprosy bacilli in artificial media, which makes it difficult to demonstrate the presence of viable infecting bacteria in test samples. Years may pass before an infected individual shows signs of disease, which makes it difficult to establish a source of infection. DNA can be detected in tissue samples and environmental specimens, which does at least provide some indication of the distribution of leprosy bacilli in the environment. However, the detection of DNA does not prove that viable infectious bacilli are present in a sample. Because some patients give no history of contact with another case of leprosy it has been suspected that transmission may sometimes occur indirectly via environmental contamination or some disease vector. A global study of soil samples showed that *Mycobacterium leprae* DNA could be detected in soil from inside the homes of leprosy patients in Bangladesh,

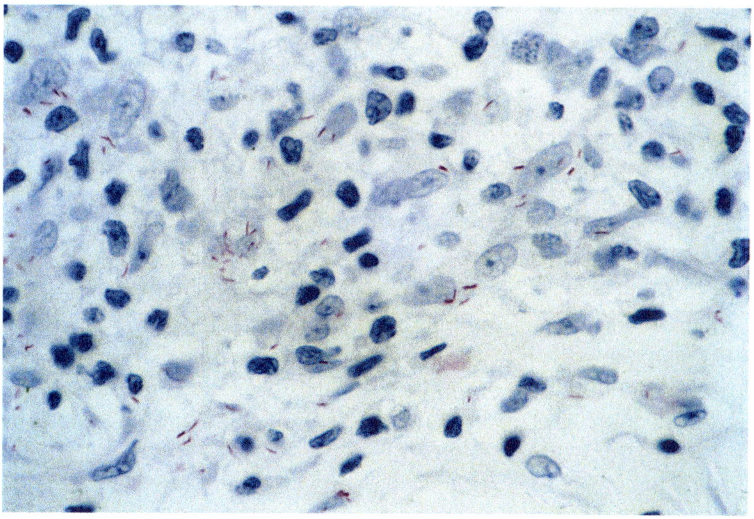

Fig. 5.41. Skin biopsy showing magenta-stained leprosy bacilli (Wellcome Collection). CC0 1.0 Universal.

from the natural habitat of infected armadillos in Surinam and from terrain frequented by infected red squirrels on Brownsea Island (Tió-Coma *et al.*, 2019). One theory advanced in the 1940s was that cockroaches might act as vectors of leprosy by feeding on the skin lesions of infected patients and by spreading the bacteria in their faeces. The faeces of German Cockroaches, *Blattella germanica*, which occurred in abundance in an African leprosy hospital were found to contain acid-fast bacilli that resembled leprosy bacilli (Moiser, 1945; Moiser, 1947).

Rat leprosy

Mycobacterium lepraemurium is the causative agent of rat leprosy. In 1902, Vyacheslav Stefansky (1867–1949), a bacteriologist based at the Odessa Bacteriological Station, noticed lesions in rats, *Rattus norvegicus*, that resembled the lesions of leprosy in humans. Stefansky was involved in a rat eradication programme in the port city of Odessa on the shores of the Black Sea. The aim of the campaign was to thwart an outbreak of bubonic plague, a bacterial infection spread by infected rats. The following year, George Dean (1863–1914), the chief bacteriologist at the Lister Institute of Preventative Medicine, published a report on rat leprosy in British rats. Dean had become aware of Stefansky's findings in Odessa and realised that he had seen similar lesions in rats collected at Elstree in Hertfordshire (Dean, 1905). It subsequently became apparent that rat leprosy was a globally distributed disease of rats (Lowe, 1937).

Mycobacterium lepraemurium shares a number of common features with the agents of human leprosy, *Mycobacterium leprae* and *Mycobacterium lepromatosis*. All of them are normally dependent on host cells for growth, although it is just about possible to culture *Mycobacterium lepraemurium* in specially supplemented artificial growth media. All three species have reduced numbers of genes compared with free-living mycobacteria and all of them cause a form of leprosy in mammalian hosts. *Mycobacterium leprae* and *Mycobacterium lepromatosis* are closely related species and it has been estimated that they diverged from a common ancestral bacterium as recently as approximately 14 million years ago (Singh *et al.*, 2015). *Mycobacterium lepraemurium* is a more distantly related species but it has followed a comparable evolutionary path to *Mycobacterium leprae* and *Mycobacterium lepromatosis* in that it too has progressively lost the genes that are necessary to allow for a free-living existence. In addition to rodents, domestic cats are susceptible to infection with *Mycobacterium lepraemurium*. The disease in cats is manifested by lumpy, ulcerating lesions on the head and limbs. It is assumed that cats acquire infection from diseased rodents through abrasions in the skin. At least two other species of mycobacteria act as agents of feline leprosy (O'Brien *et al.*, 2017). Molecular analysis of lesions has revealed the presence of two uncultured novel mycobacteria, which have been provisionally named as "*Candidatus* Mycobacterium tarwinense" and "*Candidatus* Mycobacterium lepraefelis".

Marine mammals

Whales (Order Cetacea) evolved from terrestrial hoofed animals and are distantly related to hippopotami, cows and chevrotains. In common with their herbivorous relatives, whales retain the compartmentalised stomachs of their ancestors, although unlike their terrestrial relatives, whales are carnivores. Toothed whales, such as killer whales are hunters that use echolocation to target individual prey items, including fish and marine mammals. Baleen whales (Fig. 5.42) are filter feeders that consume shoals of fish and crustaceans. The gut flora of baleen whales includes fermentative bacteria, comparable to those found in terrestrial herbivores. Whereas, a large part of the diet of herbivores consists of the polysaccharide cellulose derived from plant matter, much of the diet of baleen whales consists of the polysaccharide chitin, derived from the carapaces of marine crustacea. Chitin accounts for a significant proportion of the crustacean body mass. The compartmentalised stomachs of whales allows them to make use of this food source by accommodating fermentative bacteria that are capable of degrading chitin, in the same way that terrestrial herbivores make use of cellulose. In addition to fermentative bacteria, the gut flora of baleen whales includes bacteria similar to those found in terrestrial carnivores, which are involved in the processing of animal proteins (Sanders *et al*., 2015).

The cetacean skin flora is speciose and labile and its composition is influenced by environmental and seasonal factors. The composition of the whale skin flora varies also between individual animals and between species. Recognisably distinctive core groups of bacteria are known to occur in association with partic-

Fig. 5.42. Minke whale, a baleen whale often seen in British waters. Inner Hebrides, UK.

ular species of whale. Humpback whales, *Megaptera novaeangliae*, for example, have a characteristic skin flora, common features of which are apparent even when comparisons are made between samples taken from widely separated populations. Humpback whales occur in both the northern and southern hemispheres and have been recorded in increasing numbers from British waters in recent decades. Humpback whales migrate from relatively warm waters in winter to cooler waters situated at higher latitudes during the summer foraging season. Two genera of bacteria, *Tenacibaculum* and *Psychrobacter* were found to account for a significant proportion of the skin flora of humpback whales (Fig. 5.43) sampled from populations in the North Atlantic, North Pacific and South Pacific (Apprill *et al.*, 2014). *Tenacibaculum* species (Phylum Bacteroidota) occur widely in marine environments. One species, *Tenacibaculum maritimum*, is a fish pathogen that causes ulcerative lesions and tail rot in farmed fish. *Psychrobacter* species (Class *Gammaproteobacteria*) are cold-tolerant bacteria, associated especially with cold or cool marine and terrestrial habitats. A study of humpback whales in Antarctic waters, demonstrated seasonal variation in the whale skin flora. *Psychrobacter* decreased in abundance as water temperatures increased (Bierlich *et al.*, 2018).

Psychrobacter was found to be the most abundant member of the skin flora of captive killer whales (*Orcinus orca*) and bottlenose dolphins (*Tursiops truncatus*) (Chiarello *et al.*, 2017). The genus *Psychrobacter* contains multiple species and different kinds of cetacean are associated with particular strains or species of *Psychrobacter*. The skin flora of sperm whales, for example, includes several distinctive psychrobacters. Baleen whales are distinguished by

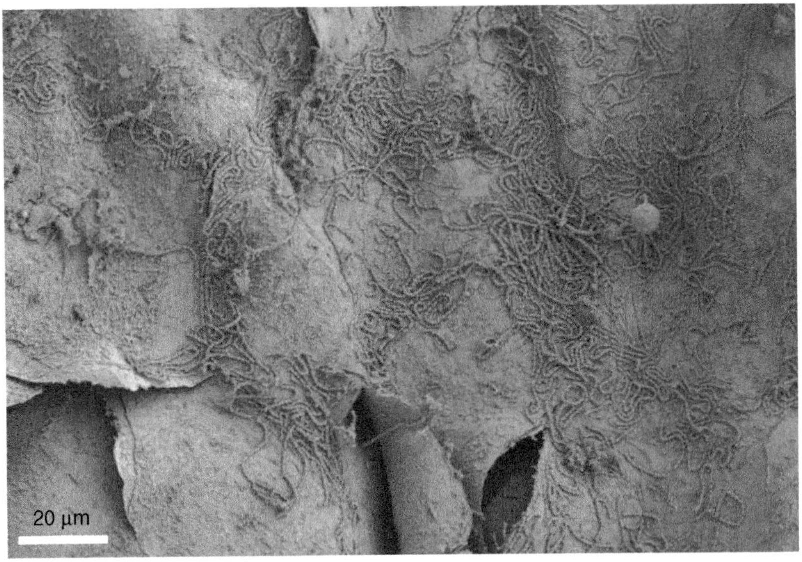

Fig. 5.43. Electron micrograph of bacteria on the surface of a humpback whale (Image cropped from figure 8 in Apprill *et al.*, 2014; *PLoS ONE*. 9(3): e90785). Creative Commons Attribution Licence.

a skin flora that includes unnamed bacteria in the family *Cardiobacteriaceae* and an as yet uncultured member of the family *Moraxellaceae.* Dolphins and pilot whales (Family Delphinidae) share broadly similar skin floras. Overall comparisons of the skin floras of different whales and dolphins make it possible to draw a tree-like diagram based on measures of microbial similarity, which broadly correlates with the phylogenetic tree for the cetaceans themselves. The congruence between the two trees suggests that cetaceans might have coevolved with their skin bacteria (Apprill *et al.*, 2020). In comparison with other cetaceans, belugas appear to have unusually variable skin floras, as measured between individual animals and between separate populations. A study of belugas in Alaska failed to detect any particular groups of skin bacteria that could be considered characteristic of belugas as a species (Van Cise *et al.*, 2020). Belugas (Fig. 5.44) are usually found at high latitudes but they are known to migrate to warmer waters to moult. Belugas are occasionally sighted in British waters.

Brucellosis in sea mammals

Brucellosis is a disease of livestock and humans caused by *Brucella* (Class *Alphaproteobacteria*). The genus was named after the microbiologist Sir David Bruce (1855–1931). Bruce chaired the Mediterranean Fever Commission (1904–1906), which established links between the bacterium, goats' milk and Malta fever, a disease which became known as brucellosis. Brucellosis has been

Fig. 5.44. Belugas. St Lawrence River, Canada.

eradicated from livestock in the UK but brucellosis still affects domesticated animals in many other parts of the world. Brucellosis can be acquired by ingesting unpasteurised milk or as a result of exposure to aerosolised bacteria. In man, brucellosis is characterised by fevers as well as a variety of signs and symptoms, including fatigue, swollen lymph nodes, abscesses, an enlarged spleen and back pain. Foci of infection may occur in the brain and in bone. In domesticated animals, brucellosis is associated with disorders of the reproductive system. Susceptible animals include sheep, goats, cattle and pigs. According to the current scheme of bacterial nomenclature, a single species, *Brucella melitensis*, accounts for infections seen in domestic livestock (Hördt *et al.*, 2020; Oren and Garrity, 2020). *Brucella melitensis* also infects wild mammals, including antelopes, buffalo and bison. Additional *Brucella* species, *Brucella microti* and *Brucella vulpis*, have been isolated from voles and foxes. Other closely related bacteria have been cultured from the root nodules of broom and lupins. As seen with other symbiotic and pathogenic alphaproteobacteria, *Brucella* has the capacity to invade and replicate inside eukaryotic cells. The bacteria stimulate the formation of membrane-bound vacuoles that resemble intracellular organelles in which the bacteria can survive and replicate.

Two *Brucella* species – *Brucella ceti*, which is associated mainly with whales and dolphins, and *Brucella pinnipedialis*, associated mostly with seals – are responsible for cases of brucellosis seen in marine mammals (Guzmán-Verri *et al.*, 2012). Most cases are detected as a result of diagnostic tests performed on sick captive animals or following post-mortem investigations of stranded animals. Brucellosis is manifested in marine mammals by focal lesions in the reproductive systems of female and male animals, in the cardiovascular system, in the central nervous system and in bones and joints. *Brucella* has been cultured from the mammary glands of whales and dolphins. Because of their reliance on echolocation, disoriented animals with lesions of the central nervous system are susceptible to stranding and cases of neurobrucellosis are diagnosed in stranded cetaceans on a regular basis.

Strandings of marine mammals are monitored by the UK Cetacean Strandings Investigation Programme. There is a hotline for reporting live and dead strandings. Regional schemes exist in Cornwall and Scotland. The Cornwall Wildlife Trust manages a Marine Strandings Network. The Scottish Marine Animal Stranding Scheme issues an annual report which includes details of post-mortem and bacteriological investigations. *Brucella pinnipedialis*, which is mainly a pathogen of seals, has been cultured from common seals, *Phoca vitulina* (Fig. 5.45) and grey seals, *Halichoerus grypus* (Fig. 5.46) and also from a stranded minke whale with neurobrucellosis (Davison *et al.*, 2021a). *Brucella ceti* has been cultured from 11 species of cetacean in the UK, including: Risso's dolphin, *Grampus griseus*; killer whale *Orcinus orca*; Sowerby's beaked whale, *Mesoplodon bidens*; Atlantic white-sided dolphin, *Lagenorhynchus acutus*; short-beaked common dolphin, *Delphinus delphis*; long-finned pilot whale, *Globicephela melas*; and minke whale (Davison *et al.*, 2021b).

Fig. 5.45. Common seal. Ravenscar, Yorkshire, UK.

Fig. 5.46. Grey seal. Ravenscar, Yorkshire, UK.

Other pathogens of sea mammals

Many of the bacteria that colonise and infect marine mammals are closely related to, or in some cases indistinguishable from, the bacteria associated with terrestrial mammals (Higgins, 2000). *Streptococcus equi*, for example, a pathogen of horses and occasionally of humans has been cultured from common seals and grey seals and also from a pilot whale with pneumonia. Some species of bacteria are specialist commensals or pathogens of marine mammals. *Streptococcus marimammalium* was described on the basis of isolates cultured from post-mortem samples obtained from a grey seal and a common seal in Scotland and *Streptococcus halichoeri* was first identified from grey seals sampled in Cornwall and Scotland. *Streptococcus phocae*, was described as a new species on the basis of cultures isolated from seals with secondary bacterial respiratory tract infection during the course of an outbreak of canine distemper virus, which affected seal populations around the coasts of the North Sea in 1988 (Numberger *et al.*, 2021). Other bacteria found in association with sea mammals include *Actinobacillus delphinicola*, which has been cultured from stranded dolphins and *Granulicatella balaenopterae*, which has been cultured from post-mortem tissue samples taken from stranded minke whales. *Mycobacterium marinum*, a bacterium found in aquatic environments, which occasionally causes skin infections in humans, known as fish tank granulomas, was isolated from the skin lesions of a captive beluga whale.

Bone worms

The carcasses of dead whales are islands of plenty on an otherwise desolate seafloor. When all the scavenging sharks and hagfish have picked the carcass clean and there is nothing left but bones, *Osedax* worms (Order Annelida) eke out a living by extracting nutrients from the bones themselves (Vrijenhoek *et al.*, 2009; Fujiwara *et al.*, 2019). *Osedax* worms (Fig. 5.47) have no mouth and no alimentary canal. Instead they possess a branching root-like system of vascularised tissue that infiltrates the bone of the dead whale. Osedax worms dissolve bone by releasing acid. The root-like tissue contains symbiotic bacteria, which enable the worms to process and absorb nutrients. Many of the bacteria found in association with bone worms, such as the genus *Neptunomonas*, belong to the *Oceanospirillaceae*, a family of bacteria associated also with marine sediments and marine molluscs (Verna *et al.*, 2010; Miyamoto *et al.*, 2017).

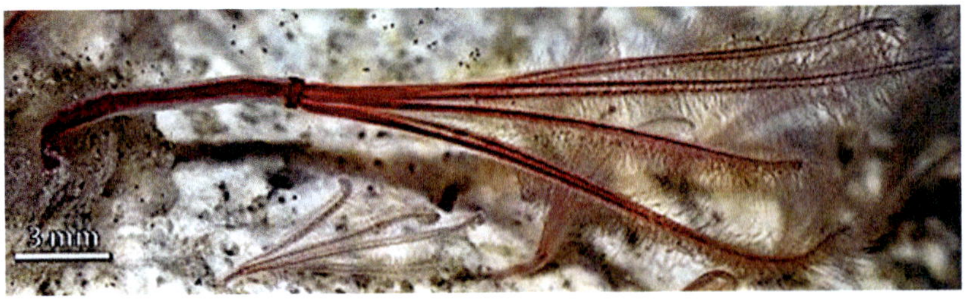

Fig. 5.47. *Osedax* bone worms on whale bone, Monterey Bay, USA (Image cropped from figure 1 in Vrijenhoek *et al.*, 2009). Creative Commons Attribution Licence.

Osedax is a speciose genus and multiple species may colonise a single whale carcass. Observations of *Osedax* were first published in 2004, following its discovery in 2002 on the carcass of a grey whale at a depth of 2891 m in Monterey Bay in the Pacific Ocean. The bones of the dead creature were found covered by a luxuriant carpet of bone worms (Rouse *et al.*, 2004). *Osedax* has since been reported from whale falls in the Indian Ocean, the Atlantic, the Mediterranean and from the Southern Ocean. Some records have been obtained by baiting the ocean floor with bones or carcasses. Cow bones deposited at a depth of 2341 m on the Arctic mid-Ocean Ridge, were found to be colonised by large numbers of bone worms when retrieved a year later (Eilertsen *et al.*, 2020). Similarly, the carcass of a stranded minke whale deposited at a relatively shallow depth of 125 m in the North Sea became colonised by *Osedax* (Glover *et al.*, 2005). It has also been shown that bone worms will colonise the bones of large fish (Rouse *et al.*, 2011).

Osedax has been feeding on the bones of vertebrate marine animals since the Cretaceous Period. Boreholes and chambers observed by scanning the fossilised bones of a plesiosaur from the Cambridge Greensand and a turtle from the Cenomanian Grey Chalk at Burham in Kent show that bone worms have been in existence for at least 100 million years (Danise and Higgs, 2015).

References

Aas, J.A., Paster, B.J., Stokes, L.N., Olsen, I. and Dewhirst, F.E. (2005) Defining the normal bacterial flora of the oral cavity. *Journal of Clinical Microbiology*, 43, 5721–5732.

Ahmed, M.Z., Araujo-Jnr, E.V., Welch, J.J. and Kawahara, A.Y. (2015) *Wolbachia* in butterflies and moths: geographic structure in infection frequency. *Frontiers in Zoology*, 12, 16.

Apprill, A., Robbins, J., Eren, A.M., Pack, A.A., Reveillaud, J. *et al.* (2014) Humpback whale populations share a core skin bacterial community: towards a health index for marine mammals? *PLoS ONE*, 9, e90785.

Apprill, A., Miller, C.A., Van Cise, A.M., U'Ren, J.M., Leslie, M.S. *et al.* (2020) Marine mammal skin microbiotas are influenced by host phylogeny. *Royal Society Open Science*, 7, 192046.

Arif, S., Gerth, M., Hone-Millard, W.G., Nunes, M.D.S., Dapporto, L. *et al.* (2021) Evidence for multiple colonisations and *Wolbachia* infections shaping the genetic structure of the widespread butterfly *Polyommatus icarus* in the British Isles. *Molecular Ecology*, 30, 5196–5213.

Atkinson, S.R., Paul, J., Sloane, E., Curtis, S. and Miller, R. (2009) The emergence of methicillin-resistant *Staphylococcus aureus* among injecting drug users. *Journal of Infection*, 58, 339–345.

Balvín, O., Roth, S., Talbot, B. and Reinhardt, K. (2018) Co-speciation in bedbug *Wolbachia* parallel the pattern in nematode hosts. *Scientific Reports*, 8, 8797.

Barnard, J.P. and Stinson, M.W. (1996) The alpha-hemolysin of *Streptococcus gordonii* is hydrogen peroxide. *Infection and Immunology*, 64, 3853–3857.

Becker, M.R., Paster, B.J., Leys, E.J., Moeschberger, M.L., Kenyon, S.G. *et al.* (2002). Molecular analysis of bacterial species associated with childhood caries. *Journal of Clinical Microbiology*. 40, 1001–1009.

Bierlich, K.C., Miller, C., DeForce, E., Friedlaender, A.S., Johnston, D.W. *et al.* (2018) Temporal and regional variability in the skin microbiome of humpback whales along the western Antarctic Peninsula. *Applied and Environmental Microbiology*, 84, e02574-17.

Billeter, S.A., Levy, M.G., Chomel, B.B. and Breitschwerdt, E.B. (2008) Vector transmission of *Bartonella* species with emphasis on the potential for tick transmission. *Medical and Veterinary Entomology*, 22, 1–15.

Bong, L.-J., Neoh, K.-B., Jaal, Z. and Lee, C.Y. (2015) *Paederus* outbreaks in human settings: a review of current knowledge. *Journal of Medical Entomology*, 52, 517–526.

Brites, D., Loiseau, C., Menardo, F., Borrell, S., Boniotti, M.B. *et al.* (2018) A new phylogenetic framework for the animal-adapted *Mycobacterium tuberculosis* complex. *Frontiers in Microbiology*, 9, 2820.

Burgess, N.R.H., McDermott, S.N. and Whiting, J. (1973) Aerobic bacteria occurring in the hind-gut of the cockroach, *Blatta orientalis*. *Journal of Hygiene* (Cambridge), 71, 1–7.

Butlin, R.K. and Hewitt, G.M.A. (1985) Hybrid zone between *Chorthippus parallelus parallelus* and *Chorthippus parallelus erythropus* (Orthoptera: Acrididae): behavioural characters. *Biological Journal of the Linnean Society*, 26, 287–299.

Cacchio, P., Ercole, C., Contento, R., Cappuccio, G., Preite Martinez M. *et al.* (2012) Involvement of bacteria in the origin of a newly described speleothem in the gypsum cave of Grave Grubbo (Crotone, Italy). *Journal of Cave and Karst Studies*, 74, 7–18.

Cagatay, N.S., Akhoundi, M., Izri, A., Brun, S., and Hurst, G.D.D. (2025) Prevalence of heritable symbionts in Parisian bedbugs (Hemiptera: Cimicidae). *Environmental Microbiology Reports*, 17(1), e70054. https://doi.org/10.1111/1758-2229.70054.

Carroll, S., McEvoy, E.G. and Gibson, R. (2003) The production of tetrodotoxin-like substances by nemertean worms in conjunction with bacteria. *Journal of Experimental Marine Biology and Ecology*, 288, 51–63.

Chiarello, M., Villéger, S., Bouvier, C., Auguet, J.C. and Bouvier, T. (2017) Captive bottlenose dolphins and killer whales harbor a species-specific skin microbiota that varies among individuals. *Scientific Reports*, 7, 15269.

Comas, I., Coscolla, M., Luo, T., Borrell, S., Holt, K.E. *et al.* (2013) Out-of-Africa migration and Neolithic coexpansion of *Mycobacterium tuberculosis* with modern humans. *Nature Genetics*, 45, 1176–1182.

Danise, S. and Higgs, N.D. (2015) Bone-eating *Osedax* worms lived on Mesozoic marine reptile deadfalls. *Biology Letters*, 11, 20150072.

Davison, N.J., Barnett, J.E., Koylass, M., Whatmore, A.M., Perkins, M.W. *et al.* (2014) *Helicobacter cetorum* infection in striped dolphin (*Stenella coeruleoalba*), Atlantic white-sided dolphin (*Lagenorhynchus acutus*), and short-beaked common dolphin (*Delphinus delphus*) from the southwest coast of England. *Journal of Wildlife Diseases*, 50, 431–437.

Davison, N.J., Dagleish, M.P., Ten Doeschate, M., Muchowski, J., Perrett, L.L. *et al.* (2021a) Meningoencephalitis in a common minke whale *Balaenoptera acutorostrata* associated with *Brucella pinnipedialis* and gamma-herpesvirus infection. *Diseases of Aquatic Organisms*, 144, 231–235.

Davison, N.J., Brownlow, A., Ten Doeschate, M., Dale, E.J., Foster, G. *et al.* (2021b) Neurobrucellosis due to *Brucella ceti* ST26 in three Sowerby's beaked whales (*Mesoplodon bidens*). *Journal of Comparative Pathology*, 182, 1–8.

De Silva, D., Peters, J., Cole, K., Cole, M.J., Cresswell, F. *et al.* (2016) Whole-genome sequencing to determine transmission of *Neisseria gonorrhoeae*: an observational study. *Lancet Infectious Diseases*, 16, 1295–1303.

Dean, G. (1905) Further observations on a leprosy-like disease of the rat. *Journal of Hygiene* (London), 5, 99–112.

Dehio, C., Sauder, U. and Hiestand, R. (2004) Isolation of *Bartonella schoenbuchensis* from *Lipoptena cervi*, a blood-sucking arthropod causing deer ked dermatitis. *Journal of Clinical Microbiology*, 42, 5320–5323.

Deng, J., Assandri, G., Chauhan, P., Futahashi, R., Galimberti, A. *et al.* (2021) *Wolbachia*-driven selective sweep in a range expanding insect species. *BMC Ecology and Evolution*, 21, 181.

Dudaniec, R.Y., Yong, C.J., Lancaster, L.T., Svensson, E.I., and Hansson, B. (2018) Signatures of local adaptation along environmental gradients in a range-expanding damselfly (*Ischnura elegans*). *Molecular Ecology*, 27, 2576–2593.

Duron, O., Bouchon, D., Boutin, S., Bellamy, L., Zhou, L. *et al.* (2008) The diversity of reproductive parasites among arthropods: *Wolbachia* do not walk alone. *BMC Biology*, 6, 27.

Edwards, R. (1997) Provisional Atlas of the aculeate Hymenoptera of Britain and Ireland. Part 1. Biological Records Centre, Huntingdon, UK.

Eilertsen, M.H., Dahlgren, T.G. and Rapp, H.T. (2020) A new species of *Osedax* (Siboglinidae: Annelida) from colonization experiments in the Arctic Deep Sea. *Frontiers in Marine Science*, 7, 443.

Fialho, R.F. and Stevens, L. (2000) Male-killing *Wolbachia* in a flour beetle. *Proceedings of the Royal Society B: Biological Sciences*, 267, 1469–1473.

Foster, G., Whatmore, A.M., Dagleish, M.P., Malnick, H., Gilbert, M.J. *et al.* (2019). Forensic microbiology reveals that *Neisseria animaloris* infections in harbour porpoises follow traumatic injuries by grey seals. *Scientific Reports*, 9, 14338.

Foster, G., Holmes, B., Steigerwalt, A.G., Lawson, P.A., Thorne, P. *et al.* (2004) *Campylobacter insulaenigrae* sp. nov., isolated from marine mammals. *International Journal of Systematic and Evolutionary Microbiology*, 54, 2369–2373.

Fujiwara, Y., Jimi, N., Sumida, P.Y.G., Kawato, M. and Kitazato, H. (2019) New species of bone-eating worm *Osedax* from the abyssal South Atlantic Ocean (Annelida, Siboglinidae). *ZooKeys*, 814, 53–69. https://doi.org/10.3897/zookeys.814.28869.

Ghose, C., Perez-Perez, G.I., Dominguez-Bello, M.G., Pride, D.T., Bravi, C.M. *et al.* (2002) East Asian genotypes of *Helicobacter pylori* strains in Amerindians provide evidence for its ancient human carriage. *Proceedings of the National Academy of Sciences of the United States of America*, 99, 15107–15111.

Gil-Turnes, M.S., Hay, M.E. and Fenical, W. (1989) Symbiotic marine bacteria chemically defend crustacean embryos from a pathogenic fungus. *Science*, 246, 116–118.

Glover, A.G., Källström, B., Smith, C.R. and Dahlgren, T.G. (2005) World-wide whale worms? A new species of *Osedax* from the shallow north Atlantic. *Proceedings of the Royal Society. B. Biological Sciences*, 272, 2587–2592.

Graffar, M. and Mertens, S. (1950) Le rôle des Blattes dans la transmission des salmonelloses. *Annales de l'Institut Pasteur* (Paris), 79, 654–660.

Grice, E.A., Kong, H.H., Renaud, G., Young, A.C., NISC Comparative Sequencing Program *et al.* (2008) A diversity profile of the human skin microbiota. *Genome Research*, 18, 1043–1050.

Grice, E.A., Kong, H.H., Conlan, S., Deming, C.B., Davis, J. *et al.* (2009) Topographical and temporal diversity of the human skin microbiome. *Science*, 324(5931), 1190–1192.

Guzmán-Verri, C., González-Barrientos, R., Hernández-Mora, G., Morales, J.A., Baquero-Calvo, E. *et al.* (2012). *Brucella ceti* and brucellosis in cetaceans. *Frontiers in Cellular and Infection Microbiology*, 2, 3.

Hammer, T.J., Le, E., Martin, A.N. and Moran, N.A. (2021) The gut microbiota of bumblebees. *Insectes Sociaux*, 68, 287–301.

Harper, C.G., Feng, Y., Xu, S., Taylor, N.S., Kinsel, M., Dewhirst, F.E. *et al.* (2002) *Helicobacter cetorum* sp. nov., a urease-positive *Helicobacter species* isolated from dolphins and whales. *Journal of Clinical Microbiology*, 40, 4536–4543.

Harris, L., Kelly, S., Hunter, M. and Perlman, S.J. (2010) Population dynamics and rapid spread of *Cardinium*, a bacterial endosymbiont causing cytoplasmic incompatibility in *Encarsia pergandiella* (Hymenoptera: Aphelinidae). *Heredity*, 104, 239–246.

Higgins, R. (2000) Bacteria and fungi of marine mammals: a review. *Canadian Veterinary Journal*, 41, 105–116.

Hördt, A., López, M.G., Meier-Kolthoff, J.P., Schleuning, M., Weinhold, L.M. *et al.* (2020) Analysis of 1,000+ Type-strain genomes substantially improves taxonomic classification of *Alphaproteobacteria*. *Frontiers in Microbiology*, 11, 468.

Howe, P.D., Bryant, S.R. and Shreeve, T.G. (2007) Predicting body temperature and activity of adult *Polyommatus icarus* using neural network models under current and projected climate scenarios. *Oecologia*, 153, 857–869.

Hurst, G.D.D., Graf von der Schulenburg, J.H., Majerus, T.M.O., Bertrand, D., Zakharov, I.A. *et al.* (1999) Invasion of one insect species, *Adalia bipunctata* by two different male-killing bacteria. *Insect Molecular Biology*, 8, 133–139.

Ilinsky, Y., Demenkova, M., Bykov, R. and Bugrov, A. (2022) Narrow genetic diversity of *Wolbachia* symbionts in Acrididae grasshopper hosts (Insecta, Orthoptera). *International Journal of Molecular Sciences*, 23, 853.

Jäckel, R., Diana M. and Dobler, S. (2013) Evidence for selective sweeps by *Wolbachia* infections: phylogeny of *Altica* leaf beetles and their reproductive parasites. *Molecular Ecology*, 22, 4241–4255.

Jahfari, S., Ruyts, S.C., Frazer-Mendelewska, E., Jaarsma, R.I., Verheyen, K. *et al.* (2017) Melting pot of tick-borne zoonoses: the European hedgehog contributes to the maintenance of various tick-borne diseases in natural cycles urban and suburban areas. *Parasites and Vectors*, 10, 134.

Kaltenpoth, M., Göttler, W., Herzner, G. and Strohm, E. (2005) Symbiotic bacteria protect wasp larvae from fungal infestation. *Current Biology*, 15, 475–479.

Kaltenpoth, M., Goettler, W., Dale, C., Stubblefield, J.W., Herzner, G. *et al.* (2006) '*Candidatus* Streptomyces philanthi', an endosymbiotic streptomycete in the antennae of *Philanthus* digger wasps. *International Journal of Systematic and Evolutionary Microbiology*, 56, 1403–1411.

Kaltenpoth, M., Winter, S.A. and Kleinhammer, A. (2009) Localization and transmission route of *Coriobacterium glomerans*, the endosymbiont of pyrrhocorid bugs. *FEMS Microbiology and Ecology*, 69, 373–383.

Kazor, C.E., Mitchell, P.M., Lee, A.M., Stokes, L.N., Loesche, W.J. *et al.* (2003) Diversity of bacterial populations on the tongue dorsa of patients with halitosis and healthy patients. *Journal of Clinical Microbiology*, 41, 558–563.

Kellner, R.L. (2000) Possible genetic basis of pederin polymorphism in rove beetles (*Paederus riparius*). *Journal of Heredity*, 91, 158–162.

Keyte, S., Abdullah, S., James, K., Newbury, H., Helps, C. *et al.* (2021) Prevalence and distribution of *Anaplasma phagocytophilum* in ticks collected from dogs in the United Kingdom. *Veterinary Record*, 188, e12.

Kidgell, C., Reichard, U., Wain, J., Linz, B., Torpdahl, M. *et al.* (2002) *Salmonella typhi*, the causative agent of typhoid fever, is approximately 50,000 years old. *Infection, Genetics and Evolution*, 2, 39–45.

Kinjo, Y., Bourguignon, T., Hongoh, Y., Lo, N., Tokuda, G. *et al.* (2022) Coevolution of metabolic pathways in Blattodea and their *Blattabacterium* endosymbionts, and comparisons with other insect-bacteria symbioses. *Microbiology Spectrum*, 10, e0277922.

Kleinschmidt, B. and Kölsch, G. (2011) Adopting bacteria in order to adapt to water – how reed beetles colonized the wetlands (Coleoptera, Chrysomelidae, Donaciinae). *Insects*, 2, 540–554.

Koch, H and Schmid-Hempel, P. (2011) Socially transmitted gut microbiota protect bumble bees against an intestinal parasite. *Proceedings of the National Academy of Sciences of the United States of America*, 108, 19288–19292.

Kolasa, M., Kubisz, D., Gutowski, J.M., Ścibior, R., Mazur, M.A., Holecová, M. and Kajtoch, Ł. (2018). Infection by endosymbiotic "male-killing" bacteria in Coleoptera. *Folia Biologica* (Kraków), 66, 165–178.

Kölsch, G. and Pedersen, B.V. (2008) Molecular phylogeny of reed beetles (Col., Chrysomelidae, Donaciinae): the signature of ecological specialization and geographical isolation. *Molecular Phylogenetics and Evolution*, 48, 936–952.

Kölsch, G. and Synefiaridou, D. (2012) Shared ancestry of symbionts? Sagrinae and Donaciinae (Coleoptera, Chrysomelidae) harbor similar bacteria. *Insects*, 3, 473–491.

Kuechler, S.M., Dettner, K. and Kehl, S. (2011) Characterization of an obligate intracellular bacterium in the midgut epithelium of the bulrush bug *Chilacis typhae* (Heteroptera, Lygaeidae, Artheneinae). *Applied and Environmental Microbiology*, 77, 2869–2876.

Kumar, P.S., Griffen, A.L., Barton, J.A., Paster, B.J, Moeschberger, M.L. *et al.* (2003). New bacterial species associated with chronic periodontitis. *Journal of Dental Research*, 82, 338–344.

Lamelas, A., Gosalbes, M.J., Manzano-Marín, A., Peretó, J. *et al.* (2011) *Serratia symbiotica* from the aphid *Cinara cedri*: a missing link from facultative to obligate insect endosymbiont. *PLoS Genetics*, 7, e1002357.

Ledda, A., Price, J., Cole, K., Llewelyn, M., Kearns, A. *et al.* (2017) Re-emergence of methicillin susceptibility in a resistant lineage of *Staphylococcus aureus*. *Journal of Antimicrobial Chemotherapy*, 72, 1285–1288.

Lewis, D. (1979) The detection of rickettsia-like microorganisms within the ovaries of female *Ixodes ricinus* ticks. *Zeitschrift für Parasitenkunde*, 59, 295–298.

Liu, G.Y., Essex, A., Buchanan, J.T., Datta, V., Hoffman, H.M. *et al.* (2005) *Staphylococcus aureus* golden pigment impairs neutrophil killing and promotes virulence through its antioxidant activity. *Journal of Experimental Medicine*, 202, 209–215.

Lo, N., Bandi, C., Watanabe, H., Nalepa, C. and Beninati, T. (2013) Evidence for cocladogenesis between diverse dictyopteran lineages and their intracellular endosymbionts. *Molecular Biology and Evolution*, 20, 907–913.

Lowe, J. (1937) Rat leprosy. a critical review of the literature. *International Journal of Leprosy*, 5, 311–328; 463–481.

Majerus, M. (1994) Ladybirds. New Naturalist. Collins, London, UK.

Majerus, M. (2006) The impact of male-killing bacteria on the evolution of aphidophagous coccinellids. *European Journal of Entomology*, 103, 1–7.

Mallet, J., Wynne, I.R., and Thomas, C.D. (2011) Hybridisation and climate change: brown argus butterflies in Britain (*Polyommatus* subgenus *Aricia*). *Insect Conservation and Diversity*, 4, 192–199.

Marden, J.N., McClure, E.A., Beka, L. and Graf, J. (2016) Host matters: medicinal leech digestive-tract symbionts and their pathogenic potential. *Frontiers in Microbiology*, 7, 1569.

Martínez-Rodríguez, P., Hernández-Pérez, M. and Bella, J.L. (2013). Detection of *Spiroplasma* and *Wolbachia* in the bacterial gonad community of *Chorthippus parallelus*. *Microbial Ecology*, 66, 211–223.

Martínez-Rodríguez, P. and Bella, J.L. (2018) *Chorthippus parallelus* and *Wolbachia*: overlapping orthopteroid and bacterial hybrid zones. *Frontiers in Genetics*, 9, 604.

Martinson, V.G., Gawryluk, R.M.R., Gowen, B.E., Curtis, C.I., Jaenike, J. *et al.* (2020) Multiple origins of obligate nematode and insect symbionts by a clade of bacteria closely related to plant pathogens. *Proceedings of the National Academy of Science of the United States of America*, 117, 31979–31986.

Mazur, M.A., Holecová, M., Lachowska-Cierlik, D., Lis, A., Kubisz, D. *et al.* (2016) Selective sweep of *Wolbachia* and parthenogenetic host genomes – the example of the weevil *Eusomus ovulum*. *Insect Molecular Biology*, 25, 701–711.

McFrederick, Q.S., Cannone, J.J., Gutell, R.R., Kellner, K., Plowes, R.M. *et al.* (2013) Specificity between lactobacilli and hymenopteran hosts is the exception rather than the rule. *Applied Environmental Microbiology*, 79, 1803–1812.

Mendell, J.E., Clements, K.D., Choat, J.H. and Angert, E.R. (2008) Extreme polyploidy in a large bacterium. *Proceedings of the National Academy of Sciences of the United States of America*, 105, 6730–6734.

Miller, W.D. (1890) The micro-organisms of the human mouth: the local and general diseases which are caused by them. S.S. White Dental Manufacturing, Philadelphia, Pennsylvania.
Miyamoto, N., Yoshida, Ma., Koga, H. and Fujiwara, Y. (2017) Genetic mechanisms of bone digestion and nutrient absorption in the bone-eating worm *Osedax japonicus* inferred from transcriptome and gene expression analyses. *BMC Evolutionary Biology*, 17, 17.
Moiser, B. (1945) Modes of transmission of Hansen's disease (leprosy). *Leprosy Review*, 16, 63–66.
Moiser, B. (1947) Hansen's disease (leprosy) and cockroaches. *East African Medical Journal*, 24, 230–236.
Moriarty, D.J.W. (1976) Quantitative studies on bacteria and algae in the food of the mullet *Mugil cephalus* L. and the prawn *Metapenaeus bennettae* (Racek and Dall). *Journal of Experimental Marine Biology and Ecology*, 22, 131–143.
Mueller, U.G., Ishak, H., Lee, J.C., Sen, R. and Gutell, R.R. (2010) Placement of attine ant-associated *Pseudonocardia* in a global *Pseudonocardia* phylogeny (*Pseudonocardiaceae, Actinomycetales*): a test of two symbiont-association models. *Antonie van Leeuwenhoek*, 98, 195–212.
Nadal-Jimenez, P., Siozios, S., Halliday, N., Cámara, M. and Hurst, G.D.D. (2022) *Symbiopectobacterium purcellii*, gen. nov., sp. nov., isolated from the leafhopper *Empoasca decipiens*. *International Journal of Systematic and Evolutionary Microbiology*, 72, 005440.
Nechitaylo, T.Y., Westermann, M. and Kaltenpoth, M. (2014) Cultivation reveals physiological diversity among defensive '*Streptomyces philanthi*' symbionts of beewolf digger wasps (Hymenoptera, Crabronidae). *BMC Microbiology*, 14, 202.
Norton, S.A. and Lyons, C. (2002) Blister beetles and the ten plagues. *Lancet*, 359, 1950.
Numberger, D., Siebert, U., Fulde, M. and Valentin-Weigand, P. (2021) Streptococcal infections in marine mammals. *Microorganisms*, 9, 350.
O'Brien, C.R., Malik, R., Globan, M., Reppas, G., McCowan, C. *et al.* (2017) Feline leprosy due to *Mycobacterium lepraemurium*. *Journal of Feline Medicine and Surgery*, 19, 737–746.
Ogston, A. (1882) Micrococcus poisoning. *Journal of Anatomy and Physiology*, 16 (4), 526–567; 17(1), 24–58.
Olsen, M.A., Blix, A.S., Utsi, T.H., Sørmo, W. and Mathiesen, S.D. (2000) Chitinolytic bacteria in the minke whale forestomach. *Canadian Journal of Microbiology*, 46, 85–94.
Oren, A. and Garrity, G.M. (2020) List of new names and new combinations previously effectively, but not validly, published. *International Journal of Systematic and Evolutionary Microbiology*, 70, 3.
Paster, B.J., Boches, S.K., Galvin, J.L., Ericson, R.E., Lau, C.N. *et al.* (2001) Bacterial diversity in human subgingival plaque. *Journal of Bacteriology*, 183, 3770–3783.
Patricelli, D., Sielezniew, M., Ponikwicka-Tyszko, D., Ratkiewicz, M., Bonelli, S. *et al.* (2013) Contrasting genetic structure of rear edge and continuous range populations of a parasitic butterfly infected by *Wolbachia*. *BMC Evolutionary Biology*, 13, 14.
Paul, J. and Bates, J. (2000) Is infestation with the common bedbug increasing? *British Medical Journal*, 320, 1141.
Penz, T., Schmitz-Esser, S., Kelly, S.E., Cass, B.N., Müller, A. *et al.* (2012) Comparative genomics suggests an independent origin of cytoplasmic incompatibility in *Cardinium hertigii*. *PLoS Genetics*, 8, e1003012.
Piel, J. (2002) A polyketide synthase-peptide synthetase gene cluster from an uncultured bacterial symbiont of *Paederus* beetles. *Proceedings of the National Academy of Sciences of the United States of America*, 99, 14002–14007.
Proemse, B.C., Eberhard, R.S., Sharples, C., Bowman, J.P., Richards, K. *et al.* (2017) Stromatolites on the rise in peat-bound karstic wetlands. *Scientific Reports*, 7, 15384.
Purcell, A., Steiner, T., Mégraud, F. and Bové, J.M. (1986) In vitro isolation of a transovarially transmitted bacterium from the leafhopper *Euscelidius variegatus* (Hemiptera: Cicadellidae). *Journal of Invertebrate Pathology*, 48, 66–73.

Qin, J., Li, R., Raes, J., Arumugam, M., Burgdorf, K.S., Manichanh, C. *et al.* (2010) A human gut microbial gene catalogue established by metagenomic sequencing. *Nature*, 464, 59–65.

Ragge, D.R. and Reynolds, W.J. (1998) The songs of the grasshoppers and crickets of Western Europe. Harley, Colchester, UK.

Regier, Y., Komma, K., Weigel, M., Pulliainen, A.T., Göttig, S. *et al.* (2018). Microbiome analysis reveals the presence of *Bartonella* spp. and *Acinetobacter* spp. in deer keds (*Lipoptena cervi*). *Frontiers in Microbiology*, 9, 3100.

Reis, F., Kirsch, R., Pauchet, Y., Bauer, E., Bilz, L.C. *et al.* (2020) Bacterial symbionts support larval sap feeding and adult folivory in (semi-)aquatic reed beetles. *Nature Communications*, 11, 2964.

Rosenbach, F.J. (1884) Microorganismen bei den Wund-Infections-Krankheiten des Menschen. J.F. Bergmann, Wiesbaden, Germany.

Roth, L.M. and Willis, E.R. (1957) The medical and veterinary importance of cockroaches. *Smithsonian Miscellaneous Collections*, 134, 1–147.

Rouse, G.W., Goffredi, S.K. and Vrijenhoek, R.C. (2004) *Osedax*: bone-eating marine worms with dwarf males. *Science*, 305, 668–671.

Rouse, G.W., Goffredi, S.K., Johnson S.B. and Vrijenhoek R.C. (2011) Not whale-fall specialists, *Osedax* worms also consume fishbones. *Biology Letters*, 7, 736–739.

Ruparell, A., Inui, T., Staunton, R., Wallis, C., Deusch, O. *et al.* (2020) The canine oral microbiome: variation in bacterial populations across different niches. *BMC Microbiology*, 20, 42.

Sabree, Z.L., Kambhampati, S. and Moran, N.A. (2009) Nitrogen recycling and nutritional provisioning by Blattabacterium, the cockroach endosymbiont. *Proceedings of the National Academy of Sciences of the United States of America*, 106, 19521–19526.

Sacchi, L., Bigliardi, E., Corona, S., Beninati, T., Lo, N. *et al.* (2004) A symbiont of the tick *Ixodes ricinus* invades and consumes mitochondria in a mode similar to that of the parasitic bacterium *Bdellovibrio bacteriovorus*. *Tissue and Cell*, 36, 43–53.

Sanders, J.G., Beichman, A.C., Roman, J., Scott, J.J., Emerson, D. *et al.* (2015) Baleen whales host a unique gut microbiome with similarities to both carnivores and herbivores. *Nature Communications*, 6, 8285.

Saylor, T.C., Casselli, T., Lethbridge, K.G., Moore, J.P., Owens, K.M. *et al.* (2022) *Borrelia burgdorferi*, the Lyme disease spirochete, possesses genetically-encoded responses to doxycycline, but not to amoxicillin. *PLoS ONE*, 17 (9), e0274125.

Seifert, B. (2021) A taxonomic revision of the Palaearctic members of the Formica rufa group (Hymenoptera: Formicidae) – the famous mound-building red wood ants. *Myrmecological News*, 31, 133–179. 10.25849/myrmecol.news_031:133.

Servín-Garcidueñas, L.E., Sánchez-Quinto, A. and Martínez-Romero, E. (2014) Draft genome sequence of *Commensalibacter papalotli* MX01, a symbiont identified from the guts of overwintering monarch butterflies. *Genome Announcements*, 2, e00128-14.

Shipley, A.E. (2015) The minor horrors of war. Smith, Elder, London, UK.

Shrivastava, A., Patel, V.K., Tang, Y., Yost, S.C., Dewhirst, F.E. *et al.* (2018) Cargo transport shapes the spatial organization of a microbial community. *Proceedings of the National Academy of Sciences of the United States of America*, 115, 8633–8638.

Smee, M.R. (2011). Population ecology and genetics of the Marsh Fritillary Butterfly *Euphydryas aurinia*. Doctoral Thesis. University of Exeter.

Smith, D., Stewart, K., and Goddard, E. (2023) The oriental cockroach *Blatta orientalis* L. recovered from Early Roman London: implications for past distribution and roman trade. *Environmental Archaeology*, 29(1), 63–70.

Smith, E.A., Anderson, K.E., Corby-Harris, V., McFrederick, Q.S., Parish, A.J. *et al.* (2021) Reclassification of seven honey bee symbiont strains as *Bombella apis*. *International Journal of Systematic and Evolutionary Microbiology*, 71, 9.

Solnick, J.V. and Schauer, D.B. (2001) Emergence of diverse *Helicobacter* species in the pathogenesis of gastric and enterohepatic diseases. *Clinical Microbiology Reviews*, 14, 59–97.

Son, Y., Luckhart, S., Zhang, X., Lieber, M.J. and Lewis, E.E. (2008) Effects and implications of antibiotic treatment on *Wolbachia*-infected vine weevil (Coleoptera: Curculionidae). *Agricultural and Forest Entomology*, 10, 147–155.

Stammer, H.J. (1935) Studien an Symbiosen zwischen Käfern und Mikroorganismen I. Die Symbiose der Donaciinen (Coleopt. Chrysomel.). *Zeitschrift für Morphologie und Ökologie der Tiere*, 29, 585–608.

Stevens, C.E. and Hume, I.D. (1998) Contributions of microbes in vertebrate gastrointestinal tract to production and conservation of nutrients. *Physiological Reviews*, 78, 393–427.

Sucháčková Bartoňová, A., Konvička, M., Marešová, J., Wiemers, M., Ignatev, N., *et al.* (2021) *Wolbachia* affects mitochondrial population structure in two systems of closely related Palaearctic blue butterflies. *Scientific Reports*, 11, 3019.

Thongprem, P., Evison, S.E.F., Hurst, G.D.D. and Otti, O. (2020) Transmission, tropism, and biological impacts of torix *Rickettsia* in the common bed bug *Cimex lectularius* (Hemiptera: Cimicidae). *Frontiers in Microbiology*, 23 (11), 608763.

Thuesen, E.V., Kogure, K., Hashimoto, K. and Nemoto, T. (1988) Poison arrowworms: a tetrodotoxin venom in the marine phylum Chaetognatha. *Journal of Experimental Marine Biology and Ecology*, 116, 249–256.

Tijsse-Klasen, E., Hansford, K.M., Jahfari, S., Phipps, P., Sprong, H. *et al.* (2013) Spotted fever group rickettsiae in *Dermacentor reticulatus* and *Haemaphysalis punctata* ticks in the UK. *Parasites and Vectors*, 6, 212.

Tió-Coma, M., Wijnands, T., Pierneef, L., Schilling, A.K., Alam, K. *et al.* (2019) Detection of *Mycobacterium leprae* DNA in soil: multiple needles in the haystack. *Scientific Report*, 9, 3165.

Tulloch, J.S.P., Semper, A.E., Brooks, T.J.G., Russell, K., Halsby, K.D. *et al.* (2019) The demographics and geographic distribution of laboratory-confirmed Lyme disease cases in England and Wales (2013–2016): an ecological study. *BMJ Open*, 9, e028064.

Van Cise, A.M., Wade, P.R., Goertz, C.E.C., Burek-Huntington, K., Parsons, K.M. *et al.* (2020) Skin microbiome of beluga whales: spatial, temporal, and health-related dynamics. *Animal Microbiome*, 2, 39.

van de Weg, C.A.M., de Steenwinkel, J.E.M., Miedema, J.R., Bakker, M., van Ingen, J. *et al.* (2019) The tough process of unmasking the slow-growing mycobacterium: case report of *Mycobacterium microti* infection. *Access Microbiology*, 2, acmi000074.

Vásquez, A., Forsgren, E., Fries, I., Paxton, R.J., Flaberg, E. *et al.* (2012) Symbionts as major modulators of insect health: lactic acid bacteria and honeybees. *PLoS ONE*, 7, e33188.

Verna, C., Ramette, A., Wiklund, H., Dahlgren, T.G., Glover, A.G. *et al.* (2010) High symbiont diversity in the bone-eating worm *Osedax mucofloris* from shallow whale-falls in the North Atlantic. *Environmental Microbiology*, 12, 2355–2370.

Viljakainen, L., Reuter, M. and Pamilo, P. (2008) *Wolbachia* transmission dynamics in *Formica* wood ants. *BMC Evolutionary Biology*, 8, 55.

Vorburger, C., Gehrer, L. and Rodriguez, P. (2010) A strain of the bacterial symbiont *Regiella insecticola* protects aphids against parasitoids. *Biology Letters*, 6, 109–111.

Vrijenhoek, R.C., Johnson, S.B. and Rouse, G.W. (2009) A remarkable diversity of bone-eating worms (*Osedax*; Siboglinidae; Annelida). *BMC Biology*, 7, 74.

Wada-Katsumata, A., Zurek, L., Nalyanya, G., Roelofs, W.L., Zhang, A. *et al.* (2015) Gut bacteria mediate aggregation in the German cockroach. *Proceedings of the National Academy of Sciences of the United States of America*, 112, 15678–15683.

Weinert, L.A., Tinsley, M.C., Temperley, M. and Jiggins, F.M. (2007) Are we underestimating the diversity and incidence of insect bacterial symbionts? A case study in ladybird beetles. *Biology Letters*, 3, 678–681.

Werren, J.H., Windsor, D. and Guo, L.R. (1995) Distribution of *Wolbachia* among neotropical arthropods. *Proceedings of the Royal Society of London B*, 262, 197–204.

Wilson, D.J., Gabriel, E., Leatherbarrow, A.J., Cheesbrough, J., Gee, S. *et al.* (2009) Rapid evolution and the importance of recombination to the gastroenteric pathogen *Campylobacter jejuni*. *Molecular Biology and Evolution*, 26, 385–397.

Wood, C., Fitt, R.N.L. and Lancaster, L.T. (2019) Evolving social dynamics prime thermal tolerance during a poleward range shift. *Biological Journal of the Linnean Society*, 126 (3), 574–586, https://doi.org/10.1093/biolinnean/bly197.

Worsfold, T.M. and Ashelby, C.W. (2008) Additional UK records of the non-native prawn *Palaemon macrodactylus* (Crustacea: Decapoda). *Marine Biodiversity Records*, 1 (e48), 1–3.

Young, B.C., Wu, C., Gordon, N.C., Cole, K., Price, J. *et al.* (2017) Severe infections emerge from commensal bacteria by adaptive evolution. *eLife*, 6, e30637

Zchori-Fein, E., Gottlieb, Y., Kelly, S.E., Brown, J.K., Wilson, J.M. *et al.* (2001) A newly discovered bacterium associated with parthenogenesis and a change in host selection behavior in parasitoid wasps. *Proceedings of the National Academy of Sciences of the United States of America*, 98, 12555–12560.

6 Plants

Canker is a disease of plants.

James Joyce (1916). *A Portrait of the Artist as a Young Man*.

Plants would not be plants without bacteria. The cells of plants contain chloroplasts, photosynthetic organelles whose antecedents were once free-living cyanobacteria. It has been estimated that the first photosynthetic eukaryotes emerged about 1.9 billion years ago when a single-celled eukaryote engulfed a free-living cyanobacterium and became a new kind of photosynthetic organism. This endosymbiotic event gave rise eventually to red algae, green algae and green plants (Sánchez-Baracaldo *et al.*, 2017). Chloroplasts contain the photosynthetic pigment chlorophyll (Fig. 6.1) that gives plants their green colouration. Chloroplasts continue to retain their own complement of DNA independently of their host cell but after hundreds of millions of years of endosymbiotic existence chloroplast genomes have become much smaller than those of free-living cyanobacteria. The closest free-living relatives of chloroplasts are thought to be members of the cyanobacterial genus *Gloeomargarita*, which inhabit freshwater lakes.

Free-living cyanobacteria continue to form symbiotic associations with certain plants and also with fungi. As well as being able to perform photosynthesis, free-living cyanobacteria are also able to fix atmospheric nitrogen. Nitrogen fixation is important because among other things nitrogen is a component of amino acids, which form the building blocks of proteins. Symbiotic relationships exist between nitrogen-fixing cyanobacteria and cycads, water ferns and gunneras.

A well-known bacteria fact is that members of the pea family, popularly known in English as legumes, accommodate nitrogen-fixing bacteria within specialised root nodules. Knowledge that leguminous plants will put the

DOI: 10.1079/9781836993933.0006

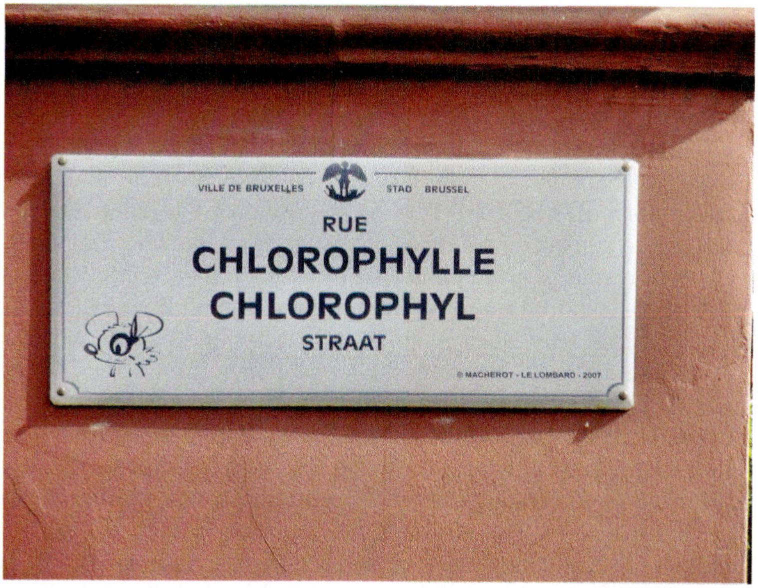

Fig. 6.1. Street sign in homage to the Belgian comic series, *Chlorophylle*. Brussels, Belgium.

goodness back into the soil forms the basis of crop rotation schemes. Less well known are the symbioses formed between nitrogen-fixing bacteria and other plant families. While members of the pea family are associated with nitrogen-fixing Gram-negative rhizobacteria, completely unrelated nitrogen-fixing Gram-positive bacteria from the genus *Frankia* form symbioses with a number of other plant families. Nodules containing *Frankia* can be found at the roots of common alder, *Alnus glutinosa*, bog myrtle, *Myrica gale* and sea buckthorn, *Hippophae rhamnoides*. *Frankia* gives these plants an edge in impoverished soils.

By chelating iron and by solubilising phosphorus, soil bacteria that live in association with the roots of plants increase the availability of these essential nutrients. The external aerial surfaces of plants provide a home for commensal bacteria. The microbes that colonise leaves, stems, flowers and fruits are known as epiphytes. Epiphytes interact with one another and with the environment. Epiphytes include nitrogen-fixing bacteria and photosynthetic cyanobacteria. Some epiphytes specialise in feeding on substances secreted by plants. Bacteria may even creep through open pores (stomata) and take up residence inside leaves and stems. They are known as endophytes. There is some evidence to suggest that nitrogen-fixing bacteria that live inside the leaves and needles of some poplar, willow and conifer species act as growth promoters and enable plants to grow in impoverished soils.

In order to capture and digest prey, some carnivorous plants, such as bladderworts and pitcher plants are equipped with specialised bladder-shaped traps or with modified leaves that function as pitfall traps. These structures are inhabited by populations of commensal bacteria. It can be argued that these enclosed spaces function in effect as plant digestive systems and also that they are miniature microbial ecosystems.

Plant diseases caused by bacteria manifest themselves as blights, galls, spots, rots, wilts and cankers. Pathogenic bacteria may cause localised damage to plant cells, resulting in spots or stripes of dead plant tissue. In cases of more extensive damage, the whole plant may die. Bacteria that invade the xylem cause tissue necrosis which leads to wilting of the plant. Necrotising bacteria can cause sunken areas of dead tissue called cankers that ooze fluids. Many cases of canker are fungal in origin but bacteria are important causes of cankers that affect fruit trees, ash and horse chestnut. Bacterial plant diseases seen in the British Isles affect many plant species and include white colour disease of creeping thistle, bleeding canker of horse chestnut, crown gall disease of cherry, lilac leaf blight, acute oak decline, ash canker, black rot of crucifers and fireblight of apples. Many of the bacterial plant diseases that commonly occur in the British Isles are caused by bacteria that belong to a relatively small number of Gram-negative genera in the Phylum *Pseudomonadota*. Variants of one particular species, *Pseudomonas syringae* are responsible for a wide range of diseases that affect a large number of different plant species (Figs. 1.9–1.12).

An understanding of plant-associated bacteria leads to a more complete appreciation of plant biology and goes some way to demystify some of the more obscure corners of natural history.

Cyanobacteria as symbionts of plants: *Gunnera* or giant rhubarb

Cyanobacteria can fix atmospheric nitrogen, enabling plants with cyanobacterial symbionts to grow in nutrient-poor soils. Plants that host symbiotic cyanobacteria include gunneras, cycads, water ferns, liverworts and hornworts. Gunneras are widely distributed in the tropics and at higher latitudes in the southern hemisphere. Many species occur at high altitudes, where they tend to frequent skeletal soils (Fig. 6.2). Others favour swampy lowland habitats. Gunneras are not native to the British Isles but are appealing to gardeners because of their large leaves and are often planted as waterside feature plants. One species in particular, *Gunnera tinctoria*, or giant rhubarb, originally from Chile, is widely planted in the British Isles (Fig. 6.3). A second South American species *Gunnera manicata* is also a popular waterside plant (Fig. 6.4). *Gunnera tinctoria* has become naturalised in Ireland, in western parts of the British Isles and in New Zealand where it poses a threat to the native flora. The sale of *Gunnera tinctoria* has been prohibited in Europe since 2018 and in the Western Isles, where the plant has become rapidly invasive a control campaign was launched. *Gunnera tinctoria* spreads by producing large numbers of seeds and displaces the native vegetation. If left unchecked *Gunnera manicata* can also become invasive in some settings. Whilst the sale of *Gunnera manicata* currently remains unrestricted, it has been listed as a non-native invasive plant in the Republic of Ireland.

Gunneras are equipped with special glands on the stem at the base of the petiole that facilitates infection with *Nostoc,* a cyanobacterium that serves as a nitrogen-fixing symbiont. When supplied with ammonium nitrate as a nitrogen

Fig. 6.2. *Gunnera insignis*, in its natural habitat. Volcan Irazú, Costa Rica.

Fig. 6.3. *Gunnera tinctoria* on the Isle of Harris, UK, where this plant has become an invasive species.

source, seedlings of *Gunnera tinctoria* fail to develop symbiotic glands. This demonstrates that the formation of the symbiotic glands is stimulated by nitrogen deprivation (Chiu *et al.*, 2005). The fossil record shows that gunneras have been around since the late Cretaceous Period. They form a rather

Fig. 6.4. *Gunnera manicata*. County Antrim, Northern Ireland.

distinctive group of plants whose closest relatives appear to be the saxifrages (Fuller and Hickey, 2005).

Cycads

Cycads make attractive houseplants. The sago palm, *Cycas revoluta*, originally from Japan, is a popular houseplant in the UK. Cycads superficially resemble palm trees but whereas palms are classified as angiosperms, or flowering plants, cycads are classified as gymnosperms, or cone-bearing plants. Cycads form a distinctive group among the gymnosperms, the Division Cycadophyta. The heyday for cycads was during the Jurassic Period. It is presumed that cycads constituted a significant part of the diet of herbivorous dinosaurs. Cycads today are widely distributed in the tropics and subtropics (Fig. 6.5). Cycads have specialised nodular coralloid roots which accommodate nitrogen-fixing cyanobacterial symbionts. Cycad cyanobacterial symbionts are acquired from the surrounding environment. The symbiont is usually a strain of *Nostoc*, although other cyanobacterial genera including *Calothrix*, *Richelia* and *Scytonema* have been found living in association with cycads (Chang *et al.*, 2019).

For many years I kept a cycad as a treasured houseplant (Fig. 6.6). When the cycad needed repotting I had an opportunity to examine the cycad's coralloid roots and to look for cyanobacterial symbionts. Surprisingly this was much easier than I expected. The coralloid roots were unexpectedly friable and the simple process of extracting the cycad from its pot caused a number of the root tips to snap off and expose the cyanobacteria. These are easily recognisable on account of their green colouration. Seen in cross section a green ring is visible,

Fig. 6.5. Zamia palm, *Macrozamia riedlei*, a cycad in its natural habitat. Perth, Australia.

Fig. 6.6. Sago palm, *Cycas revoluta*, a popular houseplant.

which is known as the cyanobacterial zone (Fig. 6.7). It seems odd that the cyanobacteria continue to produce a photosynthetic pigment, even when they dwell beneath the soil in conditions of near total darkness.

Water Ferns

Water ferns, *Azolla* spp., are the only truly aquatic ferns. In North America they are known as mosquito ferns on account of the fact that by blanketing the surface

Fig. 6.7. Coralloid roots of *Cycas revoluta,* showing green cyanobacterial zone.

Fig. 6.8. Water fern, *Azolla filiculoides.* Bramber, West Sussex, UK.

of a pond with their fronds they deter the egg-laying activities of female mosquitoes. Water ferns occur widely in the tropics. In rice growing areas, paddy fields are planted with *Azolla* between rice harvests because of its capacity to increase the availability of nitrogen. The introduced species *Azolla filiculoides*, originally from North America, is now widely distributed in the British Isles (Fig. 6.8). For much of the time *Azolla* is nowhere to be seen but occasionally massive unsightly blooms of this otherwise attractive water plant may occur (Fig. 6.9). Blooms of *Azolla* can choke waterways to the detriment of the native flora and fauna. *Azolla* is of interest to coleopterists as the food plant of a minute exotic weevil, *Stenopelmus rufinasus*. Originally from North America, the weevil was accidentally introduced to the UK along with its food plant.

Microscopic examination of *Azolla* fronds will reveal its nitrogen-fixing cyanobacterial symbiont, *Anabaena azollae* (Fig. 6.10). The cyanobacteria are readily distinguishable from the host plant's chloroplasts because of their blue-green colouration as opposed to the yellow-green hue of the chloroplasts. *Azolla* symbionts are transmitted vertically through the megaspore stage of the fern life cycle to the next generation of vegetative *Azolla*.

The *Azolla* event

About 50 million years ago during the Eocene Epoch massive *Azolla* blooms occurred in the Arctic Ocean. At that time the Arctic Ocean was encircled by land masses, was mostly stagnant with a stratified water column and was in receipt of freshwater run off from surrounding rivers. Temperatures were higher than they are today and the upper freshwater layers of the Arctic Ocean supported huge seasonal *Azolla* blooms. The dead remains of each bloom sank to the bottom of the water column to form a thick fossilised layer of *Azolla* remnants. Enormous amounts of organic carbon were buried as a result of this process; so much in fact that it has been reasoned that the Azolla event might have

Fig. 6.9. Bloom of *Azolla filiculoides*. Bramber, West Sussex, UK.

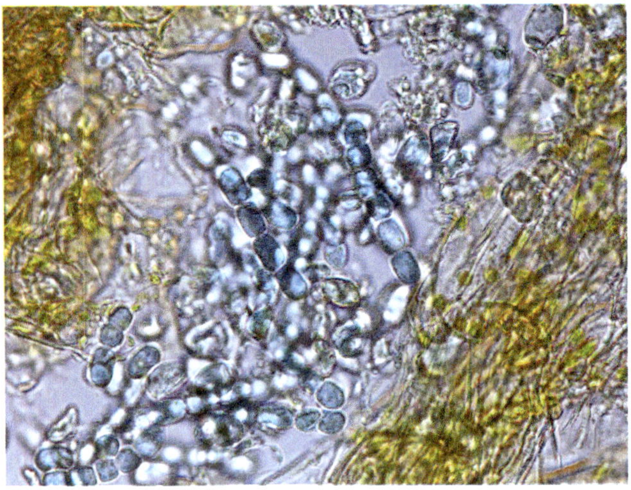

Fig. 6.10. Microscopy of water fern, showing its cyanobacterial symbiont *Anabaena azollae.*

caused a reduction in global temperatures (Brinkhuis *et al.*, 2006). There is commercial interest in the possibility of being able to exploit oil and gas reserves that may exist beneath the Arctic basin as a consequence of the *Azolla* event.

Nitrogen-fixing symbionts of legumes

The nitrogen-fixing bacteria that live in the root nodules of leguminous plants (Fabaceae, also known as the Leguminosae and Papilionaceae) (Figs. 6.11–6.12) were discovered by the Dutch microbiologist Martinus Beijerinck (1851–1931). It was known that legumes (including peas, beans and clovers) added fertility to the soil and leguminous plants were included in crop rotation schemes for that reason. The symbionts that inhabit the root nodules of legumes are informally known as rhizobacteria or rhizobia in reference to the genus *Rhizobium*. In 1889 the German microbiologist Albert Bernhard Frank (1839–1900) published the first formal description of a legume symbiont which he named *Rhizobium leguminosarum*. Numerous other nitrogen-fixing symbionts have since been discovered, including members of the genera *Bradyrhizobium*, *Mesorhizobium*, *Sinorhizobium*, *Azorhizobium*, *Methylobacterium* and *Devosia*, all of which belong to the Class *Alphaproteobacteria*. Other legume symbionts belong to a different class, the *Betaproteobacteria*. Legume symbiont genera tend to align with particular taxonomic groups of legumes (Andrews and Andrews, 2017). With regard to the British flora, members of the genera *Lathyrus* (vetchlings) and *Vicia* (vetches) have *Rhizobium* species as symbionts, whilst *Medicago* (medicks) and *Melilotus* (melilots) are colonised by *Ensifer* species (Table 6.1). Some legumes are more fussy than others. Different species in the genus *Galega*, which includes goat's-rue, *Galega officinalis*, are colonised by specific symbiotic variants of *Neorhizobium galegae*

Fig. 6.11. Mountain milk-vetch, *Oxytropis halleri*, a legume growing on coastal dunes. Invernaver, Sutherland.

Fig. 6.12. Everlasting sweet pea, *Lathyrus latifolius*, a legume growing on chalk grassland. Whitehawk Hill, Brighton, UK.

(Österman *et al.*, 2015). Clovers, in contrast, have been shown to accommodate symbionts from multiple genera.

Legume symbionts reside in the soil. They react to flavonoid compounds released by the roots of legumes by releasing compounds themselves that cause

Table 6.1. Some British legumes (Fabaceae) and their nitrogen-fixing bacterial symbionts.

Scientific name	English name	Nitrogen-fixing symbiont
Lathyrus japonicus	Sea pea	*Rhizobium*
Lathyrus aphaca	Yellow vetchling	*Rhizobium*
Lathyrus nissolia	Grass vetchling	*Rhizobium*
Lathyrus pratensis	Meadow vetchling	*Rhizobium*
Vicia cracca	Tuften vetch	*Rhizobium*
Vicia sylvatica	Wood vetch	*Rhizobium*
Vicia hirsuta	Hairy tare	*Rhizobium*
Vicia tetrasperma	Smooth tare	*Rhizobium*
Vicia sepium	Bush vetch	*Rhizobium*
Galega officinalis	Goat's-rue	*Neorhizobium*
Onobrychis viciifolia	Sainfoin	*Phyllobacterium*
Medicago lupulina	Black medick	*Ensifer*
Medicago sativa	Lucerne	*Ensifer*
Medicago polymorpha	Fimbriate medick	*Ensifer*
Melilotus alba	White melilot	*Ensifer*
Melilotus indicus	Small melilot	*Ensifer*
Melilotus officinalis	Ribbed melilot	*Ensifer*
Trifolium fragiferum	Strawberry clover	*Bradyrhizobium, Mesorhizobium, Rhizobium*
Trifolium pratense	Red clover	*Bradyrhizobium, Ensifer, Rhizobium*
Trifolium repens	White clover	*Bradyrhizobium, Ensifer, Rhizobium*
Ulex europaeus	Gorse	*Bradyrhizobium*
Cytisus scoparius	Broom	*Bradyrhizobium*

Source: Based on data from Andrews and Andrews (2017).

the roots to become receptive to entry by the bacteria. Once inside, the bacteria become incorporated inside root cells, where they continue to live in the manner of an intracellular organelle. The symbionts convert nitrogen to ammonia which is used by the plant as a source of nitrogen. The energy requirements for nitrogen fixation are met by carbohydrates supplied to the symbionts by the host plant.

Frankia

If you pull up an alder seedling it is quite likely that you will soon discover root nodules induced by *Frankia* (Fig. 6.13). *Frankia* is a nitrogen-fixing Gram-positive bacterium that enables alder trees, *Alnus* spp., to grow in nitrogen-free soils. When grown in nitrogen-free nutrient solutions alder seedlings inoculated with *Frankia* remain green and healthy, whereas uninfected seedlings fail to survive (Benson, 1982). Alders are pioneering plants that colonise nitrogen-poor soils, such as the waterlogged soils associated with bogs. Common alder, *Alnus glutinosa* often grows in the same places that are frequented by another *Frankia*-associated plant, bog myrtle, *Myrica gale*. Common alder and bog myrtle are among the few species of large woody plant species that thrive in particularly boggy locations, such as in the precarious floating bogs of the New Forest (Fig. 6.14).

Fig. 6.13. Alder seedling with root nodules.

Fig. 6.14. Common alder and bog myrtle, New Forest, UK.

Frankia (Fig. 6.15) is a widespread soil bacterium and member of the Phylum *Actinomycetota*. A survey of Scottish soils detected *Frankia* at locations where its host plants were absent as well as at sites colonised by common alder or bog myrtle (Rodriguez-Barrueco, 1968). In addition to the native common alder, several exotic alders, the hybrid alder, *Alnus* x *spaethii*, grey alder, *Alnus incana* and Italian alder, *Alnus cordata* are planted in the UK. Grey alder thrives on well-drained, nitrogen-poor soils and can sometimes be seen growing on former spoil heaps and landfill sites. Alder-associated *Frankia* species include *Frankia alni*, *Frankia torreyi* and *Frankia canadensis*. The type strain of *Frankia alni* was isolated from a root nodule of an American green alder, *Alnus viridis* subsp. *crispa* at Tadoussac in Quebec (Nouioui *et al.*, 2016). The village of Tadoussac is situated on the St Lawrence River and is a popular destination for birdwatchers and whale watchers. The village is flanked by an enormous steep-sided marine terrace composed of gritty sand dropped by glacial meltwater at the end of the Ice Age. Few plants manage to survive in such hostile terrain, apart from a few scattered bushes of American green alder (Fig. 6.16).

Sea buckthorn, *Hippophoe rhamnoides*, is the only member of the Family Elaeagnaceae that is native to Britain. The nodular roots of sea buckthorn accommodate nitrogen-fixing *Frankia* symbionts, which enable the plant to colonise nitrogen-poor coastal sand dunes (Figs. 6.17–6.18). Pollen samples show that sea buckthorn became abundant soon after the retreat of the ice at the end of the Ice Age. In modern times its distribution has become limited to the coast. Because of its ability to stabilise sand dunes, sea buckthorn has been planted at many locations around the British Isles. Sea buckthorn is an attractive plant. Its orange berries can be used to make jam and they provide food for birds. Rabbits feed on the seedlings. With the decline of rabbit populations

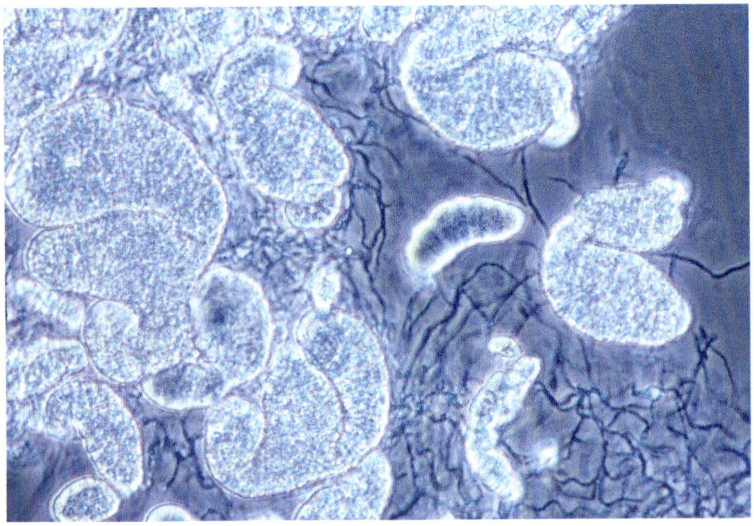

Fig. 6.15. *Frankia*, microscopy. (Image kindly provided by Philippe Normand.)

Fig. 6.16. American green alder. Tadoussac, Quebec, Canada.

Fig. 6.17. Nodular roots of sea buckthorn.

that followed the arrival of myxomatosis in the 1950s, sea buckthorn became recognised as an invasive plant that threatened to displace less vigorous members of the dune flora at some locations (Ranwell, 1972). Accordingly

Fig. 6.18. Sea buckthorn, Tyne Sands, East Lothian, UK.

control measures have been instigated at a number of sites. Clumps of sea buckthorn are cut back and seedlings uprooted. While strolling around Tentsmuir National Nature Reserve I noticed a number of saplings that showed signs of having been treated with herbicide. Sea buckthorn is a relatively recent arrival at Tentsmuir. Its rapid colonisation of the dunes poses a threat to the indigenous flora including rare orchid species.

Non-native *Frankia*-associated garden plants which can be seen in the UK include cultivars of California lilac, *Ceanothus* and oleaster, *Elaeagnus ebbingei*. Oleaster is a member of the Family Elaeagnaceae (the same family as sea buckthorn). It is used as a hedgerow plant. *Ceanothus* is a member of the Family Rhamnaceae. It is favoured by gardeners because of its blue flowers. The plants that form symbiotic relationships with *Frankia* amount to a seemingly unlikely hotchpotch of species that belong to multiple different plant families (Table 6.2). The *Frankia* strains that infect them can be grouped into a number of genetically related clades or clusters, which broadly align with the various *Frankia*-associated plant families.

One of the genetic clusters relates to strains of *Frankia* that form symbioses with members of the Family Rosaceae, as well as with some other plant families. *Frankia* is not known to form symbioses with any members of the rose family in the British Isles. In North America, however, *Frankia* can be found in association with a number of rosaceous plant species. One of them is yellow mountain avens, *Dryas drummondii*. Oddly, the closely related mountain avens, *Dryas octopetala*, which also occurs in North America, seems to be resistant to infection with *Frankia* and it does not form root nodules. *Dryas octopetala* is widely distributed in the northern hemisphere. Its attractive white eight-petalled flowers are a feature of the flora of the Burren in western Ireland.

Table 6.2. Relationships between *Frankia* strains and host plants.

Strain cluster and species	Associated plants	Distribution
Cluster 1 – 'Alnus strains' *Frankia alni* *Frankia torreyi* *Frankia canadensis*	Alders, myrtles	Worldwide (but not Australia)
Cluster 1 – 'Casuarina strains' *Frankia casuarinae*	Casuarinas (she-oaks)	Australia
Cluster 2 – 'Rosaceous strains' *Frankia coriariae*	Yellow mountain avens, California lilac, *Purshia, Chamaebatia, Datisca, Cercocarpus, Coriaria*	The Americas, New Zealand, New Guinea, Asia, Mediterranean
Cluster 3 – 'Elaeagnus strains' *Frankia elaeagni* *Frankia disciariae*	Sea buckthorn, oleaster, *Colletia, Discaria,* alders, myrtles	Europe, Asia, the Americas, Australia, New Zealand

Source: Adapted from Normand *et al.* (2007) with permission from David Benson.

The 'Casuarina strains' are associated with casuarina trees, also known as she-oaks. Casuarina strains have not been detected in soils in the absence of casuarinas. Casuarina strains have unusually small genomes and appear to have become dependent on their host plants. These strains have quite likely coevolved with their host plants in Australia since the break-up of Gondwanaland. Casuarinas are adapted to life in dry nutrient-poor soils in Australia but have since been planted in similar soils in other continents and have taken their symbiotic bacteria with them. 'Alnus strains' occur worldwide but not in Australia and appear to be able to exist as free-living bacteria in soils in the absence of host plants. 'Elaeagnus strains' also occur in soils in the absence of host plants and occur worldwide, including Australia.

Nitrogen-fixing endophytes of poplars, willows and conifers

Many of the plants that flourish in nitrogen-poor soils are species with root nodules inhabited by nitrogen-fixing bacteria. Explaining how other plants that grow in such places obtain nitrogen presents more of a puzzle, although a growing body of evidence supports a view that some poplars, willows and conifers acquire nitrogen from nitrogen-fixing bacteria that live inside their leaves or needles (von Wuehlisch, 2011). Poplars and willows can grow rapidly in waterlogged as well as in well-drained sandy soils. Poplar suckers are a common sight along the sandy banks of some rivers, where few other plant species seem able to gain a root hold. It has also been noted that nitrogen fertilisers have little effect on the growth rates of these trees.

Poplar seedlings grow more vigorously when inoculated with a nitrogen-fixing plant-associated strain of *Paenibacillus humicus* (Scherling *et al.*, 2009). Nitrogen fixation has been shown to occur in the leaves of wild poplars harvested from a river bank in Washington State. The poplar leaves were colonised by multiple species of endophytic nitrogen-fixing bacteria (Doty *et al.*, 2016). The inoculation of Douglas fir, *Pseudotsuga menziesii*, seedlings with nitrogen-fixing bacteria results in the growth of taller, bulkier plants (Khan *et al.*, 2015). Douglas fir was introduced to Britain from North America in 1827 by the botanist, David Douglas. At the time of writing the tallest tree in England is a 60-plus metre high Douglas fir at Nutcombe Bottom in Somerset. The tallest tree in Britain is a 66-m high Douglas fir at Reelig Glen, near Inverness.

Carnivorous plants

In nutrient-poor soils, carnivorous plants supplement their intake of organic nitrogen and other nutrients by digesting the decomposed remains of trapped invertebrates. Sundews and butterworts secrete enzymes in order to digest their prey. Other carnivorous plants such as pitcher plants and bladderworts capture their victims in fluid-filled traps that contain complex microbial communities (Fig. 6.19). The commensal microbes of carnivorous plants contribute to the digestive process in a manner comparable to the commensal bacteria that inhabit the digestive tracts of animals.

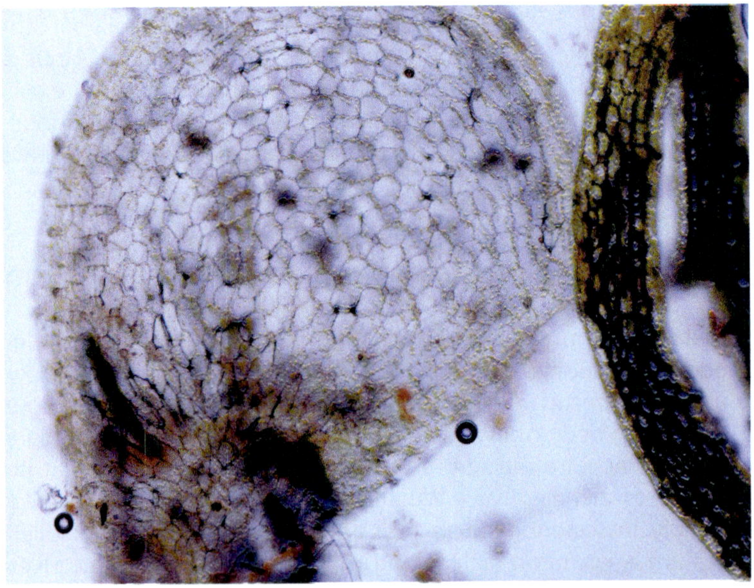

Fig. 6.19. Bladderwort bladder, microscopy. New Forest, UK.

Pitcher plants

The plants broadly known as pitcher plants belong to several phylogenetically unrelated plant families. True pitcher plants belong to the family Sarraceniaceae. They are distantly related to heathers. The tropical pitcher plants, some of which are grown as house plants, belong to the family Nepenthaceae. The Australian pitcher plant, *Cephalotus follicularis*, is a member of the family Cephalotaceae. Despite their separate evolutionary origins, true pitcher plants and tropical pitcher plants host comparably complex communities of bacteria. A molecular study of tropical pitcher plants, which involved an analysis of 16 pitcher samples, detected a total of 3260 OTUs (i.e. approximately 3260 species of bacteria) belonging to 17 different phyla, a flora comparable in the complexity to the mammalian gut flora but less complex than that of surrounding soils (Takeuchi *et al.*, 2015; Bittleston *et al.*, 2018). The microbes that colonise pitcher fluids include bacteria that degrade organic nitrogen compounds and bacteria capable of digesting the chitinous exoskeletons of arthropods. Pitcher plants host enormous numbers of bacterial cells. A sample of pitcher fluid may contain as many as 100 million bacterial cells per cubic centimetre of fluid. It has been demonstrated that the bacteria found in pitcher fluids reduce the surface tension of water. Insects that fall into pitcher fluids are more likely to sink to the bottom and drown (Armitage, 2016). Pitcher plants acquire their commensal microbes largely from the local environment. When tropical pitcher plants from Asia were planted experimentally in a bog in New England, their pitchers acquired a microbial flora similar to that of the native purple pitcher plant (Bittleston *et al.*, 2018). Pitchers remain sterile until they open. Larvae of the Pitcher Plant Mosquito, *Wyeomyia smithii*, a species that lives in the pitchers of purple pitcher plants, has been shown to acquire and retain bacteria found in pitcher fluids. Mosquitoes appear to play a role in the transmission of bacteria from one pitcher plant to another (Arellano and Coon 2022). The bacterial communities of pitchers colonised by mosquito larvae are more diverse than those found in mosquito-free plants (Peterson *et al.*, 2008).

The Purple Pitcher Plant, *Saraccenia purpurea* (Fig. 6.20) now occurs at many localities in the British Isles. Localities include Woodfield Bog in County Offaly, Peatlands Park in County Armagh and Lower Hyde Heath in Dorset. Originally from North America, pitcher plants have been grown at Kew Gardens since the seventeenth century. Pitcher plants were first found growing in the wild in Ireland in the late nineteenth century. In recent years pitcher plants have been recorded from increasing numbers of sites in Britain and Ireland. Pitcher plants thrive in acidic boggy habitats. The plants are long-lived, produce large numbers of seeds and given suitable conditions can displace the native flora. As a consequence, control measures were initiated which resulted in the eradication of colonies of naturalised pitcher plants in Somerset and the New Forest. In other places control measures have been less effective. In 2013, Wedholme Flow, part of the South Solway Mosses National Nature Reserve, was estimated to hold a population of 37,000 plants (Walker, 2014).

Fig. 6.20. Purple Pitcher Plant, *Saraccenia purpurea*. Briar Island, Nova Scotia, Canada.

Bladderworts

Bladderworts occur worldwide. Hundreds of species have been described, of which about half a dozen occur in the British Isles. British bladderworts can be readily identified as belonging to one of three species groups but identification to the actual species is more tricky. One of the more common species, Lesser Bladderwort, *Utricularia minor* is widespread in acidic areas of western Britain and is also locally common in the New Forest (Fig. 6.21). Other British species include Southern Bladderwort, *Utricularia australis* and Greater Bladderwort, *Utricularia vulgaris*. Bladderworts belong to the same family as the insectivorous butterworts, the Lentibulariceae.

Bladderworts host complex communities of microbes which are more diverse than the communities of bacteria associated with pitcher plants. Analysis of sequence data derived from specimens of *Utricularia australis* and *Utricularia vulgaris* resulted in the detection of more than 4500 OTUs (i.e. approximately 4500 species of bacteria) in the bladders and on the surfaces of bladderworts (Sirová *et al.*, 2018). Although the microbial floras of the two bladderwort species are of comparable overall composition, different bladderwort species are associated with different species of commensal bacteria. Southern and Greater Bladderwort are difficult to distinguish in the field and can only be reliably separated by examining the structure of their flowers. It may be the case that some bacteria are better than field naturalists at identifying bladderworts.

Fig. 6.21. Lesser Bladderwort, *Utricularia minor*. New Forest, UK.

A significant proportion of the bacteria found in the bladders of bladderworts are capable of digesting cellulose of plant or algal origin. In this respect, bladderwort bacteria resemble the bacteria found in the guts of ruminants. It has been noted that many of the green algae that get sucked into the bladders fail to survive and eventually decompose. Observations have also shown that bladderworts are not particularly successful at catching invertebrates. It therefore seems likely that the principal function of the bladders is to harvest and process algal and plant-derived materials and that bladderworts might better be described as omnivorous rather than as carnivorous plants.

The normal microbial flora of flora

Bacteria live in association with the roots, stems, leaves and other aerial structures of plants. Bacteria occur in large numbers inside plant tissues as well as on plant surfaces. Plant surfaces are covered in bacteria. The surface of a leaf may accommodate as many as 10 million bacterial cells per square centimetre (Lindow and Leveau, 2002). The total surface area of plant aerial surfaces worldwide has been estimated to exceed the total land surface area of the earth. Plants provide living space for many of the world's bacteria. Plant-associated bacteria are diverse. A single plant may be colonised by several thousand species of bacteria. The bacteria that live as commensals inside foliage have been shown to be more diverse than the ones that occur on the surfaces of leaves and shoots. Although diverse at a species level, most of the bacteria that live in association with plants belong to a small number of phyla, notably the *Pseudomonadota*, *Actinomycetota* and *Bacteroidetes* (Vorholt, 2012).

Plant-associated microbes include bacteria that feed on substances secreted by plants, bacteria that fix nitrogen, bacteria that act as plant growth promoters, pathogenic bacteria that cause plant diseases and also commensal bacteria that provide protection against plant pathogens.

Much of what is known about the normal microbial flora of plants has been gained as a result of detailed microbiological investigations involving a relatively small number of plant species. Most of them are species that are grown as food crops. One particular plant which has received a large amount of attention by plant microbiologists is thale cress, *Arabidopsis thaliana*, a dainty but rather weedy looking annual crucifer with small, white flowers. Thale cress readily colonises stony field margins, gravel paths and old walls. It is an unobtrusive plant and one is more likely to see through it than look at it. Nevertheless, thale cress is a much-studied plant. It is widely distributed across the northern hemisphere and has been introduced to Australia. It can be cultivated with ease, it grows rapidly and it has been adopted as a model plant for studies by geneticists and plant biologists. For these reasons more is known about the bacteria associated with thale cress than is known about the microbial flora of most other plants.

Thale cress is associated with a diverse microbial flora. A study of 196 cress plants grown outdoors resulted in the detection of more than 5000 species of bacteria from the leaves alone. An average plant was shown to accommodate more than 3000 species. The leaf flora of thale cress varies according to the genotype of the plant. This finding shows that the colonisation of thale cress plants by different kinds of bacteria is influenced by the genetic makeup of individual plants (Horton *et al.*, 2014). An advantage of studying thale cress over domesticated plants is that thale cress is a wild plant. Compared with the varieties of domesticated plants that are cultivated as food crops, wild plant species are genetically very variable. Hence, thale cress provides a good model species for the study of wild plants in general. The bacterial flora of a thale cress plant becomes more diverse over the course of time. As the plant matures, patterns of distribution become apparent in its bacterial flora. Different species of bacteria from the same genus become established in the roots as compared with the leaves of the plant (Beilsmith *et al.*, 2021). This shows that plants consist of mosaics of different microbial habitats.

Bacteria commonly found in association with thale cress include members of the genera *Sphingomonas* and *Methylobacterium* (both in the Phylum *Pseudomonadota*) and *Flavobacterium* (Phylum *Bacteroidetes*) (Bodenhausen *et al.*, 2013). Colonies of flavobacteria grown in a Petri dish are typically yellow in colour. Flavobacteria exhibit a characteristic sliding and gliding motility, which enhances their ability to colonise the roots of plants. *Flavobacterium* species are capable of degrading complex organic compounds produced by plants. Flavobacteria themselves produce substances that protect plants against fungal infection (Kraut-Cohen *et al.*, 2021).

The genus *Sphingomonas* (Phylum *Pseudomonadota*) includes bacteria that stimulate the growth of roots. Some *Sphingomonas* strains have been shown to protect plants from infection with the plant pathogen *Pseudomonas syringae* (Innerebner *et al.*, 2011). Methylobacteria are unusual in that they can subsist on a diet of single-carbon organic compounds such as methanol or formaldehyde. Methanol is released by plants during the process of cell division. It has been shown that symbiotic methylobacteria release substances that stimulate cell division, thereby increasing the supply of methanol as a food source. At the same time by stimulating cell division methylobacteria also act as plant growth promoters.

Entry of pathogens through stomata

Plant pathogens can gain entry into plant tissues through traumatic wounds, via the mouth parts of sap-feeding insects, through frost-damaged plant tissues and through stomata (Melotto *et al.*, 2006). Stomata are the microscopic pores that cover the surfaces of leaves. Stomata allow plants to transpire and to exchange gases with the atmosphere during photosynthesis. Plants respond to drought conditions by closing their stomata. The opening and closing of stomata is controlled by specialised guard cells situated near the brim of each stoma. Sensory receptors situated on guard cells allow detection of the presence of bacteria. The usual response of a plant to a bacterial pathogen is to close its stomata. Normally the stomata might not open again until several hours have lapsed. Some plant pathogens, however, have virulence factors that cause the stomata to reopen much sooner. When exposed to the plant pathogen *Pseudomonas syringae*, thale cress plants react by closing their stomata. However, *Pseudomonas syringae* releases coronatine, a virulence factor that interferes with the closure mechanism. A few of the stomata stay open and the bacteria move towards the open stomata.

Pseudomonas syringae and related species of bacteria

Pseudomonads are rather nondescript Gram-negative bacteria. They occur widely in natural environments. One species in particular, *Pseudomonas syringae*, causes a wide range of different plant diseases that affect a large number of different species of plants (Table 6.3). The species *Pseudomonas syringae* includes a number of pathogenic variants or pathovars, which broadly correlate with different plant species, or groups of plant species and with the diseases that affect them. Pathovars of *Pseudomonas syringae* are responsible for cankers of plum and cherry trees and for leaf blights of lilac (Fig. 6.22), wheat and mulberry (Hulin *et al.*, 2022). *Pseudomonas syringae* includes strains or pathovars that cause white colour disease of creeping thistle and bleeding canker of horse chestnut (see Chapter 1, Figs. 1.9–1.12). Both diseases have become a familiar sight in the British countryside. The strain associated with white colour disease of creeping thistle (strain CT99B016B) belongs to *Pseudomonas syringae* pathovar *tagetis*, a pathovar originally identified as a cause of apical blanching in marigolds. Other plants similarly affected include Jerusalem artichoke, globe artichoke and sunflowers, all of which are members of the plant family Asteraceae (Lydon *et al.*, 2011). The bacteria release a toxin which inhibits the development of chloroplasts, resulting in plant tissues which are deficient in the green pigment chlorophyll.

Table 6.3. Some plant diseases caused by *Pseudomonas syringae.*

Strain or pathovar	Plant disease	Affected plant species
aceris	Leaf spot	Sycamore
aesculi	Bleeding canker	Horse chestnut
berberidis	Leaf spot, leaf fall	*Berberis* species
tagetis	White colour disease	Creeping thistle, marigolds
delphinii	Delphinium black blotch	Delphiniums, monkshood
mori	Mulberry blight	Mulberry
morsprunorum	Canker of stone fruits	Cherry, plum
philadelphis	Leaf spot	Mock oranges
pisi	Pea bacterial blight	Peas
syringae	Bacterial shot hole	Cherry laurel
syringae	Canker	Cherry
syringae	Leaf blight	Lilac
viburni	Leaf spot	*Viburnum sargentii*

Fig. 6.22. Lilac leaves showing signs of leaf blight.

Not to be confused with ash dieback disease, which is caused by a fungus, ash canker is caused by a bacterium, *Pseudomonas savastanoi* pathovar *fraxini*. Bacteria gain entry through wounds in the bark and cause the formation of blister-like lesions. Over the course of time the blisters crack. Cracks may extend vertically down trunks for some distance. Unlike ash dieback disease, ash canker is not considered to be a particular threat to ash trees. Trees may survive for a considerable length of time following the onset of infection.

Ice-nucleating bacteria

At temperatures between about 0 and −5°C, liquid water on leaf surfaces may remain in the form of supercooled liquid water, or it may freeze to form ice crystals. The outcome depends to some extent on the number of ice-nucleating bacteria present on the leaf surface (Hirano and Upper, 2000). Frost-damaged plants may suffer from blossom blight and dieback of twigs and shoots. Wounded tissues serve as entry sites for pathogenic bacteria and fungi. Frost-damaged leaves of winter wheat, for example, are susceptible to septoria tritici blotch, a fungal disease caused by the fungus *Zymoseptoria tritici*. It has been suggested that a mutualistic relationship may exist between the fungus and ice-nucleating bacteria. By promoting frost damage, ice-nucleating strains of *Pseudomonas syringae* have been shown to facilitate the ingress of fungi into wounded leaf tissues (Fones, 2020). Fungal infection of the leaves increases the available supply of nutrients to the bacteria. Ice-nucleating proteins present on the surfaces of ice-nucleating bacteria trigger ice formation by causing supercooled liquid water molecules to assume a more ordered structure (Roeters *et al*., 2021). Ice-minus bacteria are strains of bacteria that lack ice-nucleating proteins. There is commercial interest in the possibility that ice-minus bacteria could be applied to plants of agricultural value in order to prevent frost damage.

Diseases of food crops

The countryside is full of unhealthy plants. Some plant diseases, including a few that are caused by bacteria can be recognised in the field. Often, however, it is impossible in the field to determine why a wild plant might not be looking at its best. Much of what is known about plant disease relates to commercially valuable plants (Table 6.4).

Crown gall disease affects garden and orchard plants, such as apples, pears, plums, blackberries and raspberry canes and also wild plants, such as hawthorn and willows. The causative agent of crown gall disease is the bacterium, *Agrobacterium radiobacter*. Confusingly, the synonyms *Agrobacterium tumefasciens* and *Rhizobium radiobacter* are also often applied to this bacterium. The crown gall bacterium is a member of the Class *Alphaproteobacteria* and is a close relative of the nitrogen-fixing *Rhizobium* species that live as symbionts in the root nodules of legumes. The crown gall bacterium causes knobbly or warty growths on the stems, trunks or at the root crown of infected plants. The bacteria inhabit the soil. After infecting a susceptible plant, the bacterium transfers a small portion of its DNA to the host cell, where it integrates with the host genome. The transferred DNA includes genes which encode plant hormones that stimulate the growth of nodular, warty plant tissues. Galls formed by herbaceous plants, such as beetroot, decompose and liberate bacteria back into the soil. Woody plants are much better at tolerating infection.

Table 6.4. Some pathogenic bacteria and associated plant diseases.

Pathogenic bacterium	Plant disease	Affected plant species
Agrobacterium tumefaciens	Crown gall disease	Apple, hawthorn, rose, etc.
Brenneria goodwinii	Acute oak decline	Pedunculate oak, sessile oak
Brenneria salicis	Watermark disease	White willow
Burkholderia cepacia	Slippery skin	Onion
Dickeya dadantii	Soft rot	Potato, tomato, maize etc.
Erwinia amylovora	Fireblight	Apples, pears, hawthorn
"*Candidatus* Liberibacter solanacearum"	Zebra Chip	Potato, tomato, aubergine
Phytoplasma species	Little leaf, stunting etc.	Many plant species
Pectobacterium atrosepticum	Potato blackleg	Potato
Pseudomonas savastanoi pathovar *fraxini*	Ash canker	Ash
Ralstonia solanacearum	Bacterial wilt	Potato, tomato
Rhodococcus fascians	Leafy gall	Many plant species
Streptomyces scabies	Potato scab	Potato
Xanthomonas campetris	Black rot	Crucifers
Xylella fastidiosa	Dieback	Olive trees (in Europe)

New and emerging plant diseases

An international trade in plants and plant products brings with it novel plant diseases. *Xylella fastidiosa*, a bacterial pathogen of coffee and citrus plants in the Americas was first recorded in Europe in 2013 as a disease of olive trees in Italy. Outbreaks of disease have since been reported in France, Spain and Portugal, affecting lavender plants and almond trees. *Xylella fastidiosa* is transmitted by sap-feeding insects. Although cases of the disease have not yet been reported in Britain, it is considered to pose a serious threat. Many plant species are considered to be susceptible to infection and the native meadow spittlebug, *Philaenus spumarius*, is known to be a competent vector.

Zebra chip is a bacterial disease of potatoes in New Zealand and the Americas. Features of the disease include stunted growth and blanched and purple foliage. When fried, affected potato tubers become discoloured and assume a stripy appearance. Following an analysis of 16s RNA gene sequences, a new species of bacterium, "*Candidatus* Liberibacter solanacearum" was described as the agent of zebra chip by a team of plant scientists based in New Zealand in 2009 (Liefting *et al.*, 2009). Because the zebra chip bacterium could not be isolated in culture it was named using the *Candidatus* designation, in order to conform with the conventions of bacterial nomenclature.

The zebra chip bacterium causes disease in other members of the potato family, including tomatoes, peppers and aubergines. Related strains of the bacterium cause disease in members of the carrot family, including carrots

Fig. 6.23. Carrot plants showing signs of infection with "*Candidatus* Liberibacter solanacearum". (Image cropped from figure 2 in Munyaneza *et al.*, 2016; *PLoS ONE*. 11 (8), e0161016. CCO.)

(Fig. 6.23), parsnips, celery and fennel (Munyaneza *et al.*, 2016). Specific strains also infect stinging nettles and members of the Convolvulaceae. Together these novel plant diseases are known as the zebra chip complex of plant diseases. Zebra chip bacteria live in the phloem of infected plants and are transmitted by sap-feeding psyllids (jumping plant-lice). In the UK, zebra chip has not yet been detected as a disease of potatoes. However, the strains of zebra chip bacteria associated with carrots and stinging nettles have been detected in plants and in jumping plant-lice in Scotland. A survey of carrot crops and field margins resulted in the detection of high levels of zebra chip bacteria infection in *Trioza urticae*, a psyllid that feeds on nettles and in *Trioza anthrisci*, a psyllid that feeds on members of the carrot family. Zebra chip bacteria were also detected in stinging nettles and in cow parsley, a member of the carrot family (Sumner-Kalkun *et al.*, 2020).

Acute oak decline

Dieback diseases of native oak trees have been recorded in Europe for more than 100 years. Since the beginning of the current millennium, however, a new wave of disease has been affecting British oak trees (Figs. 6.24–6.26). It is a rapidly progressing condition, which can result in the death of a tree within

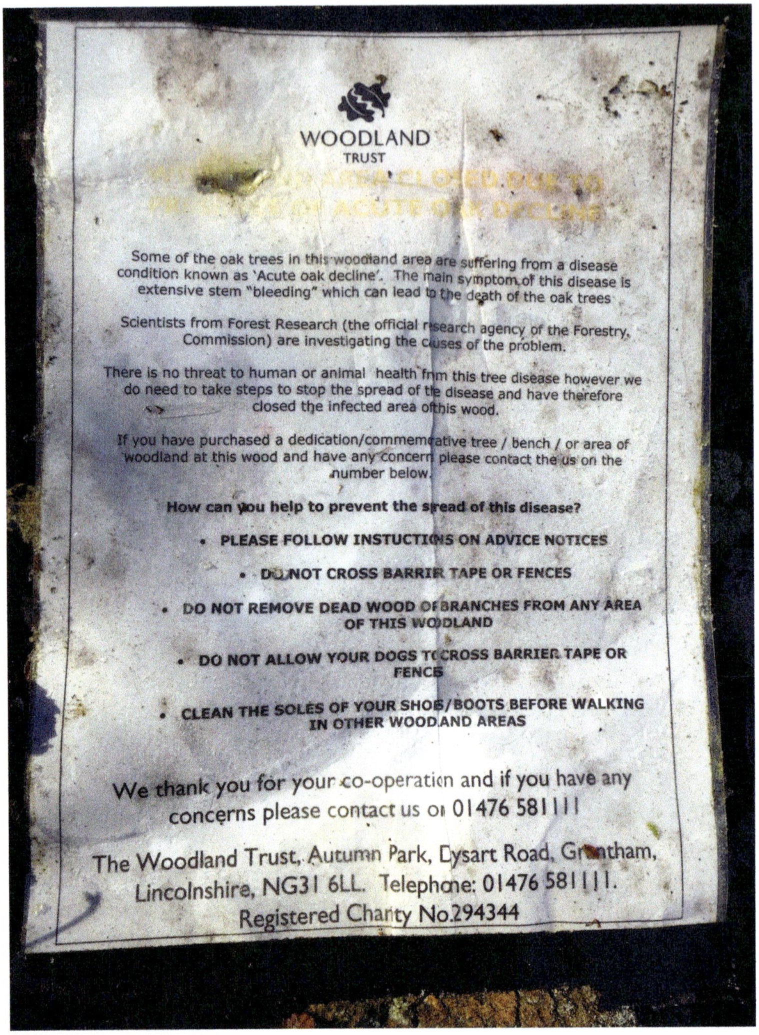

Fig. 6.24. Acute oak decline notice, Stratfield Brake, Oxfordshire, UK.

5 years of disease onset. Affected trees look unhealthy. Vertical streaks of fluid known as stem bleeds ooze down their trunks. Fluid can also be seen seeping from cracks in the bark. The canopies of affected trees become thinner. The disease was given the name acute oak decline (Denman *et al.*, 2014). In 2012, a novel species of bacterium cultured from affected trees was described as *Brenneria goodwinii*. Two other species of bacteria, *Gibbsiella quercinecans* and *Rahnella victoriana* have been found regularly in association with diseased trees. In addition it has been noted that affected trees often become infected with a bark boring beetle, *Agrilus biguttatus* (the Oak Jewel Beetle).

Fig. 6.25. Restricted access to woodland affected by Acute oak decline, Stratfield Brake, Oxfordshire, UK.

Agrilus biguttatus is a rather attractive metallic green jewel beetle (Family Buprestidae). The beetles are attracted to stressed trees. Female beetles lay their eggs on the bark of tree trunks. The beetle larvae live under the bark. The adult beetles bore their way out through the bark, leaving D-shaped holes that are characteristic sign of infestation with jewel beetles. It is currently unclear exactly what roles the beetles might have, if any, in exacerbating the disease or in transmitting potential causative agents. *Brenneria goodwinii*, the bacterium most strongly associated with acute oak decline, has been cultured from healthy as well as from diseased trees. Hence, it seems likely that a combination of factors must be necessary in order to cause trees to become unhealthy and to exhibit signs of acute oak decline. The geographical distribution of acute oak decline is currently centred on drought-prone areas of central southern England.

Fig. 6.26. Diseased and fallen trees, Stratfield Brake, Oxfordshire, UK.

References

Andrews, M. and Andrews, M.E. (2017) Specificity in legume-rhizobia symbioses. *International Journal of Molecular Science*, 18, 705.

Arellano, A.A. and Coon, K.L. (2022) Bacterial communities in carnivorous pitcher plants colonize and persist in inquiline mosquitoes. *Animal Microbiome*, 4, 13.

Armitage, D.W. (2016) Bacteria facilitate prey retention by the pitcher plant *Darlingtonia californica*. *Biology Letters*, 12, 20160577.

Beilsmith, K., Perisin, M. and Bergelson, J. (2021) Natural bacterial assemblages in *Arabidopsis thaliana* tissues become more distinguishable and diverse during host development. *mBio*, 12, e02723-20.

Benson, D.R. (1982) Isolation of *Frankia* strains from alder actinorhizal root nodules. *Applied and Environmental Microbiology*, 44, 461–465.

Bodenhausen, N., Horton, M.W. and Bergelson, J. (2013) Bacterial communities associated with the leaves and the roots of *Arabidopsis thaliana*. *PLoS ONE*, 8, e56329.

Brinkhuis, H., Schouten, S., Collinson, M.E., Sluijs, A., Sinninghe Damsté, J.S. *et al.* (2006). Episodic fresh surface waters in the Eocene Arctic Ocean. *Nature*, 441, 606–609.

Chang, A.C.G., Chen, T., Li, N. and Duan, J. (2019) Perspectives on endosymbiosis in coralloid roots: association of cycads and cyanobacteria. *Frontiers in Microbiology*, 10, 1888.

Chiu, W.L., Peters, G.A., Levieille, G., Still, P.C., Cousins, S. *et al.* (2005) Nitrogen deprivation stimulates symbiotic gland development in *Gunnera manicata*. *Plant Physiology*, 139, 224–230.

Denman, S., Brown, N., Kirk, S., Jeger, M. and Webber, J. (2014) A description of the symptoms of Acute Oak Decline in Britain and a comparative review on causes of similar disorders on oak in Europe, *Forestry: An International Journal of Forest Research*, 87, 535–551.

Doty, S.L., Sher, A.W., Fleck, N.D., Khorasani, M., Bumgarner, R.E. *et al.* (2016). Variable nitrogen fixation in wild *Populus*. *PLoS ONE*, 11, e0155979.

Fones, H.N. (2020) Presence of ice-nucleating *Pseudomonas* on wheat leaves promotes septoria tritici blotch disease (*Zymoseptoria tritici*) via a mutually beneficial interaction. *Scientific Reports*, 10, 17738.

Fuller, D.Q. and Hickey, L.J. (2005) Systematics and leaf architecture of the Gunneraceae. *The Botanical Review*, 71, 295–353.

Hirano, S.S. and Upper, C.D. (2000) Bacteria in the leaf ecosystem with emphasis on *Pseudomonas syringae* – a pathogen, ice nucleus, and epiphyte. *Microbiology and Molecular Biology Reviews*, 64, 624–653.

Horton, M.W., Bodenhausen, N., Beilsmith, K., Meng, D., Muegge, B.D. *et al.* (2014) Genome-wide association study of *Arabidopsis thaliana* leaf microbial community. *Nature Communications*, 5, 5320.

Hulin, M.T., Vadillo Dieguez, A., Cossu, F., Lynn, S., Russell, K. *et al.* (2022) Identifying resistance in wild and ornamental cherry towards bacterial canker caused by *Pseudomonas syringae*. *Plant Pathology*, 71, 949–965.

Innerebner, G., Knief, C. and Vorholt, J.A. (2011) Protection of *Arabidopsis thaliana* against leaf-pathogenic *Pseudomonas syringae* by *Sphingomonas* strains in a controlled model system. *Applied and Environmental Microbiology*, 77, 3202–3210.

Khan, Z., Kandel, S.L., Ramos, D.N., Ettl, G.J., Kim, S.-H. *et al.* (2015) Increased biomass of nursery-grown Douglas-fir seedlings upon inoculation with diazotrophic endophytic consortia. *Forests*, 6, 3582–3593.

Kraut-Cohen, J., Shapiro, O.H., Dror, B. and Cytryn, E. (2021) Pectin induced colony expansion of soil-derived *Flavobacterium* strains. *Frontiers in Microbiology*, 12, 651891.

Liefting, L.W., Sutherland, P.W., Ward, L.I., Paice, K.L., Weir, B.S. *et al.* (2009) A new '*Candidatus* Liberibacter' species associated with diseases of solanaceous crops. *Plant Diseases*, 93, 208–214.

Lindow, S.E. and Leveau, J.H. (2002) Phyllosphere microbiology. *Current Opinion in Biotechnology*, 13, 238–243.

Lydon, J., Kong, H., Murphy, C.A., Wenming, Z., Peng, G. *et al.* (2011) The biology and biological activity of *Pseudomonas syringae* pv. *tagetis*. *Pest Technology*, 5, 48–55.

Melotto, M., Underwood, W., Koczan, J., Nomura, K. and He, S.Y. (2006) Plant stomata function in innate immunity against bacterial invasion. *Cell*, 126, 969–980.

Munyaneza, J.E., Mustafa, T., Fisher, T.W., Sengoda, V.G. and Horton, D.R. (2016) Assessing the likelihood of transmission of *Candidatus* Liberibacter solanacearum to carrot by potato psyllid, *Bactericera cockerelli* (Hemiptera: Triozidae). *PLoS ONE*, 11 (8), e0161016.

Nouioui, I., Ghodhbane-Gtari, F., Montero-Calasanz, M.D.C., Göker, M., Meier-Kolthoff, J.P. *et al.* (2016) Proposal of a type strain for *Frankia alni* (Woronin 1866) Von Tubeuf 1895, emended description of *Frankia alni*, and recognition of *Frankia casuarinae* sp. nov. and *Frankia elaeagni* sp. nov. *International Journal of Systematic and Evolutionary Microbiology*, 66 (12), 5201–5210. https://doi.org/10.1099/ijsem.0.001496.

Normand, P., Queiroux, C., Tisa, L.S., Benson, D.R., Rouy, Z. *et al.* (2007) Exploring the genomes of *Frankia*. *Physiologia Plantarum*, 130, 331–343.

Österman, J., Mousavi, S.A., Koskinen, P., Paulin, L. and Lindström, K. (2015) Genomic features separating ten strains of *Neorhizobium galegae* with different symbiotic phenotypes. *BMC Genomics*, 16, 348.

Peterson, C.N., Day, S., Wolfe, B.E., Ellison, A.M., Kolter, R. *et al.* (2008) A keystone predator controls bacterial diversity in the pitcher-plant (*Sarracenia purpurea*) microecosystem. *Environmental Microbiology*, 10, 2257–2266.

Ranwell, D.S. (ed.) (1972) The management of sea buckthorn (*Hippophae rhamnoides* L.) on selected sites in Great Britain. Nature Conservancy, Norwich, UK.

Rodriguez-Barrueco, C. (1968) The occurrence of the root nodule endophyte of *Alnus glutinosa* and *Myrica gale* in soils. *Journal of General Microbiology*, 52, 189–194.

Roeters, S.J., Golbek, T.W., Bregnhøj, M., Drace, T., Alamdari, S. *et al.* (2021) Ice-nucleating proteins are activated by low temperatures to control the structure of interfacial water. *Nature Communications*, 12, 1183.

Sánchez-Baracaldo, P., Raven, J.A., Pisani, D. and Knoll, A.H. (2017) Early photosynthetic eukaryotes inhabited low-salinity habitats. *Proceedings of the National Academy of Sciences of the United States of America*, 114, E7737–E7745.

Scherling, C., Ulrich, K., Ewald, D. and Weckwerth, W. (2009) A metabolic signature of the beneficial interaction of the endophyte *Paenibacillus* sp. isolate and in vitro-grown poplar plants revealed by metabolomics. *Molecular Plant-Microbe Interactions*, 22, 1032–1037.

Sirová, D., Bárta, J., Šimek, K., Posch, T., Pech, J. *et al.* (2018) Hunters or farmers? Microbiome characteristics help elucidate the diet composition in an aquatic carnivorous plant. *Microbiome*, 6, 225.

Sumner-Kalkun, J.C., Highet, F., Arnsdorf, Y.M., Back, E., Carnegie, M. *et al.* (2020) '*Candidatus* Liberibacter solanacearum' distribution and diversity in Scotland and the characterisation of novel haplotypes from *Craspedolepta* spp. (Psyllidae: Aphalaridae). *Scientific Reports*, 10, 16567.

Takeuchi, Y., Chaffron, S., Salcher, M.M., Shimizu-Inatsugi, R., Kobayashi, M.J. *et al.* (2015) Bacterial diversity and composition in the fluid of pitcher plants of the genus Nepenthes. *Systematic and Applied Microbiology*, 38, 330–339.

von Wuehlisch, G. (2011) Evidence for nitrogen-fixation in the Salicaceae family. *Tree Planters' Notes*, 54, 38–41.

Vorholt, J.A. (2012) Microbial life in the phyllosphere. *Nature Reviews Microbiology*, 10, 828–840.

Walker, K.J. (2014) *Sarracenia purpurea* subsp. *purpurea* (Sarraceniaceae) naturalised in Britain and Ireland: distribution, ecology, impacts and control. *New Journal of Botany*, 4, 33–41.

7 Interactions with other microbes and fungi

Vivre? les serviteurs feront cela pour nous.
[Living? The servants can do that for us.]
Auguste Villiers de L'Isle Adam (1890). *Axël.*

Viruses that infect bacterial cells are known as bacteriophages (bacteria eaters), or simply as phages. Viruses that infect archaea are referred to as archaeal viruses. There is not much to a virus, apart from the virus's repository of genetic material (either DNA or RNA), which is surrounded either by a membranous envelope or by a protective proteinaceous coat, called a capsid. When they are not infecting the cells of other organisms, viruses are so inert that some scientists contend that they are not really alive. One way to look at viruses is as entities that let cellular organisms do the living for them.

After entering a bacterial cell, the bacterium's replicative machinery reads the virus's genetic code as an instruction to assemble new copies of the virus. This process usually results in the lysis and death of the bacterium and the liberation of multiple copies of the virus into the environment. Lytic phages can be regarded as predators or parasites of bacteria. Alternatively, after entering the bacterial cell the viral genome may become integrated with the chromosomal DNA of the host bacterium or it may persist inside the bacterial cell as a separate molecule of extrachromosomal DNA. In such cases the viral genome will continue to be replicated passively in step with the replication cycle of the bacterium. At some point, however, an external stimulus, such as exposure of the host bacterium to hostile environmental conditions, may trigger the lytic phase of the phage life cycle, resulting in the production of new copies of the phage and their release into the environment from the lysed host cell. Phages may provide their host bacteria with a selective advantage by encoding

DOI: 10.1079/9781836993933.0007

useful virulence factors such as toxins. Phage derived genes may even become permanently integrated into the bacterial genome.

Viruses do not fit into the three-domain classification scheme used to classify cellular organisms. Whereas, the genomes of cellular organisms, including bacteria, archaea and eukaryotes are encoded in the form of double strands of DNA, viral genomes may consist either of DNA, or RNA, which may be double- or single-stranded, depending on the virus. Different groups of viruses appear to have evolved independently on at least 20 separate occasions (Krupovic and Koonin, 2017). The discovery of viruses came about through the use of porcelain filters, which had been developed in order to filter bacteria from liquids. In 1892, a Russian botanist, Dimitri Ivanowski (1864–1920) used filters to show that extracts from plants with tobacco mosaic disease were infectious to other plants even after removal of any bacteria.

Where there are bacteria, there are bacteriophages (Duckworth, 1976). The total global population of bacteriophages has been estimated to be somewhere in the region of 10 nonillion (1 followed by 31 zeros) virus particles, which translates into 'a trillion phages for every grain of sand in the world' (Keen, 2015). Despite their abundance, the minuteness of viruses means that viruses are estimated to account for only 200 million metric tonnes of the world's biomass, measured as carbon, compared with 70 billion metric tonnes for bacteria and 7 billion metric tonnes for archaea (Bar-On *et al.*, 2018).

Bacterial toxins encoded by bacteriophages

Cholera is a potentially severe diarrhoeal illness, caused by toxigenic strains of an aquatic bacterium, *Vibrio cholerae* (Fig. 7.1). Known since antiquity, outbreaks of disease continue to occur on a regular basis in some areas of Asia and Africa. In 1973 an apparently indigenous case of infection with *Vibrio cholerae*

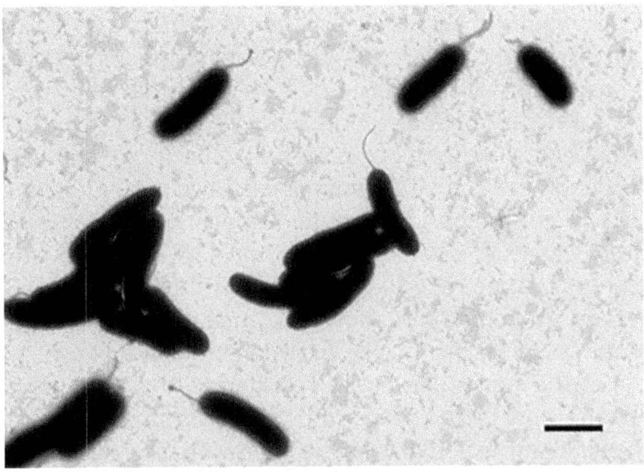

Fig. 7.1. *Vibrio cholerae*. (Image cropped from figure 6 in Giacomucci *et al.*, 2019; *PLoS ONE.* 14 (8), e0221431). Creative Commons.

was reported in Germany and the bacterium was detected in samples of river water. In 1974 four cases of infection with *Vibrio cholerae* were reported in Sweden. All four of the patients had previously been in contact with brackish water. In 1975 *Vibrio cholerae* was isolated from water samples and from ducklings at a duck farm in Denmark (Bisgaard and Kristensen, 1975). Reports of *Vibrio cholerae* in mainland Europe prompted microbiologists based at the Public Health Laboratory in Maidstone to conduct an environmental survey of water bodies in Kent. *Vibrio cholerae* was cultured from water samples collected during the summer months at multiple locations in the county, including Chetney Marshes, Minster Marshes and the Royal Military Canal (Lee *et al*., 1982). The highest colony counts were associated with bodies of stagnant brackish water. *Vibrio cholerae* was cultured also from Black-headed gulls, Greater black-backed gulls and Herring gulls. Some of the birds tested had previously been netted at various sites in mainland Europe. It became apparent that *Vibrio cholerae* was part of the natural microflora of southern England. None of the isolates investigated produced cholera toxin and it was concluded that the bacterium did not present a threat to public health. Non-toxigenic strains of *Vibrio cholerae* are known to occur in estuarine habitats worldwide. In mainland Europe, for example, the Neusiedler See, a brackish lake in Austria, is a hotspot for *Vibrio cholerae* strain diversity as well as for migrating birds (Pretzer *et al*., 2017).

The clinical manifestations of cholera are mediated by cholera toxin. Cholera toxin stimulates cells of the intestinal lining to secrete chloride ions. The build-up of chloride ions in the bowel lumen creates an osmotic gradient which draws water into the bowel, leading to watery diarrhoea and dehydration. Cholera toxin is encoded by cholera toxin phage (CTXφ), a virus which can be transmitted horizontally between different strains of *Vibrio cholerae* (Waldor and Mekalanos, 1996). After binding with the bacterium, the viral nucleic acid enters the bacterial cell and the viral genome becomes integrated with that of the bacterium. Daughter cells of the bacterium continue to incorporate copies of the viral genetic code and produce virally encoded cholera toxin. Bacterial cells can be induced to enter the lytic phase of the phage cycle by environmental triggers such as exposure to sunlight, resulting in the liberation of new infectious copies of the virus. Strains of *Vibrio cholerae* that do not produce cholera toxin, such as those found in environmental samples in Britain and mainland Europe, are known as non-cholera vibrios. They may, however, harbour other phage-encoded toxins and occasionally cause sporadic cases of diarrhoeal disease.

In the Ganges Delta of Bangladesh and India, where cholera occurs as an endemic disease, a dynamic situation exists, which involves different strains of *Vibrio cholerae*, the bacteriophages that infect them, other microorganisms and the environment. Toxigenic strains of *Vibrio cholerae*, which incorporate copies of the cholera toxin phage genome, are in turn susceptible to infection with lytic phages released by non-toxigenic strains of *Vibrio cholerae*. Blooms of toxigenic *Vibrio cholerae* and cholera epidemics correlate with episodes of high rainfall and flooding; conditions which dilute the concentration of lytic phages present in the aquatic environment (Faruque *et al*., 2005). In common with other bacteria, *Vibrio cholerae* resists hostile environmental conditions by forming biofilms (Giacomucci *et al.*, 2019). It has been shown that *Vibrio cholerae* can colonise the mucilaginous sheaths of aquatic cyanobacteria, which may serve as a reservoir for the bacterium during the intervals between epidemics (Islam, 2020).

Since the beginning of the nineteenth century, toxigenic *Vibrio cholerae* has emerged periodically from its home in the Ganges Delta and spread around the world to cause seven pandemics, four of which reached Britain during the early to mid-1800s. Cholera hospitals were established to care for victims of the disease (Fig. 7.2). During the course of the 1853–1854 epidemic, the physician John Snow (1813–1858) conducted an epidemiological study of the disease in London (Fig. 7.3). The results of his investigation caused him to reason that cholera was an infectious disease which was contracted following exposure to contaminated water supplies. The discovery of the causative bacterium *Vibrio cholerae* is credited to the Italian pathologist Filippo Pacini (1812–1883). Based on his experience of the 1854 cholera outbreak in Florence, Pacini published an account of microscopical observations of cholera and its pathology, which included observations of vibrios seen in post-mortem samples from cholera victims.

Many *Vibrio* species are susceptible to infection with phages (Fig. 7.4) and have been shown to harbour phage-encoded genetic elements (Pryshliak *et al.*, 2014; Castillo *et al.*, 2018). Whilst *Vibrio cholerae* is the best known member of its genus, other *Vibrio* species occur widely in aquatic environments. Vibrios are chitin degrading organisms which play an important role in recycling the

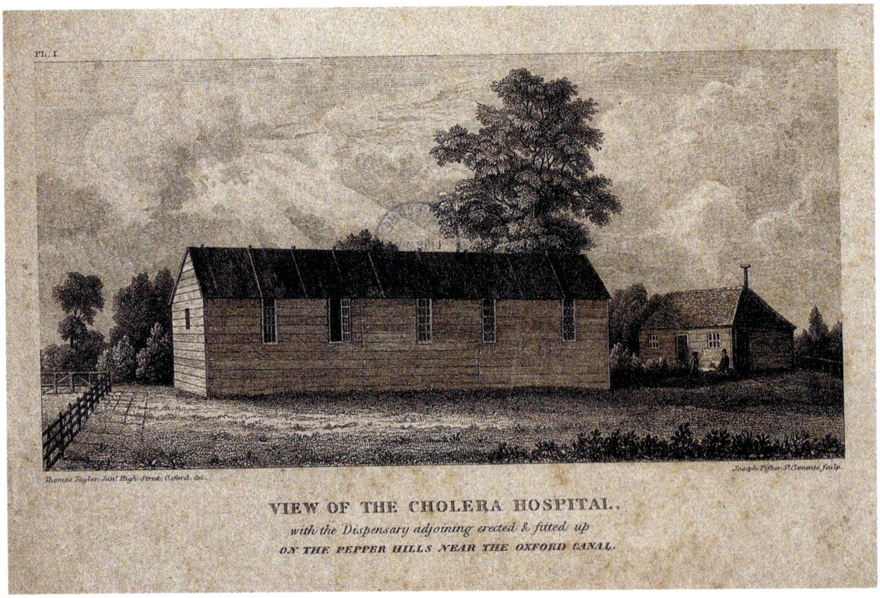

Fig. 7.2. Cholera Hospital, Oxford, near the Oxford Canal. Line engraving by J. Fisher after T. Taylor junior (Wellcome Collection). Public Domain Mark.

Fig. 7.3. The John Snow public house commemorates Snow's investigation of the 1854 Broad Street cholera outbreak in Soho, London.

chitinous exoskeletons of marine arthropods. Vibrios are particularly associated with brackish waters where they may account for a significant proportion of the microflora. *Vibrio vulnificus* and *Vibrio parahaemolyticus* are potential human pathogens. *Vibrio vulnificus* is a cause of necrotising wound infection and *Vibrio parahaemolyticus* is an agent of diarrhoeal illness. Both species thrive best in warm climates. Unusually high colony counts of *Vibrio parahaemolyticus* and *Vibrio vulnificus*, as well as *Vibrio cholerae*, were recorded at an unspecified estuarine site in south-western England during a summer heatwave in 2018 when water temperatures reached 30°C (Ford *et al.*, 2020).

Vibrio anguillarum is a fish pathogen, which causes vibriosis, a septicaemic disease of fish, including commercially important species such as farmed salmon. *Vibrio harveyi* is a bioluminescent pathogen of marine fish and crustaceans, including farmed prawns. *Vibrio anguillarum* and *Vibrio harveyi* are susceptible to lytic phages and there is interest in the possible application of lytic phages as biological control agents of pathogenic vibrios in aquaculture.

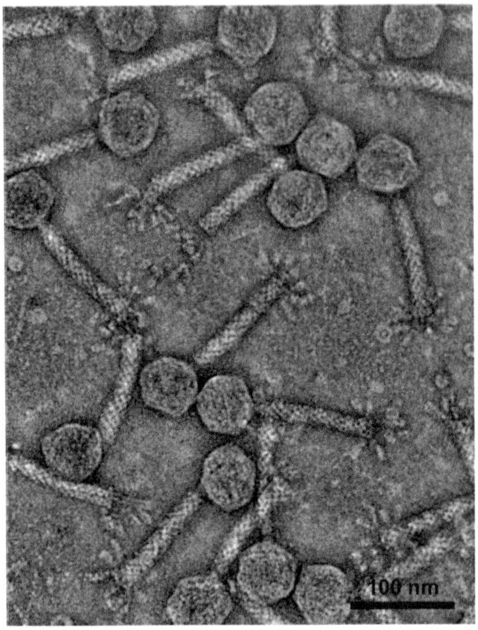

Fig. 7.4. Transmission electron micrographs of *Vibrio vulnificus* Phage PV94 particles (Image cropped from figure 1 in Pryshliak *et al*., 2014; *PLoS ONE*. 9 (4), e94707). Creative Commons.

Botulism

Botulinum toxins are neurotoxins which cause paralysis by blocking the release of neurotransmitters at neuromuscular junctions. The toxins are produced by *Clostridium botulinum* (Fig. 7.5) and six other species of *Clostridium*. Numerous subtypes of botulinum toxin have been recognised. A complex web of relationships exists between the different toxins and the bacteria that produce them. Some species of *Clostridium* have been shown to produce more than one kind of toxin and some toxins are associated with multiple species of bacteria, a pattern which indicates that toxin production is linked to the horizontal transmission of mobile genetic elements between different strains of bacteria. Genes for several subtypes of botulinum toxin are encoded by bacteriophages (Eklund *et al*., 1971; Zhang *et al*., 2022). In some strains of bacteria, the toxin genes are borne on plasmids, circular DNA molecules which exist separately from the bacterial chromosomal DNA. In others the toxin genes have become integrated within the bacterial chromosome.

Clostridia are anaerobic Gram-positive rod-shaped bacteria (Phylum *Bacillota*). Clostridia spread through the environment as spores, although their intolerance of oxygen limits vegetative forms of the bacteria to anoxic habitats, including waterlogged sediments and deposits of rotting organic matter such as the carcasses of dead animals. Toxin-producing clostridia can thrive in insufficiently treated preserved foods, such as tinned and potted meat products. Surveys of natural environments in the British Isles have demonstrated

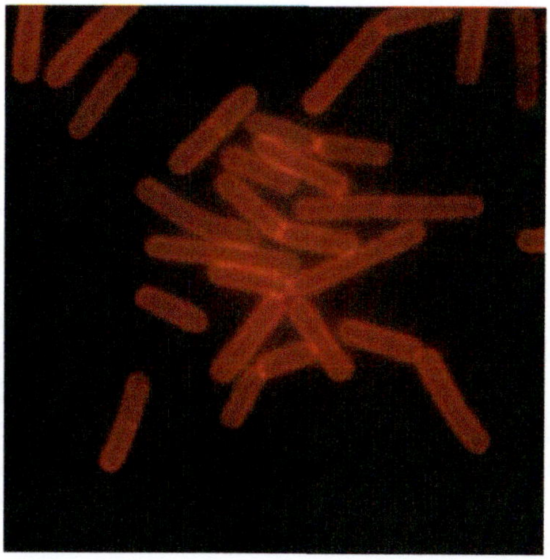

Fig. 7.5. *Clostridium botulinum* bacteria (Image cropped from figure 1 in Zhang *et al.*, 2022; *Int. J. Mol. Sci.* 23, 8391). Creative Commons.

the widespread presence of toxin-producing clostridia in sediment samples. Of 45 samples collected from the Norfolk Broads, all but one of them were found to be positive for botulinum toxin-producing clostridia (Borland *et al.*, 1977).

Cases of botulism are regularly detected in wild birds, livestock and occasionally in fish and in humans. An outbreak of botulism among wild birds at a nature reserve in Italy occurred after numbers of grey mullet, *Mugil cephalus*, died in a tidal pool frequented by water fowl (Circella *et al.*, 2019). The dead fish created an ideal microhabitat for botulinum-producing clostridia. Maggots that feed on the carcasses of dead animals accumulate botulinum toxin and serve as vectors of botulism to insectivorous birds and mammals. Landfill sites, rich in organic waste, have been recognised as a breeding ground for anaerobic bacteria and a source of botulism toxin. A survey of landfill sites in England, Scotland and Wales detected toxin-producing clostridia at 18 of the 19 sites tested (Ortiz and Smith, 1994). Botulism is a significant cause of mortality among gulls that scavenge at landfill sites. Cases of avian botulism are usually caused by type C and type D toxins, which are encoded by bacteriophages.

The earliest report of botulism in the UK concerned the 1922 Loch Maree Tragedy when eight people died of botulism after consuming toxic duck paste sandwiches (Leighton, 1923). Toxin production by the Loch Maree strain is encoded by a large circular plasmid. The plasmid contains residual phage-like DNA sequences, as do plasmids found in other strains of toxin-producing clostridia (Smith *et al.*, 2021b).

Other phage-encoded bacterial toxins

Other bacterial toxins encoded by viruses include Shiga toxin, which is produced by certain strains of *Escherichia coli*, and Panton–Valentine leucocidin, a toxin produced by some strains of *Staphylococcus aureus* (Panton and Valentine,

1932). Panton–Valentine leucocidin production is encoded by genes borne on a bacteriophage (φSa2), which integrates with the bacterial genome. The toxin is secreted as two components, one of which binds to a receptor on the surface of white blood cells. The two toxin components unite to form a pore-like structure in the cell membrane, which results in the lysis of the white cell. Toxigenic strains of *Staphylococcus aureus* cause skin and soft tissue infections such as boils and abscesses. Toxigenic strains are easily transmitted through close contact, giving toxin-positive strains of *Staphylococcus aureus* a selective advantage in close contact settings, such as occur in health care, competitive sports and military camps. Panton–Valentine leucocidin is toxic to human white cells but less so to the white cells of other mammals. Horses are susceptible to horse-adapted strains of toxigenic *Staphylococcus aureus* that produce a leucocidin (LukPQ), which is specifically toxic to the white cells of horses (Koop *et al.*, 2017). LukPQ leucocidin is encoded by a bacteriophage (φSaeq1). Cattle are susceptible to infection with cattle-adapted strains of *Staphylococcus aureus* that produce phage-encoded toxins.

Infection with Shiga toxin-producing *Escherichia coli* typically presents with severe, often bloody diarrhoea. A rare complication of exposure to Shiga toxin, haemolytic uraemic syndrome, is characterised by haemolytic anaemia and kidney damage. Shiga toxin is encoded by a phage which integrates with the DNA of the host bacterium. Toxin is produced during the lytic stage of the phage life cycle. When the host cell lyses, large numbers of new phage particles and large amounts of toxin are released. Infections are usually acquired as a result of exposure to animals or contaminated animal products. The principal hosts of Shiga toxin-producing *Escherichia coli* are cattle and other ruminants, which remain free of disease, because unlike human cells the cells of ruminants lack suitable binding sites for the toxin.

Shiga toxin is named after the Japanese microbiologist Kiyoshi Shiga (1871–1957). In 1897 Shiga identified the bacterium "*Bacillus dysentericus*" as an infectious cause of bloody diarrhoea (dysentery). The bacterium was later renamed *Shigella dysenteriae*. Shiga toxin production by *Shigella dysenteriae* is encoded by bacterial chromosomal genes derived from a phage, although a subsequent rearrangement and loss of genes means that the phage (or what might be regarded as its residual elements) is no longer capable of entering the lytic phase of the phage life cycle. Phage-encoded toxin production has been reported to occur in a related species *Shigella flexneri* (Gray *et al.*, 2014). *Shigella* species and Shiga toxin-producing *Escherichia coli* are very closely related and there is a school of thought which favours the reclassification of *Shigella* species as strains of *Escherichia coli*.

Staphylococcus aureus and *Escherichia coli* are both susceptible to infection with lytic phages. Because different strains of a single species of bacterium vary in their susceptibility to different kinds of phage, sets of phages can be used as a typing tool to distinguish between strains of bacteria. Phages manifest themselves on a Petri dish seeded with bacteria as circular plaques that mark spots where the bacteria have been lysed by the bacteriophage. Phage typing can be used to support epidemiological investigations of disease outbreaks. Phage typing systems were developed for important pathogenic bacteria, including *Staphylococcus aureus*, *Escherichia coli* and *Salmonella*

Typhi, the agent of typhoid fever. Phage typing was eventually phased out and replaced by molecular typing methods based on the comparison of bacterial DNA sequences.

Amoeba-resisting bacteria as intracellular pathogens

Amoebae are eukaryotic single-celled organisms that move about by extending their irregularly shaped cells into protrusions (known as pseudopodia) in the direction of travel. Some species, known as testate amoebae, live inside a tiny shell. Some amoebae survive adverse conditions by forming resistant cysts and spores. Amoebae are motile predators that engulf and digest bacteria. In the laboratory it is possible to observe the steady progress of a predatory amoeba as it cuts a trail across a Petri dish while grazing on a lawn of bacteria. Some bacteria, known as amoeba-resisting bacteria, resist being digested by amoebae. Amoeba-resisting bacteria can survive as long-term residents inside the cells of amoebae. Intracellular existence provides access to nutrients and offers protection from hostile and changeable environmental conditions (Kang *et al.*, 2017; Strassmann and Shu, 2017).

Amoebae have existed for more than a billion years. Amoebae have been described as 'training grounds for intracellular bacterial pathogens' (Molmaret *et al.*, 2005). Amoebae resemble the white blood cells of the mammalian immune system. In particular, free-living amoebae show a number of structural and behavioural similarities to macrophages, predatory motile white cells that roam around the tissues of the body. Macrophages recognise, envelope and digest invasive bacteria. Amoeba-resisting bacteria which normally exist as parasites in free-living amoebae also have the potential to survive inside mammalian macrophages and act as pathogens.

Phagocytosis, the process by which amoebae engulf their prey, results in the enclosure of a phagocytosed bacterium inside an intracellular vacuole, or phagosome, which then fuses with a lysosome containing digestive enzymes. Some amoeba-resisting bacteria avoid being digested by escaping from the phagosome and entering the intracellular fluid or cytosol of the amoeba. Others resist the action of the amoeba's digestive enzymes and remain inside the lysosome. Amoeba-resisting bacteria belong to disparate taxonomic groups including the phyla *Actinomycetota*, *Bacteroidota* and *Chlamydiota* and the classes *Alphaproteobacteria*, *Betaproteobacteria* and *Gammaproteobacteria*.

Legionella and Legionnaires' disease

Biofilms formed by bacteria in aquatic habitats attract predatory amoebae, which in turn become infected with amoeba-resisting bacteria. Manmade systems, including air conditioning units, cooling towers and pipework used to channel hot water and drinking water may be full of bacteria and amoebae. A study of the water supply at the University Hospital in Lausanne resulted in the detection of *Hartmannella vermiformis*, a species of amoeba commonly found in drinking water supply systems. Numerous species of amoeba-resisting bacteria were detected, including the plant symbionts *Methylorubrum extorquens* and *Bradyrhizobium japonicum* (both Class *Alphaproteobacteria*), as well as *Legionella pneumophila* (Class *Gammaproteobacteria*), the agent of Legionnaires' disease (Thomas *et al.*, 2006). *Legionella pneumophila* is an amoeba-resisting bacterium which is also able to invade and replicate within macrophages in lung tissue. Other *Legionella* species, including *Legionella longbeachae* and *Legionella bozemanni* are rarer causes of Legionnaires' disease.

Legionella pneumophila belongs to a group composed of intracellular specialists, the Order *Legionellales*. Legionellas occur naturally in fresh water and soil as free-living bacteria and as intracellular inhabitants of amoebae and ciliates. "*Candidatus* Berkiella cookevillensis" (Fig. 7.6) is the proposed name for a bacterium found in association with amoebae recovered from a biofilm in a cooling tower. It has been shown experimentally that "*Candidatus* Berkiella cookevillensis" can invade and replicate inside the nuclei of human-derived HeLa cell cultures (Chamberlain *et al.*, 2019). Other genera in the Order *Legionellales* include pathogens and symbionts of mammals and arthropods. Dozens of species of *Legionella* have been described. Amoeba-resisting bacteria are not generally amenable to isolation in culture but analysis of publicly available DNA sequence data derived from environmental samples detected sequences indicative of the existence of more than 400 uncultured and undescribed genera of *Legionellales* worldwide (Graells *et al.*, 2018).

Acid-fast mycobacteria

Mycobacteria belong to the Gram-positive Phylum *Actinomycetota*. The cells of mycobacteria are surrounded by a waxy coat composed of mycolic acids which causes them to stain poorly when stained using Gram's staining method.

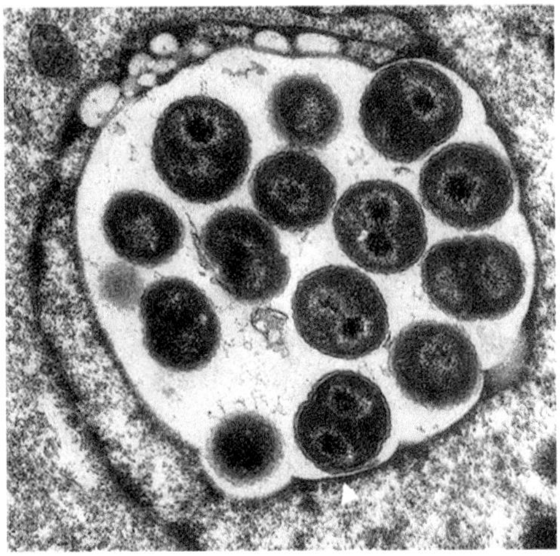

Fig. 7.6. "*Candidatus* Berkiella cookevillensis" in the nucleus of a human-derived HeLa cell. (Image cropped from figure 4 in Chamberlain *et al.*, 2019. *BMC Microbiology*. 19, 91.)

Instead of Gram's stain, the Ziehl–Neelsen staining method is used to visualise mycobacteria. The test sample is stained with carbol fuchsin and then washed in a mixture of acid and alcohol. The outer waxy coat resists the bleaching effects of the acid-alcohol mixture, causing the cells to stand out on microscopy as minute fuchsia-coloured bacilli. Mycobacteria are accordingly known as acid-fast bacteria.

Mycobacterium species occur in soils and aquatic environments. They thrive in cool, damp, acidic conditions and can be detected in a wide range of soil types. The most abundant and widely distributed soil mycobacteria detected in a global survey of 143 soils were strains related to *Mycobacterium novocastrense* (Walsh *et al.*, 2019). *Mycobacterium novocastrense* was originally described on the basis of an isolate cultured in Newcastle upon Tyne from a skin lesion on a child's hand (Shojaei *et al.*, 1997). *Mycobacterium novocastrense* is one of a large number of environmental species of mycobacteria that are occasionally detected in clinical specimens. Infections caused by environmental mycobacteria are associated with compromised host immune function and also with contaminated wounds. *Mycobacterium marinum* is an amoeba-resisting mycobacterium found in freshwater and saltwater environments. It is a fish pathogen and can cause skin and soft tissue infections, known as fish-tank granulomas, in aquarium enthusiasts. *Mycobacterium avium*, *Mycobacterium intracellulare*, *Mycobacterium chimaera*, *Mycobacterium abscessus*, *Mycobacterium chelonae* and *Mycobacterium fortuitum* are aquatic amoeba-resisting mycobacteria that occasionally cause local or disseminated disease in humans. *Mycobacterium chelonae* was originally cultured from infected sea turtles. *Mycobacterium avium* is associated with soft tissue infections in birds (Ben Salah and Drancourt, 2010).

Mycobacterium leprae, the agent of leprosy and *Mycobacterium tuberculosis*, which causes tuberculosis are specialised pathogens of humans (Fig. 7.7). Both species have retained their capacity to survive for long periods inside predatory amoebae (Mba Medie *et al.*, 2011; Wheat *et al.*, 2014).

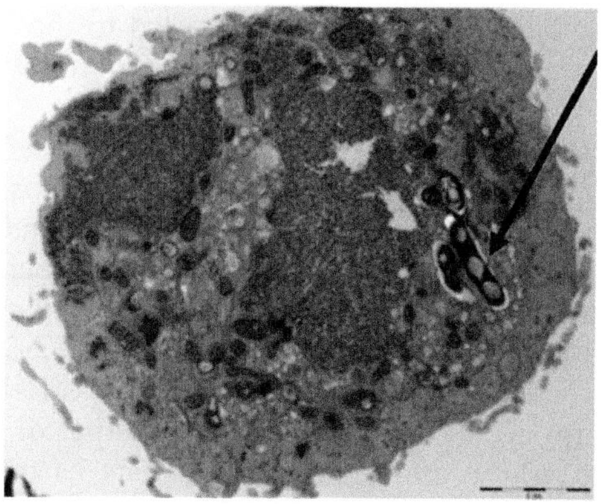

Fig. 7.7. *Mycobacterium tuberculosis* bacteria internalised by *Acanthamoeba polyphaga* (Image cropped from figure 2 in Mba Medie *et al.*, 2011; *PLoS ONE*. 6, e20499). Creative Commons.

Chlamydia and related intracellular bacteria

Members of the Phylum *Chlamydiota* are intracellular bacteria that depend on eukaryotic host cells in order to complete their life cycle, which consists of two alternating phases; an intracellular phase, known as the reticulate body and an extracellular phase, known as the elementary body, which has the capacity to infect host cells. *Chlamydiae* include symbionts of free-living aquatic amoebae and pathogens of mammals and fish (Taylor-Brown *et al.*, 2015). *Parachlamydia acanthamoebae* is a symbiont of free-living aquatic amoebae, which also has the capacity to invade human macrophages. *Chlamydia trachomatis* is an important human pathogen. Man is the only natural host. Some strains cause trachoma, an infection which leads to scarring of the eyelids and blindness. Other strains are sexually transmitted. Some sexually transmitted strains cause inflammation of the genital tract, which may lead to infertility in females. *Chlamydia psittaci* is the agent of psittacosis, a disease of birds and occasionally of humans. Compared with free-living bacteria, *Chlamydia trachomatis* and *Chlamydia psittaci* have small genomes. As intracellular parasites they have come to rely on host eukaryotic cells to supply many of their needs. The genome of *Chlamydia trachomatis* is small, being composed of only about 1 million base pairs. The genome of *Chlamydia psittaci* is not much larger. *Chlamydiae* found in association with free-living amoebae have larger genomes.

Other bacterial symbionts of amoebae

"*Candidatus* Nucleicultrix amoebiphila" (Class *Alphaproteobacteria*) is an amoeba-resisting bacterium which was discovered in amoebae recovered from sludge samples from a bioreactor used to treat nitrogenous waste. "*Candidatus* Nucleicultrix amoebiphila" was found inside the nuclei of free-living amoebae. The engorged nuclei of infected amoebae can be seen to be full of bacteria. Analysis of publicly available DNA sequence data derived from environmental sampling has shown that bacteria allied to the "*Candidatus* Nucleicultrix amoebiphila" are widely distributed in soil and freshwater environments (Schulz *et al.*, 2014).

Dictyostelium discoideum is a social amoeba that inhabits damp forest soils. When deprived of nutrients the cells of *Dictyostelium discoideum* clump together, move en masse, produce fruiting bodies and sporulate. For this reason it is sometimes referred to a cellular slime mould, although it is phylogenetically unrelated to other kinds of slime mould. *Dictyostelium discoideum* is susceptible to infection with amoeba-resisting bacteria (Fig. 7.8), including *Burkholderia agricolaris* and *Burkholderia hayleyella* (Class *Betaproteobacteria*) (Paquet and Charette, 2016; Shu *et al.*, 2018a; Shu *et al.*, 2018b; Khojandi *et al.*, 2019). Naturally infected amoebae produce as many spores as uninfected amoebae but experimentally infected naive amoebae produce fewer spores, which suggests that some strains of amoebae are more tolerant to infection than others. Some amoebae, including *Dictyostelium discoideum,* expel amoeba-resisting bacteria by wrapping them inside multiple layers of membranes known as multilamellar bodies.

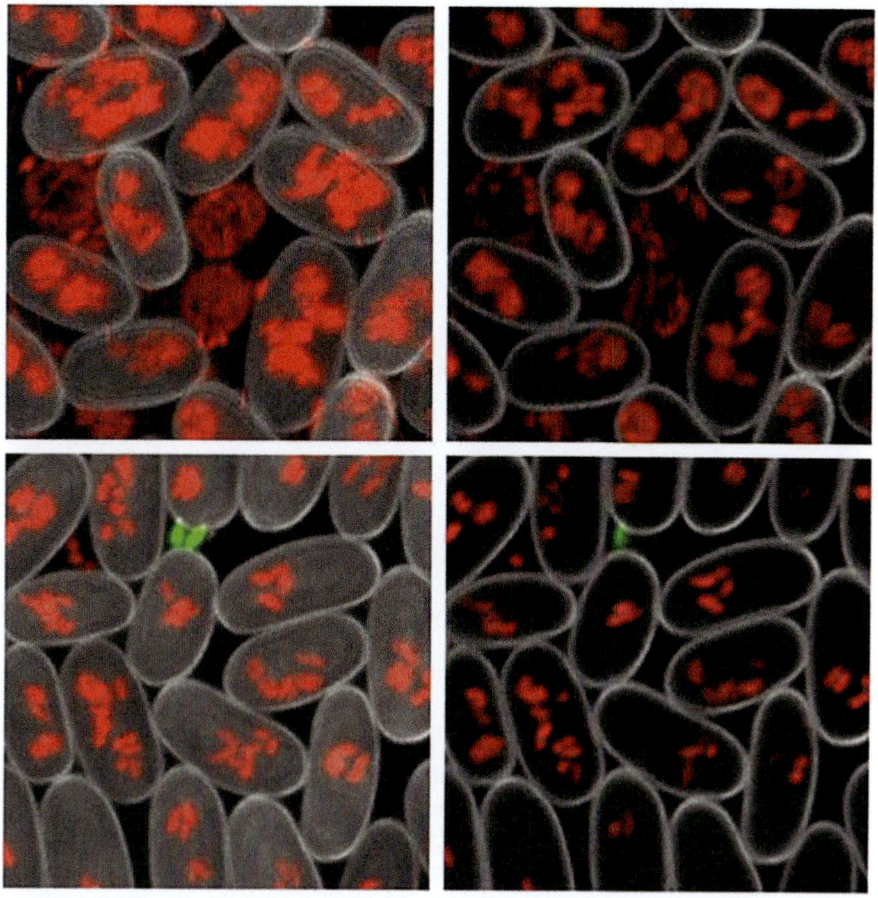

Fig. 7.8. Cells of the amoeba *Dictyostelium discoideum* infected with *Burkholderia agricolaris*, shown in red. (Khojandi *et al.*, 2019; *ISME Journal*. 13, 2068-2081. Image cropped from Figure 7). Creative Commons Attribution 4.0 International License.

Bacterial symbionts of algae and algicidal bacteria

Algal cells are enveloped by a thin dilute halo of secreted organic compounds, a zone known as the phycosphere (Bell and Mitchell, 1972). The phycosphere is a zone where bacteria and algae interact via the exchange of secreted chemical messenger molecules and nutrients. Photosynthetic algae serve as sources of organic carbon and organic sulphur compounds to heterotrophic bacteria (Amin *et al.*, 2015). Algae depend on bacteria as providers of fixed nitrogen, vitamins and iron. Algae are miniature habitats associated with communities of symbiotic bacteria that vary in complexity depending on the kind of alga involved. Pelagic algae are microscopic islands of plenty in what are otherwise largely barren oceanic waters. Algal blooms present opportunities for other organisms, including algicidal bacteria. The organisms informally known as algae include representatives of disparate phylogenetic groups, including coccolithophores, diatoms and dinoflagellates (Fig. 7.9).

Fig. 7.9. Diatoms and other eukaryotic algae surrounded by tiny bacteria. Widewater, West Sussex, UK.

Coccolithophores are planktonic photosynthetic unicellular eukaryotes. Their cells are surrounded by an arrangement of calcium carbonate plates. Coccolithophores were especially diverse and abundant during the Cretaceous Period, as witnessed by the deposits of chalk that exist today in the form of the North and South Downs, Salisbury Plain, the Chiltern Hills and the Yorkshire Wolds. Chalk is composed largely of the calcareous plates of innumerable coccolithophores. Coccolithophores are small but relatively complex microhabitats. Studies of two globally distributed coccolithophore species *Emiliania huxleyi* and *Coccolithus pelagicus* f. *braarudii* detected 85 coccolithophore-associated bacterial OTUs (i.e. approximately 85 species) from five different bacterial phyla. A single coccolithophore culture was estimated on average to host 35 species of bacteria, a richer microflora than the 9–14 species of bacteria found in association with diatoms and 17–19 species of bacteria found in association with dinoflagellates (Green *et al.*, 2015). Bacteria found in association with coccolithophores include *Roseobacter* (Class *Alphaproteobacteria*) and *Marinobacter* (Class *Gammaproteobacteria*). The coccolithophore *Emiliania huxleyi* is susceptible to 2-heptyl-4-quinolone, an antibiotic released by the algicidal bacterium *Pseudoalteromonas piscicida* (Class *Gammaproteobacteria*) (Harvey *et al.*, 2016).

Diatoms are microscopic unicellular photosynthetic eukaryotes associated with freshwater and saltwater habitats as well as damp soils. Diatoms account for much of the marine phytoplankton. They are non-motile organisms made recognisable by their silica shells that resemble miniature glass sculptures. Diatoms have been engaging in intimate relationships with bacteria for hundreds of millions of years. Diatoms release organic compounds that attract bacteria, while bacteria produce sticky extracellular polymeric substances that facilitate their attachment to diatoms.

Genomic analysis of the diatom *Phaeodactylum* indicates that a large number of diatom genes were appropriated from bacteria, including genes involved in the construction of their silica shells (Bowler *et al.*, 2008). Bacterial symbionts of diatoms include members of the genera *Sulfitobacter* and *Roseobacter* (Class *Alphaproteobacteria*) as well as members of the Phylum *Bacteroidota*. Bacteria supply diatoms with essential vitamins, fixed nitrogen and iron (Amin *et al.*, 2012; Amin *et al.*, 2015). The freshwater diatoms *Epithemia* and *Rhopalodia* harbour nitrogen-fixing spheroid bodies derived from cyanobacteria that function as intracellular organelles. The globally distributed marine diatoms *Epithemia pelagica* and *Epithemia catenata* have lost the capacity to synthesise chlorophyll and depend on cyanobacterial endosymbionts for organic carbon as well as for organic nitrogen compounds (Schvarcz *et al.*, 2022). Some diatom species are susceptible to protease enzymes released by the algicidal bacterium *Kordia algicida* (Phylum *Bacteroidota*). Algicidal bacteria such as *Kordia algicida* influence the population dynamics of marine planktonic algae (Bigalke and Pohnert, 2019). Some diatoms have been shown to release defensive antibacterial substances (Amin *et al.*, 2012).

Dinoflagellates are aquatic motile single-celled photosynthetic eukaryotes. Dinoflagellates contain the green photosynthetic pigment chlorophyll, as well as accessory orange carotenoid and yellow-brown xanthophyll pigments. Blooms of dinoflagellates known as red tides occur occasionally during the warmer months of summer. Dinoflagellates produce neurotoxins, which may harm marine life and humans who consume toxin-contaminated fish or shellfish (Fig. 7.10). It has been shown that toxin production by dinoflagellates

Fig. 7.10. Paralytic shellfish poisoning warning sign, Canada.

is modulated by symbiotic bacteria in the genus *Marinobacter* (Albinsson *et al.*, 2014). Paralytic shellfish poisoning is a clinical syndrome caused by the ingestion of shellfish contaminated with saxitoxin, a toxin produced by dinoflagellates, which can become concentrated in the tissues of filter-feeding molluscs. Paralytic shellfish poisoning is associated with numbness, paralysis and in extreme cases with respiratory failure and death. Shellfish harvested from around the British coast are monitored for the presence of toxins, the detection of which may lead to temporary harvesting restrictions and the closure of fisheries (Hinder *et al.*, 2011). During 1968, a mass die-off occurred among shags, *Phalacrocorax aristotelis*, and other seabirds on the Northumberland coast, an event which coincided with the reporting of 78 cases of paralytic shellfish poisoning in north-east England (Coulson *et al.*, 1968).

Saxitoxin, the principal toxin associated with paralytic shellfish poisoning, is produced by cyanobacteria as well as by dinoflagellates. Genomic analysis has revealed that toxin production by dinoflagellates is encoded by genes that show similarities to the genes that encode saxitoxin synthesis in cyanobacteria. This suggests a likely common origin for saxitoxin synthesis genes and that an ancestral dinoflagellate might have acquired toxin synthesis genes from a cyanobacterial source as a result of horizontal gene transfer (Orr *et al.*, 2013).

Lichens

A lichen is a symbiotic association between a fungus and one or sometimes two photosynthetic partner organisms referred to as photobionts. The fungus involved is usually a cup fungus (ascomycete). The photosynthetic organism is usually a eukaryotic alga. In some lichens (roughly 10% of species) the photobiont is a

Fig. 7.11. *Collema tenax*, a jelly lichen. Southwick, West Sussex, UK.

cyanobacterium. Some lichens incorporate two photobionts; a cyanobacterium and a green alga. Photobionts provide the fungus with a source of organic carbon. Cyanobacterial photobionts additionally act as a source of organic nitrogen compounds (Gilbert, 2001; Dobson, 2005; Sanders and Masumoto, 2021). About 20 or so of the British lichen genera feature species that include a cyanobacterial photobiont. The jelly lichens (Fig. 7.11) in the genus *Collema* are gelatinous lichens that grow on rocks, tree trunks and on sandy soils. Jelly lichens resemble their nostoc photobionts: they are blackish-green in colour; and they swell up when wet but become dry and crusty when desiccated, as do free-living colonies of nostoc. *Leptogium* is another nostoc-hosting genus of lichens. *Leptogium* species are brown or black in colour. The general demeanour of *Leptogium* species is leafy rather than jelly-like. *Lobaria pulmonaria* (Fig. 7.12) is a rather ostentatious foliose green lichen which can be seen growing on damp rocks and tree trunks in western parts of the British Isles. *Lobaria pulmonaria* is sensitive to air pollution and has disappeared from some former sites. The primary photobiont is a eukaryotic alga. *Lobaria pulmonaria* also accommodates cyanobacteria (*Nostoc*) in specialised structures known as cephalodia. Dog lichens, *Peltigera* (Fig. 7.13) can be found growing on mossy logs and damp shady lawns. *Peltigera*

Fig. 7.12. *Lobaria pulmonaria* on the bones of a Cuvier's beaked whale. Ord, Isle of Skye, UK.

Fig. 7.13. *Peltigera membranacea*. Cornwall, UK.

membranacea, a common species that occurs in grassland is greyish-green in colour with conspicuous white root-like projections known as rhizines. The photobiont of *Peltigera membranacea* is nostoc. *Sticta canariensis,* a lichen found on trees and rocks in western Britain, occurs in two forms. A brown-coloured form hosts a cyanobacterial photobiont, whereas the green form hosts a green alga. *Collemopsidium halodytes* are *Collemopsidium sublitorale* are coastal inhabitants of the littoral zone where they impart a pale brown crusty veneer to shoreline rocks and especially to the exposed surfaces of limpets and barnacles close to high water mark. The symbiont of *Collemopsidium* is reported to be a member of the cyanobacterial genus *Hyella*. Two more littoral lichens *Lichina confinis* and *Lichina pygmaea* are black-coloured bushy branching species, which together with the crustose lichen *Hydropunctaria maura* (which hosts an algal symbiont) contribute to the black zone of rocky coastal shorelines. *Lichina confinis* occurs just above and *Lichina pygmaea* occurs just below high water mark. The photobiont of *Lichina confinis* and *Lichina pygmaea* is the filamentous aquatic cyanobacterium *Calothrix*.

References

Albinsson, M.E., Negri, A.P., Blackburn, S.I. and Bolch, C.J.S. (2014) Bacterial community affects toxin production by *Gymnodinium catenatum*. *PLoS ONE*, 9, e104623.

Amin, S.A., Parker, M.S. and Armbrust, E.V. (2012) Interactions between diatoms and bacteria. *Microbiology and Molecular Biology Reviews*, 76(3), 667–684.

Amin, S.A., Hmelo, L.R., van Tol, H.M., Durham, B.P., Carlson, L.T. *et al.* (2015) Interaction and signalling between a cosmopolitan phytoplankton and associated bacteria. *Nature*, 522, 98–101.

Bar-On, Y. M., Phillips, R. and Milo, R. (2018) The biomass distribution on Earth. *Proceedings of the National Academy of Sciences of the United States of America*, 115, 6506–6511.

Bell, W. and Mitchell, R. (1972) Chemotactic and growth responses of marine bacteria to algal extracellular products. *The Biological Bulletin*, 143 (2), 265–277.

Ben Salah, I. and Drancourt, M. (2010) Surviving within the amoebal exocyst: the *Mycobacterium avium* complex paradigm. *BMC Microbiology*, 10, 99.

Bigalke, A. and Pohnert, G. (2019) Algicidal bacteria trigger contrasting responses in model diatom communities of different composition. *Microbiology Open*, 8 (8), e00818.

Bisgaard, M. and Kristensen, K.K. (1975) Isolation, characterization and public health aspects of *Vibrio cholerae* NAG isolated from a Danish duck farm. *Avian Pathology*, 4, 271–276.

Borland, E. D., Moryson, C.J. and Smith, G.R. (1977) Avian botulism and the high prevalence of *Clostridium botulinum* in the Norfolk Broads. *The Veterinary Record*, 100(6), 106–109. https://doi.org/10.1136/vr.100.6.106.

Bowler, C., Allen, A.E., Badger, J.H., Grimwood, J., Jabbari, K. *et al.* (2008) The *Phaeodactylum* genome reveals the evolutionary history of diatom genomes. *Nature*, 456, 239–244.

Castillo, D., Kauffman, K., Hussain, F., Kalatzis, P., Rørbo, N. *et al.* (2018) Widespread distribution of prophage-encoded virulence factors in marine *Vibrio* communities. *Scientific Reports*, 8, 9973.

Chamberlain, N.B., Mehari, Y.T., Hayes, B.J., Roden, C.M., Kidane, D.T. *et al.* (2019) Infection and nuclear interaction in mammalian cells by '*Candidatus* Berkiella cookevillensis', a novel bacterium isolated from amoebae. *BMC Microbiology*, 19, 1–18.

Circella, E., Camarda, A., Bano, L., Marzano, G., Lombardi, R. *et al.* (2019) Botulism in wild birds and changes in environmental habitat: A relationship to be considered. *Animals*, 9 (12), 1034.

Coulson, J.C., Potts, G.R., Deans, I.R. and Fraser, S.M. (1968) Exceptional mortality of shags and other seabirds caused by paralytic shellfish poison. *British Birds*, 61, 381–405.

Dobson, F. (2005) Lichens: An Illustrated Guide to the British and Irish Species. Richmond Publishing, Slough, UK.

Duckworth, D.H. (1976) Who discovered bacteriophage? *Bacteriology Reviews*, 40, 793–802.

Eklund, M.W., Poysky, F.T., Reed, S.M. and Smith, C.A. (1971) Bacteriophage and the toxigenicity of *Clostridium botulinum* type C. *Science*, 172, 480–482.

Faruque, S.M., Islam, M.J., Ahmad, Q.S., Faruque, A.S., Sack, D.A. *et al.* (2005) Self-limiting nature of seasonal cholera epidemics: Role of host-mediated amplification of phage. *Proceedings of the National Academy of Sciences of the United States of America*, 102 (17), 6119–6124.

Ford, C.L., Powell, A., Lau, D.Y.L., Turner, A.D., Dhanji-Rapkova, M. *et al.* (2020) Isolation and characterization of potentially pathogenic *Vibrio* species in a temperate, higher latitude hotspot. *Environmental Microbiology Reports*, 12 (4), 424–434.

Giacomucci, S., Cros, C.D., Perron, X., Mathieu-Denoncourt, A. and Duperthuy, M. (2019) Flagella-dependent inhibition of biofilm formation by sub-inhibitory concentration of polymyxin B in *Vibrio cholerae*. *PLoS ONE*, 14 (8), e0221431.

Gilbert, O. (2001) Lichens. New Naturalist. Collins, London, UK.

Graells, T., Ishak, H., Larsson, M. and Guy, L. (2018) The all-intracellular order *Legionellales* is unexpectedly diverse, globally distributed and lowly abundant. *FEMS Microbiology Ecology*, 94, fiy185.

Gray, M.D., Lampel, K.A., Strockbine, N.A., Fernandez, R.E., Melton-Celsa, A.R. *et al.* (2014). Clinical isolates of Shiga toxin 1a–producing *Shigella flexneri* with an epidemiological link to recent travel to Hispañiola. *Emerging Infectious Diseases*, 20 (10), 1669.

Green, D.H., Echavarri-Bravo, V., Brennan, D. and Hart, M.C. (2015) Bacterial diversity associated with the coccolithophorid algae *Emiliania huxleyi* and *Coccolithus pelagicus* f. *braarudii*. *Biomed Research International*, 2015, 194540.

Harvey, E.L., Deering, R.W., Rowley, D.C., El Gamal, A., Schorn, M. *et al.* (2016) A bacterial quorum-sensing precursor induces mortality in the marine coccolithophore, *Emiliania huxleyi. Frontiers in Microbiology*, 7, 59.

Hinder, S., Hays, G., Brooks, C., Davies, A., Edwards, M. *et al.* (2011) Toxic marine microalgae and shellfish poisoning in the British Isles: history, review of epidemiology, and future implications. *Environmental Health: A Global Access Science Source*, 10, 54.

Islam, M.S., Zaman, M.H., Islam, M.S., Ahmed, N. and Clemens, J.D. (2020) Environmental reservoirs of *Vibrio cholerae. Vaccine*, 38, Supplement 1, A52–A62.

Kang, S., Tice, A.K., Spiegel, F.W., Silberman, J.D., Pánek, T. *et al.* (2017). Between a pod and a hard test: the deep evolution of amoebae. *Molecular Biology and Evolution*, 34, 2258–2270.

Keen, E.C. (2015) A century of phage research: bacteriophages and the shaping of modern biology. *BioEssays: News and Reviews in Molecular, Cellular and Developmental Biology*, 37, 6–9.

Khojandi, N., Haselkorn, T.S., Eschbach, M.N., Naser, R.A. and DiSalvo, S. (2019) Intracellular Burkholderia symbionts induce extracellular secondary infections; driving diverse host outcomes that vary by genotype and environment. *ISME Journal*, 13, 2068–2081. https://doi.org/10.1038/s41396-019-0419-7.

Koop, G., Vrieling, M., Storisteanu, D.M., Lok, L.S., Monie, T. *et al.* (2017) Identification of LukPQ, a novel, equid-adapted leukocidin of *Staphylococcus aureus. Scientific Reports*, 7, 40660.

Krupovic, M. and Koonin, E.V. (2017) Multiple origins of viral capsid proteins from cellular ancestors. *Proceedings of the National Academy of Sciences of the United States of America*, 114, E2401–E2410.

Lee, J.V., Bashford, D.J., Donovan, T.J., Furniss, A.L. and West, P.A. (1982) The incidence of *Vibrio cholerae* in water, animals and birds in Kent, England. *Journal of Applied Bacteriology*, 52 (2), 281–291.

Leighton, G.R. (1923) Botulism and food preservation: the Loch Maree tragedy. Collins, London, UK.

Mba Medie, F., Salah, I., Henrissat, B., Raoult, D. and Drancourt, M. (2011) *Mycobacterium tuberculosis* complex mycobacteria as amoeba-resistant organisms. *PloS ONE*, 6, e20499.

Molmeret, M., Horn, M., Wagner, M., Santic, M. and Abu Kwaik, Y. (2005) Amoebae as training grounds for intracellular bacterial pathogens. *Applied and Environmental Microbiology*, 71, 20–28.

Orr, R.J., Stüken, A., Murray, S.A. and Jakobsen, K.S. (2013) Evolution and distribution of saxitoxin biosynthesis in dinoflagellates. *Drugs*, 11(8), 2814–2828.

Ortiz, N., and Smith, G. (1994) Landfill sites, botulism and gulls. *Epidemiology and Infection*, 112 (2), 385–391.

Panton, P.N. and Valentine, F.C.O. (1932) Staphylococcal toxin. *Lancet*, 219, 506–508.

Paquet, V.E. and Charette, S.J. (2016) Amoeba-resisting bacteria found in multilamellar bodies secreted by *Dictyostelium discoideum*: social amoebae can also package bacteria. *FEMS Microbiology Ecology*, 92(3), fiw025. https://doi.org/10.1093/femsec/fiw025.

Pretzer, C., Druzhinina, I.S., Amaro, C., Benediktsdóttir, E., Hedenström, I. *et al.* (2017). High genetic diversity of *Vibrio cholerae* in the European lake Neusiedler See is associated with intensive recombination in the reed habitat and the long-distance transfer of strains. *Environmental Microbiology*, 19 (1), 328–344.

Pryshliak, M., Hammerl, J.A., Reetz, J., Strauch, E. and Hertwig, S. (2014) *Vibrio vulnificus* phage PV94 is closely related to temperate phages of *V. cholerae* and other *Vibrio* species. *PLoS ONE*, 9 (4), e94707.

Schulz, F., Lagkouvardos, I., Wascher, F., Aistleitner, K., Kostanjšek, R. *et al.* (2014) Life in an unusual intracellular niche: a bacterial symbiont infecting the nucleus of amoebae. *ISME Journal*, 8, 1634–1644.

Schvarcz, C.R., Wilson, S.T., Caffin, M., Stancheva, R., Li, Q. *et al.* (2022) Overlooked and widespread pennate diatom-diazotroph symbioses in the sea. *Nature Communications*, 13 (1), 799.

Shojaei, H., Goodfellow, M., Magee, J., Freeman, R., Gould, F. *et al.* (1997) *Mycobacterium novocastrense* sp. nov., a rapidly growing photochromogenic mycobacterium. *International Journal of Systematic Bacteriology*, 47, 1205–1207.

Shu, L., Brock, D.A., Geist, K.S., Miller, J.W., Queller, D.C. *et al.* (2018a) Symbiont location, host fitness, and possible coadaptation in a symbiosis between social amoebae and bacteria. *eLife*, 7, e42660.

Shu, L., Zhang, B., Queller, D.C. and Strassmann, J.E. (2018b) *Burkholderia* bacteria use chemotaxis to find social amoeba *Dictyostelium discoideum* hosts. *ISME Journal*, 12 (8), 1977–1993.

Smith, T.J., Williamson, C.H., Hill, K.K., Johnson, S.L., Xie, G. *et al.* (2021) The distinctive evolution of orfX *Clostridium parabotulinum* strains and their botulinum neurotoxin type A and F gene clusters is influenced by environmental factors and gene interactions via mobile genetic elements. *Frontiers in Microbiology*, 12, 566908.

Strassmann, J.E. and Shu, L. (2017).Ancient bacteria–amoeba relationships and pathogenic animal bacteria. *PLoS Biology*, 15, e2002460.

Taylor-Brown, A., Vaughan, L., Greub, G., Timms, P. and Polkinghorne, A. (2015) Twenty years of research into *Chlamydia*-like organisms: a revolution in our understanding of the biology and pathogenicity of members of the phylum *Chlamydiae*. *Pathogens and Disease*, 73, 1–15.

Thomas, V., Herrera-Rimann, K., Blanc, D.S. and Greub, G. (2006) Biodiversity of amoebae and amoeba-resisting bacteria in a hospital water network. *Applied and Environmental Microbiology*, 72, 2428–2438.

Waldor, M.K. and Mekalanos, J.J. (1996) Lysogenic conversion by a filamentous phage encoding cholera toxin. *Science*, 272, 1910–1914.

Walsh, C.M., Gebert, M.J., Delgado-Baquerizo, M., Maestre, F.T. and Fierer, N. (2019) A global survey of mycobacterial diversity in soil. *Applied and Environmental Microbiology*, 85, e01180-19.

Wheat, W.H., Casali, A.L., Thomas, V., Spencer, J.S., Lahiri, R. *et al.* (2014) Long-term survival and virulence of *Mycobacterium leprae* in amoebal cysts. *PLoS Neglected Tropical Diseases*, 8 (12), e3405.

Zhang, Z., Douillard, F.P., Korkeala, H. and Lindström, M. (2022) Specific isolation of *Clostridium botulinum* Group I cells by phage lysin cell wall binding domain with the aid of S-layer disruption. *International Journal of Molecular Sciences*, 23, 8391.

8 What have bacteria done for us?

> I venture to suggest the name *Bacterium lactis*.
>
> Lister (1873). *A Further Contribution to the Natural History of Bacteria and the Germ Theory of Fermentative Changes.*

Among the record releases of 1986 was a compilation album of recordings by Copernicus, entitled *From Bacteria* (DMC 017). The title track's lyrics assert that human beings originated from bacteria. There is some truth in this statement because our mitochondria – organelles, found in human cells and in the cells of other eukaryotic organisms – do indeed appear to be the descendants of once free-living bacteria (Martin and Mentel, 2010). Mitochondria, which are inherited maternally, oxidise nutrients and provide the cell with a source of energy (Fig. 8.1). Eukaryotic cells are considered to represent the legacy of a primordial endosymbiotic event in which a free-living bacterium – the ancestor of mitochondria – became combined with some other kind of microbe. The partner organism was possibly an archaea-like microbe. In some respects eukaryotes more closely resemble archaea than they do bacteria.

Mitochondria have tiny genomes, consisting of about 16,500 base pairs. The genetic code of mitochondria found in most animal cells encodes only 37 genes (Chial and Craig, 2008). By comparison the genomes of free-living bacteria typically consist of several million base pairs which encode thousands of genes. The closest free-living relatives of mitochondria appear to be members of the Class *Alphaproteobacteria*. Free-living alphaproteobacteria continue to demonstrate a propensity to develop relationships with the cells of other organisms. The nitrogen-fixing symbionts which colonise the root nodules of leguminous plants are alphaproteobacteria. Some alphaproteobacteria, such as *Brucella* species, which cause the disease brucellosis, are intracellular

DOI: 10.1079/9781836993933.0008

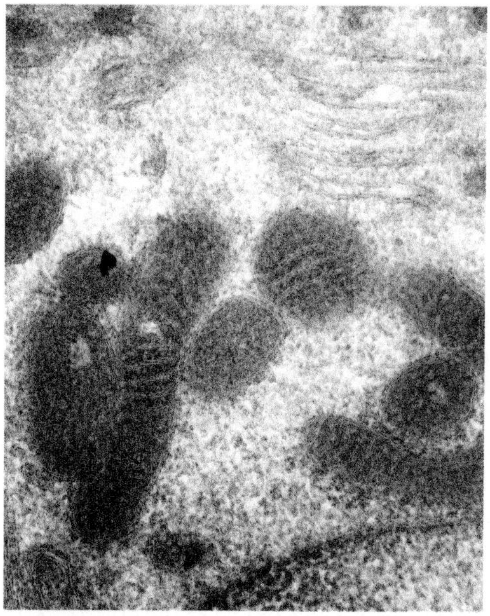

Fig. 8.1. Transmission electron microscopy, mitochondria, nucleus and membranes (Mike Kayser). CC BY 4.0.

pathogens. Alphaproteobacteria in the genus *Wolbachia* include pathogenic and symbiotic bacteria which are associated with many species of arthropods.

Bacteria-flavoured strawberries?

The methylobacteria are a group of alphaproteobacteria which can be found living in association with plants. Methylobacteria use methanol as a food source, a substance produced by plants during cell division. Methylobacteria can also be found in soil samples. One species, *Methylorubrum extorquens* (formerly known as *Methylobacterium extorquens*) has been shown to enhance the flavour of cultivated strawberries. The chemical compounds that give strawberries their distinctive strawberry flavour are known as furanones or furanoid compounds. The bacterium *Methylorubrum extorquens* readily colonises the leaves of strawberry plants. Plants react to the presence of the bacteria by increasing the production of defensive substances, including tasty furanoid compounds. *Methylorubrum extorquens* boosts the rate of furanone production by releasing lactaldehyde, a molecule utilised by strawberry plants during the synthesis of furanoid compounds. Researchers in Austria showed that strawberry plants which had been sprayed with *Methylorubrum extorquens* produced significantly larger quantities of furanoid compounds than a set of control plants which had been sprayed only with water (Verginer *et al.*, 2010).

Lactic acid bacteria and acetic acid bacteria

Lactic acid bacteria (Phylum *Bacillota*) are bacteria that produce lactic acid as a metabolite during carbohydrate fermentation. In the production of fermented foods, lactate released by lactic acid bacteria inhibits the growth of other microbes and helps to prolong shelf life (Table 8.1). Compounds produced by lactic acid bacteria modulate the textures and flavours of preserved foods. Lactic acid bacteria feature prominently in the fermentation of dairy products and in the pickling of vegetables. Silage, a fermented product consisting mainly of grass, is used widely as fodder for livestock during the winter months. The production of alcoholic drinks primarily involves the fermentation of sugars by yeasts, although lactic acid bacteria may make a significant contribution to the flavours of whisky, beer and cider. In Belgium, the production of distinctive lambic beers makes use of a fermentative process that involves environmental lactic acid bacteria.

The lactic acid bacterium *Lactococcus lactis* was the first species of bacterium to be isolated in pure culture. In '*A further contribution to the natural history of bacteria and the germ theory of fermentative changes*' the surgeon Joseph Lister (1827–1912) described experiments in which samples

Table 8.1. Lactic acid bacteria involved in food production.

Food product	Bacteria involved in its production
Sauerkraut	*Leuconostoc mesenteroides*, *Levilactobacillus brevis*, *Lactiplantibacillus plantarum*, *Pediococcus pentosaceus*
Kimchi	*Leuconostoc mesenteroides*, *Weissela koreensis*, *Pediococcus pentosaceus*, *Companilactobacillus kimchii*
Sourdough bread	*Limosilactobacillus fermentum*, *Fructilactobacillus sanfranciscensis*
Sour beers	*Lactobacillus delbrueckii*, *Lactiplantibacillus plantarum*, *Pediococcus damnosus*
Cider	*Levilactobacillus brevis*, *Oenococcus oeni*
Wine	*Oenococcus oeni*, *Pediococcus damnosus*, *Lentilactobacillus hilgardii*, *Lacticaseibacillus paracasei*
Malt whisky	*Levilactobacillus brevis*, *Lacticaseibacillus paracasei*, *Lactiplantibacillus plantarum*
Dry sausages	*Latilactobacillus sakei*, *Lactobacillus acidophilus*, *Lactiplantibacillus plantarum*, *Pediococcus pentosaceus*
Yoghurt	*Lactobacillus delbrueckii* subspecies *bulgaricus*, *Streptococcus thermophilus*
Cheeses	*Lactococcus lactis*, *Lactococcus cremoris*, *Lactobacillus delbrueckii* subspecies *bulgaricus*, *Streptococcus thermophilus*, *Lacticaseibacillus casei*, *Lactiplantibacillus plantarum*
Emmental cheese	*Streptococcus thermophilus*, *Lactobacillus helveticus*, *Propionibacterium freudenreichii*
Smear-ripened cheese rinds	*Brevibacterium linens*, *Brevibacterium aurantiacum*, *Microbacterium gubbeenense*, *Mycetocola reblochoni*, *Corynebacterium casei*, *Staphylococcus casei*

of boiled milk were inoculated with bacterial cultures taken from soured milk (Lister, 1873). Lister showed that the fermentation of milk could be brought about by a single species of bacterium. Lister made use of his "*Bacterium lactis*" (Fig. 8.2) as a model for understanding the roles of pathogenic bacteria in the aetiology of infectious diseases.

Lactic acid bacteria are microaerophiles which thrive best at subatmospheric concentrations of oxygen. They can be found in the intestinal and urogenital tracts of mammals as well as on the surfaces of leaves and flowers. Many of the lactic acid bacteria used as food preserving agents, such as those involved in the production of pickled vegetables, are strains derived from plants. The origins of the strains of lactic acid used in the production of fermented dairy products are less certain, although there are clues that at least some of them were originally derived from plants. Bacteria used in the fermentation of dairy products are often maintained as starter cultures which have become acclimatised to life in a rather narrow ecological niche. Compared with plant-associated wild strains, starter culture strains may show a reduced capacity to process plant-associated substances. Nevertheless, *Lactiplantibacillus plantarum*, one of the most widely used bacteria in the production of fermented foods, has a large complement of genes which enable it to utilise a range of different substrates. *Lactiplantibacillus plantarum* is used in the fermentation of pickled vegetables as well as dairy products.

Acetic acid bacteria (Class *Alphaproteobacteria*) are bacteria that convert alcohol to acetic acid. They are strictly aerobic organisms that require oxygen. Acetic acid bacteria are used in the commercial production of vinegar from wine, cider and malted barley. Vinegar is used as a condiment and as a preservative of pickled vegetables. In a two-step process, yeasts convert sugars to alcohol, which is converted to acetic acid by acetic acid bacteria. Acetic acid bacteria may act as spoilage organisms in the production of alcoholic drinks.

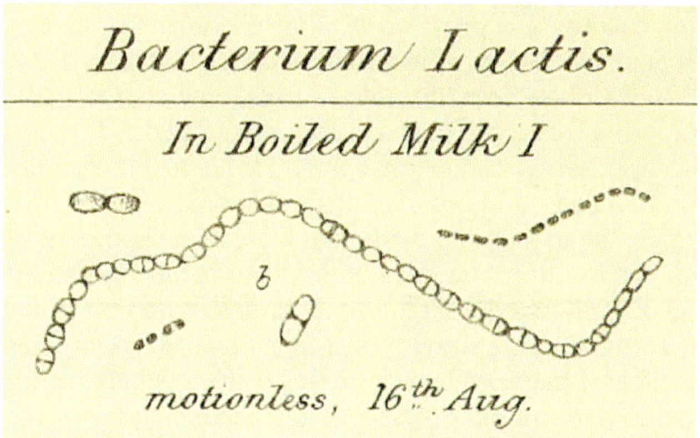

Fig. 8.2. Lister's original illustration of *Lactococcus lactis*. Image cropped from Plate XX: A further contribution to the natural history of bacteria and the germ theory of fermentative changes by Joseph Lister. (This material has been provided by the Royal College of Surgeons. Public Domain Mark.)

Louis Pasteur's earliest microbiological research involved investigations into the spoilage of wine. Microscopic investigation of spoiled wine disclosed the presence of bacteria. Pasteur showed that the bacteria could be eliminated by heating the wine: a process that became known as pasteurisation. Pasteur called his wine spoilage bacteria '*mycoderma aceti*' and referred to them as plants: '*Le* mycoderma aceti *est une des plantes les plus simples que l'on puisse imaginer*' (Pasteur, 1864). Pasteur's wine spoilage bacterium was later isolated in pure culture and formally described by the Dutch microbiologist Martinus Beijerinck as *Acetobacter aceti* (Beijerinck, 1898).

Fermented dairy products

Fermented dairy products involve the inoculation of milk with starter cultures of lactic acid bacteria. Yoghurt, an acidified milk product which has been produced for thousands of years, is made using starter cultures of *Lactobacillus delbrueckii* subspecies *bulgaricus* and *Streptococcus thermophilus* (van de Guchte *et al.*, 2006). The two species work synergistically to bring about the acidification of milk and the coagulation of milk proteins. *Lactobacillus delbrueckii* subspecies *bulgaricus* shows signs of adaptation to life as a milk-fermenting organism. Compared with other lactobacilli, *Lactobacillus delbrueckii* subspecies *bulgaricus* has a small genome. Genomic analysis shows that it has lost functional genes involved in the processing of plant-derived sugars. It has also lost the capacity to synthesise amino acids that are naturally present in milk. *Lactobacillus delbrueckii* subspecies *bulgaricus* can be regarded as a domesticated lactic acid bacterium, which has evolved to exist in a protein-rich, lactose-rich environment (El Kafsi *et al.*, 2014). Its ancestors were probably bacteria that lived in association with plant matter, possibly as residents of the intestinal tracts of herbivorous ruminants.

Additional clues regarding the origins of the lactic acid bacteria used as starter cultures are provided by the folklore associated with the production of Scandinavian fermented milk products. In *Flora Lapponica*, Linnaeus (1737) described how *tätmjölk*, a traditional soured milk product, was made by exposing fresh milk to the leaves of butterwort, *Pinguicula vulgaris* (*tätört* in Swedish, *tettegras* in Norwegian). He likened the production process to the action of a yeast and observed that additional supplies of soured milk could be generated by inoculating fresh milk with a small amount of an established *tätmjölk* culture. In his writings on Lapland he also noted that butterwort leaves would be gathered during the summer and stored for use at other times of the year. *Tätmjölk* has a stringy consistency caused by exopolysaccharides secreted by lactic acid bacteria. Leaves collected from butterwort (Fig. 3.73) have been shown to harbour exopolysaccharide-producing lactococci (Porcellato *et al.*, 2016). In the English language, *tätmjölk* and similar fermented milk products are known as ropy milk. Fermented ropy milk continues to be produced and marketed in Norway, where it is known as *tettemelk*, *tjukkmelk* or *tjukkmjølk*. The soured milk is made by adding a mixture of three traditional cultures containing the lactic acid bacterium *Lactococcus cremoris* to fresh milk.

The inoculated milk is incubated at about 20°C for 24 hours to allow the milk to coagulate and acidify. It is then refrigerated and marketed in one litre cartons.

Lactic acid bacteria used in the production of cheese include *Lactococcus lactis* and *Lactococcus cremoris* as well as *Lactobacillus delbrueckii* subspecies *bulgaricus* and *Streptococcus thermophilus*, both of which feature in the production of yoghurt. Cheese-making involves the acidification of milk and the breakdown of fats and proteins by bacteria. The flavours and textures of different cheeses are influenced by the kinds of lactic acid bacteria used as starter cultures. Starter cultures may involve single strains or multiple strains of bacteria, depending on the type of cheese. Later during the ripening process, the metabolic products of lactic acid bacteria as well as other kinds of bacteria and yeasts and moulds contribute additional distinguishing features to the cheese. Lactic acid bacteria that contribute to ripening include *Lacticaseibacillus casei*, *Lactiplantibacillus plantarum* and *Latilactobacillus curvatus* (Ercolini *et al.*, 2003). The blue veins in Stilton cheese and other blue cheeses are caused by the fungus *Penicillium roqueforti*. The characteristic cavities in Emmental cheese result from the release of carbon dioxide by *Propionibacterium freudenreichii* (Phylum *Actinomycetota*). *Propionibacterium freudenreichii* releases propionic acid, acetic acid and carbon dioxide as fermentation products of lactic acid. Starter cultures used in the production of Emmental include *Propionibacterium freudenreichii* as well as the lactic acid bacterium *Lactobacillus helveticus*.

Brevibacterium linens is an odoriferous bacterium that forms part of the normal human skin flora. In the production of smear-ripened cheeses such as the French Reblochon or the Irish Gubbeen cheese, starter cultures of *Brevibacterium linens* or *Brevibacterium aurantiacum* (Phylum *Actinomycetota*) are applied to the surface. Later during the ripening process, other bacteria (Fig. 8.3) colonise the cheese, including *Corynebacterium casei*, *Microbacterium gubbeenense*, *Mycetocola reblochoni* and *Staphylococcus casei* (Mounier *et al.*, 2005; Bora *et al.*, 2008). Smear-ripened cheeses are also known as rind-washed cheeses. During the course of the ripening process, they are washed with brine, wine or beer. Stinking Bishop, a rind-washed cheese from Gloucestershire is washed with perry. Regular washing encourages the growth of *Brevibacterium*, whose red carotenoid pigment and volatile metabolites largely account for the distinctive appearance and aroma of rind-washed cheeses.

Fermented alcoholic drinks

Malt whisky production involves the immersion of malted barley in hot water, a process known as mashing. The resulting sugary infusion is transferred to fermenting vessels where yeast is added. Environmental lactic acid bacteria contribute to the fermentation process. The numbers of bacterial cells may reach 100 million (10^8) cells per millilitre. Numerous species of lactic acid bacteria have been isolated from whisky fermentations, including *Levilactobacillus brevis*, *Lactobacillus crispatus*, *Limosilactobacillus fermentum*, *Lentilactobacillus hilgardii*, *Lacticaseibacillus paracasei*, *Lactiplantibacillus pentosus* and *Lactiplantibacillus plantarum* (van Beek and Priest, 2000). The metabolites

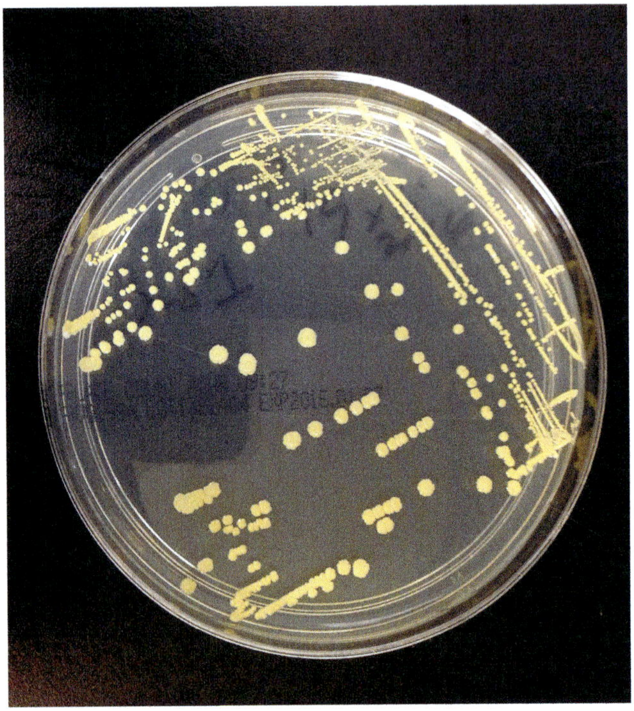

Fig. 8.3. *Staphylococcus casei* cultured from a rind-washed cheese.

of these bacteria are responsible for creating some of the distinctive features of the final product. Gamma-lactones, which are metabolites of oleic acid, a fatty acid present in barley, contribute to the perceived sweetness of malt whisky.

Yeasts are responsible for primary fermentation during the production of cider and wine. *Oenococcus oeni* and other lactic acid bacteria are responsible for a process of secondary fermentation which results in the conversion of malic acid, a sugar present in apples and grapes, into lactic acid. As lactic acid is less acidic than malic acid, secondary or malolactic fermentation reduces the acidity of the wine or cider and improves its palatability. Furthermore, the production of volatile metabolites by lactic acid bacteria improves the flavour of the final product. Nevertheless, lactic acid bacteria may also act as spoilage agents. *Pediococcus* species have been implicated in the production of distasteful metabolites, such as acrolein, a substance responsible for the odour of burnt fat. By producing exopolysaccharides, pediococci give fermented drinks an undesirable ropy consistency. Pediococci also convert amino acids to the bioactive amines histamine and tyramine, which in high concentration may cause headaches and palpitations.

Lactic acid bacilli are usually regarded as unwelcome spoilage agents of beers. Hops, *Humulus lupulus*, are added to beer to modulate the flavour and because of the antibacterial action of iso-alpha-acids released by the hops. Hop-tolerant lactic acid bacteria produce exopolysaccharides and lactic acid, resulting in cloudy, acidic beers. Lactic acid bacteria also produce diacetyl, a metabolite with a strong buttery flavour. Nevertheless, such qualities are considered responsible for the appeal of sour beers, such as Belgian Lambic beers (Fig. 8.4). Lactic acid bacteria may also contribute to the distinctive flavours of some English ales (Fig. 8.5).

Fig. 8.4. Brasserie Cantillon, Brussels, where beer is brewed with help of environmental lactic acid bacteria.

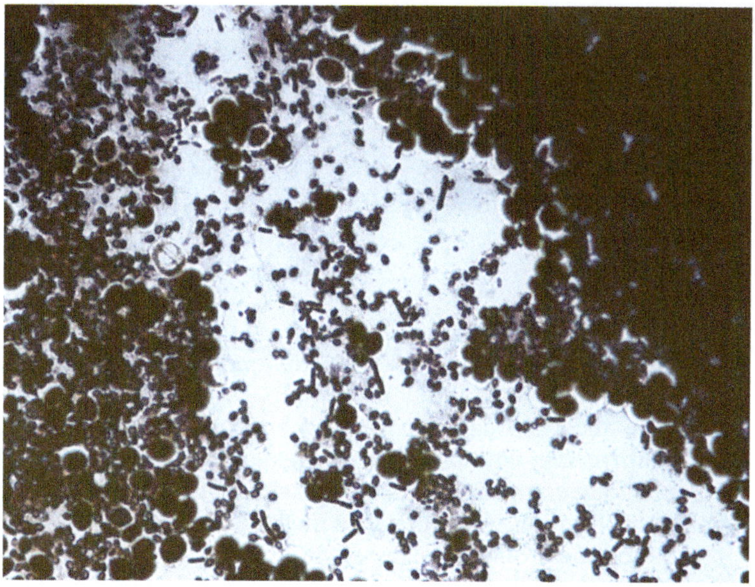

Fig. 8.5. Lactic acid bacteria dwarfed by the larger cells of brewer's yeast, cultured from the dregs of a bottled English ale.

Salt and salted foods: habitats for halophiles

Salt is a microbial habitat. The microbial flora of sea salt is dominated by halophilic archaea, including members of the genera *Halorubrum*, *Halobacterium* and *Haloarcula*, along with smaller numbers of halophilic bacteria such as *Salinibacter* and *Thalassophilus*. Halophiles have been recovered not only from salt harvested from solar salterns but also from salt made by heating seawater artificially. The halophilic archaea *Halobaculum magnesiiphilum*, *Natronoarchaeum mannanilyticum* and *Salarchaeum japonicum* were described as new species on the basis of strains isolated from samples of commercially available table salt made by heating seawater in iron pans (Shimane *et al.*, 2011). Table salt sourced as rock salt from mines may also contain halophilic microbes. The microbial flora of Himalayan pink salt, a variety of edible rock salt mined in Pakistan, was found to be dominated by the halophilic bacterium *Sulfitobacter* (Satari *et al.*, 2021).

Halophiles are involved in the production of salty foods, including fermented fish, fish sauce, brine-washed cheeses, olives and pickled vegetables. Anchovies, *Engraulis encrasicolus*, are prepared by curing them for many months in salt, after which the salted filets are tinned or packed in glass jars containing oil. The traditional method for ripening anchovies relies on the action of environmental halophilic fermentative microbes. In a blind tasting, anchovies deliberately inoculated with halophilic archaea were found to be more palatable than uninoculated control samples (Aponte *et al.*, 2010). Anchovies are an essential ingredient of Worcestershire sauce and of *colatura di alici*, a fish sauce made in Italy.

The ancient Romans referred to fish sauce as *garum*. Fermented fish sauces continue to figure prominently in the cuisines of Southeast Asia. *Natrinema gari*, *Halococcus thailandensis*, *Halobacillus thailandensis*, *Lentibacillus salicampi*, *Lentibacillus juripiscarius* and *Lentibacillus halophilus* are halophilic microbes described as new to science on the basis of strains isolated from Thai fish sauce (Namwong *et al.*, 2005). *Tetragenococcus muriaticus* is a halophilic bacterium cultured from fermented Japanese squid liver sauce. The specific names of the haloarchaea *Halalkalicoccus jeotgali* and *Natronococcus jeotgali* refer to *jeotgal*, a Korean condiment made by fermenting seafood mixed with rock salt (Lee, 2013).

The retting of flax

Linen is made from the cellulose stem fibres of flax, *Linum usitatissimum*. The process by which the fibres are separated from the surrounding matrix of pectin and other carbohydrate materials is called retting. Bundles of flax are submerged in retting tanks to create anoxic conditions in which the softer carbohydrate compounds can be broken down by the action of fermentative anaerobic bacteria. An alternative method called dew retting involves laying flax out on the ground to allow environmental fungi and bacteria to degrade the soft tissues. The retting of flax is a process that attracted the attention of some of the pioneering figures from the history of microbiology. A common finding was the discovery of Gram-positive rod-shaped anaerobic bacteria with large terminal spores; an appearance characteristic of members of the genus *Clostridium*. Working in France, Sergei Winogradsky described such organisms in a paper entitled: '*Sur le rouissage du lin et son agent microbien*' (Winogradsky, 1895). A few years later, similar organisms were isolated by Karl Störmer in Germany and by Martinus Beijerinck in the Netherlands. There is also an even earlier report by the French botanist Philippe Édouard Léon Van Tieghem, who noted that an anaerobic bacterium was involved in the retting process (Van Tieghem, 1879). The early authors used a variety of different names for these organisms, including "*Plectridium pectinovorum*" and "*Granulobacter pectinovorum*" (Störmer, 1903; Beijerinck and van Delden, 1904). The current scientific consensus is that the principal agents of pectin fermentation during the retting process are members of the genus *Clostridium*, including the species *Clostridium felsineum*, *Clostridium saccharobutylicum* and *Clostridium acetobutylicum*, accompanied by other environmental bacteria. A retting tank can easily be established by immersing a bundle of flax stems in a suitable container full of stagnant water. Environmental clostridial spores will soon germinate to produce vegetative cells and start the retting process. The bacteria can be visualised by preparing a microscope slide stained with Gram's stain (Fig. 8.6).

Antibiotics

Antibiotics are chemical compounds that inhibit or kill microorganisms (Table 8.2). The term encompasses substances naturally produced by

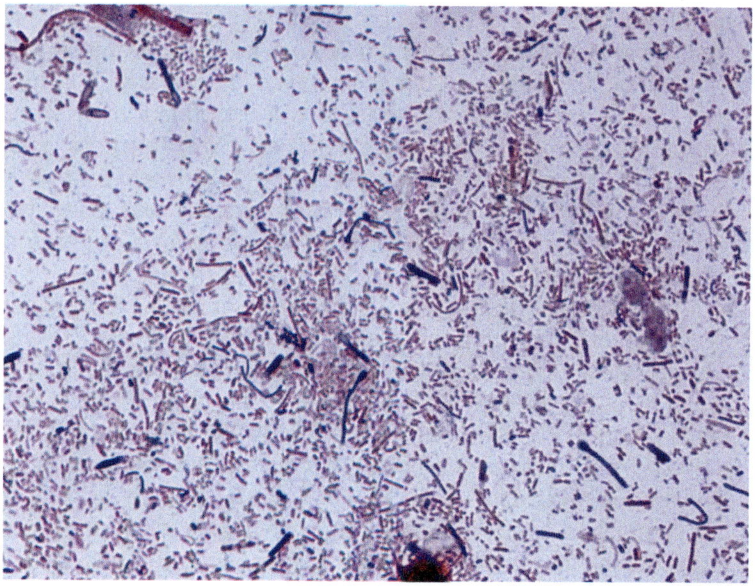

Fig. 8.6. Spore-forming clostridia and other bacteria from a retting tank. Gram's stain.

Table 8.2. Some antibiotics produced by bacteria and the bacteria that produce them.

Antibiotic-producing bacterium	Name of antibiotic	Class of antibiotic
Streptomyces griseus	streptomycin	aminoglycoside
Streptomyces kanamyceticus	kanamycin	aminoglycoside
Micromonospora spp.	gentamicin	aminoglycoside
Micromonospora inyoensis	sisomicin	aminoglycoside
Streptomyces tenebrarius	tobramycin	aminoglycoside
Pseudomonas fluorescens	mupirocin	pseudomonic acid
Streptomyces venezuelae	chloramphenicol	amide
Streptomyces lincolnensis	lincomycin	lincosamide
Streptomyces erythreus	erythromycin	macrolide
Streptomyces niveus	novobiocin	aminocoumarin
Streptomyces spectabilis	spectinomycin	aminocyclitol
Amycolatopsis rifamycinica	rifamide	rifamycins
Streptomyces aureofaciens	chlortetracycline	tetracycline
Streptomyces rimosus	oxytetracycline	tetracycline
Streptomyces roseosporus	daptomycin	lipopeptide
Amycolatopsis orientalis	vancomycin	glycopeptide
Actinoplanes teichomyceticus	teicoplanin	glycopeptide
Streptomyces candidus	avoparcin	glycopeptide
Bacillus licheniformis	bacitracin	polypeptide
Paenibacillus polymyxa	colistin	polymyxin

bacteria and moulds as well as semi-synthetic compounds derived from microbial products. Additionally, a number of synthetic substances with antibacterial properties are commonly referred to as antibiotics. Penicillin, a natural compound produced by a fungus, *Penicillium notatum*, was first administered to an infected patient in 1941. Penicillin works by binding to and inhibiting enzymes involved in cell wall synthesis. Penicillin was followed a few years later by the introduction of streptomycin, an antibiotic produced by a soil bacterium *Streptomyces griseus* (Phylum *Actinomycetota*). Other antibiotic discoveries soon followed and in combination with other antibiotics, streptomycin was used to treat tuberculosis, an infection caused by *Mycobacterium tuberculosis* (Phylum *Actinomycetota*).

The genus *Streptomyces* is a diverse genus of soil bacteria that form slow growing, sometimes filamentous colonies that resemble moulds. Streptomycetes produce many biologically active compounds including antibiotics. By inhibiting other bacteria these compounds give streptomycetes a competitive edge in microbiologically rich soil habitats. Antibiotics derived from bacteria are used to treat a wide range of infectious diseases in man and animals.

Antibiotics work by interfering with vital cellular functions. Antibiotics either kill the target cells (bactericidal antibiotics) or inhibit their growth (bacteriostatic antibiotics). Aminoglycosides, chloramphenicol, lincosamides, macrolides and tetracyclines are antibiotics that bind to the bacterial ribosome to block protein synthesis. Rifampicin is an antibiotic used in the treatment of tuberculosis which blocks protein synthesis by binding to RNA polymerase. The glycopeptides are a class of antibiotics that disrupt cell wall synthesis in Gram-positive organisms. Glycopeptides are used in the treatment of infections caused by strains of staphylococci, enterococci and other bacteria which are resistant to penicillins. Narrow-spectrum antibiotics are active against a relatively small number of bacteria whilst broad-spectrum antibiotics work against a wide range of bacteria from different taxonomic groups.

As many antibiotics are natural compounds or semi-synthetic versions of natural compounds produced by organisms that live in microbially rich and competitive habitats such as soil, it is easy to imagine that genes encoding resistance determinants to these substances will already exist in nature. Soon after the introduction of penicillin into clinical practice, penicillin-resistant strains of bacteria began to emerge and proliferate. The widespread use of antibiotics creates a strong selective advantage for resistant strains of bacteria, which undergo clonal expansion. Furthermore, bacteria readily exchange resistance genes via the transfer of mobile genetic elements such as plasmids. In addition to the acquisition of preformed resistance determinants, resistance may appear *de novo* through genetic mutation. Medically important bacteria that are resistant to multiple antibiotics have become known as superbugs (Fig. 8.7).

Large volumes of antibiotics are used in animal husbandry and veterinary practice, not only to treat infection but also as growth promoters. In the early 1950s it was discovered that antibiotics stimulated the growth of chickens. Chickens fed with waste materials derived from the production of the antibiotic

Fig. 8.7. *Superbugs*, an installation by the artist Anna Dumitriu at the Science Museum, London (Anna Dumitriu).

chlortetracycline gained weight faster than chickens raised on an antibiotic-free diet. Animal fodder supplemented with antibiotics soon became the norm. In the 1960s an outbreak of antibiotic-resistant salmonella was linked to the addition of antibiotics to animal feedstuffs. In 1969 the Swann Committee advised the UK government that antibiotics used in clinical medicine should not be fed to farm animals (Swann *et al.*, 1969). In the 1990s the European Union banned the use of the antibiotic avoparcin and several other antibiotics as growth promoters for livestock. Avoparcin is chemically similar to vancomycin, a glycopeptide antibiotic used in clinical practice. During the 1980s and 1990s outbreaks of glycopeptide-resistant enterococci occurred in hospitals (Fig. 8.8). Vancomycin resistance genes can be transferred between bacteria via mobile genetic elements. The detection of glycopeptide-resistant enterococci in livestock provided evidence of a possible animal origin for resistance genes found in clinical strains (Bates *et al.*, 1994).

Anti-cancer agents

Erwinase is an enzyme produced by the bacterium *Dickeya chrysanthemi* (previously known as *Erwinia chysanthemi*). Erwinase breaks down the amino acid asparagine. Erwinase is used in the treatment of acute lymphoblastic leukaemia. Unlike normal human cells, leukaemia cells do not synthesise asparagine and fail to proliferate when exposed to erwinase. *Dickeya chrysanthemi* is otherwise known as a plant pathogen.

Cetoniacytone is a compound with anti-cancer properties obtained from a symbiotic bacterium (*Actinomyces* sp. strain Lu9419) cultured from the intestinal flora of the rose chafer, *Cetonia aurata* (Family Scarabaeidae). Rose chafers (Fig. 8.9) are metallic green and about the same size as queen bumblebees. The distribution of rose chafers in the British Isles is oddly patchy but there

Fig. 8.8. Colonies of vancomycin-resistant enterococci growing on selective medium (Janice Bates).

Fig. 8.9. Rose chafers, *Cetonia aurata*, West Sussex, UK.

are places in southern and western Britain, such as the Isle of Scilly, Portland, Lundy and the Arundel area in West Sussex, where they are locally common. They may be seen flying around gardens or resting on flower heads. Cetoniacytone resembles a number of other bioactive compounds produced by actinomycetes. Actinomycetes, especially members of the genus *Streptomyces* are major sources of antibiotics. As soil organisms, actinomycetes presumably gain a selective advantage by producing substances that inhibit competitive bacteria and other organisms. Rose chafer larvae live in soil which is rich in decomposed vegetable matter. Any benefits that cetoniacytone might confer on *Actinomyces* sp. strain Lu9419 or on the beetles which host the bacterium are unknown. As yet, cetoniacytone has not found a role in clinical practice.

Thermostable enzymes

The enzymes and other proteins of most organisms are denatured and rendered ineffective when exposed to high temperatures. However, thermophilic bacteria that live at high temperatures and hyperthermophilic bacteria that live at very high temperatures, are equipped with thermostable enzymes. PCR (polymerase chain reaction) tests make use of thermostable enzymes to copy and amplify DNA in test samples. The classic PCR method uses the enzyme *Taq* DNA polymerase (usually abbreviated to *Taq*) from the thermophilic bacterium *Thermus aquaticus*. PCR has become a routine laboratory procedure in forensic and diagnostic laboratories. Tiny traces of DNA in a test sample can be amplified by PCR to yield enormous numbers of copies. The amplified product can then be subjected to various test procedures such as DNA sequencing to check for the presence of specific genes or to allow comparison with reference material. In addition to *Taq*, thermostable enzymes obtained from the thermophilic bacterium *Thermus litoralis* and the hyperthermophilic archaeon *Pyrococcus furiosus* are sometimes used to perform PCR tests

The industrial processing of starch involves the enzymatic breakdown of starch molecules to produce sugars. Sugars can be converted by fermentation to other compounds such as ethanol or citric acid. Sugars themselves are used also as sweetening agents, such as in the form of fructose corn syrup, which is an ingredient of some processed foods. The degradation of starch to produce glucose is conducted at high temperatures and uses thermostable enzymes from *Bacillus licheniformis* and *Bacillus stearothermophilus*. Glucose can be converted to fructose using a thermostable glucose isomerase enzyme derived from *Streptomyces murinus*. Thermostable enzymes are also used in baking to ensure the freshness and prolong the shelf life of baked bread. Left to its own devices, baked bread naturally becomes stale and brittle due to the gradual recrystallisation of is constituent starch molecules. When added to dough, a thermostable enzyme derived from *Bacillus stearothermophilus* converts the ends of starch molecules to maltose, a process which prevents the molecules from recrystallising (Vieille and Zeikus, 2001; Sarmiento *et al*., 2015).

Metal

The term bioleaching refers to industrial processes that use microorganisms to extract metals from metal ores. Bioleaching techniques are used in the commercial extraction of copper, cobalt, gold, nickel, uranium and zinc (Watling *et al*., 2015). The physicochemical conditions associated with ore deposits can be manipulated in order to create conditions that optimise bioleaching. Metal ores can be agitated and aerated and ores can be inoculated with starter cultures of bacteria. Substrates can be added that stimulate the growth of bioleaching bacteria. Bioleaching can be conducted *in situ*, in specially constructed heaps or in stirred tanks. The bioleaching of copper from copper ores involves the oxidation of iron sulphide to produce iron sulphate and sulphuric acid. The copper is leached from the ore by the acid. Acidophiles such as *Acidithiobacillus ferrivorans* and *Acidithiobacillus thiooxidans* can be added as starter cultures to promote the oxidation of sulphur and iron sulphide present in the ore. Numerous other species of bacteria and archaea have been earmarked as possible future bioleaching agents that could be used in commercial operations.

The microscopic examination of gold grains (Fig. 8.10) from an Australian gold deposit revealed the presence of microbial biofilms inhabited by the metal-tolerant bacterium *Cupriavidus metallidurans*. In nature, gold occurs as metallic gold and also in groundwater in aqueous solution in the form of soluble compounds such as gold chloride and gold thiosulphate. *Cupriavidus metallidurans* and other bacteria can catalyse the precipitation of metallic gold from solutions containing the toxic compound gold chloride. Gold precipitation by *Cupriavidus metallidurans* is thought to serve as a detoxifying mechanism.

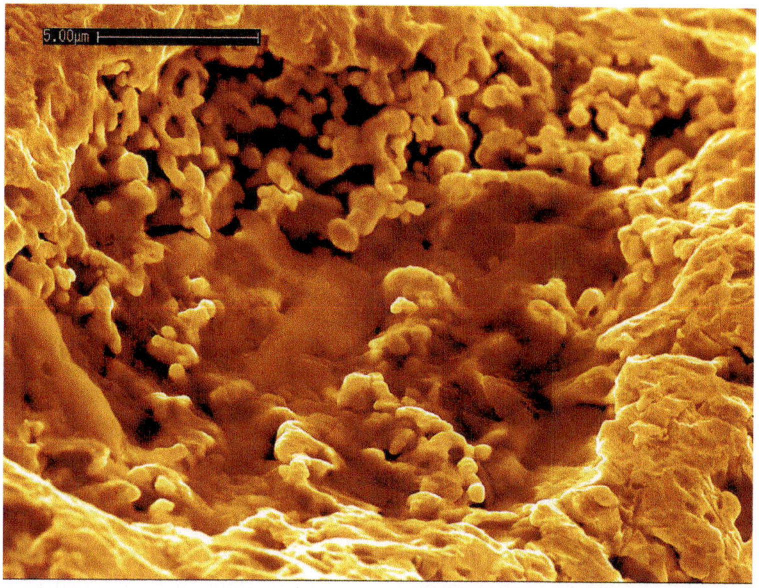

Fig. 8.10. Coloured scanning electron image of bacterioform gold on a gold grain from the Hit or Miss Mine in northern Queensland (CSIRO Scienceimage 3908). CC BY 3.0.

Gold-precipitating bacteria, including *Acidothiobacillus thiooxidans* and cyanobacteria have also been found in a South African gold mine. In New Zealand thermophilic bacteria and archaea are responsible for deposits of metallic gold and silver sinter deposits in the vicinity of hot springs (Reith *et al.*, 2007).

The Great Work of the Metal Lover is an installation created by Adam W. Brown, an artist based at Michigan State University. It features a culture of *Cupriavidus metallidurans* housed in a glass bioreactor containing toxic gold chloride. Microscopic quantities of metallic gold are produced in real time by the bacteria before the eyes of their audience.

Bioart

The term bioart was coined in 1997 by the American-Brazilian artist Eduardo Kac. His 1999 work, *Genesis*, featured a genetically modified bacterium. Kac transcribed a passage from the Book of Genesis – '*Let man have dominion over the fish of the sea, and over the fowl of the air, and over every living thing that moves upon the earth*' – into a coded base pair sequence of DNA which he then inserted into the genome of a living bacterium. The term bioart has been applied to a range of artistic practices in which artists adopt scientific methodologies and use biological materials in order to investigate different aspects of the living world. Bioart of a kind was already flourishing unobtrusively in various forms before the term was invented. Alexander Fleming used pigmented bacteria to produce Petri-dish sized miniature artworks. The practice of painting with microbes continues among members of the microbiological community to this day.

The Canadian artist Jennifer Willett directs the INCUBATOR art lab at the University of Windsor in Ontario. Her art lab fosters collaborative education, thinking and practice in the fields of art and ecology. *When microbes dream* is a bacteriocentric multimedia installation that incorporates artworks, theatre and performance art. The title piece is a mural, which represents a dream world and encourages the viewer to imagine what it might be like for the microbes. The *Baroque biology paper theatre* shows pictures of bacteria busily sharing their biotech processes with humans.

The bioartist Nurit Bar-Shai cofounded Genspace, a community biology lab in Brooklyn, New York, where non-scientists can attend classes and learn the art of microbial painting with cultures of living pigmented bacteria, or investigate their own DNA using DNA technologies. Nurit Bar-Shai collaborated with Eshel Ben-Jacob, a theoretical physicist whose research interests include intercellular communication in colonies of the bacterium *Paenibacillus vortex*, which led to his development of ideas about bacterial self-organisation and the concept of bacterial intelligence. Colonies of *Paenibacillus vortex* form intricate patterns when grown on agar in Petri dishes. Nurit Bar-Shai's bioart project *Objectivity [tentative]* includes live performances during which external stimuli are used to influence the growth patterns produced by the bacteria.

Among the projects created by the British artist Anna Dumitriu are *Normal Flora* (2004), the *MRSA Quilt* (2011), *Confronting the Bacterial Sublime* (2012), *Superbugs at the Science Museum* (2017–2019) and the *Plague Dress* (2018). The normal flora project was an investigation of bacteria and other microbes in the domestic environment. Bacteria cultured from household objects formed the basis for artworks, including *Bed and Chair Flora* (Fig. 8.11) and *Cutlery Flora* (Fig. 8.12). The *MRSA Quilt* was made from squares of cotton fabric embedded in Petri dishes containing chromogenic agar used routinely in diagnostic laboratories to detect antibiotic-resistant *Staphylococcus aureus* (Figs. 5.35–5.36). Phosphatase enzyme produced by the bacteria elicits the transformation of a chromogenic compound to yield a blue pigment which is taken up by the cotton squares. *Confronting the Bacterial Sublime* explored the philosophical ideas of eighteenth century philosopher and politician, Edmund Burke (1729–1797) regarding the sublime and the beautiful, as applied to bacteria; that an aesthetic apprehension of the sublime – which Burke describes as a sense of "*delightful horror*" – can be triggered by exposure not only to dangerous, violent and grandiose aspects of nature but also by acquaintance with the microscopic and the infinitely small.

Music

In the world of recorded music, bacteria and archaea contribute to the subject matter of recordings, the titles of tracks and albums, to the names of artists and to the names of record labels. Archaea is the name of a melodic death metal

Fig. 8.11. *Bed and Chair Flora.* Textile artworks depicting environmental bacteria cultured from household objects (Anna Dumitriu).

band based in Gothenburg, Sweden. Markus Heidarsson, guitarist in Archaea commented that apart from being a cool name, the *Archaea* were one of the three major groupings of life and also that a part of the Precambrian was called the Archaean (Markus Heidarsson, personal communication, 2009). Archaea was also the name of a death metal band based in Brisbane, Queensland. The Australian Archaea were active in the early 2000s. There was also a British Archaea whose members self-released an EP of experimental music entitled *Synaesthesia*. Writing on behalf of the trio, Steven Edmunds explained that their choice of name was inspired by reading about microbial life in one of Bill Bryson's books (Steven Edmunds, personal communication, 2009). 2006 saw the release of *Archaea* (Mutable 17525-2), an album consisting of three pieces of minimal chamber music recorded by the Dan Joseph Ensemble in New York. The composer Dan Joseph recalled that he first heard the word used in a radio interview in which *Archaea* were described as the original blocks of life. "I think it was also the quality of the word itself – its simple abstract beauty – as well as its evocation of things archaic, arcane and ancient – qualities present in my music – that made it seem an appropriate title" (Dan Joseph, personal communication, 2008). *Archaea apart* is one of the nine tracks on *The serene life of microbes* (AD55CD), an electronic ambient album recorded by the studio explorer Richard Bone. When asked about where the idea for the album came from, Richard explained that the title came to him in a dream and that he then surfed microbiology websites in search of possible song titles (Richard Bone, personal communication, 2009). In 1977, musician and performance artist Cosey Fanni Tutti gave a performance entitled 'Genetic Fear' at the ACME Gallery in London. The publicity materials for the performance included a

Fig. 8.12. *Cutlery Flora*. Engraved objects depicting environmental bacteria (Anna Dumitriu).

newspaper cutting which reported the discovery of a 'third form of life' (the archaea).

Based in London, the group Morphogenesis released a three track album, *Stromatolites* in 1998 (Vintage electronic records VER 01). Morphogenesis use homemade instruments, including a bioactivity translator which converts electrical signals obtained from plants, fungi and the human nervous system into auditory output. For two tracks on *Stromatolites*, layers of different pieces of music were combined to create a laminated sound effect. Cover artwork by Roger Sutherland has an abstract but geological feel which complements the music. Michael Prime, a member of Morphogenesis who keeps a fossil stromatolite on his studio mixing desk, explained that: "Stromatolites were fascinating to us on many levels. One aspect was the incredible amount of time they had been around on planet earth. This earliest of life forms had managed to survive every cataclysm the planet had gone through over hundreds of millions of years, and would no doubt still be around long after humans were gone. We felt that the archaic, yet enduring qualities of stromatolites shared an atmosphere with the music on the CD. We were also intrigued by the physical structure of stromatolites, and the fact that microorganisms could leave behind such complex, substantial and enduring structures" (Michael Prime, personal communication, 2010). *Stromatolite* (Musea Records FGBG 4719) is the title of an album released in 2007 by Outer Limits, a progressive rock band from Japan. Also based in Japan, the artist Taku Sakakibara (known as TaQ), released an electronic album entitled *Ongaq: stromatolite* (VAP inc. VPCC-81477) in 2004.

A number of artists have used bacteria in their album titles or as a name for themselves, sometimes to convey a whimsical sense of humour or to imbue their work with a feeling of mystery, although in many cases bacteriological terms are used in order to reflect the uneasy nature of their music. In 1997, the French progressive rock band Longshot released *The cosmic bacteria's experiences* (Target 29971), an album based on the science fiction writings of author Thierry Guilleminot. The bacteriocentric plot describes the escapades of a cosmic bacterium which reaches planet earth, having infected an escaped prisoner from Alpha Centauri. Bands with bacteria in their titles include: Bacteria Cosmica, a Latin rock band from California; Larry Simon and Groove Bacteria, a jazz fusion combo from Portsmouth, New Hampshire; Bloodred Bacteria, a grindcore band from Germany; and Red Bacteria Vacuum, an all-girl band from Japan. The late Ivor Cutler (1923–2006), Scottish songwriter, writer and artist, mentions bacteria in several of his musical compositions. *Befriend a Bacterium: Stickies* is a booklet that features messages on sticky labels that Ivor Cutler printed out and attached to various objects (Cutler and Honeysett, 1992).

Literature

Literary history is littered with the names of writers who have succumbed to infectious diseases, notably to tuberculosis (TB) caused by the bacterium *Mycobacterium tuberculosis*. Thomas Mann's (1875–1955) novel, *The Magic Mountain* is set in a TB sanatorium at Davos, situated high in the Swiss Alps. *The Magic Mountain*, which pre-dates the discovery of antibiotics, is set in an era when bed rest, fresh air and the surgical resection of diseased tissue were the only available treatments. In Thomas Mann's novel, the character Hans Castorp travels to Davos to visit a tubercular relative. Inside the sanatorium Castorp is introduced to "*Der Blaue Heinrich*" (Mann, 1924). The *Blue Henry* (2014) is an altered object, created by the artist Anna Dumitriu. The artwork features a vintage pocket sputum flask of a type once carried by expectorating TB sufferers (Fig. 8.13). Blue Henries were equipped with a spring-loaded lid. The lid of Anna Dumitriu's *Blue Henry* is engraved with a transmission diagram which represents the genomic base pair differences and inferred transmission links between *Mycobacterium tuberculosis* isolates cultured from different patients in a TB outbreak (Fig. 8.14). The *Blue Henry* is one of a number of TB-related pieces that featured in the exhibition: *The Romantic Disease: An Artistic Investigation of Tuberculosis* (Martin, 2014).

Fig. 8.13. The *Blue Henry* is an altered vintage sputum flask of a type used by tuberculosis patients (Anna Dumitriu).

Fig. 8.14. The *Blue Henry*'s lid is engraved with a transmission diagram representing the genomic analysis of mycobacterial isolates involved in a TB outbreak (Anna Dumitriu).

Albert Camus (1913–1960), who died of tuberculosis, based his novel *La Peste* on a historic cholera epidemic which affected the port city of Oran in Algeria (Camus, 1947). The spectre of cholera, an infection caused by the bacterium *Vibrio cholerae* loiters in the background of the magic realist novel, *Love in the Time of Cholera* by the Colombian writer, Gabriel García Márquez (1927–2014) (Márquez, 1985). The text of A.J. Cronin's (1896–1981) novel *The Citadel* includes a fictional account of a typhoid outbreak in a Welsh mining community. Cronin, who had firsthand experience of medical practice, describes an epidemiological outbreak investigation and measures taken to bring the outbreak to an end. The story of Mary Mallon (1869–1938), a healthy typhoid carrier most often known by her soubriquet Typhoid Mary, has become a part of popular culture. Having emigrated from Ireland to the USA she worked as a cook in New York City. Multiple cases of typhoid fever were traced back to Mary as the source of infection. Initially she refused to provide samples for microbiological testing but after having been forcibly detained, she was found to an asymptomatic carrier of *Salmonella enterica* serovar Typhi. Several theatrical productions have been based on her story, including *Forgiving Typhoid Mary* (1991) by the American playwright Mark St. Germain. Despite a promising title, *Salmonella men and planet porno*, a collection of short stories by the Japanese science fiction writer Yasutaka Tsutsui, has little to say about bacteria beside a brief statement to the effect that life on all planets evolves from bacteria (Tsutsui, 2006).

The discovery of DNA

The discovery of DNA can be traced back to the pioneering bacteriological experiments of Frederick Griffith (1879–1941). Griffith investigated the capsular variants of *Streptococcus pneumoniae* (Fig. 8.15). *Streptococcus pneumoniae*, known informally as the pneumococcus, is a Gram-positive bacterium that commonly colonises the upper airways of humans. As well as behaving as a harmless commensal organism, the pneumococcus is a potential pathogen and a common cause of bacterial pneumonia. Pneumococci can

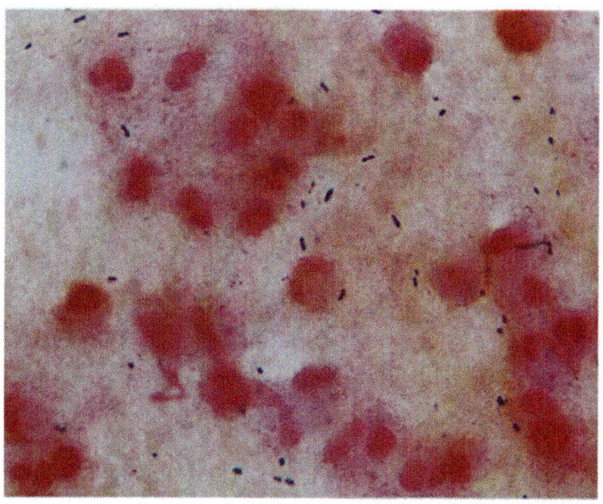

Fig. 8.15. Clinical specimen showing *Streptococcus pneumoniae*. Gram's stain.

also cause ear infections, bloodstream infections and meningitis. Some strains are more pathogenic than others and the type of polysaccharide capsule that surrounds the pneumococcus is an important virulence determinant. Some strains of pneumococci, the so-called smooth strains, do not produce a capsule. Griffith conducted a classic experiment in which a mouse was inoculated with a live strain of a non-capsulated pneumococcus plus a killed strain of a capsular pneumococcus. Subsequently, live capsular pneumococci were recovered from the mouse. Griffith deduced that the live non-capsulated pneumococcus had incorporated some substance from the dead capsulated bacteria, causing it to become capsulate. He described this as a transforming principle (Griffith, 1928). During the 1940s, Oswald Avery (1877–1955) and colleagues at the Rockefeller Institute for Medical Research determined that the transforming principle was DNA (Avery *et al.*, 1944). Avery's findings led to the discovery of the double helical structure of DNA and the molecular basis of the genetic code (Watson and Crick, 1953). The pneumococcus is a naturally transformable bacterium which can integrate DNA obtained from an external source into its own complement of DNA. Horizontal transfer of genetic information from another pneumococcus can lead to a change of capsular serotype. Human immunity to pneumococcal infection depends on an ability to recognise capsular antigen. Accordingly, pneumococcal vaccines work by priming the immune system to recognise commonly occurring capsular serotypes. By changing capsular serotype, pneumococci can evade the immune response. Serotype switching has been shown to occur in nature (Kell *et al.*, 1993). Transformation allows pneumococci to gain a selective advantage by acquiring additional useful genes, such as antibiotic resistance genes.

Griffith conducted his research at the Ministry of Health Pathological Laboratory in London. In 1941, Griffith and his colleague William McDonald Scott (1884–1941) were killed in an air raid. Scott had been appointed director of the Emergency Public Health Laboratory Service, an organisation which had been mobilised in 1939. The service continued to operate in peace time as the Public Health Laboratory Service. *Reciprocal Nuclei*, a sculpture by Mair Livingstone was commissioned in memory of Scott and Griffith by Scott's daughter Peggy Pereira. It stands inside the gates of the former Central Public Health Laboratory at 61, Colindale Avenue, London (Fig. 8.16). Mair Livingstone was the married name of the microbiologist Mair Eleri Morgan Thomas (1919–2014). In her guise as a sculptor she created artworks. The subject of one of her sculptures is *Anaerobiospirillum thomasii* (Fig. 8.17), a species of gut bacterium which had been named after her (Malnick, 1997).

Plague and pestilence

Analysis of ancient DNA has shown that the Neolithic decline, a period of population decline which occurred approximately 5400 years ago, coincided with the appearance of multiple strains of the plague bacillus *Yersinia pestis* among farming communities across ancient Europe. The archaeological record shows

Fig. 8.16. *Reciprocal Nuclei* by the sculptor Mair Livingstone, commissioned in memory of W.M. Scott and F. Griffith.

that neolithic settlements were regularly burned and abandoned, possibly in response to disease outbreaks. Genomic comparison of dated ancient DNA samples suggests that the plague bacillus may have evolved from an ancestral *Yersinia* species during a time when large densely populated settlements were being constructed in Eastern Europe, creating conditions which would have favoured the emergence of new infectious diseases. The plague bacillus, *Yersinia pestis* is closely related to *Yersinia pseudotuberculosis*, a bacterium found in wild rodents. *Yersinia pseudotuberculosis* occasionally causes infections in domestic animals and humans. From its original homeland, which was situated possibly somewhere north of the Black Sea, *Yersinia pestis* spread westwards and as far north as Scandinavia and eastwards into Asia. Ancient bacterial DNA indicative of plague has been detected in human remains from an approximately 5000-year-old neolithic grave in Sweden. The grave, containing the remains of 78 bodies, may represent the aftermath of a local plague epidemic (Rascovan *et al.*, 2019).

Bubonic plague has shaped the course of human history and brought about societal change (Fig. 8.18). Numerous references to plague exist in art, literature

Fig. 8.17. Sculpture of *Anaerobiospirillum thomasii* by Mair Livingstone (Anna Eleri Livingstone).

Fig. 8.18. Plague memorial, Winchester, Hampshire, UK.

in other cultural media and in the form of artefacts, such as plague stones, in whose vinegar-filled hollows coins were deposited as an early example of contactless payment (Fig. 8.19). When plague reached the village of Eyam in Derbyshire during the Great Plague of 1665–1666 the villagers implemented a policy of self-quarantine by restricting direct contact with outsiders. Many of the villagers died. Today Eyam is a tourist attraction where it is possible to inspect plague-related exhibits in the local museum, visit the graves of plague victims (Fig. 8.20) and search for Eyam's historic plague stone, situated at a boundary with a neighbouring plague-free village.

Written sources record a total of three historic plague pandemics. The Justinian plague, which affected the Byzantine Empire, began around AD 541 and lasted for 200 years. The Black Death, which began in AD 1346 marked the beginning of the second pandemic, which continued as a series of outbreaks and epidemics for 500 years. The last major epidemic to affect Britain was the Great Plague of 1665–1666. The third pandemic began in 1855. From a focus in China the plague spread to Europe, Africa and the Americas.

Fig. 8.19. A plague stone. Mousehole, Cornwall, UK.

Fig. 8.20. Gravestone marking the final resting place of Catherine Mompesson who died of plague in 1666, Eyam, Derbyshire

Yersinia pestis has been confirmed as the cause of all three pandemics by analysis of ancient bacterial DNA found in association with plague victims.

In common with its close relative *Yersinia pseudotuberculosis*, *Yersinia pestis* can become established among populations of wild rodents. Plague among wild animals is known as sylvatic plague. Numerous rodent species are susceptible to infection. Sylvatic plague occurs in Asia, Africa and the Americas. In the USA sylvatic plague has been responsible for a decline in prairie dog populations. *Yersinia pestis* is transmitted by flea bites or via the inhalation of aerosolised droplets.

Susceptibility to infection depends to some extent on how well the invasive bacteria are controlled by macrophages; predatory white cells which resemble amoebae. During the early stages of infection in humans, plague bacilli are consumed by and persist inside macrophages. *Yersinia pestis* is an amoeba-resisting bacterium, which can survive after being ingested by predatory amoebae. It has been suggested that soil amoebae might serve as a refuge and reservoir for plague bacilli (Benavides-Montaño and Vadyvaloo, 2017).

People living today are heirs to genetic changes wrought on human populations by the selective effects of plague. Comparison of human DNA recovered from burials in London and Denmark, dating from before, during and after the time of the Black Death, shows that plague was responsible for significant changes in gene frequencies in the human population, especially with regard to genes involved in immune responses to pathogens. One of the selected variants relates to the ability of human macrophages to kill ingested plague bacilli. Some of the gene variants that confer protection against infection with *Yersinia pestis* increase susceptibility to autoimmune diseases, including Crohn's disease and rheumatoid arthritis (Klunk *et al.*, 2022).

References

Aponte, M., Blaiotta, G., Francesca, N. and Moschetti, G. (2010) Could halophilic archaea improve the traditional salted anchovies (*Engraulis encrasicholus* L.) safety and quality? *Letters in Applied Microbiology*, 51, 697–703.

Avery, O.T., MacLeod, C.M. and McCarty, M. (1944) Studies on the chemical nature of the substance inducing transformation of pneumococcal types: induction of transformation by a desoxyribonucleic acid fraction isolated from pneumococcus type III. *Journal of Experimental Medicine*, 79, 137–158.

Bates, J., Jordens, J.Z. and Griffiths, D.T. (1994) Farm animals as a putative reservoir for vancomycin-resistant enterococcal infection in man. *Journal of Antimicrobial Chemotherapy*, 34, 507–514.

Beijerinck, M.W. (1898) Über die Arten der Essigbakterien. *Zentralblatt für Bakteriologie, Parasitenkunde, Infektionskrankheiten und Hygiene*, II (4), 209–216.

Beijerinck, M.W. and van Delden, A. (1904) Sur les bacteries actives dans le rouissage du lin. *Archives Néerlandaises des Sciences Exactes et Naturelles*, 9, 418–441.

Benavides-Montaño, J.A. and Vadyvaloo, V. (2017) *Yersinia pestis* resists predation by *Acanthamoeba castellanii* and exhibits prolonged intracellular survival. *Applied Environmental Microbiology*, 83, e00593-17.

Bora, N., Vancanneyt, M., Gelsomino, R., Snauwaert, C., Swings, J. *et al.* (2008) *Mycetocola reblochoni* sp. nov., isolated from the surface microbial flora of Reblochon cheese. *International Journal of Systematic and Evolutionary Microbiology*, 58, 2687–2693.

Camus, A. (1947) La Peste. Gallimard. France.

Chial, H. and Craig, J. (2008) mtDNA and mitochondrial diseases. *Nature Education*, 1 (1), 217.

Cronin, A.J. (1937) The Citadel. Gollancz, London, UK.

Cutler, I. and Honeysett, M. (1992) Befriend a bacterium. Pickpocket Books, Hastings, UK.

El Kafsi, H., Binesse, J., Loux, V., Buratti, J., Boudebbouze, S. *et al.* (2014) *Lactobacillus delbrueckii* ssp. *lactis* and ssp. *bulgaricus*: a chronicle of evolution in action. *BMC Genomics*, 15, 407.

Ercolini, D., Hill, P.J. and Dodd, C.E. (2003) Bacterial community structure and location in Stilton cheese. *Applied Environmental Microbiology*, 69, 3540–3548.

Griffith, F. (1928) The significance of pneumococcal types. *Journal of Hygiene* (Cambridge), 27, 113–159.

Kell, C.M., Jordens, Z., Daniels, M., Coffey, T.J., Bates, J. *et al.* (1993) Molecular epidemiology of penicillin-resistant pneumococci isolated in Nairobi, Kenya. *Infection and Immunity*, 61, 4382–4391.

Klunk, J., Vilgalys, T.P., Demeure, C.E., Cheng, X., Shiratori, M. *et al.* (2022) Evolution of immune genes is associated with the Black Death. *Nature*, 611, 312–319.

Lee, H.S. (2013) Diversity of halophilic archaea in fermented foods and human intestines and their application. *Journal of Microbiology and Biotechnology*, 23, 1645–1653.

Linnaeus, C [von Linné, C]. (1837) Flora lapponica. Salomon Schouten, Amsterdam, Netherlands.

Lister, J. (1873) A further contribution to the natural history of bacteria and the germ theory of fermentative changes. J.E. Adlard, London, UK.

Malnick H. (1997) *Anaerobiospirillum thomasii* sp. nov., an anaerobic spiral bacterium isolated from the feces of cats and dogs and from diarrheal feces of humans, and emendation of the genus *Anaerobiospirillum*. *International Journal of Systematic Bacteriology*, 47, 381–384.

Mann, T. (1924) Der Zauberberg. S. Fischer Verlag, Berlin, Germany.
Márquez, G.G. (1985) Love in the time of cholera. Editorial Oveja Negra, Colombia.
Martin, C. (2014) Interweaving the culture and science of tuberculosis. *Lancet*, 383, 773–774.
Martin, W. and Mentel, M. (2010) The Origin of Mitochondria. *Nature Education*, 3 (9), 58.
Mounier, J., Gelsomino, R., Goerges, S., Vancanneyt, M., Vandemeulebroecke, K. *et al.* (2005) Surface microflora of four smear-ripened cheeses. *Applied Environmental Microbiology*, 71, 6489–6500.
Namwong, S., Tanasupawat, S., Smitinont, T., Visessanguan, W., Kudo, T. *et al.* (2005) Isolation of *Lentibacillus salicampi* strains and *Lentibacillus juripiscarius* sp. nov. from fish sauce in Thailand. *International Journal of Systematic and Evolutionary Microbiology*, 55, 315–320.
Pasteur, L. (1864) Mémoire sur la fermentation acétique. *Annales Scientifiques de l'École Normale Supérieure Paris*, 1, 113–158.
Porcellato, D., Tranvåg, M. and Narvhus, J. (2016) Short communication: Presence of *Lactococcus* and lactococcal exopolysaccharide operons on the leaves of *Pinguicula vulgaris* supports the traditional source of bacteria present in Scandinavian ropy fermented milk. *Journal of Dairy Science*, 99, 7049–7052.
Rascovan, N., Sjögren, K.G., Kristiansen, K., Nielsen, R., Willerslev, E. *et al.* (2019) Emergence and spread of basal lineages of *Yersinia pestis* during the Neolithic decline. *Cell*, 176, 295–305.e10.
Reith, F., Lengke, M., Falconer, D., Craw, D. and Southam, G. (2007) The geomicrobiology of gold. *ISME Journal*, 1, 567–584.
Sarmiento, F., Peralta, R. and Blamey, J.M. (2015) Cold and hot extremozymes: industrial relevance and current trends. *Frontiers in Bioengineering Biotechnology*, 20, 148.
Satari, L., Guillĕn, A., Latorre-Perĕz, A. and Porcar, M. (2021) Beyond archaea: the table salt bacteriome. *Frontiers in Microbiology*, 12, 714110.
Shimane, Y., Hatada, Y., Minegishi, H., Echigo, A., Nagaoka, S. *et al.* (2011). *Salarchaeum japonicum* gen. nov., sp. nov., an aerobic, extremely halophilic member of the Archaea isolated from commercial salt. *International Journal of Systematic and Evolutionary Microbiology*, 61, 2266–2270.
Störmer, K. (1903) Über die Wasserroste des Flachses. *Chemisches Zentralblatt*, 76, 41.
Swann, M.M., Baxter, K.L., Field, H.I., Howie, J.W., Lucas, I.A.M. *et al.* (1969) Report of the Joint Committee on the use of Antibiotics in Animal Husbandry and Veterinary Medicine. HMSO, London, UK.
Tsutsui, Y. (2006) Salmonella men on planet porno. Alma Books, Richmond, UK.
van Beek, S. and Priest, F.G. (2002) Evolution of the lactic acid bacterial community during malt whisky fermentation: a polyphasic study. *Applied Environmental Microbiology*, 68, 297–305.
van de Guchte, M., Penaud, S., Grimaldi, C., Barbe, V., Bryson, K. *et al.* (2006) The complete genome sequence of *Lactobacillus bulgaricus* reveals extensive and ongoing reductive evolution. *Proceedings of the National Academy of Sciences of the United States of America*, 103, 9274–9279.
Verginer, M., Siegmund, B., Cardinale, M., Müller, H., Choi, Y. *et al.* (2010) Monitoring the plant epiphyte *Methylobacterium extorquens* DSM 21961 by real-time PCR and its influence on the strawberry flavor. *FEMS Microbiology Ecology*, 74. 136–145.
Vieille, C. and Zeikus, G.J. (2001) Hyperthermophilic enzymes: sources, uses, and molecular mechanisms for thermostability. *Microbiology and Molecular Biology Reviews*, 65, 1–43.

Watling, H.R., Shiers, D.W. and Collinson, D.M. (2015) Extremophiles in mineral sulphide heaps: some bacterial responses to variable temperature, acidity and solution composition. *Microorganisms*, 9, 364–390.

Watson, J. and Crick, F. (1953) Molecular structure of nucleic acids: a structure for deoxyribose nucleic acid. *Nature*, 171, 737–738.

Winogradsky, S. (1895) Sur le rouissage du lin et son agent microbien. *Comptes rendus de l'Académie des Sciences*, 121, 742.

9 Bacteria tourism

Alle Stromatolithe zeigen im vertikalen Schnitt deutliche Lagenstruktur …
[All stromatolites exhibit a distinctly layered structure in vertical cross section …]
Ernst Louis Kalkowsky (1908).
Oolith und Stromatolith im norddeutschen Buntsandstein.

When I first read about the stromatolites at Shark Bay in Western Australia it struck me as an odd kind of quest for anyone to want to travel halfway around the world to see rock-like structures made by microscopic organisms (Attenborough, 1979). It still does but the fact is that somehow those stromatolites got under my skin. Stromatolites exercise a strange allure that I could not resist. Eventually, decades after having first read about them, I made the journey to Shark Bay myself and realised that I had become a bacteria tourist. The pace of my bacteria tourism quickened when my attention became focussed on the need to assemble content for *Bacteria in the Natural World*. To some extent this chapter has been written to serve as something like a gazetteer of selected bacteriological tourist attractions. Its themes are arranged roughly according to the order of the main text in the manner of a series of supplements. It includes personal commentaries on some of the journeys made to specific destinations in order to gain experience and to gather material for this book. Many of the places mentioned really are tourist attractions. Some of the sites mentioned are buildings or memorials which are associated with the history of microbiology and infectious diseases. Others are well-known natural locations where particular kinds of bacteria can be seen in the wild, or where stromatolites may be seen in a natural setting. An account is provided of some of the more famous and culturally important iron springs and sulphur wells. Many of these places have been significantly modified by humans. Nevertheless, signs of their

DOI: 10.1079/9781836993933.0009

resident bacteria remain readily apparent to the bacteria tourist. This chapter concludes with accounts of visits to a number of globally important sites.

Bacteria in the natural world made me think hard about how to frame a narrative about a subject that to the uninitiated may seem to be so wilfully obscure and which is accompanied by a body of literature so steeped in technical detail as to seem almost unreadable. The task made me wonder how to write about microorganisms in a way that sits comfortably within the purlieu of an inquisitive field naturalist. Over the course of time the process started to make me see things differently. By the time I had got to the stage of sorting through photographs and wondering about what to do to fill the gaps in my photographic library, I recognised that I had acquired a new way of seeing. For example, one day I found myself facing Aust Cliff square on, with my eye to the viewfinder of my camera. I had decided to fulfil three aims with one journey: to photograph the muddy banks of the River Severn at Aust Warth as an example of an estuarine habitat; to photograph a pretty tufa mound situated near the foot of the cliff as an example of a cyanobacterial habitat; and to photograph the bands of gypsum in the cliff face as an example of an ancient evaporite deposit, once a habitat for sulphide-oxidising bacteria and other microbes. As always, I was captivated by the grandeur of the cliff face. The cliff's tableau of pigmented strata when seen through the viewfinder as a framed upright rectangle of horizontal lines reminded me of paintings by Mark Rothko. But then I discerned that without having made any special effort to do so, I was already contemplating the cliff's different layers one by one and that I was asking myself questions about the lives of the bacteria in each of those different ancient environments (Fig. 9.1).

The quest for fossilised stromatolites

Some of the greatest adventures are adventures that you can have on your own doorstep. There is really no need to travel all the way to Shark Bay to observe the living stromatolites that reside in its briny shallows: there are plenty of places in the British Isles that harbour fossilised stromatolites, which in their own way are equally as impressive and informative as the ones that live in Shark Bay. I made a decision to hunt down a comprehensive set of them (as described in Chapter 2); dating from the Precambrian to the Jurassic Period and at locations stretching from Islay to the Isle of Portland. And out of a sense of thoroughness I also visited the more recent Cretaceous chalk cliffs at Hunstanton in Norfolk, where structures present in the Hunstanton Formation have been tentatively identified as the fossilised traces of ancient microstromatolites. The search for stromatolites led me to some beautiful and remote locations. Some of these locations were already familiar to me. During the course of my earliest expedition to Orkney I had paid a visit to the sea cliffs at Yesnerby. On that occasion I arrived at Yesnerby on a bicycle hired in Stromness. I was looking for *Primula scotica*, the Scottish primrose. But at the time I was unaware of Yesnerby's fossilised stromatolites, so accordingly I did not actually recognise

Fig. 9.1. Aust Cliff, Gloucestershire, UK.

any of them as such! Similarly, when I visited Orkney for a second time I managed not to see any stromatolites, despite the fact that our campsite was only a few minutes' walk from some easily accessible specimens located along the coastline not far from Stromness (Fig. 9.2). But by the time I visited Orkney for a third time, I had become an experienced bacteria tourist: I knew all about fossilised stromatolites, what they looked like, where they were and I knew exactly how to find them (see Chapter 2).

Some of my excursions were only partially successful. On one occasion I visited Anglesey in order to investigate the acidophilic bacteria that thrive in the highly acidic water bodies on Parys Mountain. Having explored the mountain, I continued towards the village of Cemaes to search for stromatolites among the outcrops of Precambrian limestone which occur along the north coast of the island. However, I was not well prepared and in the fading light I realised that it was going to be a fruitless search. I had neglected to bring map references and in any case it was already too late in the day to locate

Fig. 9.2. Stromatolite reef, near Stromness, Orkney, UK.

the elusive stromatolites. I soon gave up looking for the stromatolites, which were supposed to be present somewhere amongst the geologically confusing hotchpotch of rocks that are exposed along the coastline. Some years later, I returned to Cemaes and set off on foot along the lanes towards Llanbadrig Church. The ancient church is situated on the coast, near a cave in which according to legend, St Patrick took shelter when shipwrecked while travelling from Iona to Ireland in 440 AD. I began to hunt for signs of prehistoric microbial life among the rocks. Cemaes and its environs are of interest to geologists because of a phenomenon known as the Gwna mélange, which features a seemingly chaotic disarray of different rock types. The mélange is thought to represent the aftermath of a gargantuan submarine landslide. Not far from the coast an entire quarry has been carved into what was once a single colossal block, or megaclast, of detached limestone. The rocks at Llanbadrig have been quarried and subjected to coastal erosion and have been cracked and crumpled during the course of geological time. Veins of calcite run across the layers of limestone. At first I found it hard to spot any signs of mineralised microbial growth among the outcrops of distressed rocks. But after a short search, I found what I was looking for (Fig. 9.3). Recent studies indicate that the stromatolites on Anglesey are among the oldest examples known from Britain. They may be as much as 800–860 million years old (Horák and Evans, 2011).

As a general rule, fossilised stromatolites occur in places that are wild, or in places that at least feel as though they are wild. Coastal erosion makes fossils visible and creates a context in which fossilised stromatolites appear uncannily as though they might still be alive. Abandoned quarry faces provide a more academic context, in which stromatolitic horizons can be examined in relation

Fig. 9.3. Stromatolite, Llanbadrig near Cemaes, UK.

to a succession of ancient environments. When I started to plan a field trip to photograph examples of Carboniferous stromatolites in south-western England I initially toyed with the idea of visiting Cheddar Gorge. The gorge is certainly a tourist attraction and would not be out of place in this chapter. Cheddar Gorge is an especially attractive locale for botanists and it is also known to harbour Carboniferous stromatolites in the Clifton Down Limestone. Then I learnt of the existence of a well-exposed stromatolite bed, also from the Clifton Down Limestone, at Barnhill Quarry, near Chipping Sodbury, which had recently been made accessible to the public. Over the course of geological time earth movements have displaced the originally horizontal stromatolite bed, such that it now lies at a jaunty angle. It currently forms part of the inclined eastern rim of a former quarry. The quarry itself has been filled in and repurposed as a space for residential housing. I found a place to park not too far from the quarry and set out on what seemed initially like a weirdly normal suburban walk through a pedestrianised housing estate. Before long I started to become acclimatised to the appearance of rows of newly constructed homes. Then I turned another corner to be confronted with the majestic apparition of a remarkably well-preserved slanting limestone rock face, the surface of which was bristling with an abundance of Carboniferous stromatolites (Fig. 9.4). A newly laid tarmacadam path leads the visitor between a row of brightly painted new houses and a geological site of special scientific interest. The stromatolite bed is situated towards the northern end of the site (ST725827). There is no mistaking the organic form of the stromatolites (Fig. 9.5). Good examples of ripple-marked rock surfaces are also present. Two interpretative boards have been erected for the benefit of visitors. One board provides information about ripple marks and the other provides information about the stromatolites. There

Fig. 9.4. Barnhill Quarry, Gloucestershire, UK, general view of stromatolite bed.

Fig. 9.5. Barnhill Quarry, Gloucestershire, UK, stromatolites.

are pictures of the stromatolites as seen in cross section and pictures of living stromatolites at Shark Bay. One of the games you can play as a bacteria tourist is to spot pictures of the Shark Bay stromatolites on websites, in textbooks and on interpretative boards. It is as though the Shark Bay stromatolites have become an unofficial benchmark against which all other stromatolites must be compared.

Visitors to the town of St Andrews have the opportunity to investigate a local tourist attraction which incorporates an actual stromatolite specimen. St Andrews has its own Geological Wall (NO506171), composed of locally occurring geological samples, including a stromatolite dating from the Carboniferous Period. The wall was constructed as recently as 2000 but its coastal aspect and exposure to the elements mean that all of its disparate geological components have already weathered to an almost uniform greyish hue, rendering the wall perfectly camouflaged against the tone of the local building stone (Fig. 9.6). Most visitors pass by the wall without a second glance. Situated

Fig. 9.6. The Geological Wall, St Andrews, UK.

Fig. 9.7. Weathered stromatolite specimen, St Andrews Geological Wall, UK.

towards the top lefthand corner of the landward side of the wall sits a solitary stromatolite (Fig. 9.7). A plaque attached to the end of the wall describes the rock as an algal limestone. Careful examination of the specimen reveals characteristic laminated concentric layers of mineralised microbial growth, although weathering of the rock's surface and a surfeit of lichenous growth make it difficult to appreciate much in the way of detail. The sample's origin is an outcrop of rocks known as the Randerston limestones which occur on the Fife coast near the village of Kingsbarns. Freshly exposed examples of the Randerston stromatolites exhibit swirling patterns in rocks that are stained red with haematite. Strictly speaking they are oncolites. Oncolites are roughly spherical structures which formed as a result of the mineralisation of successive layers of microbial growth around a nucleus, such as a shell fragment. They are surprisingly beautiful. The rock exposures near Kingsbarns are rich in fossils and it is relatively easy to find the shelly remains of molluscs and the preserved roots and stems of Carboniferous plants. The stromatolites, however, can take some finding amongst the confusing coastal melee of rock pools, shingle and slippery seaweeds. They occur only in one narrow band on the shoreline to the south of the village. It is a reminder that the growth of stromatolites is dependent on a fairly specific set of environmental conditions which favour the growth and mineralisation of microbial biofilms. As a general rule living stromatolites only flourish in places where they are not unduly disturbed by other organisms. One way to locate the Randerston stromatolite bed is to conduct a search along the pebble beach for pieces of rock which have become separated from the stromatolite bed and fashioned by wave action into stromatolitic cobble stones (Fig. 9.8).

Chalybeate springs

When I started collecting material for *Bacteria*, an aim of one of my early expeditions was to visit the Irons Well (SU229147) in the New Forest, take photographs of the habitat and obtain samples of iron bacteria for microscopic examination and photography (see Chapter 1, Figs. 1.4–1.6). It made sense to begin my acquaintance with iron-oxidising bacteria at the Irons Well, because it is an example of an iron spring that has managed to retain much of its natural character over the centuries and furthermore it is situated in the New Forest. Any naturalist will recognise the New Forest as a place that has natural history written all over it. My visit to the Irons Well was a minor adventure because the Irons Well is tucked away in a slightly obscure corner of the forest at the end of a road that leads through the village of Fritham and which eventually peters out, not far beyond Eyeworth Pond. Even as a seasoned connoisseur of the forest, it occurred to me while taking my first steps along that road, that this was a place that felt strangely *new*. The otherworldliness of the adventure continued under the shade of trees until the Irons Well came into sight. It is a place where different strands of microbiology collide. The Irons Well is an example of an iron spring, where water rich in dissolved ferrous iron comes to the surface. Iron bacteria mediate the oxidation of ferrous iron to yield vibrant rust-coloured pigments. Over an area of a few square metres bacteria

Fig. 9.8. Stromatolitic cobble stones near Kingbarns, Fife, UK.

dominate the scene through the formation of bright orange mats and streamers. The spot is marked by a rustic wooden fence which encloses the outflow. The stream is tinted bright orange by a gloopy suspension of the sheaths of the iron bacterium *Leptothrix ochracea*, the sheaths of which may be visualised easily in samples viewed under a microscope. In addition to iron, manganese is oxidised too, by *Leptothrix discophora*, which forms fragile refractive films on the water surface. As a bonus, the Irons Well is of interest to medical microbiologists and medical historians as a site purportedly once visited by lepers in search of a cure. At a later date the well was redeployed as a treatment station for mangy dogs (Shore, 1891).

Rust-coloured slimes and streamers, indicative of iron bacteria are to be seen in many wetland areas of Britain. To some extent we see only what we know, or what we care to know. A growing fondness for iron bacteria made me start to take notice of iron-rich seepages that I might otherwise have ignored. I began to wonder how the Irons Well might compare with other noteworthy springs. At Balcombe in Sussex there is an iron spring that shares a number of characteristics with the Irons Well. The spring (TQ311311) is set in woodland

Fig. 9.9. Balcombe iron spring, Sussex, UK.

and enclosed by a wooden fence (Fig. 9.9). It is situated a short walk to the north of the village. This unassuming natural feature is not even marked on the 1:50,000 scale Ordnance Survey map but one might imagine that so vivid a natural spectacle would have attracted a significant amount of human attention over the millennia. The village is served by its own railway station but despite such ease of access, the iron spring at Balcombe somehow evaded development into a full-scale tourist attraction during the heyday of mineral springs and the vogue for taking the waters during the eighteenth and nineteenth centuries.

Some iron springs have become local landmarks. One sees them marked on maps as chalybeate springs, from the Greek χάλυβας (chalybas) for steel. The Irons Well is a natural feature, which remains more or less natural, despite having achieved some cultural significance. Other culturally important chalybeate springs have been developed and enclosed within pump rooms. Spa towns have developed around some of them. In effect their resident iron bacteria have been domesticated and marginalised. When I started doing fieldwork for *Bacteria in the Natural World*, I avoided such places, assuming that there would be little that remained of bacteriological interest. However, my attitude gradually changed. I began to realise that mineral springs were not only important habitats for bacteria but also that the associated human interest in them as tourist attractions was worthy of some attention. Furthermore, I began to wonder what vestiges of nature might persist at heavily modified sites and whether or not any signs of iron bacteria might be apparent. There is a chalybeate spring at Buckhurst Park, near the village of Withyham in East Sussex, that resides under the shelter of its own Grade II listed wellhouse (TQ499352). I walked up the drive from the village, past a cricket ground and an exposure of sandstone to find the wellhouse in dappled shade. I was encouraged to see

Fig. 9.10. Buckhurst Park, iron well, Sussex, UK.

Fig. 9.11. Buckhurst Park wellhouse, Sussex, UK.

water welling up over the weedy ferruginous square of flagstones set within the spring's little enclosure (Figs. 9.10–9.11).

A few miles away, across the county boundary in Kent is the famous Chalybeate spring at Tunbridge Wells. The spring is situated within an historic pedestrianised part of the town known as the Pantiles. When I approached the site on a Sunday morning, people were outdoors, disporting themselves at cafe tables in the glare of summer sunshine but hardly anybody was taking any interest in the Chalybeate spring (TQ580387), which appeared to be closed for business at the time of my visit. The Bath House at Number 6, The Pantiles provides an early nineteenth century backdrop to the waters that emerge from twin outflows situated on either side of the entrance (Figs. 9.12–9.13). On the occasion of my visit they were clamped down by glass covers that looked like the lids from monstrous Kilner jars. On first sight the wells looked awfully dry. An orange tidemark of iron oxide indicated former water levels (Fig. 9.14). Then on further inspection I noticed an iridescent rainbow glimmer on the surface of a puddle to the side of one of the well lids. In all likelihood this signalled the presence of a surface film produced by *Leptothrix discophora*. I mused to myself that iron bacteria could continue to thrive in this most urbanised of habitats. Another chalybeate spring situated on the outskirts of Tunbridge Wells in Dunorlan Park (TQ600396) appeared to be in a healthier state from a bacteriological viewpoint. Although the park was heaving with visitors, nobody was showing any interest in iron bacteria. The chalybeate spring is set within

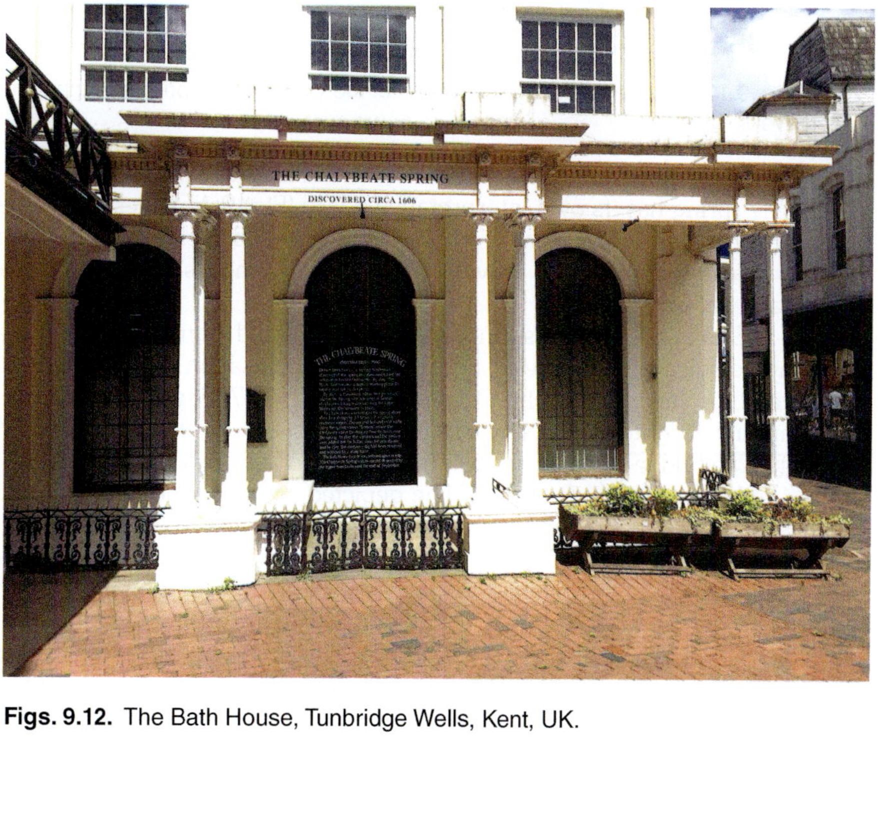

Figs. 9.12. The Bath House, Tunbridge Wells, Kent, UK.

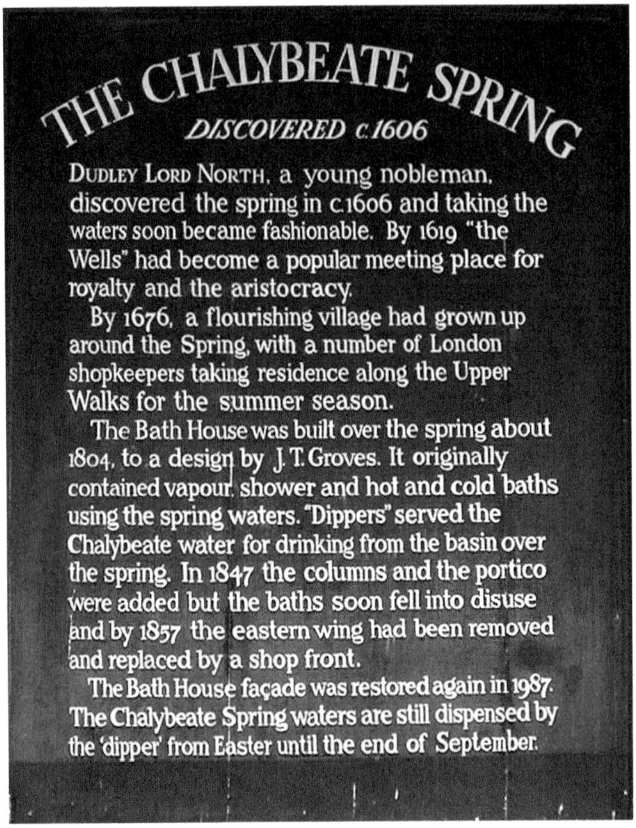

Fig. 9.13. Plaque recording the history of the Chalybeate spring at Tunbridge Wells, UK.

a bricked hollow that resembles a North African early Christian baptistry. The spring has its own interpretative board (Fig. 9.15). Despite the seasonal dryness, water was flowing up from the well and trickling into a runnel that contained sticks, dead leaves and brilliant orange slime indicative of iron bacteria.

Some chalybeate springs have disappeared completely as a result of hydrological disturbance brought about by urbanisation. In the early nineteenth century, the painter John Constable (1776–1837) depicted the Sussex coast around Brighton as a sparsely populated place. In the Fitzwilliam Museum in Cambridge there is a painting, labelled *Shoreham Bay, walk to the Chalybeate Wells Brighton*. The depicted scenery looks almost wild. When I visited St Ann's Well it was out of historic interest more than anything else as the actual well had reportedly dried up long ago. St Ann's Well is situated within the bounds of a park in Hove (TQ299049), less than a mile to the west of Brighton railway station. Although not far from the seafront and several busy roads, it is separated from the hubbub by blocks of residential buildings. During the eighteenth and nineteenth centuries people would come to take

Fig. 9.14. Iron well at Tunbridge Wells, UK.

the waters and a pump room was erected over the chalybeate spring. In 1805 the physician and author Alexander Gastard Marcet (1770–1822) published *A chemical account of the chalybeate spring, near Brighton*, noting that '*its surface is commonly found covered, sometimes entirely, sometimes only partially, with a very thin iridescent pellicle*' (Marcet, 1805). Marcet's description (Fig. 9.16) matches perfectly the appearance of a refractive surface film caused by the secretion of polysaccharides by the iron and manganese-oxidising bacterium *Leptothrix discophora*. In the early nineteenth century the site was clearly in good bacteriological health. According to a sign erected by Brighton and Hove Council, the pump house was demolished when the waters ran dry as a consequence of water extraction from artesian wells. The position of the former pump room is marked by an ersatz well-house set within a bricked seating area. However, among the shrubs not far from the seating area there were signs of life: the bottom of a dry drainage channel situated downstream from the former pump room contained a layer of desiccated ferruginous sediment and the adjacent vegetation was stained

Figs. 9.15. Iron well at Dunorlan Park, Kent, UK.

with ochre, showing that during wet weather the channel fills with water (Fig. 9.17). This indicates that iron bacteria may yet be hanging on by the tips of their fibrils in St Ann's Well Gardens.

Another historic site, the Sandrock Chalybeate Spring, near Niton on the Isle of Wight, disappeared completely in a landslip. The access road to the spring disappeared too. The road now terminates as a parking area from where a footpath leads down into the tangled vegetation of the slumping undercliff towards Rocken End. I have explored the slopes above Rocken End on a number of occasions. Many of the classic undercliff species are found here, including the Glanville Fritillary butterfly, *Melitaea cinxia* and a scarce tiger beetle, *Cicindela germanica*. Buried on those slopes somewhere are the remnants of a grotto that was built around the former spring. Visitors to the Sandrock Chalybeate Spring could find accommodation nearby at the Royal Sandrock Hotel. Queen Victoria was a guest there. The Royal Sandrock Hotel burnt down in 1984. Coastal erosion and frequent landslips ensure that the Isle of Wight undercliff is continuously reinventing itself. Despite human efforts to colonise the undercliff, the unstable local geology dictates an inevitable sequence of land slippage, followed by automatic rewilding and restoration to a more natural habitat.

By choosing one's moment with reference to the tide tables, it used to be possible to walk from Niton down to the shoreline at Rocken End and progress westwards along the beach towards Whale Chine where a set of

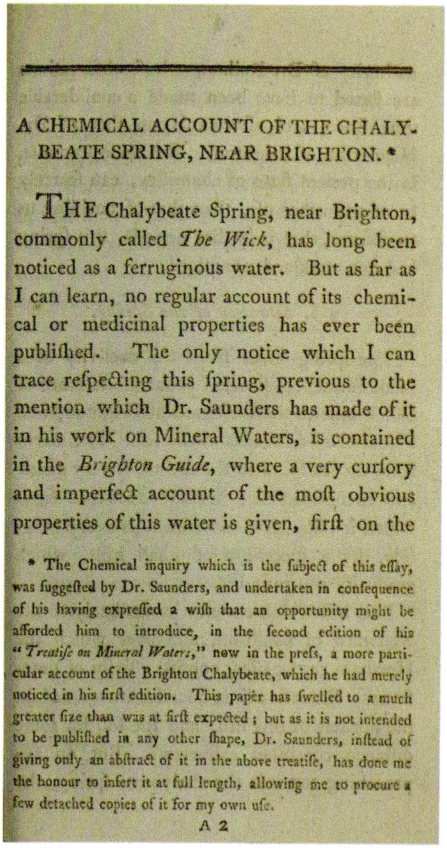

A CHEMICAL ACCOUNT OF THE CHALYBEATE SPRING, NEAR BRIGHTON. *

THE Chalybeate Spring, near Brighton, commonly called *The Wick*, has long been noticed as a ferruginous water. But as far as I can learn, no regular account of its chemical or medicinal properties has ever been publiſhed. The only notice which I can trace reſpecting this ſpring, previous to the mention which Dr. Saunders has made of it in his work on Mineral Waters, is contained in the *Brighton Guide*, where a very curſory and imperfect account of the moſt obvious properties of this water is given, firſt on the

* The Chemical inquiry which is the ſubject of this eſſay, was ſuggeſted by Dr. Saunders, and undertaken in conſequence of his having expreſſed a wiſh that an opportunity might be afforded him to introduce, in the ſecond edition of his "*Treatiſe on Mineral Waters*," now in the preſs, a more particular account of the Brighton Chalybeate, which he had merely noticed in his firſt edition. This paper has ſwelled to a much greater ſize than was at firſt expected; but as it is not intended to be publiſhed in any other ſhape, Dr. Saunders, inſtead of giving only an abſtract of it in the above treatiſe, has done me the honour to inſert it at full length, allowing me to procure a few detached copies of it for my own uſe.

A 2

Fig. 9.16. *A chemical account of the Chalybeate Spring, near Brighton,* by Alexander Marcet (1805). Facsimile provided by the Royal College of Surgeons where the originals may be consulted. Public Domain Mark.

steps once allowed access back up towards the coast road. In the 2013/2014 winter the steps were washed away and at the time of writing there appears to be no direct access to Whale Chine. The rocks at Whale Chine are rich in iron pyrite. As the cliffs erode, exposure to the air allows iron-oxidising and sulphide-oxidising bacteria to oxidise deposits of pyrite. Sulphuric acid is released as a waste product, resulting in the development of acidic conditions, which encourage the localised growth of heather. A partly sandy, partly boggy ledge to the east of Whale Chine hosts an interesting assortment of insect species, including the beautiful powder-blue dragonfly *Orthetrum coerulescens*, at its only Isle of Wight locality; although it is plentiful not many miles away in the bogs of the New Forest. Proximity to the English Channel and protection from the north wind by high cliffs promotes a long season for insect life. Common Field Grasshoppers, *Chorthippus brunneus*, for example, remain active into the autumn months long after they have died off at localities only a short distance away further inland. In 2001 I found a number of Common Field Grasshoppers, as well as a solitary Cepero's Groundhopper, *Tetrix*

Fig. 9.17. Residual traces of iron bacteria. St Ann's Well, Hove, Sussex, UK.

ceperoi, on the amazingly late date of 22 December. Iron seepages are dotted about here and there. They can be seen in Whale Chine itself and eastward at regular intervals along the coast below Blackgang Chine and towards the site of the former Sandrock Chalybeate Spring (Figs. 3.2–3.3, 9.18–9.19). When walking along this coastline, the constant changes to the landscape give the sense that this is a place where nature cannot be tamed.

Sulphur spas

Many of Britain's sulphur springs became tourist attractions because of the supposed therapeutic value of their sulphurous waters. Entire settlements have grown up around some of the more famous examples, such as at Harrogate and Llandrindod Wells. The sulphurous springs at Bath have been a tourist attraction at least since Roman times. The waters emerge from deep in the earth at a temperature of 46°C and provide a rare glimpse of the deep biosphere. I was taken to the Roman Baths as a small child and encouraged to

Fig. 9.18. Sandrock Chalybeate spring, near Niton, Isle of Wight, UK. Line engraving by G. Brannon, 1828, after himself (Wellcome Collection). Public Domain Mark.

Fig. 9.19. General view of coast east of Whale Chine showing iron seeps. Isle of Wight, UK.

drink the spring's waters. I remember that the flavour was unusual and distinctive. I am glad that I sampled the water when I did because it means that at an early age I had made contact with a rare thermophilic bacterium, *Annwoodia aquaesulis*. At the time of writing nobody has detected an exact match for

Fig. 9.20. Boston sulphur well, Anglesey, UK.

Annwoodia aquaesulis anywhere else in the world, although related organisms appear to be globally distributed in deep subsurface environments.

In search of a sulphur spring situated in a more natural setting I travelled to the Boston Sulphur Well on Anglesey (SH485860). The Boston Well is situated beside a footpath that runs across open heathland. The well itself is enclosed by a handsome Grade II listed stone wellhouse (Fig. 9.20). The weathered stones do not look out of place. Set into the stonework there is a plaque bearing the inscription: 'Sulphur Well, "Boston" 1864.' Although the Boston Well may once really have been an actual sulphur spring, disappointingly at the time of my visit, the scent of hydrogen sulphide did not greet my nostrils and I failed to find any evidence of purple sulphur or white sulphur bacteria. Instead, a carpet of ferruginous slime on the floor of the wellhouse made it immediately apparent that nowadays the Boston Well is in fact essentially an iron spring. Perhaps there has been some change in the local hydrology since 1864? In Welsh, the well is currently known as Ffynnon Goch (red spring). I concluded that the Boston Well does not currently serve as a suitable sanctuary for sulphur bacteria. The wellhouse and the surrounding heathland do, however, provide a pleasant location for the observation of iron-oxidising bacteria.

Therefore, I turned my attention to the spa town of Moffat in Dumfriesshire. The Moffat Mineral Well can be reached by following a lane that runs north from the town. The road is narrow and appears to be heading nowhere in particular as it climbs gently away from the commotion of the town centre. The well is situated in a rural location. There is a parking place and a gate. A path runs down a grassy slope to the wellhouse (Fig. 9.21), the interior of which reeks with the stench of hydrogen sulphide. A plaque attached to the wall notes that the well was discovered in 1633 by Rachael Whyteford and that the current building was repaired in 1987. Nearby, an interpretative board states that the well remained popular with visitors until the late nineteenth century, after which numbers began to decline. Most excitingly the well itself persists as a living habitat for white sulphur bacteria. Mats and bundles of their white filaments blanket the well basin (Fig. 9.22). The carpet of white bacteria, offset by the occasional patch of blue-green cyanobacteria, enmeshes stray elements of the local flora – dead leaves, twigs, needles and the odd cone. A coating of ferruginous sludge suggests that iron-oxidising bacteria reside here too. It is a delight to lean over the railings within this spartan building and observe the natural collage of bacteria, mosses and shade-tolerant higher plants which are sustained by the sulphurous spring water.

Cyanobacteria at home and abroad

Cyanobacteria are perhaps the only bacteria that really provide suitable material for serious study by amateur naturalists. They occur in terrestrial, freshwater and

Fig. 9.21. Sulphur spa wellhouse, Moffat, UK.

Fig. 9.22. White sulphur bacteria, Moffat, UK.

marine habitats. Colonies of cyanobacteria are generally readily visible to the naked eye and cyanobacteria have relatively large cells which can seen easily under a microscope. In many ways cyanobacteria bear a superficial resemblance to eukaryotic algae. British species of algae as well as cyanobacteria are covered by *The Freshwater Algal Flora of the British Isles* (John *et al.*, 2002), which also usefully includes marine species of cyanobacteria. The *Flora* includes locality data for some of the more local species and lists water bodies that are of particular interest. Careful microscopic examination of samples will be necessary in order to identify most species but some species of cyanobacteria can be recognised in the field. One of them is *Rivularia atra*, a locally distributed marine species that can be found in rock pools in the upper intertidal zone around the British Isles from Cornwall to the Hebrides. *Rivularia atra* regularly occurs during late summer in large numbers at Tyne Sands on the east coast of Scotland (John *et al.*, 2002). Tyne Sands is situated within the bounds of the John Muir Country Park, not far from Dunbar. There is a parking area at the end of a lane called Limetree Walk (NT606808), from where a track leads along the edge of woodland to the coast (NT635806). The coastline is fringed by sand dunes which have become vegetated by dense clumps of sea buckthorn, a plant that accommodates the nitrogen-fixing bacterium *Frankia* in root nodules. Colonies of *Rivularia atra* are small (c. 2 mm diameter), blackish-green in colour and roughly hemispherical in shape. Large numbers of them speckle the bottom of shallow rock pools (Fig. 9.23). They look like peppercorns (Fig. 9.24). Other *Rivularia* species, such as *Rivularia biasolettiana* and *Rivularia haematites* can be seen in freshwater habitats elsewhere in the UK (see Chapter 2).

Fig. 9.23. Rock pools colonised by *Rivularia atra*, Tyne Sands, East Lothian, UK.

Cyanobacteria can tolerate solar radiation and desiccation and are among the pioneering organisms that colonise rock faces and walls. Black vertical streaks on rock faces, known in German as *Tintenstriche* (ink streaks) indicate the presence of dense growths of cyanobacteria. These streaks mark the courses of streams of water that flow down the rock face at irregular intervals. When you have learnt to appreciate ink streaks for what they are, you start to notice them on all kinds of exposed surfaces. They are an example of a biological patina. Similar black coatings are formed by lichens and microscopic fungi (moulds and black yeasts) and in order to determine the biological agent responsible for ink streaks it may be necessary to examine scrapings under a microscope (Fig. 9.25). Streaks caused predominantly by cyanobacteria sometimes look distinctly blue-green in colour. Eukaryotic algae are prone to form films or streaks on rock surfaces that look bright green or yellow-green in colour, rather than blackish or blue-green. Ink streaks composed of cyanobacteria are most conspicuous when seen against a background of pale-coloured limestone, although pleasing compositions of streaks are sometimes encountered

Fig. 9.24. Colonies of *Rivularia atra*, Tyne Sands, East Lothian, UK.

on other materials. Figure 9.26 shows the ink-streaked rock face of a granite quarry on Lundy. The angular arrangements of hewn stone and zigzag streak patterns put me in mind of dazzle-camouflaged war ships. Ink streaks are possibly at their most impressive when seen in glaring Mediterranean sunlight against a background of white limestone. The Peñón de Ifach (Penyal d'Ifac), is a national park and tourist attraction, situated on the outskirts of Calpe between Alicante and Valencia. The Peñón is an isolated block of tertiary limestone. A tunnel and path provide access to the summit. On the way up the trail passes within easy reach of the streaks. The best way to appreciate them, however, is from a distance (Fig. 9.27).

Memorials to Robert Hooke (1635–1703)

In Oxford High Street there is a plaque set into a wall that marks the former residence of Robert Boyle and which celebrates Robert Hooke's construction of a microscope and identification of the first living cell (Fig. 9.28). When I read

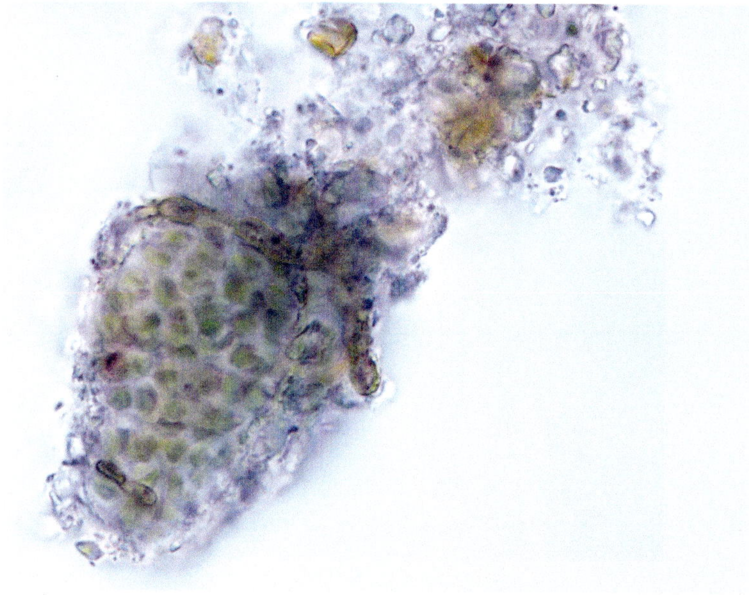

Fig. 9.25. Cyanobacterial cells and calcite crystals, Peñón de Ifach, Spain.

Fig. 9.26. Cyanobacterial ink streaks on granite, Lundy, UK.

Fig. 9.27. Cyanobacterial ink streaks, Peñón de Ifach, Spain.

Fig. 9.28. Robert Hooke memorial plaque, Oxford High Street, UK.

the inscription it had a curious effect on me. Up to that point I had not thought of Robert Hooke as being a real person. Hooke's discovery of living cells was simply a disembodied fact that I had learnt in a biology lesson. But in some way by linking this fact to an actual place, it transformed what had previously been an abstract idea into something tangible. I began to imagine what Robert Hooke might have been like when he was still alive. I started to take more of an interest in Hooke as a scientist and I discovered that Robert Hooke had been the first person in England to record microscopic observations of bacteria.

Microbiology began as a science in the seventeenth century with the fabrication of microscopes, the publication of Robert Hooke's *Micrographia* (1665) and the discovery of bacteria by Antoni van Leeuwenhoek (also Anton, Antony, Leeuwenhoeck). Robert Hooke (1635–1703) was born in Freshwater on the Isle of Wight. Hooke was apprenticed to a portrait painter, worked in Oxford alongside the chemist Robert Boyle (1627–1691) and became Curator of Experiments at the Royal Society in London. His *Micrographia* included the first depictions of a microorganism (a fungal mould) and of cells (in a sheet of cork). Antoni van Leeuwenhoek (1632–1723) was a native of Delft, the Netherlands. He learnt to grind lenses and build his own simple microscopes that consisted of a metal plate with a hole, a lens and an adjustable pin on which to mount a specimen. Although Leeuwenhoek's microscopes had only a single lens, he could obtain a magnification of several hundred times; enough to see bacteria. In 1683 Leeuwenhoek reported to the Royal Society in London that he had observed microscopic animalcules in plaque from his teeth. The organisms to which he referred as animalcules were clearly bacteria. There is an earlier account though, dating from 1676, in which Leeuwenhoek describes what may actually have been his first observations of bacteria. Having become curious about the piquant taste of pepper, Leeuwenhoek left some peppercorns in water, wondering if soaking might expose microscopic features that could account for the spiciness of the peppers (Gest, 2004). On examination of the pepper water weeks later he saw various organisms including some that were really minute. In 1677 Robert Hooke, '*being extremely desirous to examine the matter farther ...*' repeated Leeuwenhoek's experiment by soaking some peppercorns in river water for several days and observed organisms '*so exceedingly small, that millions of millions might be contained in one drop of water*'.

As well as being the first scientist in England to record observations of bacteria, Robert Hooke achieved many other things. In collaboration with Sir Christopher Wren, Robert Hooke designed the London Monument to the Great Fire of London, a 222-foot high column of Portland stone (TQ329808). It was originally envisaged that the monument would house a telescope and function as an observatory. A sign attached to the monument notes Hooke's involvement in its design. There is also a memorial inscription, set into the pavement. Other inscriptions dedicated to Hooke can be seen in Westminster Abbey and in St Paul's Cathedral and there is a memorial plaque attached to a wall near St Helen's Church at Bishopsgate in the City of London (Fig. 9.29). When he died, Hooke was first interred at St Helen's. His remains were later exhumed and translocated to a different unrecorded site elsewhere in London. Hooke's

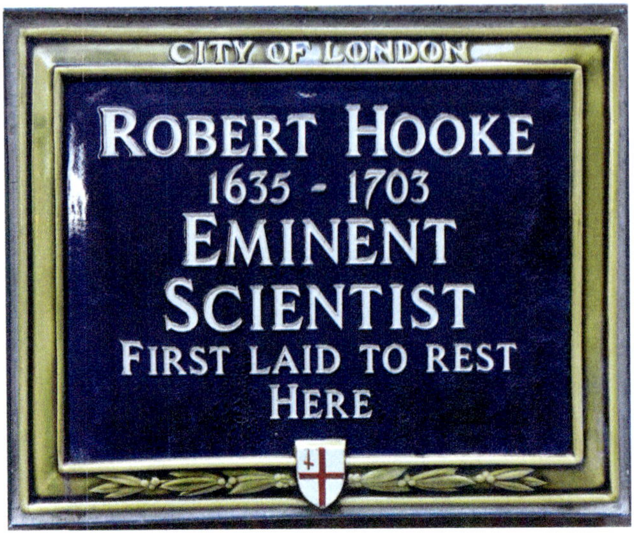

Fig. 9.29. Robert Hooke memorial plaque, St Helen's Church, London, UK.

penultimate resisting place, St Helen's Church, is within walking distance of the London Monument.

Hooke's most worthy epitaph is his exquisitely illustrated *Micrographia*. Once, while visiting an exhibition at the Tate Britain I noticed an original copy of *Micrographia* on display in a glass cabinet. The pages had been opened to reveal Hooke's engraving of a flea, an insect which happens to be a potential vector of bacterial infections. Other visitors were steadily ambling past, although nobody was taking much notice of it. It made me realise that to appreciate such things you need to become a bacteria tourist. On another occasion I peered into a display case which formed part of an exhibition at the Royal College of Physicians in London and saw a replica of Hooke's microscope. Hooke's microscope is an elegant physical object that symbolises the beginnings of microbiology.

A shopping trip to the Co-op supermarket in Freshwater on the Isle of Wight led to the chance discovery of yet another memorial to Robert Hooke. It is a roughly hewn stone cuboid (Fig. 9.30). It carries a message to the effect that the scientist and architect Robert Hooke was born nearby.

Buildings made of stone made by bacteria

Some of the most widely used building stones owe their existence, at least in part to bacteria. Dolomitic limestone, for example, cannot be readily precipitated under abiotic conditions. It is generally assumed that most if not all of the significant deposits of dolomitic limestone that exist accumulated under conditions in which microbes mediated the precipitation of the dolomite. A building made of dolomitic limestone is a statement about human ingenuity and culture.

Fig. 9.30. Robert Hooke memorial stone, Freshwater, Isle of Wight, UK.

It is also a document that offers information about ancient environmental conditions, geological time and bacteria. Dolomitic limestone is also known as magnesian limestone, or dolostone. Dolomitic limestone is sometimes also referred to simply as dolomite, because it contains the mineral dolomite, or calcium magnesium carbonate. The dolomitic limestones of northern England were deposited in a shallow salty sea during the Permian Period, more than 250 million years ago. Stagnant sediments, populated by sulphate-reducing bacteria, methanogenic archaea and other microbes provided environmental conditions that promoted the precipitation of dolomite. Exposures of dolomitic limestone can be seen today on the coast between Sunderland and Hartlepool. In some places, the limestone is associated with reefs formed by the growth of stromatolites. A narrow band of dolomitic limestone runs southwards across County Durham and through Yorkshire. Dolomitic limestone quarried from various locations was used in the construction of numerous notable buildings in northern England. An encounter with a building composed of microbially mediated materials such as dolomitic limestone presents the bacteria tourist with an opportunity to look beyond the human significance of the building and to contemplate the natural history of the building material itself. During the course of an expedition to County Durham it is possible to see stromatolites at Blackhall Rocks (Fig 9.31), visit old limestone quarries and appraise buildings that have been constructed from the quarried limestone. At Marsden Bay (NZ399648) a visiting naturalist can enjoy a dynamic coastal landscape consisting of herb-rich limestone grassland and cliffs and sea caves carved by coastal erosion in dolomitic rocks. The banks of Marsden Quarry are visible from the coast road. The landscape is dotted with buildings made from the local dolomitic limestone. Not far away in Sunderland is St Andrews Church, a grandiose building

Fig. 9.31. Coastal dolomitic limestone scenery, Blackhall Rocks, Durham, UK.

constructed from dolomitic limestone quarried at Marsden. Further south in Yorkshire, dolomitic limestone was used in the construction of notable buildings in the City of York. One of them is York Minster (SE603521).

In southern England, a visit to Kimmeridge Bay in Dorset allows the bacteria tourist to see a working beam pump and contemplate the world of petroleum microbiology and also to observe exposures of dolomitic limestone (Fig. 9.32). An outcrop of Jurassic dolomitic limestone known as the Yellow Ledge dolomitic limestone was used in the construction of various local buildings, including the sea wall at Kimmeridge Bay and the Clavell Tower, a nineteenth century folly and landmark that overlooks the bay (SY909786). At low tide a ledge of Yellow Ledge dolomitic limestone can be seen in Kimmeridge Bay. Just around the corner is a beam pump used for oil extraction (SY904792). Petroleum microbiology must be one of the least accessible facets of natural history. It concerns itself with organisms that are invisible to the naked eye that reside deep below the surface of the earth. The Kimmeridge beam pump provides a tangible link to that seemingly hostile subterranean world. According to some estimates, most of the world's bacteria are subterranean. They populate the rocks, aquifers and sediments beneath our feet.

Oolite

Oolitic limestone is composed of minute ooids consisting of concentric rings of calcium carbonate, which were precipitated under the influence of bacteria in warm shallow seas. Beaches of oolitic sand can be seen today in such places as the Persian Gulf, the Gulf of Mexico and around the Great Salt Lake in Utah

Fig. 9.32. Kimmeridge Bay, dolomite ledge with Clavell Tower in the background. Dorset, UK.

(Fig. 9.33). Oolitic limestone is favoured as a building material by stone masons because its ooid composition means that it can be cut freely in any direction. A belt of Jurassic oolitic limestone stretches across England from Dorset to Yorkshire. In some areas whole villages assume the tones of Jurassic oolite. Jurassic oolitic limestone is often richly coloured. The well-known Portland stone, from the Isle of Portland in Dorset is, however, unusually pale in colour. The glare from bright sunlight on Portland stone is almost painful. One can imagine dazzling silvery beaches of oolite sand that once must have existed during the late Jurassic Period, about 145 million years ago. One particular variety of oolite, Bath stone accounts for the singular aspect of the City of Bath. Most of the city's historic buildings are made from Bath stone which was obtained from local quarries. The Roman Baths at Bath are of special interest to bacteriologists as the home of thermophilic bacteria, including, *Thermithiobacillus tepidarius* (the Tepid Bath Bacterium), and *Annwoodia aquaesulis*, (Ann Wood's Bacterium). Visitors can sample the distinctively flavoured waters in the Grand Pump Room, a late eighteenth century building constructed from Bath stone. Bath stone was exported to other parts of Britain. Some examples of buildings that feature Bath stone include Bristol Cathedral and the Royal Berkshire Hospital in Reading. The Brighton Pavilion is dressed both with Bath stone and Portland stone. Figure 9.34 shows an arrangement of oolite blocks that form one of the outside walls of the Oxford University Museum of Natural History. Many naturalists will be familiar with this building on account of its important collections and because the museum acts as a venue for the meetings of various national natural history societies and organisations. It is also a museum where visitors can see a decent selection of fossilised stromatolites. The museum's interior design features columns made from different geological materials. Numerous other buildings in central Oxford were constructed using

Fig. 9.33. Beach composed of oolitic sand, Great Salt Lake, Utah, USA.

Fig. 9.34. Oxford University Museum of Natural History, UK, detail of oolite stonework.

oolite (Arkell, 1947; Gomez-Heras, 2010). Much of it was sourced from quarries in Oxfordshire and much of it has been exposed to the elements for long enough to show signs of weathering.

Portland stone was used extensively in the rebuilding of London after the Great Fire in 1666. Noteworthy buildings in London made of Portland stone include St Paul's Cathedral, the British Museum, Somerset House and the National Gallery. Other significant buildings made using Portland stone include the Port of Liverpool Building in Liverpool and the United Nations Headquarters in New York. The London Monument to the Great Fire of London (TQ329808) symbolises the adoption of Portland stone as the quintessential metropolitan building stone. The London Monument was built in the 1670s to a design by Robert Hooke, working in collaboration with Sir Christopher Wren. A more modern building, the London School of Hygiene and Tropical Medicine (TQ298818) dates from the twentieth century. The London School was founded as an institution in 1889. In 1929 the School occupied its present home at the junction of Keppel Street and Gower Street. The London School is an institution that concerns itself with medical microbiology and the study of infectious diseases. The building consists of a steel frame, faced with Portland stone (Fig. 9.35). Gilded bronze sculptures of the vectors of infectious diseases,

Fig. 9.35. Detail of Portland stone stonework, an oolitic limestone. London School of Hygiene and Tropical Medicine, UK.

including a number of bacterial infections, decorate the outside of the building. These vectors include a rat, *Rattus* sp., a flea (Order: Siphonaptera), a tick (Family: Ixodidae), a house-fly, *Musca* sp., a tsetse fly, *Glossina* sp., a body louse, *Pediculus humanus* and two different mosquitoes, *Anopheles* sp. and *Aedes* sp. There is also a bed bug, *Cimex* sp. Although bed bugs are not considered to act as the vectors of disease, bed bugs are of bacteriological interest because they act as the hosts of symbiotic bacteria. The London Monument and the London School of Hygiene and Tropical Medicine are two edifices that span a centuries-long tradition of using Portland stone as a building material in the capital.

The extremophiles of Richard II's palace

William Shakespeare's *Richard II* was first performed at around the same time that the first microscopes were being developed. Shakespeare did not live to see the first report of the discovery of microorganisms and he would have waited another 400 years until the application of molecular techniques revealed the presence of a complex microbial ecosystem within the decaying stonework of Richard II's Palace at Portchester Castle in Hampshire (Zanardini *et al.*, 2016; Zanardini *et al.*, 2019). Portchester Castle is a tourist attraction, managed by English Heritage. I booked a ticket and travelled to Portchester to pay a visit to the castle as a bacteria tourist. Richard II's palace was constructed in the 1390s within one corner of the extensive grounds of a former Roman fort that later became a Norman castle. The castle is situated at the head of Portsmouth Harbour in a relatively sheltered spot, which is, nonetheless, exposed to sea breezes and salt spray. The palace was built using Purbeck limestone and locally sourced green sandstone (from exactly where is unclear), which shows signs of discolouration and weathering. The ruined walls are situated partly in full sunlight and partly in shade. Some walls are sheltered, while others are exposed to the elements. The varying conditions create a mosaic of stony microhabitats. The green sandstone is both porous and nutritious. It contains iron and potassium and provides a suitable habitat for rock-dwelling microbes.

Within the broader context of environmental microbiology, there is probably nothing particularly special about the stonework at Portchester Castle. Its weathered surfaces resemble those of old buildings and natural rock exposures elsewhere in southern Britain. It happens though that the stonework at Portchester Castle has been subjected to detailed microbiological investigations. The results of these investigations reveal an amazing richness of microbial life. Research conducted by British and Italian microbiologists and specialists in culturally important stonework has shown that Richard II's palace provides a home for more than 100 different genera of bacteria, as well as a handful of archaeal genera (Zanardini *et al.*, 2016; Zanardini *et al.*, 2019). The palace's microbial inhabitants include genera associated with extreme environmental conditions, such as microbes that tolerate high salt concentrations and exposure to ultraviolet radiation, extremes of pH and high temperatures. Rod-shaped and filamentous bacteria have been seen on scanning electron microscopy of samples taken from the window of Richard II's banqueting room. Most of the bacteria that could be cultured from samples of degraded stonework produced yellow or red pigments, which suggests that microbial pigments are at least partly responsible for the discolouration of the decaying stone surfaces. The ammonia-oxidising archaeon *Nitrosphaera* and halophilic archaea belonging to the genera *Halococcus*, *Halobacterium* and *Halomarina* were also detected. The masonry is a microbial ecosystem. The decayed stonework contains carbon-fixing bacteria such as the cyanobacterium *Phormidium*. They are the primary producers. By fixing carbon they act as a source of organic carbon compounds that sustain other microbes. The stonework also provides a niche for iron-oxidising bacteria *Acidimicrobium* and *Ferrimicrobium*. As they go about their business, endolithic iron-oxidising bacteria steadily degrade the stonework. They have probably been doing this since the fourteenth century. By conducting a functional analysis of the genes associated with the various microbes detected at Portchester Castle, the scientists who investigated its microbial ecology were able to map out the biochemical pathways by which the rock-dwelling microbes participate in the cycling of nitrogen and sulphur. Nitrogen-fixing species such as cyanobacteria fix atmospheric nitrogen and act as a source of organic nitrogen compounds that can be utilised by other microbes. The ammonia-oxidising microbes *Nitrospira* (a bacterium) and *Nitrosphaera* (an archaeon) convert the nitrogenous waste product ammonia to nitrite. In turn, nitrite-oxidising bacteria and denitrifying bacteria convert nitrate back to nitrogen. Overall, the evidence suggests that the stonework of Richard II's Palace provides all that is necessary to sustain a complex community of bacteria and archaea. Sunlight provides a source of energy, the atmosphere provides carbon and nitrogen and it is likely that sea spray supplies sulphate. Over the course of time though these microorganisms are steadily wrecking their own home. Their metabolic products include organic acids as well as nitric and sulphuric acid, which dissolve the stonework. Meanwhile some of the other resident bacteria occupy their time by actively boring into the rocks.

At Portchester Castle, historic architecture and bacteria complement one another. On arriving at Portchester I passed through the walls of the former Roman fort via the landgate into the outer bailey. I crossed the former moat, passed through the gatehouse and presented my ticket to the attendant. Most of the visitors seemed to be heading directly to the great tower but I had other plans. To my left was Richard II's great hall and I found that I had the place to myself (Fig. 9.36). The unexpected moment of solitude allowed me to contemplate the passage of time since the building's construction. As a new building it must have looked splendid. It made me think of the Wilton diptych, an artwork that was painted around the same time that the palace was built. The inner panels of the diptych show Richard II kneeling before the Virgin and Child and eleven angels. The outer panels depict Richard II's symbol, the white hart. On entering the palace I located the sampling sites, which had been subjected to such detailed microbiological study. Two of the sites were exfoliated sandstone surfaces (Fig. 2.89). The gnarled stone surfaces demonstrate what microorganisms can do to solid rock if left to their own devices for 600 years. The

Fig. 9.36. Doorway at Richard II's Palace, Portchester Castle, UK.

third sampling site was different. It was in a more sheltered spot and densely coated by a black encrustation (Fig. 2.90). The two exfoliated surfaces and the encrusted surface actually represent two different kinds of microhabitat that contain strikingly different microbial floras. Whilst *Rubrobacter* bacteria and archaea identified as *Halomarina* are among the predominant microbes at the exfoliated sites, archaea related to *Halobacterium* and bacteria related to *Thermomicrobium* predominate at the encrusted site. The genus *Rubrobacter* is a member of the Phylum *Actinomycetota*. *Rubrobacter* species are tolerant of exposure to ionising radiation. The genus *Thermomicrobium* is a member of the Phylum *Thermomicrobiota* and was originally described on the basis of a strain isolated from a hot spring in Yellowstone National Park. *Thermomicrobium* bacteria are thermophiles that can tolerate high temperatures. *Halomarina* and *Halobacterium* are halophilic archaea that are more usually associated with hypersaline environments. It may not be apparent to the casual observer but Richard II's Palace has become a haven for groups of microbes whose best-known representatives are usually regarded as extremophiles. The realisation that the crumbling stonework contains such marvellous microorganisms makes me hope that English Heritage will come to value their microbes and the microbial encrustations and microbial pigments as much as the palace itself. At Portchester Castle, historic architecture and bacteria complement one another in a singularly harmonious fashion.

In search of leprosy

The red squirrel, *Sciurus vulgaris*, is widely distributed in Europe and Asia but in Britain red squirrels have been in a state of decline since the nineteenth century as a consequence of the introduction and spread of the North American grey squirrel, *Sciurus carolinensis*. Most of Britain's red squirrels now occur in forested areas of Scotland and northern England. In some parts of Scotland it is not unusual to see red squirrels on bird feeders in gardens and in nature reserves, such as the RSPB reserve at Loch Garten. In southern England, isolated populations of red squirrels persist on Brownsea Island and the Isle of Wight. Both islands remain free of grey squirrels. A glimpse of a red squirrel during the course of an expedition to either of these islands is an experience that would have seemed commonplace elsewhere in southern England only a few centuries ago. An encounter with a red squirrel is of bacteriological interest because British red squirrels are infected with leprosy. Leprosy has been recognised since antiquity as a disease of humans, *Homo sapiens*. Two related species of bacteria, *Mycobacterium leprae* and *Mycobacterium lepromatosis*, cause leprosy in humans. Only recently has leprosy been identified as a disease of squirrels.

The decline of the red squirrel in Britain has been linked to competition from the invasive grey squirrel and also to susceptibility to Squirrelpox virus (SQPV), which is carried by grey squirrels. Accordingly, red squirrel disease surveillance is conducted by a team of veterinarians and scientists based at the University of Edinburgh. In 2006 a squirrel carcass was submitted for examination showing features that were not typical of infection with Squirrelpox virus. The animal had patchy hair loss and swollen shiny lesions of the extremities. Similar cases were subsequently reported from across Scotland. The microscopic examination of stained tissue sections revealed the presence of so-called acid-fast bacteria, or mycobacteria, although cultures using routine culture media for pathogenic mycobacteria were negative. However, sequencing of an amplified fragment of DNA showed a very close match for *Mycobacterium lepromatosis*, a bacterium that causes leprosy in humans (Avanzi *et al.*, 2016).

In the 1970s it had been shown that nine-banded armadillos, *Dasypus novemcinctus*, could be experimentally infected with leprosy and it became apparent that natural infections occurred amongst wild armadillos in both the USA and in South America. Infections have occasionally been reported also in non-human primates. Apart from armadillos and humans, red squirrels are the only mammals that have been demonstrated to serve as a host species that can sustain leprosy transmission. Human cases of leprosy are seldom seen in Britain today (Fulton *et al.*, 2016). The rare cases that continue to be diagnosed in Britain can be linked to prior exposure in other parts of the world where leprosy remains an endemic disease. Most cases of leprosy in humans are caused by *Mycobacterium leprae*. *Mycobacterium lepromatosis* causes a different form of disease known as diffuse lepromatous leprosy. Most cases have been reported from Mexico, where diffuse lepromatous leprosy was recognised as a separate clinical entity in the nineteenth century. *Mycobacterium lepromatosis* was recognised as a separate species when bacterial DNA sequences were found to differ markedly from those of *Mycobacterium leprae* (Han *et al.*, 2008). Red squirrels on Brownsea Island have been shown to be infected with *Mycobacterium leprae* whilst squirrels in Scotland, Ireland and the Isle of Wight were found to be infected with *Mycobacterium lepromatosis*. Squirrels infected with either species of leprosy bacillus do not necessarily display signs of disease. *Mycobacterium leprae* was detected in all of 25 squirrels tested from Brownsea Island but only eight of them showed visible signs of disease (Avanzi *et al.*, 2016). I was taken to Brownsea Island as a child and can remember my disappointment in failing to spot a red squirrel during that first visit. Decades later I made a return visit to the island and without making any special effort encountered several red squirrels but I felt strangely disappointed not to spot any squirrels showing clear physical signs of infection with the leprosy bacillus.

Whole genome sequencing of *Mycobacterium lepromatosis* DNA from British squirrels shows that bacterial strains from different parts of the British Isles, including Ireland, are very closely related. The red squirrel was on the brink of extinction in Ireland as early as the seventeenth century. A number of factors including hunting and deforestation might have led to its decline. Red squirrels were reintroduced into Ireland from Britain during the nineteenth century. It seems likely that some of these animals would have been infected with the 'British strain' of *Mycobacterium lepromatosis.* The strain of *Mycobacterium lepromatosis* found in British squirrels resembles the strain of *Mycobacterium lepromatosis* found in Mexican leprosy patients but there are significant genetic differences. Comparison of the genetic differences between the two strains and knowledge of the mutation rate in the related species *Mycobacterium leprae* make it possible to estimate that the British squirrel strain and the Mexican human strain of *Mycobacterium lepromatosis* began to diverge from an ancestral bacterial strain approximately 27,000 years ago. The date for the arrival of leprosy in the Americas is uncertain. It seems likely that *Mycobacterium leprae*, including the strain that infects armadillos, was introduced by European colonists (Schuenemann *et al.*, 2018). The history of *Mycobacterium lepromatosis* in the Americas though is less clear and there is a school of thought that it might have been introduced as a result of the migration of Pre-Columbian people from Asia.

One of the earliest human skeletons unearthed in the British Isles that shows signs of leprosy is that of a young man who was interred in an Anglo-Saxon cemetery at Great Chesterton in Essex sometime around 500 AD. Isotope analysis of the bones suggests that he had originated from continental Europe (Inskip *et al.*, 2015). Genotyping of ancient DNA extracted from the diseased bones confirms that he was infected with *Mycobacterium leprae*, and that the infecting strain was the same as the one found in red squirrels on Brownsea Island. During the middle ages, leprosy became prevalent in Britain and more than 300 leprosy hospitals were established in England alone (Roffey *et al.*, 2012). Analysis of ancient DNA extracted from the bones of four individuals from the cemetery of a leprosy hospital near Winchester showed that at least two separate strains of *Mycobacterium leprae* were present in mediaeval England. Both strains have also been detected in bones recovered from a mediaeval cemetery in Denmark. One of the two strains matches the one from the Anglo-Saxon cemetery in Essex and the one associated with red squirrels on Brownsea Island. The dates of the Winchester burials range from around AD 1000–1200. The detection of closely related leprosy bacilli from British red squirrels and from human remains from Anglo-Saxon and mediaeval cemeteries suggests that leprosy might have been transmitted from humans to squirrels at some time in the Middle Ages when leprosy was prevalent among the human population. A survey of French, German, Italian and Swiss squirrels failed to detect any signs of infection (Schilling *et al.*, 2019). This suggests that squirrel leprosy may be a peculiarly British phenomenon. Red squirrels were kept as pets in England during the Middle Ages (Walker-Meikle, 2012).

Leprosy was prevalent in England between about AD 1050–1350 (Roffey, 2012). When case numbers began to decline, leprosy hospitals were repurposed or fell into ruin. A handful of mediaeval leprosy hospital buildings survive to the present day. All of them are situated on the outskirts of historic settlements. Surviving mediaeval leprosy hospital buildings can be seen in Cambridge, Glastonbury, Maldon, Norwich, Pamphill, Ripon and elsewhere. These ancient structures provide points of contact between the present day and a way of life which for several centuries was overshadowed by *Mycobacterium leprae*. Most of the extant buildings are chapels. One of them is St Bartholomew's Chapel, or the Bartlemas, which was founded in 1126 by Henry I (Fig. 9.37). The approach road to the Bartlemas (SP534054) is a narrow quiet lane that leads the visitor away from Oxford's Cowley Road, past some allotments and into a splintered fragment of Mediaeval England. The lane that leads to the Bartlemas is strangely cryptic. From the Cowley Road the entrance only appears to be visible when one is actively looking for it. The lane does not appear to have a name. There is another road nearby called Bartlemas Close which leads somewhere else. The tiny oasis of timeless hedgerows in which the Bartlemas exists is now hemmed in by Oxford's modern housing developments. In its day this hamlet would have felt fairly remote, situated as it was in the Oxford hinterland not far from the margins of Cowley Marsh. During the cholera epidemic

Fig. 9.37. St Bartholomew's Chapel, or the Bartlemas, site of a former leprosy hospital. Oxford, UK.

of 1832, the Bartlemas saw service as a convalescent home for recovering patients. Church services still take place in the chapel but its leprosy patients departed centuries ago.

The Leper Hospital of St Mary Magdalene was established on the eastern outskirts of Winchester around 1180. An archaeological excavation at the site uncovered the remains of the hospital's former patients and the sequencing of ancient DNA extracted from their bones confirmed that some of the patients were infected with *Mycobacterium leprae*. There is nothing for a modern visitor to see now at the hospital's original location apart from a grassy field. Nevertheless, a beautiful arched doorway from the site was transported stone by stone to the city centre to be incorporated into St Peters Church (SU481297). When I visited the church, the door was open. It now serves as a portal into an early twentieth century building. The casual visitor might feel inclined to be drawn inside without a second glance at the stonework but I felt compelled to stand back and examine the doorway itself and study it against the context of its modern surroundings. It was immediately obvious that the doorway belonged somewhere else as well as to its current location. The stones and the stonework do not match the rest of the building but the doorway does not look out of place (Fig. 9.38).

Fig. 9.38. Doorway from the Leper Hospital of St Mary Magdalene, Winchester, UK.

On the outskirts of Chichester, a secular building in Swanfield Drive preserves the remaining architectural vestiges of the St James and St Magdalene Leper Hospital (SU871053). In the eighteenth century, remnants of the former hospital were incorporated into a new building, which survives as a residential thatched cottage. The former hospital was founded in the twelfth century. Later the hospital buildings became almshouses. The excavated remains from the hospital cemetery have allowed archaeologists to document the history of the site and chart the decline of leprosy amongst the patients admitted to the hospital (Magilton *et al.*, 2008). The hospital was established not far from the main road from London to Chichester, on a site that is now a zone of residential housing and supermarkets. During the Middle Ages pilgrims would have passed the hospital while travelling towards Chichester Cathedral. I stopped at a local supermarket and like a modern pilgrim picked my way on foot along the verge of the busy Westhampnett Road. Despite its proximity to a major road, the Leper's Cottage is situated in the calm of a neighbouring residential street, made accessible from the main road by a footpath. The Leper's Cottage is a Grade II listed building. It is surrounded by modern houses. A red memorial plaque has been attached to the cottage by the City of Chichester (Fig. 9.39).

Fig. 9.39. Red plaque on wall of the Leper's Cottage, Chichester, UK.

Fig. 9.40. St Michael's Leper Hospital, chapel, Warwick, UK.

In 1135, Roger de Beaumont, Earl of Warwick, founded the St Michael's Leper Hospital. The stone chapel that served the St Michael's Leper Hospital still stands beside the Saltiford Road on the western fringes of Warwick (SP276654). The site was in a sorry state when I paid it a visit (Fig. 9.40). An accompanying timber-framed building of uncertain purpose known as the Master's House looked to be in a state of near collapse. The Master's House was in fact hardly visible beneath a supportive encasement of scaffolding and tarpaulins. The buildings of the former hospital were recorded as being in a state of disrepair as long ago as the sixteenth century. The surviving buildings were later converted into cottages. At the time of writing the former chapel and Master's House are unoccupied and the surrounding plot is fenced off and overgrown with weeds.

The Old Leper Church of St Nicholas (Fig. 9.41) at Harbledown in Kent is situated on the Pilgrim's Way on the outskirts of Canterbury (TR13581). For the mediaeval pilgrims en route to the shrine of St Thomas Becket, the leprosy hospital would have been the final landmark of their route before they entered the city. The hospital was founded in about 1084. In one form or another the site has been in constant use up until the present day and from time to time its buildings have accordingly been updated. Parts of the church date back to the hospital's foundation, not long after the Norman conquest. From the time of the reformation onwards the hospital buildings were repurposed as alms-houses. The almshouses that currently exist on site date from 1840. Bypassed by the A2050, Harbledown now has the feel of a quiet backwater. I walked

Fig. 9.41. The Old Leper Church of St Nicholas near Canterbury, UK.

down Church Hill, a road lined by houses that somehow still has the character of an old sunken lane. A green sign on a brick wall announces St Nicholas Hospital Harbledown, Church and Chantry. A flight of steps leads the visitor up from the road past a row of almshouses and towards the church. The entire site is self-evidently a special place with its own particular atmosphere.

The sanatorium archipelago

Tuberculosis is an infectious disease, primarily of the lungs. The causative agent, *Mycobacterium tuberculosis*, is a member of the same genus of bacteria as the two species that cause leprosy. Most infections in humans are caused by human-adapted strains of *Mycobacterium tuberculosis*. Humans can also become infected with bovine strains of the bacterium by drinking unpasteurised milk from infected cattle. In addition to cattle, susceptible animal species include badgers, deer, llamas and alpacas. Until the implementation of effective antibiotic treatment regimes in the 1950s, tuberculosis (TB) imposed a major burden of disease on the human population. Streptomycin, the first drug that was shown to work effectively against tuberculosis was derived from a soil bacterium, *Streptomyces griseus*, a member of the Phylum *Actinomycetota* and itself a distant relative of *Mycobacterium tuberculosis*. In the pre-antibiotic era, the available treatments for tuberculosis included surgery, bedrest and fresh air, despite which many patients succumbed to their disease. A network of TB sanatoria was created. Generally these institutions were established in relatively remote country areas where the air was considered to be more healthy. With the rapid decline of cases of infection, former TB sanatoria were either repurposed or demolished. An archipelago of former TB sanatorium sites lies scattered across Britain. Quietly neglected for decades, these nearly forgotten institutions are now deemed to be of historic interest. Informative plaques and interpretative boards have started to spring up, meaning that these places are beginning to attain the status of minor tourist attractions. Many of them are situated in places that are attractive to field naturalists. From time to time while out and about exploring the countryside one happens upon the site of a former sanatorium. A major sanatorium complex, known as the Royal National Hospital, once existed on the Isle of Wight near Ventnor. The hospital was demolished in 1969 and the site now accommodates the Ventnor Botanical Gardens. Hardly a trace of the once extensive former sanatorium now remains. I had walked past the site on many occasions unaware of the existence of the TB hospital until one day in 2013 when I noticed a plaque attached to a low inconspicuous Victorian building beside the coast path (Fig. 9.42). The plaque stated that the building once served as a site for the incineration of medical

Fig. 9.42. Surviving outbuilding of the now demolished TB hospital, Ventnor, Isle of Wight, UK.

waste from the former TB hospital. On another occasion while following a footpath beside the Gynack Burn above Kingussie, I chanced upon an interpretative board which featured a facsimile of an old advertisement for the former Grampian Sanatorium. The building still exists as an NHS community hospital. And while out walking in a Gloucestershire woodland, a chance conversation led to the discovery that a neighbouring field marked the site of the former Cotswold Sanatorium. I later learned that the writer George Orwell (Eric Arthur Blair, 1903–1950) completed his novel *Nineteen Eighty-four* (Orwell, 1949) whilst receiving treatment there. By coincidence at around the same time while walking through the village of Sutton Courtney I ventured into the grounds of All Saints Church and saw a headstone with Eric Blair's name on it. Coins and flowers placed around the stone indicated that a good many people must go out of their way to visit Orwell's final resting place. The list of famous creative tuberculosis victims is a long one. A seemingly disproportionately large number of novelists, poets, painters and composers shared the misfortune of being afflicted with tuberculosis. Their illness became very much a part of their personal histories. Anyone who likes to visit the places associated with these individuals has in some small way become a practicing bacteria tourist tracing a path, unwittingly perhaps, along one of the many tangents of bacteria tourism.

Monuments to penicillin

Over the course of time, history begins to seep into our cultural fabric. Partly forgotten historic personages are rediscovered, reassessed and reinterpreted and incorporated into our cultural heritage. In 1945 a Nobel Prize was awarded to Sir Alexander Fleming, Ernst Boris Chain and Sir Howard Walter Florey for the discovery of penicillin. The award recognised Fleming's observations regarding the antibacterial action of a *Penicillium* mould (Fleming, 1929) and the roles of Florey and Chain in the development of penicillin as a therapeutic agent against bacterial infection (Chain *et al.*, 1940). Fleming made his observations, more or less in isolation, as an independent researcher in his laboratory at St Mary's Hospital, Paddington, London. Florey led a multidisciplinary research team based at the Sir William Dunn School of Pathology in Oxford.

In 1952, a stained glass window, designed by Arthur Edward Buss (1905–1999) was installed in St James's Church, Paddington, London, as a replacement for glass that had been destroyed by bomb damage during the Second World War. The window depicts Sir Alexander Fleming, dressed in a white laboratory coat and holding a Petri dish (Fig. 9.43). In 1953, a simple, understated cuboid block of grey stone, inscribed with the names of eight penicillin researchers, was installed not far from the entrance to the Oxford Botanic Garden. The names of the eight researchers, which include those of Florey and Chain, are listed in alphabetical order (Fig. 9.44). An additional inscription records the fact that the memorial was presented by the Albert and Mary Lasker Foundation, New York. The stone has been exposed to the elements for long enough to appear slightly weathered. The memorial itself has attained a kind of historic importance and has become a habitat for bacteria.

Fig. 9.43. Fleming with Petri dishes in stained glass, St James's Church, Paddington, London, UK.

During the 1960s a new home was built for the then Ministry of Health at Elephant and Castle in London. The building was named Alexander Fleming House. The architect who designed it was the Hungarian émigré, Ernö Goldfinger (Warburton, 2003). The style of the building has been described as modernist and constructivist. It is a building that shows you what it is. Ernö Goldfinger later designed a number of acclaimed brutalist buildings, including the 26 storey Balfron Tower in Poplar, East London and another high-rise residential tower block, Trellick House, situated in Kensal Town, West London. Alexander Fleming House became unpopular with the civil servants who occupied it. It was deemed to be suffering from sick building syndrome. Eventually, the Department of Health found a new home at Skipton House, a building completed in 1993 and whose offices overlook the former Alexander Fleming House. On occasions when I visited Skipton House, I used to gaze out of the window and marvel at the stark industrial immensity of the former Alexander Fleming House. I had no knowledge at the time of the building's name or history and I had no idea that it had once served as a monument to penicillin but nonetheless for some reason I felt compelled to take a photograph of it

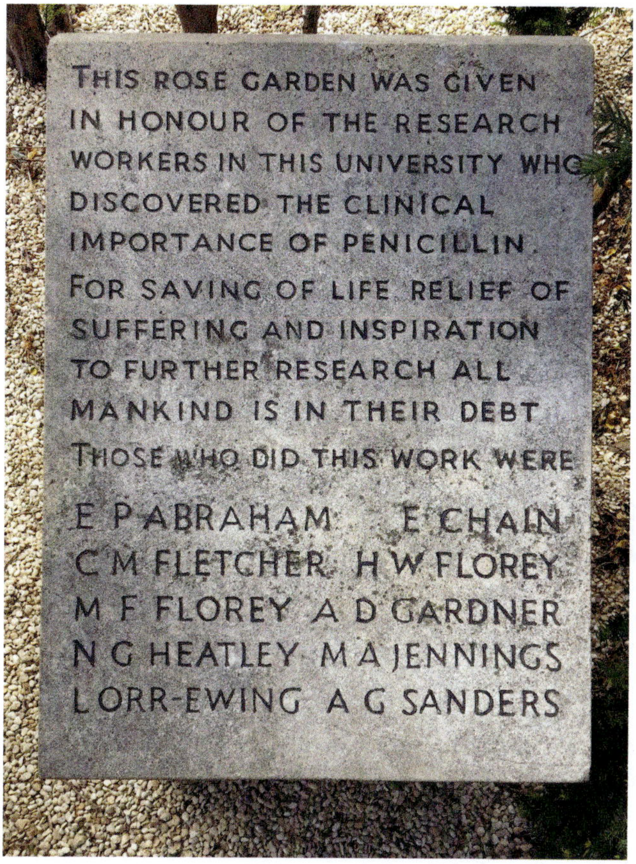

Fig. 9.44. Lasker Foundation stone commemorating penicillin research, Oxford, UK.

(Fig. 9.45). Alexander Fleming House was converted into residential flats and renamed Metro Central Heights. In 2013, Metro Central Heights was declared a listed building. The novelist Ian Fleming (1908–1964) appropriated the name Goldfinger for the character Auric Goldfinger in his novel *Goldfinger* (Fleming, 1959).

The practice of marking the former homes and work places of notable people with blue plaques originated in London in the nineteenth century. Blue plaques are engagingly minimalistic signifiers that codify the meaning of people and events within a single sparse line of prose. Blue plaques find their way into guidebooks and feature in walking tours of towns and cities. They become *de facto* tourist attractions. They take on a life of their own. In 1981, the Greater London Council attached a blue plaque to Sir Alexander Fleming's former residence, 20a, Danvers Street, Chelsea, London, although it was not until 1993 that the space occupied by Fleming's former laboratory at St Mary's Hospital, Paddington, became designated as a museum. A commemorative plaque can be seen attached to an external wall of the building in Praed Street, situated a short walk from Paddington Railway Station. Between 2010 and

Fig. 9.45. A view of the former Alexander Fleming House from Skipton House, London, UK.

2018 penicillin-themed blue plaques appeared on walls all over Oxford. One of them honours Norman Heatley (1911–2004), a biochemist who developed a method that facilitated the bulk production of penicillin. The plaque is attached to Heatley's former residence at 12, Oxford Road, Old Marston. In 2018, blue plaques appeared on the wall of the Sir William Dunn School of Pathology (Fig. 9.46), where penicillin was first produced in sufficient quantities to be used as a therapeutic agent and on the western wall of the former Radcliffe Infirmary, where penicillin was first used to treat a bacterial infection. A blue plaque appeared on the wall of Dorothy Hodgkin's former home at 94, Woodstock Road. Dorothy Hodgkin determined the structure of penicillin by using X-ray crystallography. In 1964, she was awarded the Nobel Prize for Chemistry for determining the structures of vitamins, proteins as well as antibiotics, including penicillin. In 2014, the Royal Society of Chemistry recognised her achievements by designating her former laboratory as a national chemical landmark. A hexagonal blue plaque can be seen on the wall of the Inorganic Chemistry Laboratory in South Parks Road (Fig. 9.47).

In 1941, Florey and Heatley travelled to America, with the aim of winning support from the American pharmaceutical industry to produce penicillin on an industrial scale as part of the war effort. Meanwhile in Britain, a penicillin factory was established at the Royal Navy Medical School at 5, Elton Road, Clevedon in Somerset. Freeze-dried penicillin was produced in Clevedon for use by the armed forces (Houlberg *et al.*, 2019). After the war, the facility came under the aegis of the Medical Research Council. Antibiotic research continued at 5, Elton Road until 1961. The former laboratories have since been converted into flats. A blue plaque installed by the Clevedon Civic Society commemorates

Fig. 9.46. Blue plaque commemorating the production of penicillin for clinical use. South Parks Road, Oxford, UK.

the building's history (Fig. 9.48). Between 1941 and 1953, modest quantities of penicillin were produced for civilian use by Kenneth White, an enterprising pharmacist based at Ripley in Surrey. A blue plaque marks the site of his former pharmacy where he produced penicillin. In the post-war years, industrial production of penicillin began at a purpose-built factory on the outskirts of Ulverston in Cumbria on a site previously occupied by an ironworks. Discarded slag from the ironworks continues to dominate the local landscape in the form of an enormous coastal slag bank. The slag bank is attractive to solitary bees, including *Osmia parietina*, a scarce northern species, which is more often found nesting in areas of limestone pavement. It is possible to access the slag bank by walking along the coast from a public carpark at Canal Foot. I visited the slag bank in bright May sunshine. It was a delightful experience to stroll along the shoreline and find the slag bank's special bees while at the same time listening to the cooing calls of the local eider ducks. The slag bank provides a vantage point for viewing the GlaxoSmithKline factory, where antibiotics have been made since 1948.

Fig. 9.47. Blue plaque commemorating Dorothy Hodgkin's research on the structure of penicillin. South Parks Road, Oxford, UK.

Bacteria abroad: Shark Bay

An image of the stromatolites at Shark Bay (Fig. 9.49) has become something of a prerequisite for books on geomicrobiology and also for interpretative boards at fossilised stromatolite sites. No respectable publication on ancient life fails to mention them. Anyone familiar with David Attenborough's 1979 television series, *Life on Earth*, or the accompanying book, will have seen images of these structures in their remote salty backwater. It was something of a moment when, 28 years after watching the television broadcast, I found myself strolling *en famille* into the Hamelin Pool Telegraph Station compound to start the trail to the beach (Fig. 9.50). In addition to a trickle of tourists who stop briefly to walk the trail and experience the place before heading to the resorts of Denham and Monkey Mia, there is an even smaller trickle of visitors (mostly geologists rather than microbiologists or naturalists it seems) who reach the location after long and purposeful voyages. The top of the beach features the stone-like domes of dead stromatolites, showing that the water level was once higher. Many of these domes are coated with an orange-red patina. These are especially attractive at sunset. A boardwalk leads one seaward from the

Fig. 9.48. During the Second World War, penicillin was made at the Royal Navy Medical School in Clevedon: blue plaque, Clevedon, Somerset, UK.

beach over the colony, allowing one to view the stromatolites without causing damage to these delicate structures (Fig. 9.51). Below high water mark there is an interesting area where cart ruts are clearly visible against the blackish background of cyanobacteria (Fig. 9.52). No wagons have rolled over the beach for more than half a century, showing what a fragile environment this is. In deeper water are the classic columnar and mushroom-shaped greenish-black stromatolites, as depicted on postcards and in microbiology and geology textbooks. There are plenty of interpretative signs along the boardwalk, reminding one of such information as the fact that cyanobacteria are responsible for enriching the early atmosphere with oxygen and drawing one's attention to the occasional bubbles that float up to the surface from the stromatolites. The old Telegraph Station is now a museum, containing telegraph memorabilia and also specimens of recent stromatolites from Shark Bay and elsewhere, as well as fossil stromatolites from Western Australia. It is possible to sit down and watch a short stromatolite video. A stromatolite tank contains living specimens. The captive stromatolites are supplied with electric light and regular doses of

Fig. 9.49. Stromatolites, Hamelin Pool, Shark Bay, Australia.

Fig. 9.50. Telegraph Station, Hamelin Pool, Shark Bay, Australia.

sediment. Compared with the free range ones, the captive stromatolites look a little too green for comfort, suggesting that eukaryotic algae may be flourishing at the expense of the original cyanobacteria. It is wonderful to spend a night at the Telegraph Station campsite as it grants one the opportunity to

Fig. 9.51. Viewing platform, Hamelin Pool, Shark Bay, Australia.

Fig. 9.52. Cart ruts, Hamelin Pool, Shark Bay, Australia.

walk down to the beach and watch the sun setting over the stromatolites and experience the site's ambience in relative solitude. It is possible to look out to sea over the domes of the stromatolites and wonder if the Precambrian scene was really something like this. After dark, when walking to the toilet block, one has the opportunity to pass the museum and gaze through the window at the eerie stromatolite tank illuminated by a fluorescent lamp. There is even a shop, which during opening hours sells a range of bacteria tourist souvenirs, including hats, shirts, tea-towels, car stickers and sediment globes.

The stromatolites at Lake Thetis

The Shark Bay stromatolites reside in the shallows of Hamelin Pool, a sheltered salty inlet of one of Shark Bay's southern extremities. The growth of marine stromatolites depends on a steady supply of calcium carbonate and a lack of disturbance by grazing organisms and large animals. Apart from Hamelin Pool, marine stromatolites occur also in the Bahamas, especially on Highborne Cay. Highborne Cay is privately owned and although open to researchers, not readily accessible to the bacteria tourist. It happens that Hamelin Pool and the Bahamas have marine environments that are unusually saline, restricting colonisation by many other organisms. In addition to the marine stromatolites at Hamelin Pool there are also some less well publicised colonies of stromatolites in several of Western Australia's saline lakes. One that may be visited conveniently en route from Perth to Shark Bay is at Lake Thetis near Cervantes. Pleasantly situated on the coast, Cervantes provides facilities for visitors to the Pinnacles Desert in the Nambung National Park. Most visitors seem to head straight for the park and miss the side road signposted to Lake Thetis and stromatolites (Fig. 9.53). This makes it possible for the microbe connoisseur

Fig. 9.53. Lake Thetis, stromatolite road sign, Australia.

to walk along the lake shore and view the site undisturbed by other visitors (Figs. 9.54–9.55). Above the shoreline on dry land are the rocky domes of stromatolites, left over from a time when the water level was higher. Although they are protected, it is interesting to note that a fence post has been cemented into one of the dry domes beached above the current water level, showing just how solid these structures are. There are other saline lakes with stromatolites in Western Australia including Lake Richmond and Lake Clifton south of Perth and Rottnest Island near Fremantle.

Museum of Western Australia: fossilised stromatolites

Although the tourist industry may not yet have woken up to the concept of bacteria tourism, the term geotourism already exists. It describes a touristic niche centred on sites of geological interest. There is a certain amount of overlap between geotourism and bacteria tourism. What makes living stromatolites of such special interest, apart from their rarity in the modern world, is that they suggest a vision of life on earth from past ages and support the interpretation of some of the oldest known fossils from the early Precambrian as the remains of once living stromatolites. It happens that Western Australia is home not only to the best known living stromatolites but also to some of the oldest

Fig. 9.54. Lake Thetis, information board, Australia.

Fig. 9.55. Lake Thetis, stromatolites, Australia.

stromatolite fossils. Famous examples may be seen in the Museum of Western Australia in Perth (Fig. 9.56). The museum features a display of material that allows comparison of specimens from the extant colony at Shark Bay with fossils from various ages. The collection includes a remarkably well-preserved 3.4 billion year old 'egg carton' slab of conically shaped stromatolites that provides one of the most compelling pieces of evidence of life in the early Precambrian. The specimen was discovered in 1997. In order to protect the specimen from erosion, souvenir hunters and commercial collectors, a team of scientists visited the site in 1999 to document the specimen *in situ*, before removing it for transportation to the museum. Despite the remoteness of the locality, it was noted that some tourists were present when the scientists arrived there. The specimen now resides in its own glass case in Perth. In addition to stromatolites, other evidence of early life is on display. There is a chunk of banded ironstone from Greenland whose carbon profile provides evidence that living organisms might have existed as long as 3.7 billion years ago. Also on display is a fragment of the Murchison meteorite, a body of extraterrestrial origin that contains amino acids, the constituent molecules of proteins in living cells. Before leaving the museum we ventured into the gift shop where I purchased a copy of a popular guide to stromatolites.

Pink lakes

Halophilic microbes thrive best in highly saline habitats. Many halophiles are photosynthetic organisms. The red, orange and pink photosynthetic pigments

Fig. 9.56. Museum of Western Australia.

of halophiles are responsible for the peculiar aesthetic qualities of hypersaline environments. Halophiles flourish in places where the climate is dry and sunny. The blooms of halophilic microorganisms that occur in solar salterns on the Mediterranean coast and in salt lakes in the more arid parts of the world are among the great spectacles of bacteria tourism. In the British Isles, halophiles do not really get much of a chance to express themselves. Nevertheless, although halophiles do not feature prominently in Britain's microflora, with the application of new molecular methods for the detection of microorganisms, it is becoming increasingly apparent that halophilic bacteria and archaea occur more widely than was formerly supposed. Halophiles related to *Haloferax mediterranei* (the Mediterranean Salt-bug) and *Natrialba magadii* (the Magadi Soda-bug) have been detected in a salt marsh at Colne Point in Essex and halophiles have also been detected within the salty interstices of decayed stonework on the Hampshire coast. Halophilic archaea have been cultured from samples of rock salt obtained from salt mines in Cheshire and Yorkshire and there is a school of thought that populations of these organisms might have survived in rock salt deposits over the course of geological time. The halophiles recovered from rock salt deposits may be the descendants of organisms that once flourished in ancient salt lakes during the arid Triassic and Permian Periods. First-hand experience of the halophilic organisms that live in salt lakes today, makes it easier to imagine what life might have been like in those ancient environments. Direct acquaintance with halophilic microorganisms facilitates one's understanding of halophile biology and the adaptations that allow halophiles to live in saline environments.

Santa Pola

One of the most intensively studied sites in the world for halophiles is the Braç del Port solar saltern at Santa Pola (Fig. 9.57). The saltern is situated on the Mediterranean coast not far from Alicante in a part of Spain that is popular with holidaymakers. It is one of the known localities for *Haloquadratum walsbyi* (Walsby's Square Archaeon), a square-shaped halophile that thrives only in nearly saturated salt solution. Originally discovered in a brine pool in Sinai, Walsby's Square Archaeon has since been detected at a number of widely separated locations around the world. It has been cultured from samples collected from a crystalliser pond used for commercial salt production at Braç del Port. The commercial enterprise at Braç del Port is surrounded by a nature reserve, the Parc Natural de les Salines de Santa Pola. There is also a salt museum (Museo de la Sal) on the outskirts of Santa Pola, which houses an exhibition about commercial salt production and the wildlife of the reserve (Fig. 9.58). The exhibits include pictures of halophilic microbes and provide information about halophile biology and microbial nutrient cycles.

The salterns at Santa Pola were established in the late nineteenth century. Seawater flows through a system of ponds of increasing salinity, resulting in the sequential precipitation of different salts: initially calcium carbonate, followed by gypsum (calcium sulphate) and halite (sodium chloride). This method of salt production is dependent on a rate of evaporation that exceeds precipitation.

Fig. 9.57. Braç del Port, Spain, home of Walsby's Square Archaeon.

Fig. 9.58. Santa Pola Salt Museum, Spain.

During the summer, the pools are tinted with shades of pink and purple by pigmented microbes. The nature reserve is bisected by a public road. There are parking areas from which the visiting naturalist can observe the colourful pools and the birdlife. Also visible from the roadside are great heaps of salt and the machinery used for salt production. Having enjoyed the view from the road, we set off down a side road in order to explore the Santa Pola hinterland. Beyond the perimeter of the nature reserve new holiday apartments were springing up along the roadside but in the middle of a little patch of halophilic vegetation I was pleased to find a colony of *Heteracris adspersa*, an attractive species of speckled grasshopper that is associated with saline environments.

Lake Magadi

The beds of red sandstone that furnish the landscape of Orkney with its distinctive appearance were deposited about 390 million years ago in a shallow lake known to geologists as Lake Orcadie. Devonian fish once flourished in Lake Orcadie and extensive stromatolite reefs developed in the lake's shallows. At Stromness, stromatolites occur in association with nodules of chert. The chert at Stromness has been interpreted as being a Magadi-type chert (Parnell, 1987). Chert is a kind of rock, composed of quartz (silica or silicon dioxide). Well-known varieties of chert include flint and jasper. Magadi chert is a variety of chert found at Lake Magadi, an alkaline (pH 10), hypersaline soda lake, which lies deep in the arid Rift Valley of southern Kenya in East Africa. A visit to Lake Magadi makes it possible to appreciate what Orkney might have been like 390 million years ago. Lake Magadi has no outward drainage. The lake receives water from seasonal rainfall and from hot springs, around which the water temperature may reach 86°C. Lake Magadi in Kenya and Lake Natron in Tanzania, have been described as the most alkaline environments on earth. In roadside pools, the pH may reach 12, a value equivalent to the alkalinity of domestic bleach. As well as being corrosively alkaline, the lake's waters are hypersaline. A crust of trona ($Na_3(CO_3)(HCO_3)\cdot 2H_2O$), which develops on the surface, is commercially extracted and treated on site to make soda ash (sodium bicarbonate), a substance used in various industrial processes. Lake Magadi contains relatively large amounts of silica, which is soluble in highly alkaline conditions. Silica precipitation leads to the formation of chert deposits. Magadi chert contains microfossils, which suggests that the induction of silica precipitation by microbes accounts at least partly for the origin of Magadi's chert deposits (Behr and Röhricht, 2000). The lake is rich in nutrients and supports an abundance of birdlife. There is even a species of fish, the Magadi tilapia, *Alcolapia grahami*, which has become adapted to the extreme conditions. Flourishing populations of extremophiles, including halophilic alkaliphilic archaea and purple sulphur bacteria exist in the lake. During the dry season, their pigments permeate the lake with rich tones of pink and purple. Examples of the alkaliphilic halophiles that have been discovered in Lake Magadi include the archaeon *Natrialba magadii* and the bacteria *Halomonas magadiensis*, *Tindallia magadiensis* and *Natranaerobaculum magadiense*.

During the early 1990s, when I was working in Nairobi, a weekend excursion down the Magadi Road to Lake Magadi was one of my preferred escape routes from life in the city. The countryside between Lake Magadi and the city was sparsely populated and alive with birds and insects. It was unusual to see traffic because the only real purpose of the Magadi Road was to serve the Magadi Soda Company. Although known to birdwatchers and Nairobi residents, Magadi does not tend to feature prominently in the itineraries of regular tourists. Bacteria tourists can enjoy excellent views from a causeway that crosses the lake, or alternatively from another road that follows the eastern shoreline of the lake (Fig. 9.59). On one occasion, a colleague and I drove all the way down to see Lake Natron. Beyond Magadi, the road becomes a dirt track, which gradually peters out. Eventually, Lake Natron comes into view. The lake is surrounded by a lunar landscape composed of volcanic grit. Lake Natron marks the national boundary with Tanzania but there are no roads, no discernible national frontier and no border post. Lake Natron is a breeding ground for Lesser flamingos, *Phoeniconaias minor*. Alkaliphilic cyanobacteria make up part of their diet. Carbonate stromatolites have been found on the slopes above Lake Natron. The stromatolites date from a time when the water level was much higher than it is today.

Fig. 9.59. Lake Magadi, Kenya, home of alkaliphile organisms.

Chott el Djerid

The chotts of North Africa are a series of ephemeral salt lakes, fed by artesian springs and seasonal rain. Low rainfall and high temperatures lead to the accumulation of evaporite deposits. As well as salt (sodium chloride), there are extensive deposits of gypsum (calcium sulphate). A visit to the Plateau of the Chotts makes it possible to imagine what environmental conditions might have been like when evaporite deposits of salt and gypsum were laid down more than 200 million years ago during the Triassic Period in what is now England. A thick red layer of mudstone (previously known as Keuper marl) at the foot of Aust Cliff in Gloucestershire shows that England was once a desert country. Horizontal bands of gypsum that run through the mudstone mark the former existence of transient shallow evaporite ponds, once inhabited by halotolerant microbes.

One of North Africa's more accessible sites is Chott el Djerid, a huge salt pan situated in southern Tunisia on the northern edge of the Sahara. A road runs across the length of Chott el Djerid. I travelled there on a conventional tourist bus. Chott el Djerid is a desolate but strangely beautiful tourist attraction (Fig. 9.60). The microbiology of Chott el Djerid has been well documented, which makes it attractive to bacteria tourists. The local microflora includes halotolerant and halophilic archaea and bacteria and eukaryotic algae (Abdallah *et al.*, 2016). A sediment sample from Chott el Djerid yielded *Halanaerobaculum tunisiense*, a halophilic bacterium that was new to science (Hedi *et al.*, 2009). The gypsum crusts at Chott el Djerid have been studied as analogues of evaporite deposits that might exist on Mars. Chott el Djerid

Fig. 9.60. Chott el Djerid, Tunisia, evaporite habitat.

has a long history of gypsum deposition, which makes it possible to compare fossilised traces of the microbes that are entombed deep down in the crust with the bacteria that are currently living on the surface and in so doing to develop expertise in the interpretation of microfossils found in ancient evaporite samples.

Geothermal features

Thermophilic bacteria and archaea do occur in and around the British Isles but they are not readily apparent. Thermophiles have been recovered from heaps of smoking coil spoil in Wales and Warwickshire and from the hot water supply of a hospital in London. Hyperthermophiles have been recovered from petroleum reservoirs under the North Sea, where temperatures may exceed 100°C. Moderate thermophiles exist in the moderately hot (46°C) thermal spring water at Bath. Thermophiles probably occur more widely in Britain than current records suggest. Thermophiles cannot be cultured from environmental samples unless a special effort is made to incubate them at high temperatures. Thermophiles can also be detected by sequencing DNA extracted from environmental samples. Recently, bacteria identified as members of the thermophilic genus *Thermomicrobium* were discovered in stonework at Portchester Castle. In coastal sediments, thermophiles exist unobtrusively as dormant spores. When incubated at 60°C, a sample of sediment from Woodhill Bay near Portishead on the bank of Severn Estuary yielded a thermophilic sulphate-reducing bacterium. Thermophilic sulphate-reducing bacteria flourish in deep subsurface environments. Oceanic currents must wash enormous numbers of their spores around the world's seas.

In order to become properly acquainted with thermophilic bacteria and archaea it is necessary to venture beyond the British Isles and explore naturally hot environments, such as the geothermal environments of Iceland, Italy or the USA. These places are, in effect, hot islands. They are spots on the surface of earth that reach unusually high temperatures. Although such places are few and far between, they share similar communities of closely related specialised thermophilic microbes. Very high temperatures exclude most other forms of life, making it possible to observe the lives of bacteria and archaea, undistracted by other kinds of organisms.

Iceland

Some of Iceland's geothermal features are tourist attractions but many of them have been put to use as sources of geothermal energy. Reykjavik, the capital city, derives its hot water supply from geothermal boreholes in the local area, which have been drilled down to depths of more than a kilometre and where water temperatures may exceed 100°C. Hot water is piped to users from a storage facility called Perlan (the Pearl), situated on a hill called Öskjuhlíð that overlooks Reykjavik. Perlan consists of a set of huge storage tanks on top of which is a restaurant (Fig. 9.61). Because of its geothermal origin, hot tap water in Reykjavik smells slightly sulphurous. Reykjavik's hot water supply makes it possible to study the organisms that exist in the subsurface deep beneath the city (Marteinsson *et al.*, 2001). In Reykjavik, a hot water sample is in effect a

Fig. 9.61. Perlan, Iceland.

sample from the hot deep biosphere. Thermophilic sulphate-reducing bacteria have been detected in Reykjavik's hot water supply as well as species of bacteria belonging to the Phylum *Aquificota*. *Aquifex* and allied genera obtain energy by oxidising hydrogen and sulphur compounds. They produce water as a waste product.

A popular Icelandic tour bus route known as the Golden Triangle includes a visit to the Geysir geothermal area. The principal attraction is a geyser called Strokkur. There are also hot pools and runoff channels which are adorned with colourful bacterial mats (Fig. 9.62). Thermophilic bacteria known to occur in the Geysir geothermal area include *Thermocrinis albus* (the White Water-bug) and *Geobacillus stearothermophilus* (the Autoclave Bug) (Tobler and Benning, 2011). *Thermocrinis albus* (Phylum *Aquificota*) grows as dense bundles of white hair-like filaments in hot outflow channels. *Geobacillus stearothermophilus* is a bacterium that produces heat-resistant spores. The spores are used to test the efficacy of autoclaves and heat sterilising processes. *Geobacillus stearothermophilus* was originally identified as a spoiling agent of insufficiently heated canned foods. The Strokkur geyser is the type locality for *Chloroflexus islandicus*, a yellowish-green photosynthetic green non-sulphur bacterium. The Golden Triangle tour includes a visit to Þingvellir, the location of the original Icelandic parliament. Þingvellir is visibly situated in a volcanic rift valley at the junction of the American and European tectonic plates. Iceland's geothermal features and the organisms that depend on them are sustained by

Fig. 9.62. Geysir geothermal area, Iceland, a thermophile habitat.

the heat released from magma that rises towards the surface at the junction between the two plates. Tour buses also stop briefly at Hveragerði, where there is a geothermally heated greenhouse with a cafeteria. A whole day can be usefully devoted to Hveragerði's varied geothermal features and associated thermophilic microflora.

The town of Hveragerði is situated in a geothermal area. Hot water, steam and volcanic gases rise through fissures and emerge at the surface to create a suite of varied geothermal features. Hveragerði is about 25 miles (about 40 kilometres) from Reykjavik. A regular bus service connects the town with the capital city. One day, during the course of a family holiday to Iceland, we caught the early morning bus to Hveragerði. In the middle of the town there is a small geothermal park. There is a kiosk that provides information about the site, although when we visited Hveragerði on a freezing day in February the kiosk was shut and the streets were deserted. Meanwhile, the park's attractions continued to steam gently in the chilly atmosphere. Not far from the town's northern boundary there is a geothermal hot water plant, which receives hot water from a borehole and a minor road that heads northwards into a scenic valley called Reykjadalur. Along the roadside there are hot springs and fumaroles. Hveragerði's microflora has been systematically investigated for several decades. Acidic mud holes (c. 80°C) provide suitable habitat for *Sulfolobus islandicus* (the Icelandic Sulphur-lump), an archaeon that thrives in highly acidic, sulphurous habitats. Hot alkaline pools and channels (c. 75°C) contain hair-like streamers of *Thermocrinis albus* (the White Water-bug). Cooler temperatures favour the growth of colourful mats composed of cyanobacteria and *Chloroflexus* (green non-sulphur bacteria). Exuberant bacterial mats and streamers decorate a wastewater channel from the local geothermal hot water plant (Fig. 9.63). Bacteria found in the channel include *Thermocrinis albus* (the White Water-bug) and *Thermus scotoductus* (the Hot Water Pipe Bacterium).

Fig. 9.63. Hveragerði, wastewater channel.

Yellowstone National Park

When I began to gather material for *Bacteria in the natural world*, bacteria holidays soon became the norm. One summer a trip to Yellowstone became our family holiday. Yellowstone is one of those rare places on earth where bacteria are writ large. The thermophilic bacteria and archaea at Yellowstone were the first to be studied and characterised. Thomas Brock's *Thermophilic Microorganisms and Life at High Temperatures* (Brock, 1978) provides an account of the pioneering research into thermophilic organisms that took place at Yellowstone during the 1960s. Brock's studies redefined the upper limits of temperature for life and established the concept of boiling water as a natural habitat. Yellowstone is the world's oldest national park. Its status as a park has afforded protection to its geothermal features and preserved the natural integrity of its ensemble of thermal habitats. Yellowstone National Park includes numerous geothermal features that lie in and around the collapsed caldera of a super-volcano. The park covers an area of almost 9065 square kilometres (3500 square miles; roughly one-twentieth the area of the island of Great Britain). The many geysers, hot springs and run-off channels feature a wide range of physicochemical conditions that encourage a diversity of prokaryotes. Yellowstone's microbes have been studied for decades and are unusually well-documented, making it possible for the visitor to target specific locations and habitats in the knowledge that particular species of thermophilic bacteria and archaea are likely to be present. Geothermal habitats are rendered even more beautiful by brightly pigmented microorganisms that flourish at high temperatures. Colourful bacterial mats and streamers are visible in many of Yellowstone's hot springs. A well-illustrated guidebook, *Seen and Unseen: Discovering the Microbes of Yellowstone* (Sheehan *et al.*, 2005), provides maps and information about the microbes found in different areas of the park. There is also a good amount of information about Yellowstone and its prokaryotes in *A Field Guide to Bacteria* (Dyer, 2003). The park itself is replete with interpretative boards and signs warning visitors not to molest the bacterial mats (Fig. 9.64).

Many of the park's geothermal features can be found in a series of geyser basins (Fig. 9.65). Each geyser basin contains pools, where the water may be close to boiling point, as well as cooler marginal areas and runoff channels that are often decorated with colourful bacterial mats and streamers composed of orange and green cyanobacteria (Fig. 9.66), which can thrive at temperatures of up to 70°C, as well as members of a separate group of photosynthetic thermophiles, the Phylum *Chloroflexota*, or green non-sulphur bacteria. Green non-sulphur bacteria are photosynthetic scavengers. They use light energy to assemble more complex hydrocarbons from small organic molecules that have been discarded by other microbes. At Octopus Spring (Fig. 9.67), in the Lower Geyser Basin, the predominant green non-sulphurs are members of the genus *Roseiflexus*, which are actually reddish-brown in colour (Reysenbach *et al.*, 1994; Huber *et al.*, 1998; Schaffert *et al.*, 2012). At temperatures between 50–70°C the predominant cyanobacteria are members of the genus *Synechococcus*. Different strains of *Synechococcus* occur at different sites

Fig. 9.64. Yellowstone, USA: do not mark bacterial mats.

Fig. 9.65. Norris Geyser Basin, Yellowstone, USA.

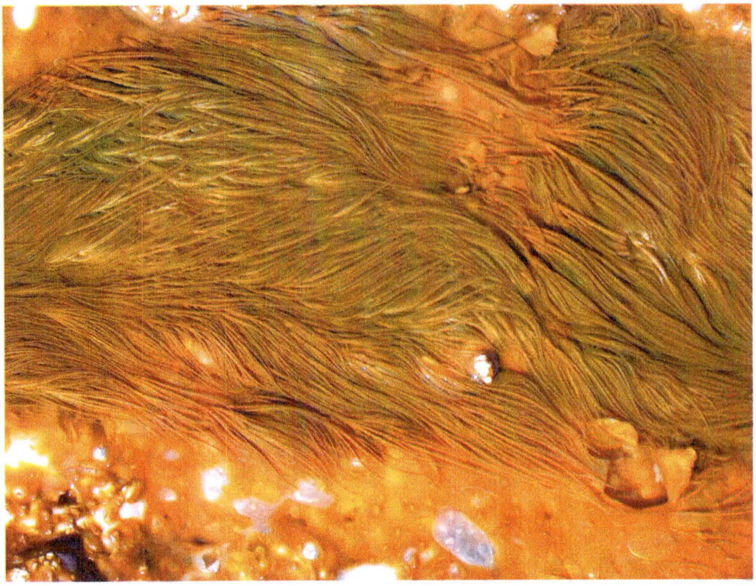

Fig. 9.66. Bacterial streamers, Yellowstone, USA.

Fig. 9.67. Octopus Spring, Yellowstone, USA.

along the temperature gradient, each strain being optimally adapted to a particular temperature range (Pedersen and Miller, 2017; Rosen *et al.*, 2018). Another genus of cyanobacterium, *Phormidium*, tolerates water temperatures that range between 30–60°C. *Phormidium* can form extensive tough rubbery orange mats around hot springs. In hot shallow pools *Phormidium* mats can develop into conical structures that resemble stromatolites. A third genus of thermophilic cyanobacterium, *Mastigocladus*, can be found growing in the same temperature zone as *Phormidium*. Filamentous cyanobacteria belonging to the genus *Calothrix* favour slightly cooler temperatures that range between 25–50°C. *Calothrix* forms greyish brown mats that become encrusted with silica.

Slightly alkaline pools, where the water approaches boiling point, are present in many areas of the park. These hot pools provide typical habitat for hyperthermophilic bacteria that belong to the Phylum *Aquificota* (water-bugs). The *Aquificota* are thermophiles that can obtain energy by oxidising hydrogen, a reaction that generates water as a waste product. Some members of the *Aquificota* can also oxidise sulphur and arsenic compounds. One of them is *Thermocrinis ruber* (the Ruby Water-bug), whose raspberry-coloured streamers can be seen in the outflow channels of very hot, slightly alkaline pools. *Thermocrinis ruber* has been recorded from Black Pool (85°C) in the West Thumb Geyser Basin and from the outflow channel (83°C) at Octopus Spring. Microbiological investigations at Octopus Spring began in the 1960s (Brock, 1978). Octopus Spring is one of the most studied sites in Yellowstone and has an unusually well-characterised microbial flora. *Thermus aquaticus* is also present at Octopus Spring. It favours slightly lower temperatures (c. 70°C). *Thermus aquaticus* was originally discovered not far from Octopus Spring at Mushroom Pool in 1969. Heat-stable polymerase enzymes from *Thermus aquaticus* form the basis of diagnostic polymerase chain reaction (PCR) tests.

In addition to the geyser basins (Fig. 9.65) there are some highly acidic sites, including the Mud Volcano Area and Roaring Mountain, which provide suitable habitat for *Sulfolobus acidocaldarius* (the Hot Sulphur Archaeon). The Mud Volcano Area hosts a suite of very hot (80°C), highly acidic (pH 1.2) mud pools (Fig. 9.68). The so-called Roaring Mountain is an example of a

Fig. 9.68. Mud Volcano, Yellowstone, USA.

solfataric field. Solfataras are hot sulphurous environments named after the Solfatara di Pozzuoli, a volcanic crater situated in Italy near Naples. *Sulfolobus acidocaldarius* was first found when microscope slides were placed in the outflow channel of a solfataric field at Yellowstone (Brock, 1978). The slides became coated with the cells of an unusually shaped lobular microbe. The Hot Sulphur Archaeon converts hydrogen sulphide to sulphuric acid, which dissolves rocks leading to the creation of hot acidic mud baths. Sulphuric acid bleaches colour from the landscape and erodes rocks to generate thick clay deposits. *Sulfolobus acidocaldarius* and related sulphur-oxidising archaea occur globally in acidic hot springs and solfataric fields and have even been recovered from the smouldering waste of a uranium mine in Germany (Fuchs *et al.*, 1996).

A colourful feature of some acidic hot springs is *Cyanidium caldarium*, a eukaryotic alga that tolerates both high temperatures and acidic conditions. *Cyanidium caldarium* and related members of the Cyanidiales flourish in the hot acidic channels that drain solfataric fields. Their lime green photosynthetic pigments provide a beautiful contrast to the drab grey clay pools that are frequented by *Sulfolobus acidocaldarius*.

Another distinctive part of Yellowstone is the area around Mammoth Hot Springs, which is rich in carbonate and where there is active travertine formation (Fig. 9.69). Travertine deposition is particularly associated with calcareous geothermal springs. Travertines generated at geothermal sites are known as thermogene travertines (Pentecost and Whitton, 2000). Tivoli in Italy is such a site: its waters remain at a constant 23°C (Gandin and Capezzuoli, 2014). The ancient Romans used travertine to construct the coliseum in Rome. Known to

Fig. 9.69. Mammoth Hot Springs, Yellowstone, USA.

them as *lapis tiburtinus*, travertine was quarried near Rome at Tibur, a settlement that later became Tivoli. The steaming pools at Mammoth Hot Springs in Yellowstone National Park reach temperatures of 65–72°C (Dong *et al.*, 2019). The extensive travertine terraces are inhabited by thermophilic bacteria and archaea. Having passed along a fault line through limestone, the sulphurous hot water that emerges from the ground is saturated with calcium carbonate. Loss of carbon dioxide to the atmosphere triggers the precipitation of calcium carbonate and the patterns of carbonate deposition are influenced by filamentous thermophilic bacteria that act as templates for precipitation. The microbial flora is dominated by members of the genus *Sulfurihydrogenibium* (sulphur gobblers) that oxidise sulphide and fix carbon dioxide (Dong *et al.*, 2019). Milky white streamers of *Sulfurihydrogenibium* (Phylum *Aquificota*) crowd the steaming channels. The bundles of bacterial streamers become entombed in crystals of calcium carbonate to create wavy ridge-like patterns in the developing layers of travertine. As the water cools and the chemistry changes other microbes become more prominent. On the cooler slopes, cyanobacteria and green sulphur bacteria become apparent (Fouke *et al.*, 2003). The beauty of the shining mounds of travertine at Mammoth Hot Springs is subtly enhanced by the pigments of photosynthetic bacteria.

Artis-Micropia: Amsterdam's museum of microbes

Natura Artis Magistra (nature is the teacher of art), or Artis for short, is a complex of buildings and gardens in Amsterdam which includes a zoo, an arboretum, a natural history museum, a planetarium, an aquarium and a microbe museum. Set over two floors, Artis-Micropia (Fig. 9.70) is itself something of a microcosm in that it combines the features of a museum, a mini-zoo and an art instillation. Flasks of cyanobacteria, Winogradsky columns, Petri dishes and various other cultures of living microbes are displayed alongside microbially fermented foodstuffs, interactive exhibits and glass sculptures of viruses, bacteria and diatoms. At the end of a gallery, a plate glass window reveals the world of a working microbiology laboratory. Microbiologists in white lab coats venture out and mingle with the crowds, check that the microscopes are in working order and talk to visitors. When I explored Artis-Micropia I was impressed by the fact that most of its visitors were young people.

The tomb of Louis Pasteur, Paris

Microbiology is a relatively young science. It did not really gather momentum until the nineteenth century, when scientists such as Robert Koch (1843–1910), Fanny Hesse (1850–1934) and Julius Petri (1852–1921) in Germany and Louis Pasteur (1822–1895) in France, developed new laboratory methods, which made it possible to establish links between particular species of bacteria and disease. One of those pioneering scientists, Louis Pasteur, spent the last seven years of his life living in an apartment at the Pasteur Institute, a research

Fig. 9.70. Amsterdam's museum of microbes, Artis-Micropia, the Netherlands.

centre established in 1887 and named in honour of Pasteur (Fig. 9.71). The Pasteur Museum (25, rue du Docteur Roux, Paris) preserves Pasteur's living quarters, furnishings, photographs and personal effects as well as the laboratory equipment used in his famous experiments. A guided tour takes the visitor through the various rooms of the building. At the end of the tour, visitors are led down a flight of marble steps to Louis Pasteur's final resting place in the Pasteur chapel and mausoleum, where his remains lie entombed in a granite sarcophagus, surrounded by symbolist artworks (Fig. 9.72). Pasteur's tomb was designed by the architect Charles-Louis Girault (1851–1932) and was based on the mausoleum of Galla Placidia in Ravenna. Galla Placidia was the daughter of the Roman emperor, Theodosius the Great. The walls are richly adorned with paintings and mosaics by the artists Luc-Olivier Merson (1846–1920) and Auguste Guilbert-Martin (1826–1890). Depictions of grazing sheep symbolise Pasteur's interest in anthrax, a deadly disease of livestock, caused by the bacterium *Bacillus anthracis*. Vines and grapes allude to the pasteurisation of wine. Chickens alert the visitor to Pasteur's investigations of chicken cholera, a disease of chickens caused by the bacterium *Pasteurella multocida*. There are mosaics that depict rabid dogs and images of mulberry trees, silkworms and silk moths that represent the silkworm diseases, *pébrine* and *flacherie*. At the far end of the chapel is an altar, beneath which lie the remains of Madame Pasteur (1826–1910).

Fig. 9.71. Pasteur Institute, Paris, France.

Fig. 9.72. The tomb of Louis Pasteur, Institut Pasteur, Paris, France.

References

Abdallah, M.B., Karray, F., Mhiri, N., Mei, N., Quéméneur, M. *et al.* (2016) Prokaryotic diversity in a Tunisian hypersaline lake, Chott El Jerid. *Extremophiles*, 20, 125–138.

Arkell, J.W. (1947) Oxford stone. Faber and Faber, London, UK.

Attenborough, D. (1979) Life on earth. Collins/BBC, London, UK.

Avanzi, C., Del-Pozo, J., Benjak, A., Stevenson, K., Simpson, V.R. *et al.* (2016) Red squirrels in the British Isles are infected with leprosy bacilli. *Science*, 354, 744–747.

Behr, H.-J. and Röhricht, C. (2002) Record of seismotectonic events in siliceous cyanobacterial sediments (Magadi cherts), Lake Magadi, Kenya. *International Journal of Earth Science*, 89, 268–283.

Brock, T.D. (1978) Thermophilic microorganisms and life at high temperatures. Springer-Verlag, New York.

Chain, E., Florey, H.W., Gardner, A.D., Heatley, N.G., Jennings, M.A. *et al.* (1940) Penicillin as a therapeutic agent. *Lancet*, II, 226–228.

Dong, Y., Sanford, R.A., Inskeep, W.P., Srivastava, V., Bulone, V. *et al.* (2019) Physiology, metabolism, and fossilization of hot-spring filamentous microbial mats. *Astrobiology*, 19, 1442–1458.

Dyer, B.D. (2003) A field guide to bacteria. Cornell, Ithaca, New York.

Fleming, A. (1929) On the antibacterial action of cultures of a *Penicillium*, with special reference to their use in the isolation of *B. influenzæ*. *British Journal of Experimental Pathology*, 10, 226–236.

Fleming, I. (1959) Goldfinger. Jonathan Cape, London, UK.

Fouke, B.W., Bonheyo, G.T., Sanzenbacher, B. and Frias-Lopez, J. (2003) Partitioning of bacterial communities between travertine depositional facies at Mammoth Hot Springs, Yellowstone National Park, U.S.A. *Canadian Journal of Earth Science*, 40, 1531–1548.

Fuchs, T., Huber, H., Burggraf, S. and Stetter, K.O. (1996) 16S rDNA-based phylogeny of the archaeal order *Sulfolobales* and reclassification of *Desulfurococcus ambivalens* as *Acidianus ambivalens* comb. nov. *Systematic and Applied Microbiology*, 19, 56–60.

Fulton, N., Anderson, L.F., Watson, J.M. and Abubakar, I. (2016) Leprosy in England and Wales 1953–2012: surveillance and challenges in low incidence countries. *BMJ Open*, 6 (5), e010608.

Gandin, A. and Capezzuoli, E. (2014) Travertine: distinctive depositional fabrics of carbonates from thermal spring systems. *Sedimentology*, 61, 264–290.

Gest, H. (2004) The discovery of microorganisms by Robert Hooke and Antoni Van Leeuwenhoek, fellows of the Royal Society. *Notes and Records of the Royal Society of London*, 58, 187–201.

Gomez-Heras, M., Smith, B.J. and Viles, H.A. (2010) Oxford stone revisited: causes and consequences of diversity in building limestone used in the historic centre of Oxford, England. *Geological Society of London, Special Publications*, 333, 101–110.

Han, X.Y., Sizer, K.C., Thompson, E.J., Kabanja, J., Li, J. *et al.* (2009) Comparative sequence analysis of *Mycobacterium leprae* and the new leprosy-causing *Mycobacterium lepromatosis*. *Journal of Bacteriology*, 191, 6067–6074.

Hedi, A., Sadfi, N., Fardeau, M.L., Rebib, H., Cayol, J.L. *et al.* (2009) Studies on the biodiversity of halophilic microorganisms isolated from El-Djerid Salt Lake (Tunisia) under aerobic conditions. *International Journal of Microbiology*, 2009, 731786.

Hooke, R. (1665) Micrographia: or, Some physiological descriptions of minute bodies made by magnifying glasses. With observations and inquiries thereupon. J. Martyn and J. Allestry, London, UK.

Horák, J.M. and Evans, J.A. (2010) Early Neoproterozoic limestones from the Gwna Group, Anglesey. *Geological Magazine*, 148, 78–88.

Houlberg, K., Wickenden, J. and Freshwater, D. (2019) Five centuries of medical contributions from the Royal Navy. *Clinical Medicine* (London), 19, 22–25.

Huber, R., Eder, W., Heldwein, S., Wanner, G., Huber, H. *et al.* (1998) *Thermocrinis ruber* gen. nov., sp. nov., a pink-filament-forming hyperthermophilic bacterium isolated from Yellowstone National Park. *Applied and Environmental Microbiology*, 64, 3576–3583.

Huber, H., Hohn, M.J., Rachel, R., Fuchs, T., Wimmer, V.C. *et al.* (2002) A new phylum of *Archaea* represented by a nanosized hyperthermophilic symbiont. *Nature*, 417, 63–67.

Inskip, S.A., Taylor, G.M., Zakrzewski, S.R., Mays, S.A., Pike, A.W.G. *et al.* (2015) Osteological, biomolecular and geochemical examination of an early Anglo-Saxon case of lepromatous leprosy. *PLoS ONE*, 10, e0124282.

John, D.M., Whitton, B.A. and Brook, A.J. (eds.) (2002) The freshwater algal flora of the British Isles: an identification guide to freshwater and terrestrial algae. Cambridge University Press, Cambridge, UK.

Magilton, J., Lee, F. and Boylston, A. (eds.) (2008) "Lepers outside the Gate": Excavations at the Cemetery of the Hospital of St James and St Mary Magdalene, Chichester, 1986–87 and 1993. (Chichester Excavations, 10; CBA Research Report 158.) Council for British Archaeology, York, UK.

Marcet, A.G. (1805) A chemical account of the Brighton chalybeate. Phillips and Fardon, London, UK.

Marteinsson, V.T., Hauksdóttir, S., Hobel, C.F., Kristmannsdóttir, H., Hreggvidsson, G.O. *et al.* (2001) Phylogenetic diversity analysis of subterranean hot springs in Iceland. *Applied and Environmental Microbiology*, 67, 4242–4248.

Orwell, G. (1949) Nineteen eighty-four. Secker and Warburg, London, UK.

Parnell, J. (1987) Mineralized magadi-type cherts in the Devonian of Scotland; support for a model of sulfide deposition in low-temperature alkaline conditions. *Economic Geology*, 82, 1053–1056.

Pedersen, D. and Miller, S. (2017) Photosynthetic temperature adaptation during niche diversification of the thermophilic cyanobacterium *Synechococcus* A/B clade. *ISME Journal*, 11, 1053–1057.

Pentecost, A. and Whitton, B.A. (2000) Limestone. In: The ecology of cyanobacteria. Eds. Whitton, BA and Potts, M. Kluwer, Dordrecht, Netherlands.

Reysenbach, A.-L., Wickham, G.S. and Pace, N.R. (1994) Phylogenetic analysis of the hyperthermophilic pink filament community in Octopus Spring, Yellowstone National Park. *Applied Environmental Microbiology*, 60, 2113–2119.

Roffey, S. (2012) Medieval leper hospitals in England: an archaeological perspective. *Medieval Archaeology*, 56, 203–233.

Rosen, M.J., Davison, M., Fisher, D.S. and Bhaya, D. (2018) Probing the ecological and evolutionary history of a thermophilic cyanobacterial population via statistical properties of its microdiversity. *PLoS ONE*, 13, e0205396.

Schaffert, C.S., Klatt, C.G., Ward, D.M., Pauley, M. and Steinke, L. (2012) Identification and distribution of high-abundance proteins in the Octopus Spring microbial mat community. *Applied and Environmental Microbiology*, 78, 8481–8484.

Schilling, A.K., Avanzi, C., Ulrich, R.G., Busso, P., Pisanu, B. *et al.* (2019) British red squirrels remain the only known wild rodent host for leprosy bacilli. *Frontiers in Veterinary Science*, 1, 8.

Schuenemann, V.J., Avanzi, C., Krause-Kyora, B., Seitz, A., Herbig, A. *et al.* (2018) Ancient genomes reveal a high diversity of *Mycobacterium leprae* in medieval Europe. *PLoS Pathogens*, 14, e1006997.

Sheehan, K.B., Patterson, D.J., Dicks, B.L. and Henson, J.M. (2005) Seen and unseen: discovering the microbes of Yellowstone. Falcon, Helena, Montana.

Shore, T.W. (1891) Springs and streams of Hampshire. *Proceedings of the Hampshire Field Club*, 2, 33–58.

Tobler, D.J. and Benning, L.G. (2011) Bacterial diversity in five Icelandic geothermal waters: temperature and sinter growth rate effects. *Extremophiles*, 15, 473–485.

Walker-Meikle, K. (2012) Medieval pets. Boydell and Brewer, Rochester, New York.

Warburton, N. (2003) Ernö Goldfinger: the life of an architect. Routledge, London, UK.

Zanardini, E. May, E., Inkpen, R., Cappitelli, F., Murrell, J.C. *et al.* (2016) Diversity of archaeal and bacterial communities on exfoliated sandstone from Portchester Castle (UK). *International Biodeterioration and Biodegradation*, 109, 78–87.

Zanardini, E., May, E., Purdy, K.J. and Murrell, J.C. (2019) Nutrient cycling potential within microbial communities on culturally important stoneworks. *Environmental Microbiology Reports*, 11, 147–154.

10 Bacteria, who are you?

> With offspring mapped below in lineage,
> Till the tangles troubled me,
>
> Thomas Hardy (1916). *The Pedigree.*

A singularly rewarding facet of natural history is the process of finding an organism, working out what it is and putting a name to it. In James Joyce's posthumously published draft novel *Stephen Hero* (1944), Stephen determines that *claritas* (clarity or radiance) equates with *quidditas* (quiddity or whatness). Stephen's conclusion could be interpreted as meaning that things become clear when you recognise what something is. As soon as you know what something is you can consult the published literature and find out what is known about it. When you know what kind of organisms exist in a given habitat you can start to imagine how they might relate to one another. The reliable identification and naming of organisms makes it possible to investigate their significance in the wider world. In recent decades, advances in molecular methods of detection, high throughput sequencing and genomic analysis have made it possible at long last to see what bacteria really are and what they are doing.

At the time of writing, validly published names exist for more than 40 bacterial and archaeal phyla. These groups contain all of the well-known and well-characterised species of bacteria and archaea. Analysis of DNA sequence data obtained from environmental samples indicates that many more additional major taxonomic groups exist in nature (Hug *et al.*, 2016). As yet, representatives of these candidate phyla have not been isolated in culture and relatively little is known about them. Nevertheless, functional analysis of their genes provides clues about the roles that these organisms might occupy in nature. Many of these

DOI: 10.1079/9781836993933.0010

obscure bacteria are associated with subsurface groundwater environments and many of them have unusually small cells and small genomes.

The basic principles of taxonomy are classification, nomenclature and identification. An ideal scheme of classification reflects the evolutionary history of the organisms that are being classified. But because bacteria and archaea can exchange chunks of genetic material, even between distantly related species, family trees based on the comparison of limited numbers of particular genes have tangled branches. Even the term species can become problematic when applied to bacteria because the sharing of similar genes by unrelated species of bacteria may confound attempts to define exactly where one species of bacterium ends and where another one begins. One can easily be drawn into a kind of labyrinth by discussions about bacterial species concepts, species definitions and the validity of the various approaches used to determine the phylogenetic histories of different groups of bacteria. But this chapter will avoid such debates. What follows is a series of brief accounts of some of the major taxonomic divisions of bacteria and archaea, illustrated by examples of species that belong to them.

A note about the names of phyla

Taxonomy is a restless endeavour. Schemes of classification and scientific names hardly seem to sit still for very long. Instead they continue to shift incrementally ever closer towards an ideal. In 2021, the International Committee on Systematics of Prokaryotes (ICSP) decided that the names of all phyla should end in *-ota*, thus making it possible for the first time to recognise the names of phyla as such, simply by looking at their endings (Oren and Garrity, 2021). Hence, the Phylum *Cyanobacteria* was renamed *Cyanobacteriota* and *Actinobacteria* was renamed *Actinomycetota* (Table 10.1). An unhelpful consequence of the new nomenclature, however, is that some of the new names look completely unlike their precursors. The phylum *Firmicutes* was renamed *Bacillota* and the phylum *Proteobacteria* was renamed *Pseudomonadota.* As members of the latter are commonly referred to according to their constituent classes – the names of which all end in *-proteobacteria*: Class *Alphaproteobacteria*, *Betaproteobacteria* and so on – the use of the name *Pseudomonadota* is likely to cause confusion. Furthermore, the new nomenclature runs the risk of losing cohesion with a vast body of existing microbiological literature. Another consideration is that some of the names that were proposed in 2021 for new phyla are currently regarded as synonyms. It would be easy to become sidetracked by discussions about nomenclature. To avoid this the text of *Bacteria in the Natural World* simply follows the nomenclature used in the *List of Prokaryotic Names with Standing* (LPNS) as posted at the time of writing. LPNS is published online by the Leibniz Institute DSMZ (Deutsche Sammlung von Mikroorganismen und Zellkulturen – German Collection of Microorganisms and Cell Cultures GmbH).

Table 10.1. Some of the better-known phyla in the Domain *Bacteria.*

ICSP nomenclature (2021)	Traditional name	Informal names and notes
Gram-positive bacteria		
Actinomycetota	*Actinobacteria*	Filamentous Gram-positive bacteria
Bacillota	*Firmicutes*	'Low G + C' Gram-positive bacteria
Gram-negative bacteria		
Acidobacteriota	*Acidobacteria*	Found in acidic soil habitats
Bacteroidota	*Bacteroidetes*	Commensals of animals and plants
Fusobacteriota	*Fusobacteria*	Animal commensals and pathogens
Pseudomonadota	*Proteobacteria*	'Ordinary' Gram-negative bacteria
Green photosynthetic bacteria		
Chlorobiota	*Chlorobi*	Green sulphur bacteria
Chloroflexota	*Chloroflexi*	Green non-sulphur bacteria
Cyanobacteriota	*Cyanobacteria*	'Blue-green algae'
Thermophilic bacteria		
Aquificota	*Aquificae*	Hot springs and hydrothermal vents
Deinococcota	*Deinococcus-Thermus*	Deinococci and some thermophiles
Dictyoglomerota	*Dictyoglomi*	Hot springs and industrial settings
Thermodesulfobacteriota	*Thermodesulfobacteria*	Hot springs
Thermotogota	*Thermotogae*	Petroleum reservoirs
Some miscellaneous phyla		
Chlamydiota	*Chlamydiae*	Intracellular parasites of animals
Chrysiogenota	*Chysiogenetes*	Arsenate-reducing bacteria
Deferribacterota	*Deferribacteres*	Animal commensals; also thermophiles
Elusimicrobiota	Termite group 1 phylum	Environmental and commensal
Lentisphaerota	*Lentisphaerae*	Environmental and human gut
Mycoplasmatota	*Tenericute*	Parasitic bacteria that lack cell walls
Nitrospirota	*Nitrospirae*	Spiral shape, nitrite-oxidising
Planctomycetota	*Planctomycetes*	Aquatic environments
Spirochaetota	*Spirochaetes*	Spiral shape; some are pathogens
Verrucomicrobiota	*Verrucobicrobia*	'Wart-like' bacteria

Phylum *Bacillota* (previously *Firmicutes*): the 'low G + C' Gram-positive bacteria

The Phylum *Bacillota* is one of the two major divisions of Gram-positive bacteria (Table 10.2). Gram-positive bacteria have thick cell walls composed of peptidoglycan, which retain crystal violet stain when stained using Gram's staining method. Traditionally, the two divisions of Gram-positive bacteria were separated according to the ratios of guanine plus cytosine (G + C) versus adenine plus thymine (A + T) base pairs in their DNA. Hence, the *Bacillota* are sometimes referred to informally as the 'low G + C' Gram-positive bacteria.

Members of the genus *Staphylococcus* are Gram-positive cocci that typically appear in clusters when viewed microscopically (Fig. 5.33). Many of the described species are commensal bacteria that colonise the body surfaces of

Table 10.2. Examples of bacteria belonging to the Phylum *Bacillota* (formerly *Firmicutes*).

Species of bacterium	Notes
Bacillus amyloliquefaciens	Soil bacterium with heat-resistant spores (up to 420°C)
Bacillus anthracis	Soil bacterium and agent of anthrax in animals and humans
Bacillus aryabhattai	Soil bacterium, originally discovered in the stratosphere
Bacillus thuringiensis	Used to control Blandford fly, *Simulium posticatum*
Clostridium botulinum	Found in sediment and soil; agent of botulism
Clostridium felsineum	Pectin-degrading bacterium, used in the retting of flax
Clostrium tetani	Toxin-producing soil organism; agent of tetanus
Clostridioides difficile	Toxin-producer; agent of antibiotic-associated diarrhoea
"*Ca.* Desulforudis audaxviator"	A deep subsurface sulphate-reducing bacterium
Enterococcus faecalis	A common gut bacterium of animals
Enterococcus faecium	A common gut bacterium of animals
"*Epulopiscium fishelsoni*"	A giant (600 µm) symbiont of surgeonfish
Granulicatella balaenopterae	From post-mortem samples from minke whales
Lactococcus lactis	A lactic acid bacterium; a Gram-positive coccus
Lactobacillus delbrueckii	A lactic acid bacterium; a Gram-positive rod
Listeria monocytogenes	Environmental bacterium and intracellular pathogen
Pediococcus pentosaceus	A lactic acid bacterium
Staphylococcus aureus	Commensal and pathogen of humans and animals
Staphylococcus caledonicus	Isolated from dogs in Scotland
Staphylococcus epidermidis	A common skin commensal
Staphylococcus lutrae	Isolated from otter carcasses in the Hebrides
Staphylococcus piscifermentans	From fermented fish in Thailand
Staphylococcus succinus	From inclusions in 25–35 million year old amber
Streptococcus agalactiae	Commensal and pathogen of humans and cattle
Streptococcus phocae	Isolated from seals with respiratory infection
Streptococcus pneumoniae	Human commensal and pathogen
Streptococcus pyogenes	Human commensal and pathogen
Veillonella denticariosi	Anomalous Gram-negative member of *Bacillota*

mammals. Members of the genus *Streptococcus* and related genera are Gram-positive cocci that typically appear as long chains of bacteria when viewed microscopically (Fig. 10.1). The genus *Streptococcus* includes oral and upper respiratory tract commensals as well as potential pathogens of mammals. The lactic acid bacteria consist of a group of plant and animal-associated genera that produce lactic acid as a fermentation product. Some of the lactic acid bacteria are rod-shaped (Fig. 10.2); others are coccoid. *Bacillus* and related genera are relatively large, square-ended Gram-positive rods. They can be cultured from soils and other environmental samples (Fig. 10.3). Some *Bacillus* species are toxin producers. *Bacillus thuringiensis* is an insect pathogen. *Bacillus cereus* is an agent of food poisoning and *Bacillus anthracis* is the agent of anthrax. *Bacillus* species produce resilient spores which can survive in hostile environmental conditions. *Clostridium* and related genera are strictly anaerobic spore-forming Gram-positive rods (Figs. 7.5, 8.7) found in anoxic environments such as sediments and damp soils. Some of them are toxin producers.

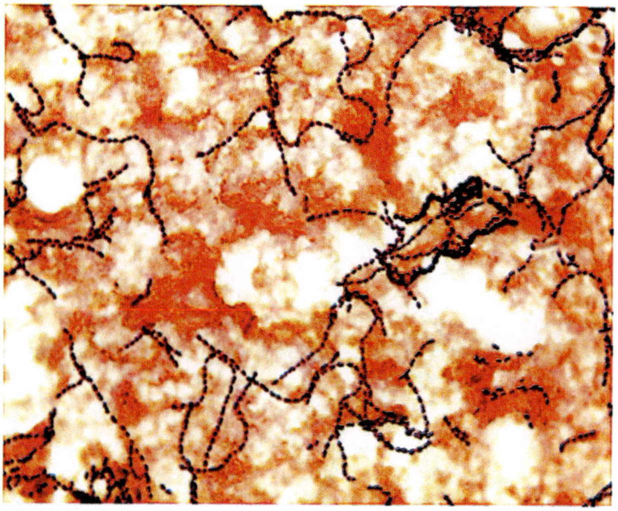

Fig. 10.1. Characteristic chains of a *Streptococcus* species as seen on microscopy. Gram's stain.

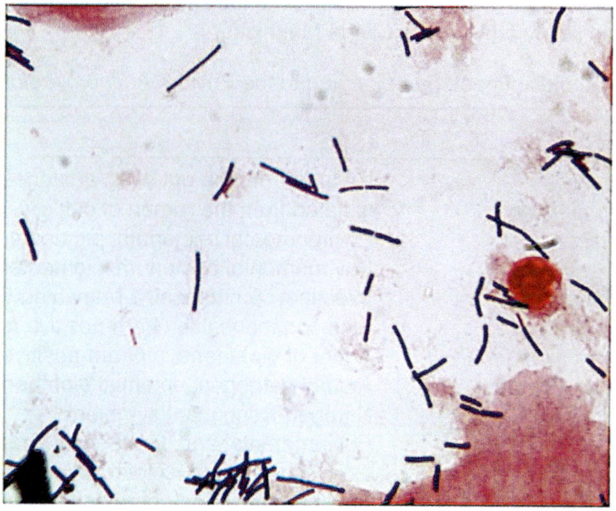

Fig. 10.2. Gram's stain of lactobacilli.

Phylum *Actinomycetota* (formerly *Actinobacteria*): the 'high G + C' Gram-positive bacteria

The Phylum *Actinomycetota* (Table 10.3) contains Gram-positive bacteria whose DNA includes a high proportion (more than 50%) of the bases guanine and cytosine. Hence as a group they are sometimes called the 'high G + C' Gram-positive bacteria. Many genera, including the soil bacteria of the

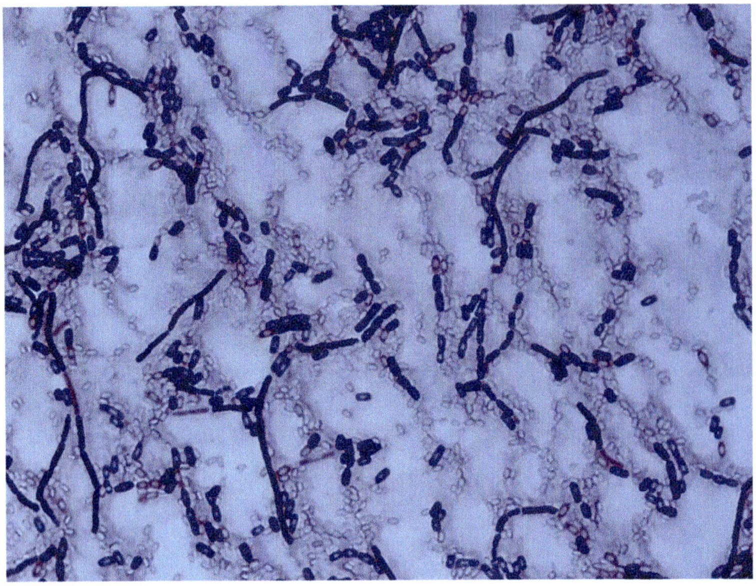

Fig. 10.3. Gram's stain of an environmental *Bacillus* species cultured from soil. (Isolate kindly provided by Minna Männistö.)

Table 10.3. Examples of bacteria belonging to the Phylum *Actinomycetota*.

Species of bacterium	Notes
Actinomyces haliotis	Isolated from the gut of an abalone
Actinomyces ruminicola	Isolated from the rumen of cattle
Arthrobacter agilis	Environmental bacterium; pigmented colonies
Arthrobacter roseus	Environmental bacterium; pigmented colonies
Bifidobacterium longum	Intestinal commensal; a branching Gram-positive rod
Brevibacterium linens	Skin commensal; a Gram-positive rod
Corynebacterium diphtheriae	Agent of diphtheria; a Gram-positive rod
Corynebacterium ulcerans	Animal pathogen; potential diphtheria agent in humans
Frankia alni	Nitrogen-fixing plant symbiont
Kocuria rosea	Environmental and commensal Gram-positive coccus
Microbacterium murale	Gram-positive rod isolated from a mouldy wall
Micrococcus luteus	Environmental and commensal Gram-positive coccus
Mycobacterium angelicum	Isolated from aquarium angelfish
Mycobacterium leprae	An agent of leprosy
Mycobacterium salmoniphilum	A pathogen of salmon
Mycobacterium sphagni	Isolated from sphagnum bogs
Mycobacterium tuberculosis	Agent of tuberculosis
Nocardia asteroides	Filamentous soil bacterium and potential pathogen
Nocardia camponoti	Isolated from an ant, *Camponotus japonicus*
Nocardia farcinica	Agent of farcy, a disease of horses
Nocardia salmonicida	A pathogen of salmon
Streptomyces albidoflavus	A filamentous soil bacterium
Streptomyces alni	Isolated from an alder root
Streptomyces coryli	Isolated from hazelnut orchard soil
Streptomyces griseus	The official microbe of the State of New Jersey

genus *Streptomyces* grow as a mycelium of branching filaments or hyphae. They are known informally as actinomycetes (meaning ray-fungi, in reference to the radial, fungoid appearance of colonies grown on agar). They occupy the same ecological territory as soil fungi, which they resemble in appearance and lifestyle. The mycelial growth habit shared by actinomycetes and soil fungi appears to be an example of convergent evolution. Filamentous cells may offer soil organisms a selective advantage: hyphae allow cells to bridge the gap between soil particles, whilst the production of aerial hyphae facilitates the dispersal of spores. Actinomycetes and soil fungi both produce antibiotics which inhibit the growth of other soil microbes. Many antibiotics used in medicine and veterinary practice were developed from substances isolated from cultures of soil-dwelling *Actinomycetota*.

Some *Actinomycetota* have waxy coats and are known as acid-fast bacteria because they retain pigment when stained and treated with acid. Acid-fast genera include *Nocardia* and *Mycobacterium* (Figs. 5.40–5.41), many species of which occur in soil or aquatic habitats. Some mycobacteria are important pathogens. *Nocardia* species are soil organisms, some which cause skin and pulmonary infections in mammals. *Nocardia* was named in honour of the French veterinarian Edmond Isidore Etienne Nocard (1850–1903) (Fig. 10.4). Bacteria in the genus *Actinomyces* and related genera are anaerobes that live on the mucosal surfaces of mammals and as part of the invertebrate gut flora.

Members of the genus *Corynebacterium* are Gram-positive rods. Some of them occur on the human skin as harmless commensals. *Corynebacterium diphtheriae* is the agent of diphtheria (Fig. 10.5). *Corynebacterium ulcerans* is another toxin-producing bacterium which infects cattle and occasionally humans. *Micrococcus* species are environmental Gram-positive cocci, some of which produce colourful pigments.

Fig. 10.4. Street sign in honour of veterinarian Edmond Nocard, Avignon.

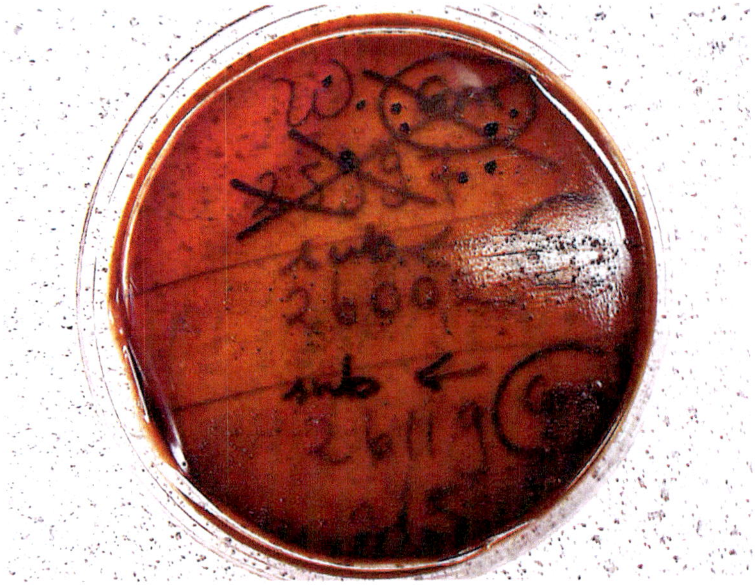

Fig. 10.5. Black colonies of *Corynebacterium diphtheriae* growing on tellurite agar.

Phylum *Mycoplasmatota*: mycoplasmas and spiroplasmas

Mycoplasmas and their allies are phylogenetically related to Gram-positive bacteria in the Phylum *Bacillota*. *Mycoplasma* and allied genera are unusual parasitic bacteria that lack cell walls (Fig. 10.6). They act as infectious agents of plants, insects and vertebrates (Table 10.4) and are dependent on their host cells for a supply of essential nutrients. *Mycoplasma pneumoniae* is an agent of respiratory tract infection in humans. *Mycoplasma genitalium* is a sexually transmitted bacterium which causes infections of the urethra in men and infections of the genital tract in women, although many cases of infection are asymptomatic. *Mycoplasma genitalium* shows high rates of resistance to antibiotics that are routinely used to treat sexually transmitted diseases (Daubenspeck *et al.*, 2020).

The *Pseudomonadota* (formerly *Proteobacteria*): a phylum of Gram-negative bacteria

A large proportion of the Gram-negative bacteria that can be cultured easily in routine microbiology laboratories belong to the Phylum *Pseudomonadota*. Numerous medically important Gram-negative bacteria are members. The group is split into classes, distinguished by Greek alphabetical prefixes: *Alpha-*, *Beta-*, *Gamma-*, *Deltaproteobacteria* and *Epsilonproteobacteria* (Table 10.5). The Class *Alphaproteobacteria* includes intracellular pathogens

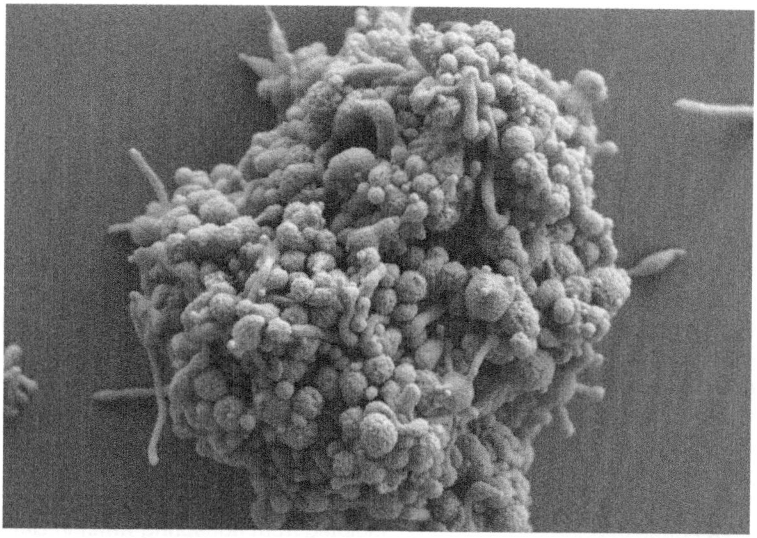

Fig. 10.6. *Mycoplasma genitalium.* Electron micrograph. (Image cropped from figure 1 in Daubenspeck *et al.*, 2020; *Frontiers in Microbiology*. 11, 585524). Creative Commons Attribution Licence.

Table 10.4. Examples of bacteria belonging to the Phylum *Mycoplasmatota*.

Species of bacterium	Notes
Entomoplasma freundtii	Isolated from a green tiger beetle (species not stated)
Mesoplasma grammopterae	Isolated from a longhorn beetle, *Grammoptera* species
Mesoplasma syrphidae	Isolated from a hoverfly
Mesoplasma tabanidae	Isolated from a horse fly
Mycoplasma anseris	Isolated from geese
Mycoplasma anserisalpingitidis	Isolated from a goose with an inflamed oviduct
Mycoplasma buteonis	Isolated from buzzards
Mycoplasma capricolum	Isolated from goats
Mycoplasma falconis	Isolated from falcon
Mycoplasma genitalium	An agent of sexually transmitted infection in humans
Mycoplasma pneumoniae	An agent of respiratory tract infection in humans
Mycoplasma haemofelis	Isolated from cats
Mycoplasma phocidae	Isolated from seals
Mycoplasma phocoeninasale	Isolated from a harbour porpoise
Spiroplasma apis	Isolated from honey bees
Spiroplasma citri	A plant pathogen
Spiroplasma culicicola	Isolated from the mosquito *Aedes sollicitans*

and commensals of animals (Fig. 5.3) and plants, as well as iron-oxidising, nitrite-oxidising and methanotrophic bacteria. The Class *Betaproteobacteria* includes environmental bacteria (Figs. 3.8–3.9) with diverse nutritional strategies as well as commensals and pathogens of mammals. *Escherichia*, *Salmonella* and *Shigella* are enteric bacteria that belong to the Class *Gammaproteobacteria*.

Table 10.5. Examples of *Pseudomonadota* (formerly *Proteobacteria*).

Species of bacterium	Notes
Class *Alphaproteobacteria*	
Agrobacterium radiobacter	Plant pathogen; agent of crown gall disease
Anaplasma phagocytophilum	Animal pathogen that invades white blood cells
Bartonella henselae	Agent of cat scratch disease
Brucella pinnipedialis	Intracellular pathogen of sea mammals
Mariprofundus ferrooxydans	Iron-oxidising bacterium; deep sea vents
Methylocystis heyeri	Methanotrophic bacterium found in peat bogs
Methylorubrum extorquens	Plant commensal; uses methanol as food source
Magnetospirillum magneticum	Magnetotactic aquatic bacterium
Nitrobacter winogradskyi	Nitrite-oxidising bacterium
Rhodomicrobium vannielii	Photosynthetic purple non-sulphur bacterium
Rickettsia prowazekii	Agent of louse-borne typhus
Wolbachia pipientis	Intracellular symbiont of arthropods
Class *Betaproteobacteria*	
Annwoodia aquaesulis	Isolated from hot spring water in Bath
Gallionella ferruginea	Iron-oxidising bacterium
Leptothrix ochracea	Iron-oxidising bacterium
Neisseria gonorrhoeae	Agent of gonorrhoea
Nitrosomonas europaea	Ammonia-oxidising organism
Variovorax paradoxus	Soil bacterium that uses multiple food substrates
Class *Gammaproteobacteria*	
Acidiferrobacter thiooxydans	Iron and sulphur oxidising bacterium
Beggiatoa alba	White sulphur bacterium
Escherichia coli	Gut commensal of mammals
Salmonella enterica	Gut commensal and pathogen
Serratia marcescens	Pigmented environmental bacterium
"*Thiomargarita magnifica*"	The world's largest known bacterium
Pseudomonas syringae	Plant pathogen
Chromatium vinosum	Purple sulphur bacterium
Thiocapsa roseopersicina	Purple sulphur bacterium
Photobacterium phosphoreum	Bioluminescent commensal of anglerfish
Photorhabdus asymbiotica	Bioluminescent pathogen of insects
Methylomonas methanica	Methanotrophic bacterium
Vibrio anguillarum	Fish pathogen
Class *Deltaproteobacteria*	
Bdellovibrio bacteriovorus	Predator of other bacteria
Desulfobacter postgatei	Sulphate-reducing bacterium
Desulfovibrio sulfuricans	Sulphate-reducing bacterium
Geobacter metallireducens	Iron- and uranium-reducing bacterium
Class *Epsilonproteobacteria*	
Campylobacter lari	Isolated from gulls; human pathogen
Campylobacter pinnipediorum	Seal pathogen
Campylobacter vulpis	Isolated from red foxes
Helicobacter pylori	Colonist of the human stomach
Sulfuricurvum kujiense	Isolated from crude oil

Other gammas include *Serratia marcescens* (a bacterium whose pigmented colonies (Fig. 10.7) can sometimes be seen in damp domestic environments), plant pathogens (Fig. 1.12), the photosynthetic purple sulphur bacteria (Figs. 3.32–3.33) and the giant white sulphur bacteria. *Bdellovibrio bacteriovorus* (Class *Deltaproteobacteria*) is a predator of other Gram-negative bacteria. *Bdellovibrio bacteriovorus* digests its prey from the inside and subsequently replicates to release daughter cells. *Bdellovibrio* is, however, not an obligate parasite of other bacteria. It is possible to maintain pure cultures of *Bdellovibrio* in nutrient broth in the absence of other bacteria. The Class *Epsilonproteobacteria* includes the genera *Campylobacter* (curved Gram-negative bacteria) and *Helicobacter* (helical Gram-negative bacteria) both of which are associated with the gastrointestinal tracts of animals (Table 10.6). There is an argument in favour of treating these organisms as a separate phylum (*Campylobacterota*).

Phylum *Acidobacteriota*: acid-bugs

Although acidobacteria (Figs. 3.48, 10.8; Table 10.6) commonly occur in soils and sediments, relatively few species have been isolated and named. *Acidobacterium*

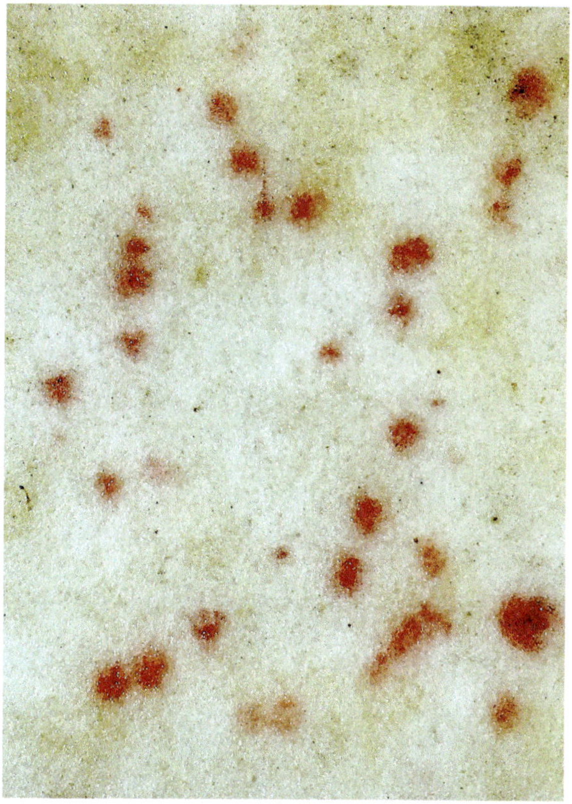

Fig. 10.7. Pigmented colonies of *Serratia marcescens* growing on a damp plastic foam tile.

Table 10.6. Major groups of Gram-negative bacteria other than *Pseudomonadota.*

Species of bacterium	Notes
Phylum *Acidobacteriota*	
Acidobacterium capsulatum	A soil bacterium
Geothrix fermentans	From freshwater sediments
Holophaga foetida	Anoxic sediments; a dimethylsulfide producer
Granulicella mallensis	Isolated from tundra soil
Terriglobus saanensis	Isolated from tundra soil
Phylum *Bacteroidota*	
Mucilaginibacter lappiensis	Isolated from lichen in a pine forest
Mucinivorans hirudinis	From leach intestinal flora
Bacteroides thetaiotaomicron	From human intestinal flora
Capnocytophaga canimorsus	From canine oral flora
Prevotella denticola	From human oral flora
Sphingobacterium species T2	Lignin-degrading bacterium
Tenacibaculum maritimum	Marine fish pathogen
Phylum *Fusobacteriota*	
Cetobacterium ceti	Isolated from a porpoise and a minke whale
Fusobacterium equinum	Isolated from equine oral flora
Fusobacterium gastrosuis	Pig stomach colonist
Streptobacillus moniliformis	Agent of rat bite fever

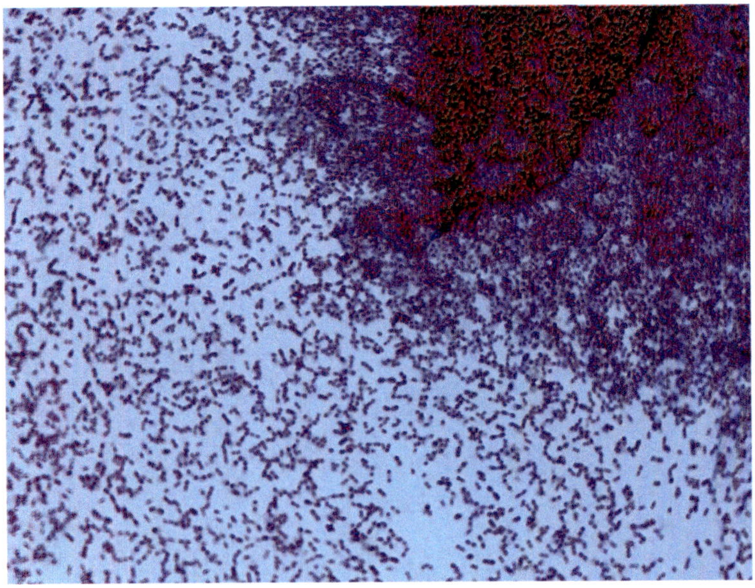

Fig. 10.8. An acidobacterium *Granulicella mallensis* cultured from tundra soil. (Isolate kindly provided by Minna Männistö.)

capsulatum, the first species to be formally described, was originally isolated from acid mine drainage in Japan but has since been detected in soil samples worldwide.

Phylum *Bacteroidota*: intestinal commensals, sliders and gliders

Bacteroides species are anaerobic bacteria that account for a large proportion of the mammalian intestinal flora. The related genera *Prevotella* and *Porphyrimonas* are members of the oral flora. *Flavobacterium* (Fig. 10.9) and *Sphingobacterium* are environmental organisms, which are often found in association with plants. Members of the genus *Mucilaginibacter* (Fig. 10.10) are slime-secreting bacteria that produce extracellular polymeric substances as a defence against desiccation and to stabilise their environment. They occur widely in soils and sediments. *Capnocytophaga* species occur as oral commensals of mammals. When grown on agar plates, colonies of *Capnocytophaga* exhibit so-called gliding motility.

Phylum *Fusobacteriota*

Fusobacterium species are thin, fusiform Gram-negative anaerobic bacteria with tapered ends (Table 10.6). Fusobacteria are mammalian oral commensals. They inhabit mucosal surfaces where they subsist by fermenting carbohydrates. Some *Fusobacterium* species are potential pathogens. Necrobacillosis, also known as Lemierre's syndrome, is a disease which is characterised by abscess formation, venous thrombosis and septicaemia. The causative agent is *Fusobacterium necrophorum*. Vincent's angina is a term used to describe

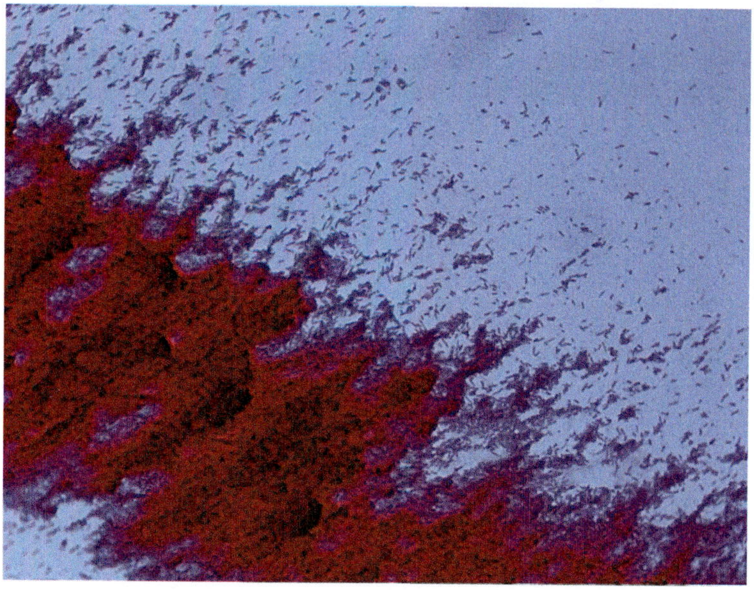

Fig. 10.9. *Flavobacterium* species cultured from soil. (Isolate kindly provided by Minna Männistö.)

mixed infections of the tonsils and pharynx involving a *Fusobacterium* and a spirochaete, sometimes referred to as "*Treponema vincenti*". These are the so-called 'Vincent's organisms' (Fig. 10.11). Vincent's organisms are also associated with necrotising ulcerative gum disease.

Streptobacillus moniliformis is a rat commensal and a cause of rat bite fever. It is an unusual bacterium which spontaneously produces cell-wall-deficient variants known as L-forms (named after the Lister Institute in London).

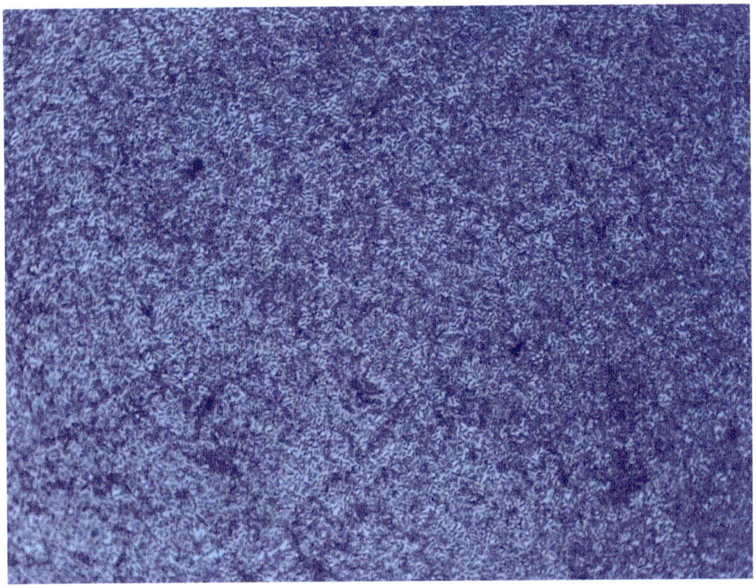

Fig. 10.10. *Mucilaginibacter lappiensis* cultured from lichen in a pine forest. (Isolate kindly provided by Minna Männistö.)

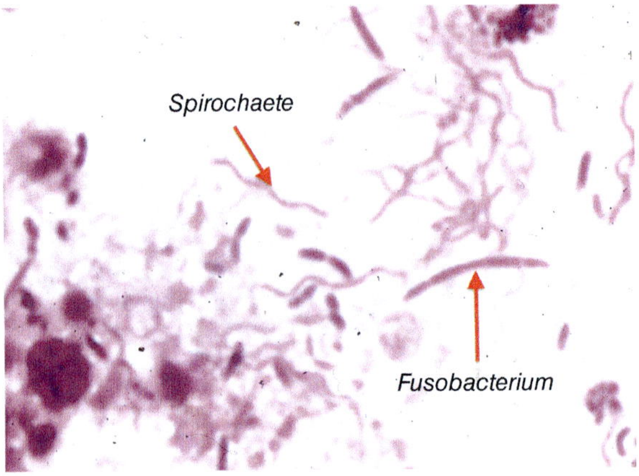

Fig. 10.11. Gram's stain of a sample from a case of oral infection showing 'Vincent's organisms' (a mixture of spirochaetes and fusobacteria).

Phylum Chlorobiota: green sulphur bacteria and allied organisms

The Phylum *Chlorobiota* includes a group of bacteria known as the green sulphur bacteria. They are called green sulphurs because they are pigmented with green bacteriochlorophyll and because they perform a version of photosynthesis that involves the oxidation of hydrogen sulphide (Table 10.7). Some of them, however, oxidise ferrous iron instead of sulphide.

Phylum *Chloroflexota*: green non-sulphur bacteria

The Phylum *Chloroflexota* includes organisms known as the 'green non-sulphur bacteria' (as distinct from the green sulphur bacteria, Phylum *Chlorobiota*). Members of the phylum include photosynthetic as well as non-photosynthetic

Table 10.7. Green photosynthetic bacteria.

Species of bacterium	Notes
Phylum *Chlorobiota*	
"*Chlorobium ferrooxidans*"	An iron-oxidising photosynthetic bacterium
Chlorobium limicola	A sulphide-oxidising photosynthetic bacterium
Chlorobium phaeoferrooxidans	An iron-oxidising photosynthetic bacterium
Ignavibacterium album	Terrestrial hot springs
Phylum *Chloroflexota*	
Chloroflexus aurantiacus	Terrestrial hot springs; photosynthetic
Chloronema giganteum	A photosynthetic freshwater bacterium
Phylum *Cyanobacteriota*	
Aphanothece stagnina	Floating colonies in fresh and brackish waters
Croftalania venusta	Microfossil discovered in the Rhynie Chert
Chroococcus limneticus	Planktonic in lakes and ponds
Chroococcus minor	Shallow ponds and streams
Eohalothece lacustrina	Microfossil from Precambrian rocks in Scotland
Gloeocapsa alpina	Limestone rocks in upland Britain
Gloeocapsa magma	Non-calcareous rocks in upland Britain
Hyella caespitosa	Intertidal rocks and shells
Homoeothrix crustacea	Calcareous streams
Lyngbya gracilis	Filamentous, intertidal cyanobacterium
Microcoleus vaginatus	Moist soils in association with mosses and algae
Nostoc commune	Colonies on limestone and calcareous soils
Nostoc pruniforme	Floating in fresh water
Nostoc verrucosum	On rocks in streams
Oscillatoria limosa	Fresh and brackish waters
Oscillatoria rupicola	On wet rocks
Rivularia atra	Upper intertidal zone
Rivularia biasolettiana	Stones in calcareous streams
Rivularia haematites	Stones in calcareous streams
Schizothrix calicola	On wet rocks

bacteria (Table 10.7). *Chloroflexus* species are filamentous thermophilic bacteria that use gliding motility to form dense yellow, orange or green aggregations or mats in hot springs. *Chloronema giganteum*, a green filamentous bacterium with gas vesicles, was discovered amongst the plankton of a freshwater lake.

Phylum *Cyanobacteriota*: blue-green algae

Cyanobacteria (see text and figures in Chapter 3) contain the photosynthetic pigment chlorophyll and perform oxygenic photosynthesis (Table 10.7). Cyanobacteria frequently occupy the same ecological spaces as eukaryotic algae and the two groups of organisms can often be found together in freshwater, terrestrial and saltwater habitats. Some species of cyanobacteria are solitary planktonic organisms, while others are colonial. Cyanobacteria exhibit tolerance to desiccation, salinity and extremes of temperature.

Phylum *Aquificota*: water bugs

Aquifex and its relatives are thermophilic or hyperthermophilic bacteria that occur in geothermal habitats. They obtain energy by oxidising sulphur, thiosulphate or hydrogen. The oxidation of hydrogen produces water as a waste product; hence the generic name *Aquifex*, meaning water-maker. Related organisms have been detected in hot compost samples.

Phylum *Deinococcota*: radiation-resistant and thermophilic bacteria

The *Deinococcota*, formerly known as the *Deinococcus–Thermus* phylum, is a clade that contains two phylogenetically related but phenotypically different groups of bacteria. *Deinococcus* species exhibit an unusual level of tolerance to ionising radiation and desiccation. During the course of experiments conducted at Corvallis, Oregon in the 1950s to investigate radiation as a means of preserving tinned meat, a bacterium was recovered from a spoiled tinned meat sample, despite its exposure to high doses of gamma rays. A coccoid organism was cultured aerobically that formed red-pigmented colonies. Initially it was named *Micrococcus radiodurans* but was later given a new generic name as *Deinococcus radiodurans*. Similar isolates were later recovered from ground meat, sausages and the carcasses of various animals. The isolation of *Deinococcus* species from environmental samples can be facilitated by exposing the sample to ionising radiation, which kills other kinds of faster-growing bacteria. *Deinococcus radiopugnans* was recovered from soil near Nottingham, using a technique that involved treating soil samples with ultraviolet light. Although deinococci have structurally Gram-negative cell walls, their unusual thickness causes them to retain crystal violet stain. Multiple copies of the genome in each cell coupled with an unusually effective DNA repair protein (RecA) allow deinococci to restore genomic integrity after radiation damage. Many of the described species were first isolated in hot desert environments.

Thermus species are thermophilic organisms (Fig. 4.8; Table 10.8). Species include *Thermus aquaticus*, which is found in hot springs and *Thermus scotoductus* which occurs in manmade hot water systems as well as in natural geothermal habitats.

Phylum *Dictyoglomerota*: thermophilic thread-ball bacteria

This phylum is represented by a single genus of anaerobic thermophiles (Table 10.8). *Dictyoglomus thermophilum* was isolated from a hot spring in Japan. A second species, *Dictyoglomus turgidum* was described from material collected at the Uzon Volcano on the Kamchatka Peninsula. A strain representing an undescribed species that uses xylan as its sole food source was isolated from wood pulp from a paper pulp factory in Finland.

Phylum *Thermodesulfobacteriota*: sulphate-reducing thermophiles

Thermodesulfobacterium species are sulphate-reducing thermophiles (Table 10.8). *Thermodesulfobacterium thermophilum* was isolated from a petroleum reservoir and *Thermodesulfobacterium hydrogeniphilum* from a marine hydrothermal vent. Other species inhabit terrestrial hot springs.

Table 10.8. Radiation-resistant, thermophilic and hyperthermophilic bacteria.

Species of bacterium	Notes
Phylum *Aquificota*	
Aquifex pyrophilus	Isolated from a marine hydrothermal vent
Balnearium lithotrophicum	Deep sea hydrothermal vents
Hydrogenobacter thermophilus	Terrestrial hot springs
Persephonella hydrogeniphila	Deep sea hydrothermal vents
Thermocrinis ruber	Hot springs, Yellowstone and Iceland
Phylum *Deinococcota*	
Deinococcus aetherius	Isolated from dust in the stratosphere
Deinococcus radiodurans	Cultured from tinned irradiated meat
Thermus aquaticus	Geothermal habitats
Thermus scotoductus	Manmade and geothermal habitats
Truepera radiovictrix	A radiation-resistant thermophile
Phylum *Dictyoglomerota*	
Dictyoglomus thermophilum	Isolated from a hot spring in Japan
Dictyoglomus strain B1	Isolated from a paper pulp factory in Finland
Phylum *Thermodesulfobacteriota*	
Thermodesulfobacterium commune	A sulphate-reducing thermophile
Phylum *Thermomicrobiota*	
Thermomicrobium roseum	Terrestrial hot springs
Phylum *Thermotogota*	
Petrotoga mobilis	Isolated from a North Sea oil reservoir
Thermotoga maritima	Geothermal seafloor habitats

Phylum *Thermotogota*: boiler-suited hyperthermophiles

These hyperthermophilic bacteria have an outer coat or 'toga'. They utilise organic carbon compounds as food sources. *Petrotoga mobilis* was isolated from a North Sea oil field. A related species, *Petrotoga mexicana* was isolated from an oil reservoir in the Gulf of Mexico. *Thermotoga maritima* was first isolated from hot seafloor mud at Isola Vulcano in Italy.

Phylum *Chlamydiota*: *Chlamydia* and allied genera

Members of the *Chlamydiota* are intracellular symbionts or pathogens of eukaryotic cells, including those of freshwater amoebae and animals (Table 10.8). They lack cell walls and cannot be seen with routine staining techniques. *Chlamydiota* are dependent on their host cells for their energy requirements and cannot be grown in cell-free culture media. A distinctive feature of *Chlamydia* and allied bacteria is a biphasic life cycle, consisting of an extracellular infectious phase, known as the elementary body and an intracellular non-infectious phase, known as the reticulate body (Fig. 10.12).

Phylum *Chrysiogenota*: arsenate-reducing, nitrite-reducing and selenite-reducing bacteria

Chrysiogenes arsenatis was isolated from the arsenic-contaminated wastewater of a gold mine in Australia. *Chrysiogenes arsenatis* is an anaerobe that uses arsenate

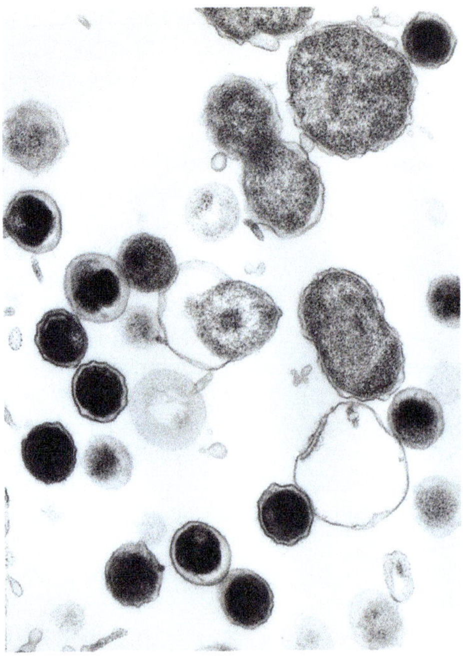

Fig. 10.12. Electron micrograph showing dark elementary bodies and larger reticulate bodies of *Chlamydia trachomatis*. (Image kindly provided by Ian Clarke.)

as an oxidising agent to extract energy from organic substrates. *Desulfurispirillum alkaliphilum*, a sulphur and nitrite-reducing bacterium, was discovered in a bioreactor in the Netherlands. *Desulfurispirillum indicum*, a selenate and selenite-reducing organism, was discovered in an estuarine canal in India.

Phylum *Deferribacterota*

Deferribacter thermophilus was isolated from an oil reservoir in the British sector of the North Sea. Related species have been found around hydrothermal vents in the Atlantic and Pacific. *Geovibrio ferrireducens* is an iron-reducing organism found in habitats that are rich in hydrocarbons. *Mucispirillum schaedleri* is a spiral anaerobic organism isolated from the gastrointestinal tract of a mouse.

Phylum *Dehalococcoidota*: bioremedial bacteria

"*Dehalococcoides ethenogenes*" is distinguished by its ability to reduce chlorinated solvents, yielding non-toxic ethene as a waste product. Chlorinated solvents are used for dry cleaning, in industrial processes and in products such as solvent-based glue. Cultures of "*Dehalococcoides ethenogenes*" are used for the bioremediation of polluted groundwaters in the vicinity of industrial plants.

Phylum *Lentisphaerota*: slime-ball bacteria

This phylum was proposed in 2004 to accommodate two recently characterised genera as well as a number of uncultured taxa detected in environmental samples. *Lentisphaera araneosa* was isolated from seawater near the coast of Oregon. *Victivallis vadensis* was isolated from a faecal sample from a healthy volunteer in the Netherlands.

Phylum *Nitrospirota*: nitrospires and allied taxa

Nitrospires are spiral-shaped bacteria found in wastewater treatment plants, oxygenated river sediments and in aquaria. Nitrospires play a role in the nitrogen cycle by oxidising nitrite to nitrate. "*Candidatus* Magnetobacterium bavaricum" is a magnetotactic bacterium which was discovered in lake sediment in Bavaria (Fig. 3.13).

Phylum *Planctomycetota*: planctomycetes, bud-bugs and plankton-bugs

The *Planctomycetota* are strange bacteria whose DNA is bounded by an intracytoplasmic membrane, similar to that of the nuclear membrane of eukaryotes (Table 10.8). This feature was first noted in *Gemmata obscuriglobus* (Fig. 10.13), a freshwater bacterium discovered in Queensland, Australia

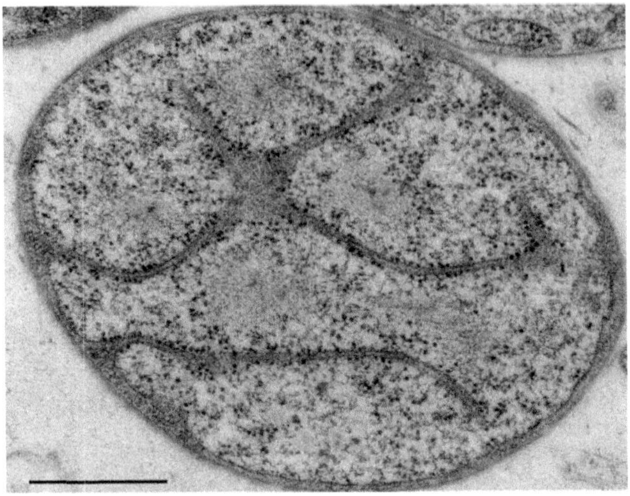

Fig. 10.13. *Gemmata obscuriglobus* showing intracytoplasmic membrane (Image cropped from figure 3 in Santarella-Mellwig *et al.*, 2010; *PLOS Biology*. 8, e1000281). Creative Commons Attribution Licence.

(Santarella-Mellwig *et al.*, 2010). Whereas other bacteria reproduce by binary fission, planctomycetes reproduce by producing yeast-like buds. *Planctomyces bekefii* which was discovered in a pond in Hungary exhibits an unusual colonial morphology, consisting of rosettes of stalked cells. High densities of planctomycetes have been recorded at Esthwaite Water in the English Lake District (Thorpe, 2022).

Phylum *Spirochaetota*: spirochaetes

Their spiral shape allows spirochaetes to corkscrew their way through viscous substrates (Fig. 10.11). Some members of the phylum live as commensal bacteria in the gastrointestinal tracts of animals. Others are important pathogens. *Borrelia burgdorferi* (Fig. 5.29) is the agent of Lyme disease. *Treponema pallidum* is the agent of syphilis. *Leptospira* species cause leptospirosis or Weil's disease. Other spirochaetes are free-living environmental organisms.

Phylum *Verrucomicrobiota*: wart-bugs and slime-eaters

Verrucomicrobium is a genus of free-living heterotrophic bacteria associated with fresh water and soil (Table 10.9). Their cells have peculiar spiny or wart-like projections. *Akkermansia muciniphila* (Fig. 10.14) is a mucin-eating bacterium found in the human intestine (Ottman *et al.*, 2017). Mucin which consists of glycoproteins forms a protective slime layer that coats the lining of the intestine.

Table 10.9. Miscellaneous bacterial phyla.

Species of bacterium	Notes
Phylum *Chlamydiota*	
Chlamydia trachomatis	A pathogen of humans
Parachlamydia acanthamoebae	A symbiont of freshwater amoebae
Phylum *Chrysiogenota*	
Chryiogenes arsenatis	From arsenic-contaminated waste-water
Desulfurispirillum alkaliphilum	From a sulphide-removing bioreactor
Phylum *Deferribacterota*	
Deferribacter thermophilus	From a North Sea oil reservoir
Geovibrio ferrireducens	Iron-reducing bacterium
Mucispirillum schaedleri	From mouse gut flora
Phylum *Dehalococcoidota*	
"*Dehalococcoides ethenogenes*"	Degrades chlorinated solvents
Phylum *Lentisphaerota*	
Lentisphaera araneosa	Isolated from seawater
Victivallis vadensis	Isolated from human faeces
Phylum *Nitrospirota*	
Nitrospira marina	Spiral-shaped nitrite-oxidising bacterium
Leptospirillum ferroxidans	An iron-oxidising bacterium
"*Ca*. Magnetobacterium bavaricum"	A magnetotactic sediment-dwelling bacterium
Phylum *Planctomycetota*	Yeast-like budding mode of reproduction
Gemmata obscuriglobus	Freshwater habitats, discovered in Australia
Planctomyces bekefii	Freshwater habitats, discovered in Hungary
Phylum *Spirochaetota*	Helical morphology
Borrelia burgdorferi	An agent of Lyme disease
Treponema pallidum	Agent of syphilis
Phylum *Verrucomicrobiota*	
Akkermansia muciniphila	Slime-eating member of the gut flora
Verrucomicrobium spinosum	Found in soil and fresh water

Some other phyla in the Domain *Bacteria*

The Phylum *Calditrichota* accommodates *Caldithrix abyssi*, a thermophile isolated from a hydrothermal vent on the Mid-Atlantic Ridge. *Gemmatimonas aurantiaca* described in 2003 belongs to the Phylum *Gemmatimonadota*. *Gemmatimonas aurantiaca* was isolated from sludge obtained from a waste-water treatment reactor. *Gemmatimonas groenlandica* (Fig. 10.15) and *Gemmatimonas phototrophica* are photosynthetic bacteria (Zeng *et al*., 2021). The Phylum *Synergistota* accommodates *Synergistes jonesii*, a bacterium isolated from the rumen of a goat and capable of degrading mimosine, a toxic compound present in the foliage of leguminous trees. Numerous other bacterial phyla are known to exist. The Superphylum "*Candidatus* Patescibacteria" consists of several dozen candidate phyla. Representative members of these phyla have been characterised by molecular methods but as yet have not been isolated in culture.

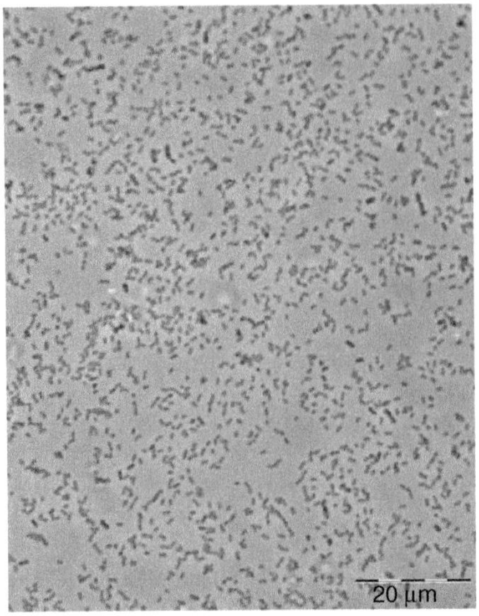

Fig. 10.14. Cells of *Akkermansia muciniphila*, a bacterium that feeds on intestinal mucus (Image cropped from figure 4 in Ottmann et al., 2017; *PLoS One*. 12 (3), e0173004). Creative Commons Attribution Licence.

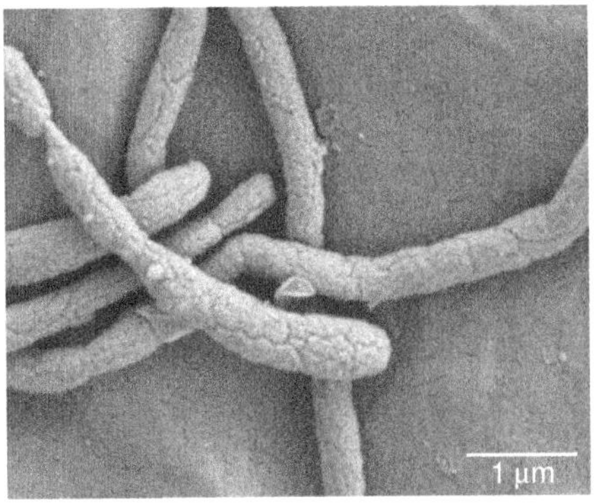

Fig. 10.15. *Gemmatimonas groenlandica*, a photosynthetic member of the Phylum *Gemmatimonadota* (Image cropped from figure 1 in Zeng et al., 2021; *Frontiers in Microbiology*. 11, 606612). Creative Commons Attribution Licence.

The Domain *Archaea* (the archaebacteria)

At the time of writing the archaea are grouped into five phyla (see Table 10.10) which have validly published names as well as a number of additional candidate phyla whose members have yet to be cultured and fully characterised. The Phylum *Microcaldota* is currently represented by a single species, *Microcaldus variisymbioticus*, an oblicate symbiont of larger free-living archaea, which was co-cultured with a thermophilic host archaeon from a hot spring in Indonesia (Sakai *et al.*, 2022).

Table 10.10. Major groups and some species in the Domain *Archaea.*

Taxonomic group	Notes
Phylum *Methanobacteriota* (previously known as Phylum "*Euryarchaeota*")	
Class *Archaeoglobi*	
Archaeoglobus fulgidus	Sulphate-reducing hyperthermophile
Class *Halobacteria*	
Halobacterium salinarum	Halophile
Haloquadratum walsbyi	Halophile, square morphology
Class *Methanobacteria*	
Methanobacterium formicicum	A methanogen associated with ruminants
Class *Methanococci*	
Methanococcus vannielii	An aquatic methanogen
Class *Thermococci*	
Thermococcus celer	A sulphur-reducing thermophile
Class *Thermoplasmata*	
Thermoplasma acidophilum	Thermophile discovered in coal refuse heaps
Phylum *Thermoproteota*	
Class *Nitrososphaeria*	
Nitrosopumilus maritimus	Ammonia-oxidising marine archaeon
Nitrososphaera viennensis	Ammonia-oxidising soil organism
Conexivisphaera calida	Sulphur and iron-reducing thermophile
Class *Thermoprotei*	
Ignicoccus hospitalis	Sulphur-reducing hyperthermophile
Sulfolobus acidocaldarius	Sulphur-oxidising acidophilic hyperthermophile
Thermoproteus tenax	Sulphur-reducing hyperthermophile
Phylum *Nanobdellota* (previously "*Nanoarchaeota*")	
"*Nanoarchaeum equitans*"	Obligate symbiont of *Ignicoccus hospitalis*

Despite its name the Phylum *Methanobacteriota* (previously known as the "*Euryarchaeota*") contains a diverse assemblage of organisms which are grouped into nine separate classes. Two classes (Class *Thermococci* and Class *Thermoplasmata*) are populated by thermophiles. Methanogens account for five classes. Additionally, the Class *Thermoplasmata* includes a number of thermophilic methanogens. The Phylum *Methanobacteriota* also includes the haloarchaea (Class *Halobacteria*).

The Phylum *Thermoproteota* contains three classes of archaea. The Class *Thermoprotei* provides a home for the sulphur-oxidising acidophilic hyperthermophilic genus *Sulfolobus* (Fig. 10.16), as well as a number of thermophilic archaea. The Class *Methanosuratincolia* contains a single species, *Methanosuratincola petrocarbonis*, a methanogenic microbe recovered from an oil field. The Class *Nitrososphaeria* (previously ranked at the level of a separate phylum, the Phylum *Nitrososphaerota*) takes its name from an ammonia-oxidising soil organism, *Nitrososphaera viennensis* which was discovered in

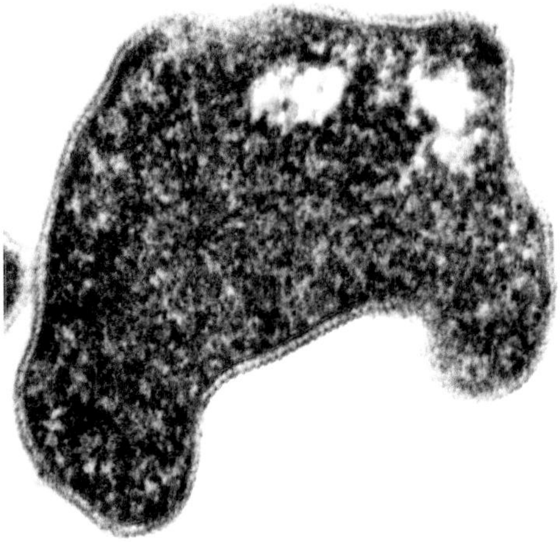

Fig. 10.16. *Sulfolobus acidocaldarius* (Image kindly provided by Thomas Brock, with permission.)

a garden in Vienna. Another ammonia-oxidising archaeon, *Nitrosopumilus maritimus*, which occurs in seawater also belongs to this class.

The Phylum *Nanobdellota* (previously known as the Phylum "Nanoarchaeota") is represented by archaea with tiny cells that live as obligate symbionts of larger free-living archaea. The first of these organisms to be discovered was "*Nanoarchaeum equitans*" (Fig. 1.34), a minute (400 nm) spherical archaeon found attached to its host *Ignicoccus hospitalis* in a sample from a hydrothermal vent near Iceland (Huber *et al.*, 2002). Another example of this group, *Nanobdella aerobiophila* was co-cultured with its host *Metallosphaera sedula* from a hot spring in Japan. A related species of nanoarchaeon, "*Candidatus* Nanopusillus acidilobi" was found living in association with a *Sulfolobus* species in a hot spring in Yellowstone National Park.

References

Daubenspeck, J.M., Totten, A.H., Needham, J., Feng, M., Balish, M.F. *et al.* (2020) *Mycoplasma genitalium* biofilms contain poly-GlcNAc and contribute to antibiotic resistance. *Frontiers in Microbiology*, 11, 585524.

Huber, H., Hohn, M.J., Rachel, R., Fuchs, T., Wimmer, V.C. *et al.* (2002) A new phylum of *Archaea* represented by a nanosized hyperthermophilic symbiont. *Nature*, 417, 63–67.

Hug, L.A., Baker, B.J., Anantharaman, K., Brown, C.T., Probst, A.J. *et al.* (2016) A new view of the tree of life. *Nature Microbiology*, 1, 16048.

Joyce, J. (1944) Stephen Hero. Jonathan Cape, London, UK.

Oren, A. and Garrity, G.M. (2021) Valid publication of the names of forty-two phyla of prokaryotes. *International Journal of Systematic and Evolutionary Microbiology*, 71 (10).

Ottman, N., Reunanen, J., Meijerink, M., Pietilä, T.E., Kainulainen, V. *et al.* (2017) Pili-like proteins of *Akkermansia muciniphila* modulate host immune responses and gut barrier function. *PLoS ONE*, 12 (3), e0173004.

Sakai, H.D., Nur, N., Kato, S., Yuki, M., Shimizu, M. *et al.* (2022) Insight into the symbiotic lifestyle of DPANN archaea revealed by cultivation and genome analyses. *Proceedings of the National Academy of Sciences of the United States of America*, 119 (3), e2115449119.

Santarella-Mellwig, R., Franke, J., Jaedicke, A., Gorjánácz, M., Bauer, U. *et al.* (2010) The compartmentalized bacteria of the planctomycetes-verrucomicrobia-chlamydiae superphylum have membrane coat-like proteins. *PLoS Biology*, 8 (1), e1000281.

Thorpe, A.C., Anderson, A., Goodall, T., Thackeray, S.J. Maberly, S.C. *et al.* (2022) Sedimentary DNA records long-term changes in a lake bacterial community in response to varying nutrient availability. *Environmental DNA*, 4 (6), 1340–1355.

Zeng, Y., Nupur, Wu N., Madsen, A.M., Chen, X. *et al.* (2021) *Gemmatimonas groenlandica* sp. nov. is an aerobic anoxygenic phototroph in the Phylum *Gemmatimonadetes*. *Frontiers in Microbiology*, 11, 606612.

Index

Note: Page numbers in **bold** and *italics* refer to tables and figures, respectively.

Printed and bound by CPI Group (UK) Ltd, Croydon, CR0 4YY

24/08/2026

14947852-0005